**Construction Operations
Manual of Policies
and Procedures**

ABOUT THE AUTHORS

SIDNEY M. LEVY has 40 years of experience in commercial, institutional, and public works construction. For the past ten years he has been a private consultant working with owners, general contractors, and developers assisting in contract administration and compliance and dispute and claims resolution. Mr. Levy is the author of 30 books relating to project management, international construction, construction materials and their application, and infrastructure development, including *Total Construction Project Management*, Second Edition and *Project Management in Construction*, Sixth Edition.

ANDREW CIVITELLO, JR., owns his own construction management consulting company, with offices in Florida and Connecticut, specializing in project management, scheduling, and construction claims. Previously, he was President of Civitello Building Company in Connecticut. Mr. Civitello is the author of *Construction Manager* and *Complete Contracting*.

Construction Operations Manual of Policies and Procedures

Sidney M. Levy

Andrew M. Civitello

Fifth Edition

New York   Chicago   San Francisco
Athens   London   Madrid
Mexico City   Milan   New Delhi
Singapore   Sydney   Toronto

Cataloging-in-Publication Data is on file with the Library of Congress.

McGraw-Hill Education books are available at special quantity discounts to use as premiums and sales promotions, or for use in corporate training programs. To contact a representative please visit the Contact Us page at www.mhprofessional.com.

Construction Operations Manual of Policies and Procedures, Fifth Edition

1 2 3 4 5 6 7 8 9 0   DOC/DOC   1 2 0 9 8 7 6 5 4

ISBN 978-0-07-182694-5
MHID 0-07-182694-7

The pages within this book were printed on acid-free paper.

Sponsoring Editor
Michael McCabe

Proofreader
Cenveo Publisher Services

Acquisitions Coordinator
Bridget Thoreson

Production Supervisor
Pamela A. Pelton

Editorial Supervisor
David E. Fogarty

Composition
Cenveo Publisher Services

Project Manager
Harleen Chopra,
Cenveo® Publisher Services

Art Director, Cover
Jeff Weeks

Copy Editor
Brooke D. Graves

Contents

Section 3. General Contracts

Preface

The first edition of this book was published in 1994. Since then, the world of design and construction has experienced a revolution in the approach to project development, design, and execution. Design professionals using Building Information Modeling (BIM) have either eliminated, or significantly reduced, many of the problems that formerly plagued both their professions and the contractors they work with. Coordination of the work of the various design disciplines, and steep declines in the number of conflicts among architectural, structural, and mechanical components, have sharply reduced misunderstandings, vagaries, change orders, and temper tantrums.

There is a new awareness that working together and sharing risks and rewards is a better approach to achieving the goal of each member of the team. New contract formats reflect this trend; by stressing collaboration and a desire to work together to solve problems, the process has become memorialized.

Today, the computer is ubiquitous: in the home office, in the field office, and in the pocket of the Project Manager and Superintendent. The speed of communication among all key members of the construction process is limited only by the time it takes to tap out a few words, snap a photograph, and press the "Send" key. Problems encountered in the field can be easily and quickly passed on to the appropriate parties; solutions can be discussed, agreed upon, and set in place, in many cases within hours of the problem being brought to light.

But one thing has not changed over these past 20 years, and that is the role of the Project Manager and the Field Superintendent in managing ever more complex construction projects and bringing them to completion "within budget and on time." One can add to that duo: "while ensuring an end project of high quality."

This aspect of the Project Manager or Field Superintendent's job remains one of *control:* control of each one of those previously listed objectives of budget, time, and quality. And these objectives or goals require full-time attention and always expecting the unexpected.

Today's managers of construction must be technically competent and possessed of managerial skills (or the ability to learn them quickly), so that they can direct teams of skilled tradespersons and legions of subcontractors who must interact effectively and cooperatively. But Project Managers and Field Supervisors must also have a smattering of accounting and have both feet in cost control. They must be aware of the legal aspects involved in nearly everything they do, from interpreting intricate construction contracts to dealing with the mosaic of local, state, and federal rules and regulations, and to the ongoing documentation that details the daily events taking place on a construction site. These details can play an important role in settling the disputes that may arise whenever a complex endeavor such as a construction project travels from start to finish.

This fifth edition of *Construction Operations Manual of Policies and Procedures*, updated to fit into today's design and construction environment, contains a series of new contract formats, checklists, forms, sample letters, and advice and suggestions gleaned from the coauthors' decades-long experience in the construction industry. We hope that this fifth edition of *Construction Operations Manual of Policies and Procedures* will provide you with some new insights to assist you in managing your new construction project more effectively.

SIDNEY M. LEVY

Introduction

Throughout the decade of the 1990s, there was an increasing awareness among construction professionals at every level of the need for efficient, coordinated management methods, procedures, and techniques that will effectively deal with the incredible complexity of our special brand of business. As we have progressed as an industry, the focus necessarily was forced to divide between these proactive management ideals and that dubious category of managerial responsibilities—"risk management."

Simply stated, the management of risk as far as an ongoing construction or construction management firm is concerned begins with an acknowledgment of the complex relationships among the various parties to every construction contract that must be aggressively pursued on a daily basis. We don't have "cracks" for things to fall through in our business—we have open crevices into which things get sucked into a vacuum if you let them. And so the "management of risk" begins with focusing our attention on not only the development of procedures and methods that address all of the activities that need to be performed, but also on those forms, procedures, follow-up, and accountability mechanisms that will create and encourage a discipline within the organization that will encourage all of these activities to regularly be performed—completely—every day.

From there, effective operations management combined with an appropriate "risk management" perspective should not limit itself to the management cliches that are offered throughout too many "authoritative" books on the subject. We need to grab the reality quickly and deal with all of the pleasant and unpleasant realities of our work environment. In other words, it does no good to formulate our management procedures and our daily operations with the idea of "construction in heaven"—as it "should" be. We cannot plan and operate our companies anticipating a work environment as we wish it to be, rather than as we know it truly is. We need to approach our work environments with a significant—and genuinely healthy—consideration of the realities of our business relationships. In this way, if we actually proceed with the expectation of routine misfires, miscommunication, and plain mistakes, our management systems will automatically be designed to include safety nets, follow-up, and instruction or guidance that will work to close all those crevices (or at least close them to the size of normal "cracks").

This is not at all to say that we must all proceed through our work days with cynicism and pessimism. This is only to say that if we proceed with the *expectation* that some number of our business associates, subordinates, and contracting parties do deal in wishful thinking, possess wide ranges of knowledge and competence, and operate with "diverse" motives, we simply will see the problems coming and deal with them all more in stride. Just as a pilot anticipates various potential malfunctions and each phase of flight—and gives each possibility very specific attention as he or she prepares and executes each phase—we as managers should learn to do the very same thing in the management of our construction contracts. We should recognize at the onset that these types of acknowledgments of human nature as applied to the construction industry do not need to be just one more source of pressure, or a new layer of stress. Instead, we must realize that the higher up the ladder of managerial responsibility that we may aspire to, the greater percentage of our work day will be devoted to dealing with exceptions to management routine. At the lower rungs of the project administration ladder, greater percentages of the complete work are characterized by routine (or at least "normal") process and procedure. Submittals are checked against contract requirements, various log forms are produced and maintained, and so on. As our responsibilities move up the managerial ladder, our jobs have an increasing requirement to introduce judgment into the process. Our judgment is presumably based upon our "experience." It follows that the only reason why "judgment" and "experience" are needed at higher-level positions are to allow one to deal with the inevitable exceptions to the smooth-running processes outlined in the procedure manual for construction in heaven.

Because of these forces that work every day against our carefully planned efforts, it has become absolutely crucial that all categories of field information be reported accurately, quickly, and completely. Administrative activities must be orchestrated, implemented, monitored, and adjusted. Late starts, mistakes, omissions, and inappropriate actions must be exposed and dealt with quickly and decisively. The success or failure of this entire effort translates directly into either containment of costs and maximization of profit, or into huge financial losses.

Dramatic and even exponential increases in the cost of doing business combined with intensifying competitive pressures dictate that managerial, administrative, and supervisory operations must be streamlined for maximum efficiency. Despite our increasing information control and reporting requirements, duplication of effort must be eliminated if overhead is to be controlled.

What This Manual Will Do for You

This fifth edition of the *Construction Operations Manual of Policies and Procedures* is designed to be your complete reference into the insights and intuition of the most successful examples of construction project management. It spans the gap between the haves and have-nots with respect to skill, experience, and resources. It will help you level the playing field if you are currently behind, and push ahead fast if you are currently at or near the forefront of your particular area of the business.

If your company has either private or public construction contracts in lump sum, construction management, or design-build formats, this manual will help you to squeeze every dollar of profit out of them. But it won't stop there. It will help you to hang on to each of those dollars and keep them away from those who would otherwise prefer to have their hands in your pockets. Every aspect of your operation will be given the tools to become more profitable. Whether or not you choose to focus your administrative processes in either manual or computerized formats, or in some combination of the two, you will be given economical, effective systems for planning, operating, and controlling. Weak areas in your operation will be flagged, and the most effective action is explicitly described in a step-by-step, how-to program. Every employee within your organization will be given fast reference to all company procedures. Each will be given the data, the tools, and the instruction that allow him or her to get it all done—the first time—every day.

The fifth edition of the *Construction Operations Manual of Policies and Procedures* is your complete:

- Office Administration Manual
- Project Manager's Handbook
- Project Engineering Operations Guide
- Field Superintendent Procedures Manual
- Project Safety and Loss Control Manual
- Contract Management Guide

- Risk-Management Manual
- Comprehensive Project Management System

In each of these areas the *Construction Operations Manual of Policies and Procedures* provides you with:

- Ready-made, full-size forms, including full descriptions and complete instruction.
- Step-by-step direction that will help you anticipate every situation and implement action at the earliest possible moment.
- Complete checklists that consolidate every important consideration of the given issue into clear action plans.
- Word-for-word letters that communicate clearly, accurately document, and compel desired responses. Each example in the Operations Manual is time-tested. Each has proven itself to be the most effective for its situation. Each is loaded with obvious and subtle language designed to orchestrate and control both the short-term and long-range aspects of each issue.
- Form letters that greatly simplify and speed-up efficient handling of common and routine processes.
- An extensive chapter on Design-Build explaining this dynamic project delivery system that is being embraced by more and more public and private owners.
- A how-to chapter on Change Orders providing guidelines for the preparation and presentation of the change order process.
- Information about Building Information Modeling and Green Buildings—two important twenty-first century advances.

Company Organization and Quality Assurance Program

1.1 Section Description and Use

This first section of your company Operations Manual is important for a number of reasons. It establishes the form of communication between the company and the employee. It is intended not only to establish a rapport with the employee, but also to initiate the employee to the communication format and the manner in which information is intended to be transferred between the company and the employee.

This first section therefore begins with a clear identification of who the company is. It should describe the people and policies that comprise the corporate identity and should provide each employee with the fundamental corporate ideals. Employees should be able to embrace each of those ideals when they are wrestling with the difficult day-to-day decisions that must be made.

The purpose of this section is to provide each employee with a concrete idea of visibility of that employee's fit within the company's operational network, and to see how that employee's daily work product interacts with all others in order to achieve project and company objectives on a daily basis. In this way, if each employee can understand the relationship of his or her work product with the work products of the project team and of the company, such an understanding can further clarify for that employee the manner in which the details of each task can and should be accomplished in order to maximize the benefit to the organization.

1.2 Company Statement of Operations

1.2.1 Company Purpose

The fundamental nature of business and the purpose of the organization is to create profits for its stockholders and to create stable and profitable employment to the organization's team members at every level. The primary method of accomplishing these goals in this industry is to provide construction and construction management services of the highest professional standards and to perform these services to the benefit of our clients. We in this industry have recognized that if we consistently serve our clients, perform each of our jobs effectively, and operate with integrity in good faith, the organization will be consistently profitable by every definition. The resulting client satisfaction will be the primary catalyst that will perpetuate our existence and encourage the organization's steady growth.

As the company continues to operate with the objective of generating profits that are consistently at or above the industry average, we must operate in the way that always adds value to the process. To this end, every executive, manager, supervisor, administrator, and employee should guide his or her day-to-day operating decisions with specific consideration of performing each task and solving each problem in ways that:

- Are expedient
- Finish the item completely—the first time
- Resolve all open issues

- Solve problems for ourselves, our clients, and our associates
- Positively influence the reputation of the company
- Add to the stature of the team

While there may be times when for given situations these objectives may seem to be mutually exclusive, striving for their attainment in every situation will consistently result in efficient work, creative solutions, respect, and personal satisfaction. When our objectives are clear, our work will be focused, our clients will be satisfied, and our operations will be profitable.

1.2.2   Operational Objectives

The Operational Objectives of the company address those processes and procedures that will become the mechanism by which the company purpose will be achieved. Because of the nature of our business, the specific operations through which we will accomplish our goals focus on the work product and professional output. To be sure, the output of the organization ultimately is the physical work-in-place constructed in the field. Into that final outcome, all else flows.

The field production is directed by the company's on-site superintendents. The field staff is directed by, coordinated with, and assisted by both the on-site and home-office project management functions. Together, these two functions combine to create our total output. In a very real sense, the rest of the organizational effort supports these two activities.

While it might be argued that the business development activities and our marketing professionals are the company team members who create the projects that we will ultimately construct, this function really only provides the company with the opportunity. The on-site and off-site project management and supervisory staff make it happen for the entire organization. All other employees not only support the business development and project management activities, but truly function each day as company ambassadors to the outside world.

1.3   Organizational Structure and Corporate Staff Functions

1.3.1   Corporate Office

The location of the corporate office is listed in the Construction Operations Manual Transmittal provided to each employee in the introductory section of the company manual.

The functions of the corporate staff include:

- Determining company objectives and corporate policy
- Conducting the centralized disciplines of business development, corporate finance, accounting, human resources, and administrative staffs and support

- Coordinating all production operations and integrating all such activities with all other corporate disciplines
- Developing company procedures and providing guidance and assistance to all company members
- Providing training, education, and professional development opportunities for all company employees
- Resolving issues that may be beyond the ability or authority of the individual project staff members
- Developing synergy within the organization through which maximum utilization is gained from the particular strengths of each individual, while continuing personal and professional development is provided for in ways that will improve our individual weaknesses

Each Jobsite Field Office is considered a satellite extension of our corporate office. Accordingly, it is expected that each Jobsite Field Office is established and maintained in a professional, clean, and proper setting that is fitting for our field staff.

1.3.2 Project Management Staff and Functions

In total, "project management" includes the following functions:

- Project management
- Site superintendence
- Project engineering
- Project scheduling
- Project accounting
- Administrative support

Project management functions are divided between "project management" and "site superintendence." Project management essentially embodies the organizational structure of the project itself, the higher authority for decision making, and the center for project administration. Site superintendence is the extension of project management directly into the field. The instruction and election of project management is carried out by the field staff, who interact at the firing line with the armies of subcontractors, suppliers, and our own trade employees who directly generate the work of our projects.

1.3.3 Operational Objectives

The specific organizational structures in the following exhibits have been designed with objectives of effectiveness, efficiency, and elimination of redundancy.

Comprehensive treatment of the complex issues faced each day in the industry can lead to complicated, detailed activities that can span many operational levels.

It is important for each individual throughout the organization to appreciate that all company procedures have been established, not for their own sake, but to achieve specific objectives. How an individual's function fits within the overall objective may not always be clear.

The ideas of "teamwork" and "cooperation" cannot be treated as the clichés they can appear to be. They must consistently be applied in the truest sense if the company objectives are to be met, and if the individuals meeting those objectives are to develop personally and professionally.

1.4   Organization Charts

1.4.1   General Description

The organizational structures that follow display the individual components of the respective operation. In large companies, each function will have individuals or even staffs assigned to it. Smaller organizations will necessarily have fewer individuals involved in administration, but the *functions* performed should be the same. The functions as listed will accordingly be combined in the responsibilities of those individuals. The formulas determining those combinations will be most often based on the specific person's capabilities, inclinations, and initiatives. A project manager (PM) in one company, for example, may also do project engineering, or in another company, may do estimating or purchasing. In still another company, the PM may also serve as the site superintendent.

Observe the functional relationships, and visualize the appropriate individual responsibility assignment within your own company.

Specific jobsite structures will be treated in Section 1.5.

1.4.2   Corporate Organization Chart

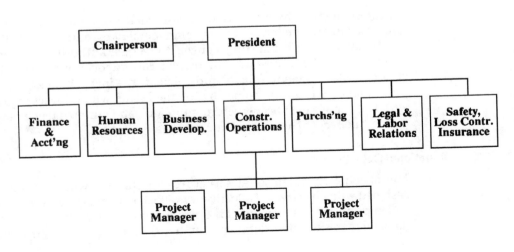

1.5   Jobsite Structures—Large and Small Projects

1.5.1   General Description

The detailed descriptions in this section assume a large-project structure for ease in clarifying the individual relationships of the particular line and staff operations. A large and/or complex project may have an individual and even staff assigned to each function—depending strictly upon the needs and responsibilities of the respective assignment.

Smaller projects will require fewer individuals, but the complete list of functions performed will be the same. Examples of ways in which duties may be combined include:

1. Project Manager/Site Superintendent/Project Engineer/Scheduler in a single individual.

2. Project Manager/Purchasing Agent working with a Site Superintendent/Project Engineer, each working with centralized scheduling.

3. Project Manager/Project Accountant working with Site Superintendent/Project Engineer/Scheduler.

1.5.2   Typical Project Organization Chart—Large Projects

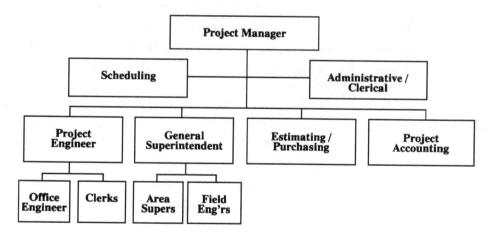

1.5.3   Concept, Organization, and Attitude

Construction Operations is the mechanism by which all company objectives as described in the company's operating statement will be achieved. It is what we sell in our business development efforts, and is the force that generates *all* revenues. Individual project success as defined by cost/profit, time of completion, and quality thus translates directly to success of the company and of the individual. Accordingly, *the purpose of virtually every position throughout the organization is to support construction operations.*

The summary descriptions (1.5.4 through 1.5.9) of the individual and/or staff responsibilities of the various project functions display their relationships. Their specific activities and responsibilities are treated in great detail throughout the Manual.

1.5.4   The Project Manager

The Project Manager (PM) is the individual charged with responsibility for the complete and satisfactory execution of the entire project. In very large organizations divided into regional areas, the PM may report to an executive or other person who would in turn report to the Operations VP. In most companies, however, the PM reports directly to the Operations VP or to the company President. In many small companies, the PM may *be* the Operations VP or the President.

In any case, the Project Manager is the company representative who deals with the outside world (owner, design professionals, vendors). His or her actions, as such, will be regarded as the company's. However good or bad that person appears to be, the company itself appears to be.

General duties of the Project Manager include:

- Assisting in development of the project staff
- Coordinating on-site with corporate office activities
- Organizing and overseeing jobsite administration
- Organizing and coordinating field supervision
- Assisting in the procurement of subcontractors and suppliers
- Developing, monitoring, updating, and communicating the progress schedule and its periodic revisions
- Managing the direct labor force and maintaining labor relations
- Managing subcontractor schedules, quality of work, coordination with other trades, and payments
- Coordinating cost-progress targets with production
- Creating and maintaining a safe/secure jobsite environment
- Identifying and resolving all changes
- Establishing and maintaining relationships with the Owner, design professionals, building officials, local businesses, and police and fire departments

1.5.5   The Superintendent

The Superintendent is responsible for the supervision of all field activities related to the physical construction. He or she reports to, and carries out the direction of, the Project Manager with respect to field operations, and thereby directs the daily progress of the work. The jobsite workforce is made up of an army of specialists. The Superintendent must ensure that the combination of their components results in a cohesive product that achieves the required quality and is completed in the shortest possible time. To that end, the Superintendent must continually work to ensure adequate staffing of the workforce, sufficient supply of materials, and complete information as necessary to assemble the product, and must plan for all these things enough in advance so as not to interfere with the progress of any one component. General duties of the Superintendent include:

- Generating, securing, or otherwise confirming all information needed to create, monitor, and modify the progress schedule on a continuing basis
- Developing the progress schedule with the Project Manager
- Participating in scope reviews of the various bid packages to properly coordinate their respective interfaces and ensure that nothing is either left out or bought twice
- Working with the Project Manager to develop and administer the site utilization program, site services, security arrangements, and other facilities and arrangements necessary for appropriate service to the construction effort
- Identifying field-construction and work-sequence considerations when finalizing bid package purchases
- Monitoring actual versus required performance by all parties
- Determining whether subcontractors are providing sufficient workforce and hours of work to achieve performance commitments
- Monitoring the performance of the company's Purchasing and Project Engineering functions to ensure that all subcontracts, material purchases, submittals, deliveries, clarifications, and changes are processed in time to guarantee jobsite arrival by, or before, the times needed
- Directing any company field staff
- Being thoroughly familiar with the requirements of the general contract, thereby identifying changes, conflicts, etc., that are beyond the scope of responsibility
- Preparing daily reports, job diaries, narratives, and all other regular and special documentation as determined by the company and by the project needs

1.5.6   The Project Engineer

As the Superintendent manages the physical construction of the project in the field, the Project Engineer in a sense builds it first—on paper. The Project Engineer is responsible for coordinating the complete construction and administrative requirements of the various bid packages; organizing the complexities into a management system that will establish, monitor, and follow up on each vendor's compliance; and orchestrating the information flows needed for each component and system to be incorporated into the project—all within the time frames needed by the progress schedule. Strict attention to detail, a profound appreciation for time constraints, and a sober understanding of the dividends reaped through proper documentation all add up to an effective Project Engineer. Duties include:

- Working with the Project Manager, Superintendent, and the estimating/purchasing effort in the final development of the various bid packages and the baseline construction schedule
- Working with the Project Manager and Superintendent to develop site utilization programs
- Establishing, maintaining, conducting, and policing detailed procedures for the submittal, review, coordination, approval, and distribution of shop drawings, samples, etc.
- Establishing and maintaining all project engineering files relating to subcontract and bid package records, plans, specifications, changes, clarifications, and as-built documents
- Reviewing all vendor schedules of values and preparing the general schedule of values as coordinated with the Project Manager
- Managing the periodic requisition procedure for review, submittal, and payment
- Expediting vendor estimates and proposals and preparing appropriate company estimates and proposals for changes to be submitted to the Owner
- Determining appropriateness and preparing subcontractor change orders to be processed through the Project Manager
- Evaluating subcontractor payment requisitions relative to actual work performed
- Working with the Project Manager, Superintendent, and Project Accountant to prepare the general requisition and follow it through to payment

1.5.7   The Project Accountant

The function of the Project Accountant can vary from strict application of technical accounting to filling a variety of roles important to the success of project administration. On large projects an individual assumes the role, with the Project Manager or Project Engineer assuming it on smaller projects.

The Project Accountant is the financial watchdog over project administration. He or she will be involved with monitoring cost performances, verifying cost information from all internal and external sources, maintaining all appropriate records, and reporting both to accommodate the technical requirements of the central accounting effort and to quantify performance. Because of its nature, the position naturally lends itself to that of Office Manager. In this role, the Project Accountant's responsibility can expand to that of auditor of company procedure. Responsibilities include:

- Establishing all vendor files relating to contracts, billings, payroll and labor records, and material costs

- Maintaining job-cost information on a current basis and producing cost-progress reports, highlighting areas potentially exceeding estimates in time to implement corrective action
- Enforcing compliance with payroll and EEO reporting and securing lien waivers from vendors of all tiers as condition of payment
- Performing periodic audits of office procedure to monitor company performance
- Reporting project financial information to central accounting

1.5.8 Planning and Scheduling

Too many people equate the scheduling function to that which simply produces the finished reports. Actually, planning, scheduling, monitoring, and updating constitute *the* fundamental activity through which all others follow. Sequences affect estimates, schedules become contract and subcontract commitments in their purchasing and performance phases, extended delays dramatically affect multiple trades, and the time of completion affects everyone's wallet.

Chronically missed schedules cause extremely disproportionate costs and interferences resulting in economic loss, problems that permeate multiple organizations, and bad reputations that are difficult to shake off. In contrast, those companies that actually give the scheduling function the effort it deserves enjoy higher profits, better relationships, and improved reputations as direct results.

From that standpoint, planning and scheduling remain *everyone's* responsibility. The time-status of every component must habitually become the focal point around which all other information is arranged. The potential effect of *every* issue on the progress schedule must always be a key consideration throughout each issue's resolution.

Duties include:

- Identifying each major construction activity, its relationships with other activities, and all necessary support
- Correlating the activity list with the contract documents and the schedule of values
- Soliciting and confirming all information from the best combination of sources, incorporating it into the plan, and distributing it in a timely manner
- Monitoring actual progress relative to planned progress continually, assessing its actual and potential impacts, displaying cause and effect relationships, and determining necessary corrections
- Monitoring the plan's implementation and maintaining all documentation relative to good and bad performances of all parties in a manner that is complete, correct, and well correlated and can be used most effectively by the project management, finance, and legal efforts

1.5.9   Project Management Services, Staff, and Support

The recurring theme throughout the Manual is that the success of the production effort *is* the success of the company. To accommodate that reality, all other company operations exist to support production efforts. The individual line-and-staff operations themselves are treated exhaustively throughout the Manual. Here, it is sufficient to emphasize that all processes are equally important, and that none exist for their own sakes.

The prevailing attitude throughout the company should be one that encourages—even rewards—early warning of potential problems and conflicts. There's no better guarantee that small problems will become large ones if they are not dealt with early, or when staff members fear a "kill the messenger" disposition or penalty in any form for bringing bad news to the attention of superiors.

With the complexities, pressures, and problems that are thrown in our faces each day, we must all constantly remind ourselves that we're all on the same side and that the clock never stops. The concept of a complete team approach must be taken as seriously as it deserves to be. Those who ignore that concept will plateau quickly, and their careers will soon follow.

1.6   Document Generation, Signing Authorities, and Communication

1.6.1   Document Authority

Strict lines of formal authority can be detailed in every organization in a straightforward manner. In the construction industry, large networks of informal authority structures can develop if they are allowed to. Those in turn can be further complicated by incorrect *perceptions* of both formal and informal authority.

Every act, deed, and remark made by a wide range of individuals carries some potential to commit the organization. Examples range from the clear language in a contract executed by a company officer to a remark by any member of the project team made at a job meeting or while touring the site. Clearly defined procedures can be thwarted when the *actions* of the parties constructively alter the "real" relationships. If, for example, a Change Order is not formally executed but the work is completed—and paid for—it can dramatically alter the legal application of the entire Change Clause and even the Dispute Resolution Clause of the general contract.

1.6.2   Document Types and Signing Authorities

Following is a list of various routine documents between the company and the outside world that are typically generated repeatedly throughout each project. Beside each document are those respective personnel authorized to generate, modify, or sign (internal document procedures are detailed elsewhere throughout the Manual).

Contracts, Subcontracts, and Contract Modifications

Owner/Company (GC, CM)	President
Company/Subvendors	Corporate Officer
Change Orders—To Owners	Corporate Officer
Change Orders—To Subvendors	Corporate Officer

Project Management, Engineering, Accounting

Schedule of Values Submission	Project Manager
Payment Applications (To Owner)	Corporate Officer
Overbudget Authorization	Corporate Officer
Subvendor Schedule of Values Approval	PM/PE
Subvendor Requisition Approval	PM/PE
Process Subvendor Requisition	PA
Shop Drawing Reviews/Transmittals	PE
Schedule Transmittal/Notifications	PM/PE

Documentation and Correspondence

Job Meeting Minutes	PE
Routine Correspondence	PM/PE/PA
Backcharge Notices	PM/PE/Super
Delay Notice—To Owner	PM
Delay Notice—To Subvendor	PM/PE/Super
Notification of Pending Change Order	PM/PE
Change Order Proposal	PM/PE
Notification of Work under Protest	PM
Work Stoppage	Corporate Officer
Notification of Claim	Corporate Officer

1.7   Quality Concept and Quality Policy

The three critical elements of a construction project are:

- Schedule
- Costs
- Quality

Scheduling and costs are discussed in subsequent sections on project management and project engineering, but the quality concept is one that ought to be rooted in the company's organization and statement of operations. Quality is an all-encompassing endeavor firmly embraced by top management and each segment of the corporation's organizational chart.

1.7.1   Quality Control versus Quality Assurance

Although used interchangeably, these two terms have distinct and separate meanings.

Quality Control (QC). The standards to which the construction or assembly of a building component have been incorporated into the project's design

Quality Assurance (QA). The process that verifies that these standards have been achieved

1.7.2   Total Quality Management

Total Quality Management (TQM) encompasses not only the field operations and the actual construction cycle, but also includes all aspects of the company's operations, for example, quality of the estimating process, quality of the Accounting Department operations—that monthly requisitions are sent to the owner on time, that they have the correct figures and all required documentation. The quality of the close-out process is also of importance. All Operation & Maintenance manuals and warranties must be assembled promptly. Punchlist is to be completed and signed off, and lien waivers obtained from subcontractors quickly. This movement focusing on Total Quality Management has spawned a number of quality-oriented processes.

1.7.3   Benchmarking

Benchmarking is a procedure where the company seeks out and studies the best practices in the industry in order to produce superior performance at home. For example, internal benchmarking of self-performing work can compare the company's performance on a project where quality levels were highest and costs lowest. Then the company can investigate and memorialize the practices that created this quality, cost-effective job.

1.7.4   International Standards Organization

The International Standards Organization (ISO) was founded in 1946 to promote voluntary manufacturing and trade communication standards. ISO 9000 and its variants ISO 9001–9004 relate to the basic quality of construction management practices.

1.7.5   Six Sigma

The Motorola Corporation created Six Sigma. The Six Sigma approach is to train a small cadre of in-house quality leaders. These leaders are called Black Belts, and they train lower supervisors in a somewhat less intense environment; these supervisors are called Green Belts. At the head of all trainer personnel is the Master Black Belt. In order to improve an operation, these Six Sigma teams employ something known as DMAIC:

D—Define the goals.

M—Measure the existing system to establish a metric by which progress toward the goals can be tracked.

A—Analyze the system to identify ways to eliminate the gap between the existing performance and the desired goals.

I—Improve the system.

C—Control the new system.

1.7.6 Quality in the Construction Process

Quality begins with the preparation of the contract plans and specifications by the design consultants; here the coordination process plays a critical role. A careful review by the design team may uncover inconsistencies, omissions, or items requiring more clarity. Although this review is the responsibility of the architect and engineers, the general contractor and its subcontractors must also perform their own scrutiny of the plans and specifications, as many bid documents and contract specifications require the contractor to perform this review and notify the owner of any errors, omissions, and inconsistencies.

The attention to quality issues will continue through construction and end after the termination of the builder's warranty period.

Quality Control and the Specifications.   Various specification sections contain references to professional and trade organization quality standards. The more common ones are the American Concrete Institute (ACI), American Society for Testing and Materials (ASTM), and the American Institute of Steel Construction (AISC).

A typical cast-in-place concrete specification includes references to the ACI and ASTM, and since most general contractors do not have these extensive and expensive manuals, they must rely on the integrity of their subcontractors and vendors to furnish materials to meet these standards.

There are, however, several trade organizations that offer low-cost or no-cost quality standard information for the products their organization represents:

Steel Door Institute. Provides information on hollow metal doors and frames, including proper installation procedures, tolerances allowed, and some repair and remedial advice.

Southern Pine Inspection Bureau. Provides specifications on various species of southern pine framing lumber.

Western Wood Products Association. Offers specifications and mill tolerances for fir, spruce, hemlock, hem-fir, and various other species of western softwoods.

Building a library of these types of quality standards, over time, will provide both the Project Manager and Project Superintendent with baseline material specifications and enhance the company's ability to provide quality service to its clients.

1.7.7 The Preconstruction Conference—QC and QA

Preconstruction conferences are often stipulated in the contract specifications, but whether they are a contract requirement or not, the use of such meetings can greatly enhance the goal for high quality levels.

Experienced installers in one trade may offer suggestions to improve their performance and the performance of predecessor activities or those that follow. Issues of constructability and achievable tolerances can be raised, reviewed, and discussed with the design consultants at these meetings.

Preconstruction conferences are particularly effective where subcontractors involved in exterior wall and building envelope components are concerned. Curtain wall or brick veneer construction involves a number of trades; the performance of each one will impact the weather-tight integrity of the building's skin. By bringing all parties together, possibly in the presence of the architect, details are either confirmed or modified in order to create the most effective exterior barrier. Rather than halt an operation in progress, problems are raised and resolved before work commences—a clear exercise of Quality Control and Quality Assurance.

Preconstruction conferences are especially useful when working with the following construction operations or components:

- Sedimentation control and maintenance where subcontractors are alerted to the provisions to be taken to preserve and avoid damage to these installations
- Cast-in-place concrete work, particularly where extreme weather conditions can be anticipated
- Precast concrete deck, wall, architectural panels requiring embedments, and attachment of other structural or architectural elements
- Structural steel and metal deck installation
- Waterproofing, commencing at the foundation and progressing up to and including the roof membrane
- Exterior wall assemblies: stucco, exterior insulation finish system (EIFS), masonry, precast concrete, metal or vinyl siding, metal panels
- Windows, louvers, ducts, and other penetrations through exterior walls
- Roofing, skylights, hatches, MEP penetrations, roof curbs, equipment platforms
- Joint sealants and compatibility to adjacent surfaces
- Interior gypsum drywall assemblies where in-wall blocking or other types of reinforcement are required
- Kitchen and bath installations where MEP coordination is required and in-wall blocking requirements need to be confirmed
- Finish wall items such as signage, window treatments, artwork installation (with lighting) to confirm the location, type, and extent of in-wall reinforcement, if required

1.7.8 Sample Panels and Mock-ups

Again, whether required by contract or not, the construction of a sample panel to document minimum quality standards and mock-ups to incorporate multiple trade work can prove invaluable. The issues of constructability and tolerances can be acknowledged or debated prior to the actual start of work. Each participant's input will confirm that the design is proper or that certain changes need to be made and approved by the design consultant in order to achieve the desired results.

These efforts taken before production begins can not only provide for a smooth flow of work, but, again, increase or at least maintain expected quality levels. Whenever these preconstruction conferences take place or panel mock-ups are prepared, the event should be memorialized to document everyone's comments, responsibilities, and agreement on the issues presented.

1.7.9 Above-Ceiling Inspections

The production of coordination drawings for above-ceiling work will generally be a contract requirement, and work should not start until an approved set of coordination drawings has been submitted to, and approved by, the design consultants. Prior to installation of work above the ceiling, all trade contractors involved in the work should be assembled in a meeting where the nature of their work and the sequence of their work is reviewed and approved by the project manager. As each trade completes its work, the work should be inspected by the Project Superintendent or his or her designee for the following quality assurance items:

- The work conforms to the contract documents with respect to materials, equipment, and "good workmanship" practices.

- The installation will not encroach on other trade's work to the point that access to the other trade's work is impaired.

- All access panels for filter replacement and belt drive replacements for fans and heating, ventilation, and air conditioning (HVAC) equipment are accessible.

- Equipment preventative maintenance work, lubrication points, etc., can be accessed without removing another trade's work.

- A statement from each trade certifying that their work has been completed, has been signed off by any inspecting authority, and is ready for close in.

- Photographs of areas to be permanently enclosed should be taken for inclusion in the Owner's O&M manuals for that particular trade, a copy of each to be placed in the general contractor's Close Out file.

1.7.10 The Punchlist and Quality Assurance

The punchlist may actually be the final test of a productive Quality Assurance program. Certainly a zero punchlist says something about the contractor's full compliance with the architect's Quality Control standards. While a zero punchlist is the goal, it is very difficult or nearly impossible to achieve, but to get closer to that goal, try the following:

- Have the Project Superintendent periodically inspect each trade and issue "pre-punchlists," but notifying the subcontractor that this is in addition to and not a substitute for the design consultant's formal punchlist.

- Reinspect again to ensure that these punchlist items have been corrected.

- If the work was not completed satisfactorily, send a letter to the subcontractor's office advising the subcontractor of the need to inspect and correct any substandard or missing work on a regular basis.

- Prior to a subcontractor demobilizing from the project, or substantially reducing its crews toward the end of the project, the project superintendent should "walk" the site or building with the subcontractor's foreman to inspect, prepare punchlists, and insist that the work be completed within a specified time frame. If these requests are not met or are met with indifference, threats of withholding payment may be the next step.

1.7.11   Developing a Company Quality Control/Quality Assurance Program

Also known as a Construction Quality Control Plan (CQCP), the formulation and implementation of a Company Quality Control/Quality Assurance Program will vary according to the size, staffing, and sophistication of the general contractor. But whatever the case, there needs to be an appointment of a Program Manager. This manager, once the program is developed and explained to all field and office personnel, will be the individual responsible for the functioning and monitoring of the program in much the same way that the Safety Director implements and monitors the company's safety program. The CQCP Manager may elect to appoint several deputies to carry out specific portions of the program or may place primary field responsibility with each project superintendent, reporting directly to the CQCP Manager or to his or her respective project managers.

1.7.12   The CQCP's Four W Approach

There are several ways in which to approach the development of a program, and by asking some basic questions—Who, What, When, Where—participation by key members of the construction organization will begin to develop the basic framework.

Who? Who will be responsible for QC/QA control for a specific project? The person selected must have received ample training and be able to pass much of his or her knowledge on to the various foremen, subcontractors, and other supervisors he or she will oversee. It must be made clear to all concerned that this technician has the authority of the CQCP Manager and the Project Manager to implement company QC/QA policy on specific jobsites.

What? What will the designated technician's responsibilities be? Will the technician be required to be on site continuously, one day each week, or only on an as-needed basis to perform in situ inspections and subsequently prepare reports for submission to the CQCP Manager or his or her project manager? Will they be responsible for coordinating inspections by government authorities and the owner's design consultants? How are they to report QC/QA violations: verbally to the affected party followed up with a written memo to the Project Manager?

When? When will some of these QC/QA activities take place: upon delivery of materials and equipment to the site? And what about components such as piping, duct, and electrical conduit risers that will be enclosed in shaftwalls and fire-rated enclosures? Since timely inspections are required, will it be the subcontractor's responsibility to notify when such inspections are to take place?

Where? Will the QC/QA technician be required to visit a subcontractor's fabrication plant (i.e., steel fabricator) to comment on the quality of fabrication equipment and processes?

The CQCP should include, as a minimum, the following components:

1. An organization chart showing where the CQCP Managers fit within the corporate management hierarchy, who they are responsible to, and who is responsible to them.
2. The proposed personnel and support staff making up the entire QC/QA program.
3. A list of specific duties and responsibilities for each assigned personnel.
4. The operating procedures—how tests and inspections will be conducted, and the documentation that will be generated prior to, during, and upon completion of each test.
5. Rework procedures to identify and record QC/QA deficiencies, notify responsible parties, and reinspect for compliance.
6. Develop a subcontractor and vendor rating system so that supervisors can evaluate and report on subcontractor/vendor performance and attention to quality issues.
7. The role the CQCP Manager will play in consultation with the Purchasing Department in selecting competent, quality-oriented subcontractors and vendors.
8. Develop a training program to familiarize office and field personnel with quality issues.
9. Develop a series of Field Inspection Reports to assist field personnel in their inspection procedures.

1.7.13 Field Inspection Reports

By developing a series of Field Inspection Reports similar to those in Sections 1.7.14 through 1.7.18, the CQCP Manager can provide the Project Manager and Project Superintendent with simple checklists that serve several purposes, including:

Providing a record of periodic inspections highlighting acceptable work and items requiring corrective action

Affirming that quality is a concern

Alerting subcontractors that periodic inspections of their work will be an ongoing activity, thereby instilling the Quality concept in their crews

1.7.14   Concrete Reinforcement Inspection Checklist

Inspection Checklist

Concrete Reinforcement   Project No._____

1. Shop drawings are approved and on site._____
2. Grade of steel delivered as required._____
3. Spacing coordinated to suit masonry/concrete units._____
4. Required clearance of steel from forms provided._____
5. Length of splices and staggered splices as required._____
6. Bends within radii and tolerance are uniformly made._____
7. Additional bars at intersections, openings, and corners provided._____
8. Bars cleaned of materials that affect bond._____
9. Dowels for marginal bars at openings._____
10. Bars tied and supported to avoid displacement._____
11. Spacers, tie wires, chairs as required._____
12. Conduit is separated by 3 conduit diameters minimum._____
13. No conduit or pipe placed below rebar material except where approved._____
14. No contact of bars is made with dissimilar metals._____
15. Bar not near surface which may cause rusting._____
16. Adequate clearance provided for deposit of concrete._____
17. Verify that contractor has resolved conflicts with embeds._____
18. Verify that contractor has coordinated for anchors, piping, sleeves._____
19. Special coating as required._____
20. No bent bars and tension members installed except where approved._____
21. Unless approved, boxing out is not approved for subsequent grouting out._____
22. Rules for bar splices: For 24d lap—multiply bar size by 3 = lap in inches
                                     For 32d lap—multiply bar size by 4 = lap in inches
                                     For 40d lap—multiply bar size by 5 = lap in inches
23. Agency/Engineer inspection is performed, if required._____

General Notes: _____

_____

_____

_____

_____

_____

Inspected by:_____ Date:_____

1.7.15   Concrete Placement Inspection Checklist

Inspection Checklist

Concrete Placement   Project No._____

1. Shop drawings approved and on site. _____
2. Verify correct psi ordered from plant. _____
3. Chutes, elephant trunks required? _____
4. Verify approval of forms and rebar prior to pour. _____
5. Requirements for testing, mix design, ingredients. _____
6. Test lab notified and tests required. _____
    Slump _____
    Number of cylinders _____
    Temperature/truck waiting time _____
7. Testing required at plant. _____
8. Vibrators to be used during pour. _____
9. Temporary form openings O.K.? _____
10. Arrange for specified curing and saw cut joints. _____
11. Arrange for cold weather protection. _____
        -or
12. Arrange for hot weather protection. _____
13. Embeds available for insertion in pour. _____
14. Box-out properly installed in form work. _____
15. Verify finishes—smooth troweled, broom. _____
16. No troweling while bleed water is on surface. _____
17. Slopes to drain properly designated. _____
18. Wet spray or curing compound adequately performed. _____
19. Traffic over area controlled. _____
20. Preparations for repairs at hand. _____

General Notes:_____

_____

_____

_____

_____

Inspected by:_____Date:_____

1.7.16   Finish Hardware—General and Butts and Hinges Inspection Checklist

Inspection Checklist

Finish Hardware—General and Butts and Hinges   Project No._____

1. Hardware schedule, product data, samples, approved and on site. _____

2. Hardware installed in accordance with manufacturer's templates. _____

3. Finishes are as required and finishes match in each area. _____

4. Hardware is removed or protected during painting operations. _____

5. Recommended order of inspection:

   In hardware storage area before installation _____

   Door butts and hinges during and after installation _____

   Locksets, latchsets, and exit bolts during and after installation _____

   Door closers, after installation _____

   Door stops, holders, push-pulls, kickplates, after installation _____

Butts and Hinges

6. Ball bearing, iolite or nylon type, is provided as required. _____

7. Solid brass, bronze, aluminum, or stainless steel is provided as required. _____

8. Fire door hinges are steel with ball bearings or as otherwise approved for a labeled assembly. _____

9. Mortise type hinges are mortised flush. _____

10. Mortise hinges on door leaf to ¼" from stop side of door and jamb leaf 5/16" from stop (3/8" and 7/16" on very thick doors) unless otherwise noted. _____

11. Unless other noted, top hinge is mounted 5" below finish door frame and bottom hinge is mounted 10" above finish floor. Intermediate hinges are spaced and mounted equidistant from top and bottom hinges and from each other. _____

12. Sufficient throw is provided to clear trim and leaf can swing functionally as required. _____

13. One-half surface hinges are used on composite doors. _____

General Notes:_____

_____

_____

_____

Inspected by:_____Date:_____

1.7.17   Hollow Metal Doors and Frame Inspection Checklist

Inspection Checklist

Hollow Metal Doors and Frames   Project No._____

1. Shop drawings and schedule are approved and on site. _____
2. Doors are as approved: type, design, material, accessories, etc. _____
3. Check panel, lights, louvers, other features. _____
4. Check for defects: dents, buckles, wraps. _____
5. Check fabrication for construction and workmanship. _____
6. Smooth edges, joints, finish, and straightness. _____
7. Additional reinforcement as required for hardware. _____
8. Observe backing plates during drilling operations. _____
9. Observe that closure channels are provided. _____
10. Provisions to receive hardware are adequate. _____
11. Backset matches with hardware requirements. _____
12. Stile edges, astragals required for pairs of doors. _____
13. Fire-rated doors have labels and proper identification. _____
14. Fire-rated frames have labels and proper identification. _____
15. Fabrication and construction of frames as required. _____
16. Frames are prebraced as required. _____
17. Extra reinforcement on frames at head, corners, and hardware. _____
18. Proper type and number of anchors are provided. _____
19. Verify adequate anchorage made during installation. _____
20. Sound-deadening treatment provided, if required. _____
21. Provide features such as silencer holes. _____
22. Frame is grouted during installation, if required. _____
23. Frame is caulked, if required. _____
24. Frames adequately braced where "built-in". _____
25. Provide spreaders when installed in masonry walls. _____
26. Protect threads of hinge plates in buck. _____
27. Fusible link holders provided, if required. _____
28. Observe installation of door and proper clearances. _____
29. Doors are hung straight, level, and plumb. _____
30. Door functions smoothly and easily. _____
31. Hardware is properly adjusted. _____
32. Observe glazing operation. _____
33. Wire glass is provided, if required. _____
34. Factory prime is touched up. _____
35. Surfaces are adequate to receive finish. _____
36. Report doors that cannot be properly cleaned. _____
37. Bumper buttons installed. _____
38. Protection provided to avoid marring. _____

General Notes: _____

_____

_____

_____

Inspected by:_____Date:_____

1.7.18   Metal Framing/Gypsum Drywall Inspection Checklist

Inspection Checklist

Metal Framing—Gypsum Drywall   Project No._____

Metal Framing

1. All submittals, samples, shop drawings are approved and on site. _____
2. Material is stored in a dry location. _____
3. Material galvanized as required. _____
4. Studs are doubled up at jambs, unless otherwise required. _____
5. Structural and/or heavy gauge studs as required. _____
6. Studs allow for movement, slab deflection. _____
7. Studs securely anchored to walls, columns, and floors. _____
8. Soundproofing provided at floor and walls as required. _____
9. Observe location, layout, plumbness. _____
10. Channel stiffeners are provided as required. _____
11. Special fastening and connections are observed. _____
12. Anchor blocking, plates, other equipment provided. _____
13. Cut studs for openings are properly framed. _____
14. Observe size, gauge of runner, and furring channels. _____
15. Hangers are saddle tied, bolted, or clipped as required. _____
16. Tie wire for channels to runners properly tied. _____
17. Elevation and layout of furring is understood. _____
18. Observe that surfaces are plumb and level. _____
19. Observe that long single lengths are used. _____
20. Control joints are installed per contract requirements. _____
21. Requirements for adjoining surfaces of different materials are accommodated. _____
22. Seating provided for sound or thermal insulation. _____
23. Spacing and construction are as specified. _____
24. Observe location of all blocking, bracing, and nailers. _____
25. Type, thickness, length, and edges are as required. _____
26. Type fastener, length, and spacing as required. _____
27. Installation complies with manufacturer's requirements. _____
28. Special type suited for damp locations if required. _____
29. Special lengths are provided as required. _____
30. Verify if horizontal or vertical application is required. _____
31. Wall board is installed with staggered application. _____
32. Internal and external metal/plastic corners as required. _____

1.7.18
Metal Framing/Gypsum Drywall Inspection Checklist *(Continued)*

33. Number of coats of compound required is provided. _____

    Level 1—All joints and interior angles have tape embedded in
        one layer of joint compound. _____

    Level 2—All joints and interior angles taped & receive two
        coats of taping compound. _____

    Level 3—All joints and interior angles taped & receive three
        coats of taping compound. _____

    Level 4—Apply Level 3 plus skim coat of joint compound
        over entire surface. _____

34. Sanding between coats is performed. _____

35. Feathering is out 12" to 16". _____

36. Provide air circulation with adequate dry heat. _____

37. If fire rated, recesses over 16" are boxed in. _____

38. Penetrations tight and sealed as required by code. _____

39. Verify contractor has coordinated cut-outs and outlet boxes correctly to avoid patching. _____

40. Wallboard is held up from floor 3/8" minimum. _____

41. Vertical joints are aligned with door jambs. _____

42. Damaged sheets are not used and are removed. _____

43. Observe minimum piecing or joining. _____

44. Nonmetallic cable, plastic, or copper pipe is not damaged. _____

45. Check for bubbles and dimples. _____

46. Curing time is adequate for subsequent finishes. _____

General Notes:_____

_____

_____

_____

_____

_____

_____

Inspected by:_____Date:_____

1.7.19   A Few Quality Tips for Field Supervisors

✓ Inspect for quality on every walkthrough and, at a bare minimum, once a week.

✓ Promptly notify a subcontractor or vendor of any unacceptable work, material, or equipment, and set a date for replacement or rework. Just as important, follow up and if that date is not met, call the subcontractor or vendor and notify them.

✓ If substandard work or repeated lack of quality continues, call the subcontractor's owner, request they come to the site, observe the poor quality, and obtain a verbal commitment to improve. Follow up with a written memorandum of the meeting. If a change in crew supervision is deemed necessary, make the request. The sample letter in Section 1.7.20 can be used for both purposes.

✓ With respect to materials and equipment purchased by the General Contractor, if deliveries include damaged or substandard products, notify the vendor, verbally or by e-mail, and request a visit to the site by its representative to inspect the poor-quality items.

✓ Prior to a subcontractor demobilizing from the site, conduct a walkthrough with its foreman and send a copy of the results of that walkthrough to the subcontractor's office. Advise that any punchlist developed during this inspection must be completed prior to demobilization, and let them know that future payments may be jeopardized if the remedial work is not done promptly. Also mention that this does not preclude the issuance of a punchlist by the architect, but is in addition to that document. The sample letter in Section 1.7.21 can be used for that purpose.

The quest for quality is unrelenting and requires the full backing of top management and the daily attention of both Project Manager and the Field Supervisors, but the concept of "Do it Right the First Time" is the essence of QCQA.

1.7.20   Sample Notification to Subcontractor of Substandard Work (*page 1.27*)

1.7.21   Sample Notification to Subcontractor of Pre-Punch Out Work (*page 1.28*)

1.8   Business Development Approach and Project Participation

"Official" business development efforts are directed from the corporate office. Nowhere else but in construction, however, are the activities of so many individuals throughout the organization so important to the sustained ability of the company to secure new contracts.

1.7.20
Sample Notification to Subcontractor of Substandard Work

Contractor's Letterhead

Addressed to: Subcontractor

Attention: Subcontractor's Owner or Partner

Re: Project name and number

Subject: Substandard work

Dear Mr. (Ms.) (   ),

At our job meeting this morning (afternoon), we inspected your work in place at (indicate the area in which the inspection took place).

We were in mutual agreement that the following items of work require rework or replacement:
   (List the substandard items discussed.)

We trust that this work will be satisfactorily completed by (date) and that you will instruct your field supervisor to pay closer attention to the performance of his or her work.

With best regards,

Project Manager (or Project Superintendent)

Most clients recognize that construction can be a difficult process riddled with inevitable problems and conflicts. The way in which we all organize and execute our work; the cooperative, optimistic attitude displayed in our approaches to problem resolution; and our effectiveness in keeping our objectives targeted at client satisfaction while preserving the other project objectives will stand like a neon sign confirming our professionalism. These elements must be second nature, *primary* considerations in the performance of every task, every debate with Owner representatives and design professionals, and the final position in the resolution of every conflict.

The successful project, satisfied client, and relationships maintained with design professionals are our best advertisements—our best sales efforts.

As our projects are completed in this manner, the primary roles of our "official" business development effort will be the demonstration that our project personnel are the company's greatest advantage that any new client can secure.

1.7.21
Sample Notification to Subcontractor of Pre-Punch Out Work

Contractor's Letterhead

Addressed to Subcontractor

Attention: Subcontractor's Owner or Partner

Re: Project name and number

Subject: Pre-Punch Out Inspection

Dear Mr. (Ms.)   (   ),

We have conducted a pre-punch out list pertaining to your firm's work that requires rework or replacement. A copy of this list is attached. It is requested that this work be completed prior to (date).

If you take issue with any items on this list, please contact me or (name of project superintendent). If not, we will expect all these items to be corrected.

[*If this work is not done after this letter is sent, add the following here and resend the letter:*

A recent inspection of your work reveals that the pre-punch list submitted to your office on (date) has not been completed. If this work is not corrected by the time you demobilize, we may, at our option, engage another firm to complete your work and backcharge your account for related costs.]

Please be advised that this list is not a substitute for the Owner's formal punchlist but is an attempt to keep that formal punchlist to a minimum.

With best regards,

Project Manager

1.9   Marketing Services and Support

1.9.1   Marketing Ideals

The primary objective of the marketing effort is to increase and maintain favorable exposure to potential buyers of construction services in the public, private, and institutional sectors. We thereby multiply our opportunities to be considered for construction assignments throughout the industry.

The marketing ideals of the company are to:

- Maintain corporate identity and image
- Operate with and promote professionalism
- Display competence, leadership, and effectiveness
- Communicate integrity

A consciousness of these ideals should be made part of everything we accomplish, and the philosophies should manifest themselves in our actions.

This section describes the objectives and responsibilities of all non-Business Development professionals throughout the company. Summaries of marketing aids available, some suggestions on their use, and your responsibilities follow.

1.9.2   The Corporate Brochure

The corporate brochure has been designed to create a favorable climate and feeling about the company in the minds of potential users of our construction services. It promotes professionalism, diversification, relevant experience, management depth, and integrity.

From time to time, select distributions of brochures will be made, but the most effective presentation is *personal contact*. Be sure that you have an adequate supply of brochures on hand and personally distribute them to appropriate people at every opportunity.

Logical recipients include:

- Current clients
- Facility planners
- Architects and engineers
- Business executives
- Building or Building Advisory Committee members
- Developers

1.9.3   Business Cards

A sample of your business card is shown below. As with any other marketing aid, it does no good in your pocket. Use it at every opportunity.

Place
your business card
here

1.9.4   Contacts

Develop "secondary" personal contacts as sources of lead information on available and upcoming projects in concept, planning, design, and bid phases. Some sources of these contacts may include:

- Architects and engineers
- Subcontractors
- Material suppliers
- Core boring and other testing companies
- Planning and zoning agencies
- Trade unions
- Government and industry sources
- Local and state development agencies
- Banks and lending institutions
- *Dodge Reports* and other project information services
- National publications and magazines
- Newspapers
- Trade/industry magazines and journals
- Business owners and executives

1.9.5   Some Dos and Don'ts

1. Speak positively about your company always, but don't overstate your capabilities.
2. Set aside a predetermined number of hours each month devoted exclusively to client contact. Activities can include phone calls, lunches, and visits.
3. Know your clients' industries. Know and use their buzz words. Get their trade journals.
4. Participate in industry trade seminars.
5. Join organizations—as many as possible.
6. Know people on development boards.
7. Invite clients to see current projects. Then ask questions that you lead.
8. Never submit a bid at the last minute without being certain that you're not putting the architect on the spot. He or she may be depending on you.
9. Be on time for appointments.
10. Drop clients' names.
11. If you see a client's name or photo favorably publicized, cut it out and send it.

1.9.6 Suggested Proposal Index

The suggested Proposal Index scheduled below is presented in order to give the Project Manager and company executives a list of possible topics to address in a project proposal that might favorably impress a potential client to further consider the company for a particular construction assignment. The information is designed to promote competence and professionalism. The proposal can include:

- Transmittal letter
- Bid format
- Company qualifications
  - Experience on comparable projects
  - Corporate management and organization
  - Statement of financial capability and bonding capacity
- Area considerations
  - Labor market
- Project approach
  - Management and labor approach
  - Preconstruction plan and program
  - Jobsite operations plan
- Project organization and résumés
  - Project staffing and organization
  - Position descriptions
  - Résumés of key personnel
- Project schedule
  - Overall plan and program
  - Review of options and considerations
- Subcontractor policy and list
  - Subcontract policy
  - Subcontract approach
  - Subcontract listing
- Construction equipment and tools
- Insurance
  - Meet or exceed minimum coverages
- Company brochures, newsletters, and other promotional materials

Company and Project Administration

2.1   Section Description

Company and Project Administration as applied in this section of the Operations Manual refers to the internal control system and housekeeping of the entire organization. It establishes the basic procedures for use of the Operations Manual, staff organization, the basic arrangement of various files and facilities, and overall logistics of communications.

The overriding objectives of this section are to establish, implement, and maintain coordinated systems for information and document management and control. Each administrative component is intended to operate as an efficient, stand-alone system and is designed to be carefully coordinated with all other administrative components. Files, records, and documentation are not only to be stored, but also are to be arranged in a manner that will facilitate retrieval and use. When these administrative systems are implemented, we will have a place to put the things developed in the other sections of the Operations Manual.

2.2   Use of the Operations Manual

2.2.1   Overall Approach

The overall organization of the Operations Manual has been designed to make its information as usable as possible to the greatest number of construction professionals at every level of company and project administration. In general, it may appear that the apparent number of differences in the way that construction and construction management companies operate may work against the idea that a standardized approach can be applied to most operations. The encouraging truth, however, lies in the fact that the real functions performed by every one of these companies are actually very similar.

Large companies will necessarily have specialized staffs with limited ranges of responsibilities. Those who purchase, for example, only purchase. Estimators only estimate, and so on. Small and midsized companies will necessarily combine any number of specialized activities in the same individual or individuals. Those who estimate and bid a project, for example, are likely to be heavily involved with purchasing. In some companies, those same individuals may then go on to become the project manager. Individuals with primary superintendent responsibilities may also be called upon to purchase, estimate, and even assist with other administrative functions. Human Resource administration may be centralized, or each private manager may have the authority to hire his or her own staff. The project manager on one project may simultaneously be an estimator for another project.

What all this simply means is that within our own company and within all those companies with which we deal, the specific combinations of functions and job assignments may vary greatly—and even change day to day. The individual functions and assignments themselves to be carried out each day, however, are remarkably consistent company to company and project to project. Every project needs the same things to be done relatively dependably. It's not a matter of "what," but a matter of "how much"—matters of degree—for a particular project, given its size, type, complexity, and dollar value. The issues and operating ideals affecting construction and its management are therefore really very similar for nearly all of us.

The construction contract on a $1 million project will be nearly identical to that on a $20 million one. The plans and technical specifications will be different, and the general conditions might have twice the detail, but the language, procedures, relationships, rights, responsibilities, and decision theories are virtually the same. Every project has a project manager, superintendent, estimator, and project engineering function or component—whether or not the particular people involved in these functions realize that this is what they're doing. The specific mix of responsibilities might have more to do with particular individuals' disposition, talent, and experience than any formal organization chart or job description. Every project has an owner, a design responsibility, a contract sum, the time of completion, and a construction force ranging from a single company to a virtual army of separate subcontractors and suppliers.

And so, as it turns out, the functions that must be performed are really very much the same for both the $1 million project and the $20 million project. The larger project (and larger organization) may need more elaborate (translate: "voluminous") file management, documentation, and record retrieval systems, but operational ideals of these physically larger systems remain virtually the same as those for the smaller, even manual (yes, manual) information control systems that are still very effective in this computer age.

This Operations Manual is intended to be used by all managers, executives, and professionals throughout the company. Employees should thoroughly familiarize themselves with the entire manual in order to secure a firm understanding of the power of effective administration and of the relationship of each function within the overall organizational structure. The Operations Manual should be referred to each time any question as to appropriate procedure arises and to confirm one's understanding of the synergy of one activity with all others.

Discussions, procedures, letters, and forms have been designed to accommodate the most effective approach to the respective issue. Each of these items has been carefully coordinated with all related company activities, the obligations of the constructor with respect to the particular project agreements, and the need to exhibit minute-to-minute control over all project information, both within the organization and without. At any given moment, at least some of this synergy and coordination may not be immediately apparent. It is therefore important that information generated and procedures for distribution of the resulting work product be followed as described. If for any reason your experience begins to suggest an alternative approach, reread Section 1.7 on Quality and proceed with your own company administrative procedures to have the Operations Manual requirements officially changed. Be sure to implement the official procedure to have those changes communicated to all those within the company who will potentially be affected by them. Do not operate as an independent maverick. Even if you have the company's best interest at heart, uncoordinated (and uncommunicated) actions and responses will almost guarantee misfires, miscommunication, and mistakes.

2.2.2 Procedure

The subject of each major section of this Operations Manual is a distinct administrative discipline within that particular area of construction management. Each subject has been arranged in a manner that will facilitate the way in which various functions must be performed. Generally, these arrangements have been determined because of operating practicalities; in some cases historical arrangement may have played some role, but only if it still makes good sense in our contemporary contracting world.

As its own discipline, each section of the Operations Manual can be considered as an individual manual for the specialist involved with that particular subject matter. Its relation to the other sections of the Operations Manual are carefully interfaced. In this way, for example, the "Purchasing" section of the Operations Manual will detail the specific procedures and recommendations for procuring the subcontract for a particular situation and thereby provide the specific "how-to" instruction for the Purchasing Agent and his or her staff. It will then go on to relate those purchasing efforts to the original estimating function—which was a source of the original purchasing budget in the first place.

The Project Engineer and Site Superintendent will be interested not only in their own specialist disciplines, but also in how the purchasing efforts handle both boilerplate and specific contract requirements throughout the subcontract terms and conditions that they'll have to live with for each project's duration. Both of these individuals should be intimately familiar with the company's standard operating documents (such as subcontracts, purchase orders, etc.). The manner in which specific purchases may have deviated from the company's "standard" must be communicated to these managers in ways that are quick, clear, and complete. And so every action that may appear far removed from another person's specific and immediate operating problem can actually be affecting them in dramatic, intimate, and persistent ways.

Each functional area within the company should have its own copy of operating procedures. It is not sufficient to have only one copy of the company's Operations Manual in the company "library," however active that particular bookcase might be. The procedures determined by company management to be critical to its operation must be immediately and simultaneously available to everyone who needs that information. Each jobsite must have atleast one copy, depending upon the actual size of the field staff. Compliance with the stated company procedures must be a primary objective of every company member, and the enforcement by every person with any supervisory capacity must be considered to be a primary responsibility. The company should carefully consider numbering each copy of the Operations Manual and formally issuing a copy to key members of management and staff.

2.3   Correspondence

2.3.1   Objectives of Effective Written Communication

This section is coordinated with Section 2.4, Files and File Management, and Section 4, Project Engineering. Effective communication is a fundamental prerequisite for all other management activities. Effective written communication goes well beyond this. It not only "communicates," but can succeed—or fail—at documenting, presenting, convincing, and complying with contract terms.

If we're not careful, the energy and emotions swirling around every jobsite can cause our communications to become confused and overly complex. At its worst, correspondence can too often be allowed to degenerate into a busy stream of innuendo, insults, or other unprofessional documents that at best will not exactly represent the company in a "favorable" light and at worst can unnecessarily expose the company to unexpected liabilities.

All correspondence in communication should be consistent within each project and throughout the company. Each communication must be clear, concise, efficient, and focused. Each must be made in the way that will make it possible for anyone unfamiliar with the original file to be able to review, correlate, arrange, and support an issue, and to coordinate the particular item with the complexities of the project record. A standard form of communication for various routine items should be adopted by the company, and each manager and professional with the company should be responsible to be sure that they comply with these ideals. Components of each communication must be closely coordinated with company routine and special file structures. Information must be factual, and conclusions must be logical. Opinions must be minimal, and emotional expressions must be controlled.

The design of effective correspondence must begin with the idea of the particular document's fundamental purpose. If its purpose is simply to document, it may be sufficient to only state the supportable fact(s) and conclude the letter. In other cases, it may be more appropriate to clearly identify the issue, refer to a source of notice or other origination, create a paper trail, prove history, and compel some action. Its success in achieving these objectives will directly translate to the success of resolving a given issue. Its efficiency in meeting these goals will directly relate to the time and cost of arriving at a favorable end result. The speed with which all information relating to a claim issue can be assembled, for example, will go directly to the time needed for research and preparation (legal and administrative fees). The coordinated and comprehensive use of information will go directly to effectiveness of any analysis and minimization of lost information and lost opportunity.

If, for example, documentation and files are so arranged as to be easily retrievable and instantly correlated, the small amount of work involved in retrieving the complete, coordinated information will instantly put a manager in a good position to negotiate an issue with a very minimal amount of preparation time. In contrast, documentation and files that are arranged in ways that require heavy research and reorientation to old issues in order to prepare for battle carry with them very high probabilities that the hard work necessary to prepare for such negotiation might not be accomplished effectively—if at all—prior to meetings with opponents. This type of scenario—which is far more common—is fundamentally the definition of "lost information" and "lost opportunity."

Each time a piece of correspondence is composed, it must be created with the idea that when the item is researched at a much later date (and possibly by different people), all required pieces of information will be physically present in the respective file or that clear references to related information (and their specific locations) are included. There should never be a need to recall circumstances without the benefit of an efficient paper trail and some method to get to the source quickly and completely.

2.3.2   Rules of Effective Project Correspondence

Correspondence is generally divided into three levels. The first level is routine, formalized processes, such as transmittal of forms, logs, etc. The second level is those communications that, although "customized" for a particular project, are actually substantially routine. These types of communications lend themselves to "form-letter" treatments that essentially retain the same content issue-to-issue and project-to-project. The third level is specific, custom communications, perhaps involving large amounts of unique and specialized detail.

The first two types of communication are treated in great detail throughout this Operations Manual. Each such communication has been designed to take advantage of the ideas of this section in that they are clear, complete, and address very specific objectives. These types are generated daily by nearly everyone in the organization. In order to measure their effectiveness, each one should be subjected to the set of rules that follow in order to best address content, correlation, efficiency, and clarity.

The first objective of any effective correspondence, then, is to communicate information. As obvious as this may seem, too many examples of correspondence lack obvious purpose, or at the very least are confusing in their apparent objectives. The final objective of each communication is to cause or compel an action on the part of the recipient.

In getting from the primary to the final objective, there may be any number of intermediate steps in the process. We are all familiar with communications that are long-winded, rambling, and confusing and that combine different subjects. It is not uncommon to see letters that contain sentences that are so "run-on" that an entire letter may contain only one period.

Following the rules below is one way of ensuring that your communications are direct, complete, and serve the intended purpose:

1. Keep each project separate.

It is not uncommon for contractors to have multiple projects with the same owner. In such cases, never combine discussions of two different projects in the same correspondence. Even if the same or similar issue occurs and for very practical reasons applies equally and in the same periods of time to different projects, keep them apart. The first reason for this practice is to avoid possible confusion, both on the part of the recipient and in the manner that the documentation will be filed and correlated with the project record. From there, the reason for this separation is to keep future developments focused and efficient. If the issue begins to develop in one way on one project, for example, those developments will not be allowed to confuse the second project. In contrast, if issues are allowed to be combined in a single letter, every discussion will remain unnecessarily complicated. If an owner representative, design professional, or anyone else should be included in distribution of the letter or by some other means should receive a copy, information regarding other projects is needlessly displayed. Finally, file and record management becomes cumbersome and more prone to error.

2. **Confine the subject of the letter to a single issue, or to a small group of closely or logically related issues.**

And keep your letters focused to the target issue. Keep separate or unrelated items in separate letters. Separate the items in order to allow direct focus and quick understanding of each issue, without unnecessarily causing distraction, confusion, and complication. The overall result will be a better chance of faster, more direct response, and reduced probability that the response itself will be confused or misdirected.

In contrast, a letter that contains several unrelated items may initially confuse or otherwise distract its reader. In beginning to deal with such a letter, its subjects must be separated from each other and individually treated. As each issue is individually researched, the separate reactions must be applied into an equally confusing response—unless the responder has the sense to separate the items into individual correspondences. Otherwise, both your letter and its response, each containing multiple unrelated items, must then be filed in each of the files corresponding to each unrelated issue. Separately filing the individual issues can become cumbersome and overly complex, with a very disproportionate risk that even the filing will not be properly done. Any resulting improper or confused filing is almost sure to cause some lost opportunity at a later time, thereby significantly increasing the risk that issues will not be resolved in the most optimal manner.

Finally, the worst condition will involve documents that contain multiple examples of unrelated information that wind up as evidence in litigation or arbitration that involves only one of the listed items. At the very least, expensive time will be consumed explaining why all those unrelated items do not apply to the discussion at hand. At worst, you may find yourself having to introduce information into an argument that you otherwise would prefer to leave out of the particular discussion.

3. **Develop and use a consistent document format.**

Be sure that every person in your organization develops correspondence that is consistent in appearance. This is not only beneficial from a company-identity standpoint, but also facilitates fast reference, efficient file and document management, and thorough research.

The "Reference" of each document should be simply the identification of the project, including appropriate contract numbers and any other necessary formality. The Reference should then be followed by "Subject" area. The Subject area is used for a number of purposes that are dealt with in detail throughout this Operations Manual. For purposes of this discussion here, the Subject area will at a minimum contain a concise description of the particular issue, and may also contain file references or other means to correlate the document with the project record. Keep the Subject description consistent from document-to-document, in order to facilitate quick research and understanding among the multiple parties.

4. **Confine letters to a single page, if at all possible.**

This correspondence rule is perhaps the most difficult of all to achieve (and may be most difficult to see the benefit of), but is actually possible in many circumstances. Remember that if your purpose is to compel action, the quickest way to achieve that result is through clear communication. Think of your letter as an executive summary written to influence the mind of the decision makers. Consider the attention span of busy individuals who must understand and respond to your letter. Consider your own reaction when you receive a long, laborious letter, and the work that you've got to go through just to make sure that you've understood the main points. Make your point(s) in your own letters clear. Get to those points fast. If the issue is complex, consider outlining the basics and steer the letter to its conclusion. Include separate attachments that support your contentions in an organized backup package.

Consider what happens in the other organization when it receives your letter. Different levels of authority may become involved with different issues. As your letter is passed from person to person and from level to level, the important, salient points distill down to the problem and (your) solution. As your letter moves up the chain in the other organization, executives at progressively higher levels are correspondingly less interested in the detail and more in the conclusion. Leave the detail in supplementary packages that support your position, but do not allow your own letter to ramble on with your intermediate supporting arguments within the body of the document.

5. **Avoid redundancy and unnecessary references and repetitions.**

If your letter format follows through with the recommendations of item #3 above, it will have included a clear project reference and the subject description that together create an unmistakable summary of the issue. There is no need to waste additional space and attention on repeating the same information in an introductory paragraph. Consider, for example, these two statements introducing the same letter:

> "This letter is in response to your letter dated June 17, 2000 in which you advised that you will not accept the pricing that was submitted on June 10, 2000 for the Vacuum Sludge Dewatering System in the north part of the Regional Waste Treatment Facility Contract #56009."

And

"This letter is in response to your letter dated June 17, 2000 regarding the subject."

The first version will put you well on your way to writing a multiple-page letter. If instead, the "Reference" in your letter included the "Regional Waste Treatment Facility Contract #56009" (complete project description), and the "Subject" referred to in the second version included "Vacuum Sludge Dewatering System Change Order Pricing" (summary description of the issue), the second version is much more clean. It properly ties your letter to the correspondence chain in an organized manner but does not unnecessarily encumber the document with extraneous language that will dilute the main point.

The second common problem with redundancy is simply repeating a point at the conclusion of your letter that was made in the early part of the document. Instead, organize your thoughts, list your points in priority sequence, and make them once and in logical order.

6. Use outline form.

Admittedly, there are times when the "single page" objective discussed above is difficult if at all possible to achieve. Many types of communications do demand logical support for conclusions drawn in the body of the document. In these and other types of circumstances that ultimately demand a longer and more complex letter, you must still do your best to break down the document into manageable and understandable components. Even short letters that contain multiple issues should be treated in this manner.

After your short and clear introduction, organize your letter into a logical developmental sequence. Consider numbering your points and summarizing each with a short title. If the letter is a chronological development, for example, you might itemize it by date or event. Clearly identify each individual item. If you're confirming several facts, separate each of them out of the body copy and number them as separate paragraphs.

If you are drawing several conclusions requiring two or more specific actions, reconsider the entire content of the letter. Look for ways to break the separate issues into individual components. List each item in outline form and number them. You may discover that it will be more appropriate to construct multiple letters from the information once all of the details have been so clarified.

Following the suggestions above will at first crystallize your own thought process. From that point, your representation of the facts should be much more understandable to everyone who will become involved either initially or at some later point in time. As the final advantage to organizing your letters in this way, research conducted at a later time or by others will be greatly facilitated. Attention will be allowed to be focused on the points without the need to wade through large and confusing amounts of extraneous information.

7. Keep each discussion simple.

This idea is much more important than simple application of the "KISS" formula. It begins with the recognition of the fundamental principle that the absolute value of total understanding drops dramatically as the number of people who must understand the particular issue increases. If, for example, ten people must be involved in the resolution of any given issue, it is all but guaranteed that at least three or four of these individuals will completely misunderstand the issue in their first consideration of it. Reasons for this effect may vary, ranging from the simple lack of attention to genuine lack of ability to grasp the concept. Whatever the true reason, however, the effect is very real.

The second idea is based in simple dilution of the issue. If the final decision maker, for example, is higher up (several levels above) than the initial recipient of your letter in their organization, that individual may be less inclined to (or interested in) understand and specifically address the detail—the minutiae—of the particular issue. The lower your issue is in terms of cost and organizational impact, the more distracted decision makers may be. Because of these two effects, it is necessary to make the issue as easy to grasp as possible in order for those busy, distracted individuals to see your point fast, recognize that it is backed up, and acknowledge that your position is supported by the project record. To the extent that all these things can be achieved with your short, clear correspondence, the point will be made to the higher-level bureaucrat that it will be a lot of work to even argue with you, much less prevail.

8. Use cause-effect style.

Develop the habit of requiring every word of your correspondence to have a purpose. Ask yourself if each sentence or phrase contributes to understanding or support, or simply adds verbiage. Arrange short, simplified statements as part of your outline presentation into logical sequence that methodically arrives at its conclusion. Don't jump around, and don't confuse your train of thought. Spend the time necessary to reconsider your draft to be sure that there isn't a more direct route to the conclusion. Whenever you've had difficulty in working with longer sentences or otherwise clarifying confusing points, show your draft to an individual in your company who is removed from the issue. Work your language until that individual clearly understands the issue, your position, and your conclusion.

9. Stay factual.

A letter (or any other written document for that matter) is no place to display emotion, sarcasm, innuendo, or irrelevant criticism. In most instances, it is also no place for opinionated statements. Instead, keep each letter confined to the facts, accurate statements, and sensible, logical conclusions. Above all, be professional.

Leave out "facts" that really aren't.

If it had for any reason become necessary to make some assumption in order to develop a position or to prepare a work product, clearly state the assumption made and the reason for it. Beyond this specific condition, do not speculate.

All of the above things may or may not have a place in verbal and off-the-record discussions, but they certainly have no place in a written bombshell that will be distributed around the planet to people with wide variations in understanding and dispositions—all without you there to explain it. Picture yourself one week from now sitting in a room with several people reviewing your written statements. How comfortable do you feel? How "professional" do you consider the letter? Do you see any justification for the other party to be insulted? Have you given your opponent any opportunity to "grandstand"—at your expense?

10. Avoid personal attack.

It's been said that calling a person a liar will not make him or her honest. Don't question the other party's motivation, intentions, or competence. Don't make accusations. Don't make threats. Even if you do believe the other person's motivations are "questionable," open accusations or insinuations will more likely drive you further apart—at least initially. Avoid these unnecessary and overemotional delays by leaving this type of language out of your letters altogether.

11. Stay cool.

Unfortunately, it is not all that unusual to receive a letter or otherwise be subjected to something that frustrates and angers you. Despite your hard work and continuing effort to be professional, you may simply be faced by an individual who does not respect the rules identified in the preceding section. Anger, insult, disgust, or any other emotion may too quickly divert your attention from the main point. If this happens, these emotions may cause you to respond inappropriately and at least delay resolution of your issue by allowing yourself to succumb to the diversion. In the worst cases, you may be prompted to fire off a hot return volley if you allow yourself to lose control.

If you do allow yourself to respond in such an uncontrolled manner, the unfortunate truth is such that your response will appear short-sighted and unprofessional. At worst, you may blurt out things that you will later regret. Somewhere in the middle of this range, your overemotional response is sure to anger at least one of your "opponents," at one or more levels of your opponent's organization. The only result (besides perhaps a short-lived feeling of satisfaction) is added delay to the resolution of your issue. You'll have given your opponent something to "prove."

In most cases, this type of response will never resolve the problem. You may get some degree of satisfaction from having "told them," but you will have shoved them a little farther away from your position before you realize that now you've got to start bringing them back.

If after considering all these reasons you still feel the need for such a response, by all means write it. Get it down on paper in all its agonizing detail. But don't send it. Set it down, and let it cool. Don't ever leave it lying around on the top of your desk. Put it in your top drawer (get it off your desk in case some well-meaning associate sends it for you). Leave it in your drawer for a day (or overnight). If the letter looks as good in the morning as it did the moment that you wrote it, send it. My bet, though, is that the light of the new day (after you've had time to cool yourself off) will show you that the letter is too long, complicated, redundant, confusing, and maybe at least a little insulting. Rewrite the letter focusing on resolving the problem. File the first one (clearly noting on its face that it had not been sent); it probably does contain a number of useful pieces of information that you have researched during its preparation that may become helpful later.

12. Get to the decision maker.

Most often, the formal route of your correspondence distribution is determined by your contract working procedure. All letters will have been required to be addressed to a particular individual, with copies to certain others. If the highest authority (. . . the Commissioner . . .) is the addressee, the routine may provide that one of the cc's (the Clerk of the Works) actually will deal with a letter first. If the addressee is the Clerk, however, the higher authority will be one of the cc's.

Understand the specific internal procedure used in your opponent's organization to review and respond to your correspondence. Know and understand the levels, if any, of cost or other issues that cause an item to be forced to another level in your opponent's organization. Know who will make the final decision in order to allow you to determine the most effective format for your particular correspondence.

If a shorter time to the decision is needed, or if overturning a lower-level decision is desired, find a way to shift your letter out of the "routine correspondence" category in order to get to the decision maker's attention directly. The contract itself may turn out to be the greatest help here. The dispute resolution clause may, for example, contain a procedure that ". . . the Commissioner will interpret the plans . . ." There may be a procedure that requires the movement of an issue through a prescribed program on the precise timetable. If such a program exists in your contract, follow it precisely and compel your opponent to do the same. Use the specific language in your contract to bring the decision makers directly to the plate, and address your problem directly if you can't get the results you want at the lower levels.

13. Respect your contract.

Far too often, professionals, managers, and staff members create or respond to an issue because of assumptions made in the relationships, rights, and responsibilities between the parties. At the jobsite level, every job "feels" like most others administratively because each will have an architect, owner representative, and construction force. People will accordingly have a great tendency to deal with a problem the way it was done on the last project.

Contemporary construction contracts, however, deal with many varied forms of project delivery systems. Lump Sum General Contracting (GC), Agency Construction Management (ACM), Construction Management with Guaranteed Maximum Price (CM w/GMP), and Design-Build (DB) are the four basic forms of project delivery. From each basic contracting form, it seems that each new drafter of construction contracts has felt the need to customize these "standard" arrangements to significant degrees. Because of all of these things, the specific rights, responsibilities, and relationships in your particular contract are very likely to be somewhat different than they were in your last contract. The General Conditions of the Contract will be customized by the Supplementary General Conditions. The entire process will (supposedly) be orchestrated by the Working Procedure or other similarly titled section of the specifications. Add to all this the state statutes and federal public contracting procedures, and it becomes easier to see that you'd better be aware of the specific set of rules for each particular project. Don't risk having your eloquent presentation being dropped into the circular file, or having to suffer the embarrassment of having to retract a position for being told that "you should have read your contract," and that "you should have known better." Because you should have known better.

14. Guide the decision.

Before you begin to draft a letter, know your issue clearly, and know your desired outcome. Don't simply state your problem and close the letter. Just stating your problem and asking for an answer isn't any better. Stating your problem and asking for the answer that you prefer may seem to be better, but it really isn't—they already expected that you wanted that answer.

Your job in drafting the most effective correspondence is to state the issue or identify the problem and then give the contractually supported reasons why the answer must be as you see it. From there, all you should need is a confirmation of the proper interpretation from the appropriate authority.

Don't, for example, simply state that:

> "Both the steel and masonry subcontractors take exception to providing the welded portion of the masonry anchors. Please advise which subcontractor should perform the work."

Instead, indicate after the first sentence that:

1. "Section 04200 3.3A provides that the Mason subcontractor include the fixed portion of all anchors.
2. Detail 6/S3 on the structural drawings shows the welded portion of the anchor on the columns.
3. General Conditions Paragraph 12.3 states the specifications take precedence over the plans.

"Therefore, in accordance with General Conditions Article 3 (the clause that says that the owner is the one who's stuck with the responsibility to make the decision) please confirm that the subject work is included in Section 04200."

Have your logical progression clearly supported by the contract specifics. In the case of a conflict within the plans and specifications, call attention to the contractual mechanism for resolving such complexity and apply it in your logical development. Use the contract to force the responsible parties to make (and remain responsible/liable for) the decisions. Don't allow your opponent to force you to make decisions that you otherwise are not contractually responsible for.

15. Require specific action by a specific date.

After you've taken the steps necessary to guide the action toward the decision you need, state the specific decision as simply and as clearly as possible. After that, state the absolute date by which the action is required and pin onto it some kind of consequence for failing to meet that date.

Examples of effective remarks include:

- "Response after that date will interfere with the masonry work and delay the project."
- "No action by that date will cause the contract work to progress beyond the work in question. In that event, cost and time needed to rework the area will be added to the cost of this additional work."
- "Response after that date will cause the steel to be delivered to the site without the subject anchors. Even if the work is determined to be contractually the responsibility of the steel subcontractor, they will then be entitled to a change order to cover the extra cost of providing the anchors in the field, instead of fabricating the anchors in the steel shop."

Above all, whatever you determined to be the consequence of late action, be certain that it is accurate and legitimate. Never use idle remarks or threats that you either cannot or will not be able to follow through on. Don't risk your stature as a respected professional over this type of issue.

If the noted date passes without the required action, react quickly. Send a notice, second request, or whatever other type of correspondence is necessary to prompt the decision. Because your original remarks were accurate, including your follow-up notice that the event has occurred, it should be made clear that your opponent had better get moving now before it gets worse. As an example, you might say something like: "As of today, we have received no response to my letter of (date). The subject anchors have accordingly not yet been ordered, and the masonry work is now delayed by this issue. The total impact on the schedule will be analyzed after your response is finally received, and you will be advised at that time of related extra costs."

16. Follow-up/follow through.

Before considering any type of deadline as recommended in item 15 above, think your entire scenarios through as clearly and as thoroughly as possible. Know the problems that you will face in all their specific detail if an adequate and complete response is not received in time for you to take your own needed action. Put yourself in that position on that future date and understand your new set of problems that exist without an appropriate response. Know the difference between effects and impact that "must" occur and those that only "may" or "can" occur in the event of an incomplete, improper, or untimely response.

After you have clearly identified and quantified these potential impacts, you will be in a better position to isolate those conditions and circumstances that will occur in the event of such improper or untimely response.

Before committing such consequences to paper, however, have your end-run decision made in advance. Do not indicate that a consequence will occur unless you are absolutely certain that it is so. If the consequence is an outside event that is beyond your control, so be it. If, however, the consequence is a specific action that you and your company will take in the event of any improper or untimely response, be absolutely certain that you and your company are thoroughly prepared to take the specific action unquestionably and without hesitation. If there is any question as to the resolve and/or your ability to take such action, do not threaten any such response.

Once having so prepared your response, keep close attention to the deadline indicated in your letter. When that deadline occurs, be sure to react immediately and decisively. Be sure to create and perpetuate your own reputation as one who says what he means and means what he says.

2.3.3 Correspondence Distribution

All correspondence must follow up with any procedure that may be described in the respective contract Working Procedure or as established in the preconstruction or initial job meeting. These instructions will incorporate the addressee along with routine distribution.

Whether or not instructed to do so officially, other distribution of your correspondence is routinely necessary. This includes:

1. The architect. Copies of all correspondence to the owner, any outside agency, other design professionals, or any other entity involved with the design or the construction of the project must be sent to the designer of record. This is not just good courtesy, but is necessary to ensure that the requirements for timely notice of all parties has been met.

2. The owner representative. Copies of all correspondence to the architect, engineers, other design professionals, outside agencies, and any other entity having anything to do with the private design, function, permits, and any other issues that are not otherwise considered to be privileged communications must be sent to the owner's representative. Each individual must be made aware of every issue on a current basis, and each must be made aware that the other has been so put on notice.

3. Each person definitely or potentially involved with the issue. If, for example, you are requesting an owner interpretation of the specification potentially involving the steel and masonry subcontractors, those individuals and companies must be copied in your correspondence. If you document any action or statement

by any individual in any company, that person must be put on distribution. This is not only good business practice, but also serves as an important notification function. Beyond that, it gives a healthy amount of legitimacy to your statement by showing to the world immediately that you're not afraid of holding your remarks up to the light. If you're putting words in a person's mouth, you're not afraid of that person seeing exactly that you know what has been said.

4. File instruction. The writer of the letter—not the administrative assistant or file clerk—knows all files potentially affected by the issue and the particular document. At the time that the letter is written, decide all files that must receive a copy and indicate such in the final file designation. For example, include a "cc: File: x, y, z" where x, y, and z are specific file folders. In this way anyone can be given the filing task without the risk of misfiling.

2.3.4 Correspondence Checklist and Desk Display (*page 2.18*)

Admittedly, it may oftentimes be a challenge to maintain the objectives of the preceding section with respect to the preparation and preservation of the most effective correspondence. The large number of issues, long hours, and other factors too often combine to perpetuate an environment that works against our effectiveness.

The checklist in Figure 2.3.4 summarizes the objectives of the preceding section as a reminder to help each of us to apply these principles in every situation. Photocopy the exhibit and cut out the checklist. Either post it near your work area, or fold on the perforation and stand it up on your desk. Keep it close as a constant reminder. Refer to it and keep your correspondence focused on achieving timely objectives—and not on keeping fires burning.

2.4 Files and File Management

2.4.1 Overall File Structure

The focus of this Operations Manual is on project management. The central accounting files of contracts, subcontracts, purchase orders, Accounts Payable, Accounts Receivable, job cost, general ledger, and tax applications are beyond the scope addressed in this book. Discussion of general accounting will be detailed as necessary for comprehensive treatment by the project management function. Accordingly, this section deals with the specific project files only.

Project management files for each project will consist of the following general sections:

- The Contract Documents
- The General Project File
- Duplicate Correspondence File Books
- Clarifications/Changes Log and Books
- Subcontractor Summary and Telephone Log
- Jobsite Subcontractor Performance Summary and Telephone Log
- Submittal Log

2.3.4
Correspondence Checklist and Desk Display

(Stand Backing)

(Fold)

Effective Correspondence:
- Keep each project separate.
- Use single subject, or small group.
- Develop consistent document format.
- Use single page, if at all possible.
- Avoid redundancy.
- Use outline form.
- Keep each discussion simple.
- Use cause-effect style.
- State factual.
- Stay cool.
- Get to the decision maker.
- Respect your contract.
- Require specific action by a specific date.
- Guide the decision.
- Follow-up, follow-through.

2.4.2 The Contract Documents

The complete set of Contract Documents typically includes the items listed below. The Bid Document and Agreement components of the Contract will be filed and managed with the General Project File. Here, we will begin with the setup and maintenance of the Plans, Specifications, Changes, and the various portions of the tactical files themselves.

The Contract Documents

Bid Documents

- Instructions to Bidders
- Prebid Meeting Minutes
- Prebid Jobsite Review/Inspection Record
- "Informational" Data
- Prime Contractor Bid
- Subcontractor Bids
- Separate Contractor Bids
- Allowance Items and Conditions
- Special and/or Supplementary Information

The Agreement

- Owner/Prime Contractor
- Owner/Design Professional
- Prime Contractor/Design Professional
- Prime Contractor/Subvendors
- Owner/Separate Contractors
- Surety Bonds
- Project Insurance
- Personal and Third-Party Guarantees
- Changes Executed after the Original Agreement

Plans

- Architectural
- Site Development
- Site Utilities
- Structural
- HVAC/Climate Control

- Fire Protection
- Electrical
- Specialty Designs
- Shop Drawings
- Product Data
- Clarifications
- Changes

Specifications

- General Conditions
- Supplementary General Conditions
- Special Conditions/Special Provisions
- Technical Specifications
- Referenced Product Standards
- Referenced Technical Standards
- Referenced Legal Standards
- Building and Other Codes
- State and Federal Regulations
- Labor Laws and Standards
- Environmental Laws and Standards

2.4.3 Setup and Maintenance of Plans, Specifications, and Changes

The discussion that follows applies to the project files to be managed during the construction phase of projects constructed under the Lump Sum General Contracting and various forms of Construction Management methods of project delivery. It is also to be used for projects constructed under the Design-Build delivery format, with the distinction that the design function has been added to the Design-Build Contractor's total responsibility. Accordingly, projects constructed under the Design-Build arrangement will have a separate but coordinated file that will accommodate the activities of the design function, as well as its interaction between the designer, the owner, and the construction force. Beyond this distinction, the design entity should be considered distinct from the construction operation—even in the Design-Build project delivery format for internal management purposes. If the relationships are preserved in this manner, the file management system described in this section and throughout this Operations Manual will continue to apply to virtually every project delivery method.

1) Initial Setup
   a) Consolidation of Drawings
      i) Generally, the bid documents will contain a provision that identifies the number of complete sets of Contract Documents that will be given to the Prime Contractor upon execution of the Agreement. Immediately after any contract is awarded, arrange to secure all such copies of the contract documents for distribution to the various subvendors for the project.
      ii) Arrange a single set of plans marked "Office" to be maintained at the central office. Use a single plan stick if possible for a small project, or separate the plans into their major sections to be arranged on several organized plan sticks in the case of a larger project.
      iii) Arrange two (2) duplicate sets of the plans for use at the jobsite. One set will be clearly marked "Jobsite," and the other set will be clearly marked "As-Builts." The uses of the sets will be described in the respective sections of this Operations Manual.
      iv) Place three (3) complete copies of the specifications including the General Conditions, Supplementary General Conditions, Working Procedure, and Technical Specifications into hard-cover three-ring binders. Each binder will be clearly labeled and distributed with the respective matching set of plans as described above. These will later be matched with the file tabs of other binders for the project in order to keep the color-coding of all project documentation consistent.
      v) Both the "Office" and "Jobsite" copies of the plans and specifications must be maintained in those respective locations and never removed. They will be managed from this point forward in a way that will help coordinate all relevant information properly during the construction period. The "As-Built" set of plans and specifications will be maintained at the jobsite and be developed throughout the life of the project. The set will serve as a complete record of all actual work and will be maintained in a form that will include at a minimum any specific requirements that may be identified in the Contract Documents. Beyond those minimum requirements, the As-Built set will be maintained as described in the appropriate section of this Operations Manual. At the appropriate time, a complete As-Built set of documents will be turned over to the owner at the conclusion of the project.
   b) Post Each Addendum.
      i) Prior to any other activity being performed on the contract, post each addendum on both the "Office" and "Jobsite" set of plans and specifications. Photocopy each addendum onto a light-colored paper (preferably pink or yellow). Physically cut out each addendum note, including its numerical reference. Paste or tape the

note at the specific area of plan or specification altered by it. Do not cover or otherwise hide the original requirement, but add a clear reference to the addendum change. Immediately adjacent to each such posting, note which addendum number it came from. Short notes or very minor changes can be made directly in red pen, with the appropriate addendum reference. This process is crucial to ensuring that no change made during the bidding process will be overlooked during the buyout, submittal, and construction processes on the project as they proceed. Failing to post the addendum items in the manner indicated here will exponentially increase the amount of work necessary to ensure that each addendum is properly accommodated every time any actions are taken, and it will greatly increase the probability that an addendum change will be overlooked sometime during the life of the project.

It will be sufficient on the "As-Built" set of plans and specifications to simply attach a copy of a complete set of addenda for the record.

ii) The posting of each individual addendum remark in each item's correct location is absolutely essential to the avoidance of serious and costly problems resulting from simple oversight of officially changed items and conditions. The method indicated here is the simplest, most secure, and time-tested procedure that will virtually eliminate the need throughout the project's life for constant hunting through an entire set of an addendum every time any specification or plan note is referenced in order to be sure that it is correct and had not been officially changed.

2) Maintenance
  a) Post Changes
    i) As the job progresses, there will be numerous sources of modifications to the physical construction. Change Orders, clarifications, job meeting discussions, and even telephone conversations will officially alter the contract requirements. Each of these changes will have its own development history and paper trail to varying degrees of complexity, and proper documentation of each is covered in detail in the respective section of this Operations Manual.
    ii) The most important idea here is simply to be certain that every modification—however simple or complex—is clearly marked in color on both the "Jobsite" and "Office" sets of Contract Documents, along with appropriate references to the correct source of the official change. Such forms of these alterations may include:

      - "SK" drawings and other clarification sketches
      - Job meeting discussions
      - Telephone and other conversations

- Requests for Information (RFIs)
- Construction Change Directives (CCDs)
- Any manner of written correspondence: letters, faxes, etc.

iii) "SK" drawings or other clarification sketches should be added directly to the "stick" sets of both "Office" and "Jobsite" Contract Documents. Clarification sketches of construction details not involving changes in clause or time can often occur as a matter of course throughout the construction phase of a project. As they occur, highlight all areas affected on both sets of plans, noting a clear reference to the new sketch or clarification. Tape the photocopy of the change sketches directly on the plan adjacent to the changed area if there is room directly on the document. If there is no room for the added clarifications sketch (as is more commonly the case), tape the sketch on the opposite side (left) sheet of the plans (onto the back of the previous plan page). At the highlighted area of the original plan, include an appropriate reference to note the presence of the sketch, such as "refer to SK-7 taped opposite."

Many times, changes and clarifications that develop during discussions—and especially subsequent to those discussions—can become more complex than may have originally been expected. In these cases, many designers resist taking the time to make sketches that properly accommodate the changes. If anyone involved with proper coordination (beginning with yourself) is at all confused with the change descriptions, or if you feel at any time that there is any potential at all for confusion now or later, insist that the designer properly issue a complete written clarification, and treat the document as described above. Follow the instructions in Section 2.4.6 Clarification/Change Log in order to keep a chronological record of all changes as they may develop.

b) Keep job meeting discussions properly coordinated.

i) Job meetings and their minutes are critically important communication and documentation vehicles, not only of project effects and impacts, but of their chronologies. They catalog the conduct of the parties and spotlight who is and who is not living up to their contractual responsibilities. If conducted and documented as described in the appropriate section of this Operations Manual, the job meeting record details will be clear and properly correlated.

The logistical difficulty in the timely incorporation of individual items occurring in any single job meeting lies not only in the number of job meeting items, but also in the time that is commonly consumed in the preparation and distribution of the "official" meeting minutes. Unfortunately, it is most common that the earliest one can hope to see the official job meeting minutes is the day of the following job meeting. If we wait until receipt of the meeting minutes before we take action on those items for which we are responsible—or to expect action on the parts of others for those items for which they are responsible—our projects will never move to completion on any timely basis. As the number of items dealt with in our job meetings inevitably increases, the risk of failing to treat one or more items in a timely manner can rise dramatically.

Because of these realities, it is critically important that we keep our own meticulous project record notes, act on our own items as quickly as possible after the meeting, and expect the other contracting parties to do the same. The best way to compel action by the other contracting parties prior to receipt of any "official" meeting minutes is to secure commitment from the responsible party to perform a specific action by a particular date, and to require that the commitment (and that performance date) be included in the official meeting minutes.

Most job meetings are held at the jobsite. If they are not being held at the jobsite on your project at the present moment, get the location of the meeting changed to the jobsite immediately. Accordingly, have the "Jobsite" set of plans and specifications available at every job meeting. Immediately upon the resolution of any item that results in a change to the documents, consider marking the set of plans and/or specifications right at the meeting, noting the job meeting number and date as the source of the change. In certain cases, it may be very possible (and certainly well advised) to have the meeting participants initial such changes directly on the jobsite copy of the plans. This will be an easy and powerful reference that will be supported by the respective meeting minutes if the particular changes are questioned at a later date. This procedure will completely eliminate all possibility that the change would be forgotten or undone—at your expense. In every case for any change that has been so initialed at the jobsite, immediately photocopy that section of the plan or specification containing the initials and distribute to the home office for appropriate filing and posting on the "Office" set of documents.

As meeting attendees continually observe this procedure as a matter of course, they may in turn become more comfortable with the legitimacy of the notations on the official project documents. They will for the most part have witnessed how the notes were made and will be hard-pressed to force any objection to these types of documented changes at a later date.

c) Adequately Document Telephone and Other Conversations

    i) Apparently minor changes and clarifications can become a daily routine (". . . there's no dimension . . ."). Simple conversation with an owner representative or design professional can get a fast answer that is acted upon immediately:

Question: . . . "I'm laying out the wall now; what's the dimension?"

Answer: . . . "the owner's furniture is 3' wide; make it 3' 3"."

In such cases, note the new dimension directly on the plans, indicate the source of information by name, and date it. Ideally, have that individual sign or initial the remark right on the drawings. At the very least, immediately write a confirming memo and send a copy to be so noted on the "office" set of plans. Refer to the section in this Operations Manual that deals with Requests for Information (RFIs), and consider using that procedure if the situation demands it.

If the instruction gets complex, or you feel that the manner in which the instruction was given is unclear or is otherwise confused, insist on the preparation of a clear sketch and/or some other adequate documentation and treat it as described in 1) above.

2.4.4   General Project File

Immediately upon notice of the award of any contract, the project executive will direct the preparation of the General Project File. As part of the initial process, the preparation will include two nearly identical sets of files. One of the sets will serve as the primary project file and be maintained at the central office. The second set—nearly identical to the first—will be arranged and maintained at the jobsite. In all cases, the central office file will be considered by the project to be the primary file, and the jobsite file will be considered to be the "working copy" file. The jobsite file is characterized as being "nearly identical" because it will have both sensitive information and records that are otherwise unnecessary for the construction of the work omitted.

The reasons for the omission of this data from the jobsite file are in no way to preclude access to this information by the senior jobsite company representatives, but to preserve the security of sensitive information and to guard against access by unauthorized individuals. In addition, the jobsite file is spared much of the bulk of preliminary records and transactions such as those involving multiple submittal and approval documents that are generated prior to final documentation that is authorized for use in construction. Managed in this way, the jobsite files will ideally retain only those documents that are to be used for coordination of the work and for actual construction.

For these reasons, the central office file is to be considered at all times to be the primary file and, as such, the integrity of the central office file must never be compromised.

Choose a label color that is coordinated with the specification binder as described in Section 2.4.2 of this Operations Manual—a color that is different from all other projects falling within the responsibility of the single project executive and/or project manager. This will distinguish the project and its file system from all other projects. As all files are developed, the files, file books, log books, and all other project records will be arranged in colors consistent with their individual projects.

If you or your company is using a computer-aided file management system that incorporates document key-word or other search/retrieval facilities, you should be able to correlate such systems with the physical document file structure as indicated in this section of the Operations Manual. Creating and preserving the record structures as indicated here will in no way alter the manner in which those electronic record identification systems will be used. Instead, the principles expressed in this section of the Operations Manual should be used as a guide for designing and organizing the actual forms and formats that will be implemented within the electronic system. Using these principles will greatly facilitate the subsequent research that will be required to retrieve the necessary records once so identified.

The overriding objectives of the document management system discussed here include to logically assign each document into a carefully coordinated set of individual files, to arrange those files in a way that will keep all associated issue files together, to preserve clear relationships among related files, and to be able to serve them up to managers when needed and in total. Note that many of the items identified in the file system, such as the Submittal Log, the Clarification/Change Log, etc., are treated in detail in their respective sections of this Operations Manual. They are not further described in this section and are only identified here for organizational and file management purposes.

The General Project File will be divided between Project Administration and Technical Construction. The individual file folders will be titled and arranged at a minimum as follows:

Project Administration

- Bid Documents
- Company Proposal, Competitive Proposals (if available), Proposal Tabulation (if available)
- Prebid Site Investigation
- Owner/Contractor Agreement
- Owner/Designer Agreement
- Contractor/Designer Agreement (Design-Build Contract)
- Payment and Performance Bonds (Prime Contractor)
- Payment and Performance Bonds (Subvendors)
- Project Insurance Policies and Certificates (Prime Contractor)
- Project Insurance Policies and Certificates (Subvendors)
- Job Cost Summary and Project Estimate
- Schedule of Values (Prime Contract)
- Schedule of Values (Subvendors)

- General Applications for Payment
- Cost Progress Reports
- Subvendors Approvals (if required)
- Jobsite Utilization Program
- Purchase/Award Schedule
- Building and Special Permits
- Daily Field Reports
- General Notifications—Owner
- General Notifications—Architect
- General Notifications—Subvendors
- Progress Photographs
- Special Photographs
- Monthly Narratives (if used)
- Baseline Progress Schedule
- Periodic Schedule Revisions (Separate file for each periodic schedule update)
- Job Meeting Minutes
- Special Meeting Minutes
- Requests for Information (RFIs)
- Nonconformance Notices (NCNs)
- Change Order Summary
- Change Order Files (Separate file for each individual potential/actual Change Order)
- Subvendor Backcharge Summary
- Subvendor Backcharge Files (Separate file for each individual subvendor)
- Substantial Completion
- Guarantees/Warranties
- As-Built Documents
- Maintenance and Operating Manuals
- Certified Payroll Reports (if used)
- Project Deliverables
- System and Facility Training/Owner-Instruction
- Certificate of Occupancy
- Punchlist/Final Completion
- Lien Releases/Waivers of Claims

- Owner Claim Summary
- Owner Claim Files (Separate file for each actual or potential claim)
- Subvendor Claim Summary
- Subvendor Claim Files (Separate file for each actual or potential claim)
- All other files as appropriate; refer to the specific project General Conditions, Supplementary General Conditions, Special Provisions, and other appropriate information as sources for any special file requirements.

Technical Construction Divisions. The file structure identified above is generally formulated based upon the structure of Division 01 of the General Requirements of the contract for construction. In 2004, the Construction Specifications Institute (CSI) reissued their Standard Specification Division numbering system adding many more sections beyond the previous Division 17.

This new system is divided in groupings as follows:

Procurement and Contracting Requirements
   Division 00: Procurement and Contracting Requirements

Specifications Group
   Division 01: General Requirements
   Division 02: Existing Conditions
   Division 03: Concrete
   Division 04: Masonry
   Division 05: Metals
   Division 06: Wood, Plastics and Composites
   Division 07: Thermal and Moisture Protection
   Division 08: Openings
   Division 09: Finishes
   Division 10: Specialties
   Division 11: Equipment
   Division 12: Furnishings
   Division 13: Special construction
   Division 14: Conveying Systems
   Division 15: Reserved
   Division 16: Reserved
   Division 17: Reserved
   Division 18: Reserved
   Division 19: Reserved

Facility Services Subgroup
   Division 20: Reserved
   Division 21: Fire Suppression
   Division 22: Plumbing
   Division 23: Heating, Ventilating, and Air-conditioning
   Division 24: Reserved

Division 25: Integrated Automation
Division 26: Electrical
Division 27: Communications
Division 28: Electronic Safety and Security
Division 29: Reserved

Site and Infrastructure Subgroup

Division 30: Reserved
Division 31: Earthwork
Division 32: Exterior Improvements
Division 33: Utilities
Division 34: Transportation
Division 35: Waterway and Marine Construction
Division 36: Reserved
Division 37: Reserved
Division 38: Reserved
Division 39: Reserved

Process Equipment Subgroup

Division 40: Process Integration
Division 41: Material Processing and Handling Equipment
Division 42: Process Heating, Cooling and Drying Equipment
Division 43: Process Gas and Liquid Handling, Purification
Division 44: Pollution Control Equipment
Division 45: Industry-Specific Manufacturing
Division 46: Reserved
Division 47: Reserved
Division 48: Electrical Power Generation
Division 49: Reserved

Each of these Divisions is further divided into Specification Sections that further define the detail included within each major design area. Among these individual Specification Sections, there is little consistency among Specification Divisions with respect to the manner in which the work of the individual Specification Sections is ultimately divided into specific subvendor bid packages. Because of this, it is necessary to arrange the General Project File to include separate file sections organized around the individual Bid Packages—and not necessarily around individual Specification Divisions and/or Specification Sections.

While it may be true that a Specification Division or Specification Section may coincidentally parallel that of a particular subvendor Bid Package, such a coincidence may not be the common case.

Bid Package File Organization. Once separate bid packages are defined and organized according to the criteria of the previous section, each of the separate bid package files will be physically divided between Contract/Correspondence and Submittals/Approvals. These two file components are separated beginning with the idea that submittals and shop drawings are bulky and become a physical inconvenience when working with the nonsubmittal items of the bid package. Because of this, the Submittals/Approval file component is merely kept as a separate file section within the bid package.

The Contract/Correspondence section of the bid package file will contain at a minimum the following components (as appropriate for the particular bid package):

- Subvendor proposal(s)
- Subcontract work Purchase Order negotiation records
- Subvendor Agreement
- Subvendor Reference Form (refer to this section of the Operations Manual)
- Correspondence (filed in reverse-chronological order)
- Subvendor Schedule of Values
- Subvendor Applications for Payment, and associated payment records
- Subvendor Requests for Information (RFIs)
- Subvendor Nonconformance Notices (NCNs)
- Subvendor Change Orders
- Subvendor Baseline and Revised Schedule Requirements
- Subvendor Bonds and Insurance
- Photographs
- Guarantees/Warranties
- As-Built Documents
- Lien Releases/Waivers of Claims
- Maintenance and Operating Manuals/System Instruction
- Substantial Completion/Certificate of Occupancy
- Punchlist/Final Completion
- Certified Payroll Reports
- Backcharges
- Subvendor Evaluation Form

Subvendor Subfiles. Within each Contract/Correspondence section of each bid package, it is important to make the additional effort to consolidate developing records for individual issues. If we simply allow our files to consolidate documents as they are generated with no further effort to distinguish among them, the resolution of every issue will require a research effort that must take us through not only the entire bid package file itself, but through all definitely—and possibly—related project records as well.

For example, if we are trying to resolve an issue related to an electrical change that may affect the placement of conduit in concrete that is currently being delayed, our research would necessarily take us through the entire "correspondence" portion of the electrical bid package, possibly through the "correspondence" portion of the concrete bid package, and even through the respective files relating to the relevant Construction Progress Schedule Updates. But unfortunately, it does not stop there. In order to be thorough, we would also be forced to review every remark in every job meeting and special meeting record, at a minimum, going back to a time preceding our current estimated date of the earliest identification of the issue. After all that work was done and our issue was (presumably) resolved, the files would be returned to their original locations. When the next issue arises, however, involving, say, changes to electrical lighting that may impact underground conduit placement, we would be forced to begin the entire process all over again from scratch.

In order to avoid this labor-intensive scenario, and work to prevent it from repeatedly occurring to the point where this type of research impacts our ability to perform proactively as project managers, we can arrange our files so that they contain all information relevant to every issue on the current bases. Such an arrangement begins with an acknowledgement that the small amount of work involved in creating these types of files is inconsequential when compared with the major amount of work involved in repeatedly researching every issue as described in the above example.

And so it begins with the principle that any issue generating at least two pieces of correspondence, along with the potential of generating more, will justify the preparation of the separate subfile within the particular subvendor bid package file. As other project records are developed (such as updated progress schedules, meeting minutes, etc.), they are identified contemporaneously, with a copy filed in the newly assigned location. Maintained in this way, one can retrieve such a file with the confidence that all of the information is already included, and that the massive effort otherwise necessary to conduct thorough research as described above is not likely to be necessary.

An example of the issue subfile as described above is as follows:

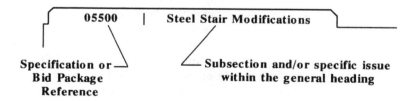

Duplicate "Correspondence File" Books.   The file record structure as described above is intended to create a thoroughly comprehensive and efficiently organized means of retaining every project record in the way they will facilitate our ongoing management of the project. Beyond these objectives, we must further address the issue of file security. We will find additionally that having adequately addressed the file security issue, we will have also created an additional supplementary file retrieval system that will improve our ability to retrieve records, depending upon the information that we have available.

Beginning with the recognition of the disastrous effects that would occur if all or a portion of the official project record files are lost or destroyed, responsible project management must provide for not only security of the original project files, but also the ability to reconstruct those files in the event of such a disaster. Even without considering a major disaster, one's imagination does not have to work very hard to consider the ways in which one single lost file can certainly constitute a "minidisaster."

The Correspondence File Book system described below addresses these issues. At the same time that the General Project File is being prepared as described in the previous section, prepare "Book #1" of the "Correspondence File." This book will be a 3″ or 4″ hard-cover three-ring binder in the color that matches that selected for the General File folders and specification binders as described above. The binder is the first of a series that will be prepared and maintained chronologically. Each binder will be identified with the following information:

- Project Name
- Company Project Number
- Owner Project Number
- Book Number
- Beginning Date
- Reserved area to insert the book's ending date (this information will be provided upon preparation of the next book following in chronological order).

The use of the Correspondence File Books begins with the idea that virtually every piece of communication demands at least one duplicate copy. Typically, the original of any document (both incoming and outgoing) will at the very least be automatically assigned a "Correspondence File" copy, which will be physically designated as the "CF" copy. As these documents are generated and/or received, the original is filed into the respective file or subfile as arranged in accordance with the previous section. Additional copies of the document may be made as necessary to be placed in other issue files and related files—also as required by the arrangement of the previous section. When all other necessary file copies have been prepared, one final "Correspondence File" ("CF") copy is prepared for insertion into the Correspondence File Book. This copy serves as the "emergency" file copy that can be used to reconstitute other files if such an effort becomes necessary.

The "CF" file copies are to be inserted into the respective Correspondence File Book in reverse chronological order (with the most recent date on top), per the date of the document (not per the date of receipt). Maintained in this way, the Correspondence File Books provide a second means of issue research that is strictly related to document dates. In this way, the Correspondence File system provides an often convenient supplemental way of record review and retrieval.

In addition to these benefits, the chronological Correspondence File record system also provides a convenient review forum that allows a manager to periodically review everything that has occurred within the particular time frame in a manner that will allow an effective double-check of his or her own follow-up activities. For example, a Monday-morning review of the previous one or two week's activities is very conveniently possible by simply flipping through the top pages of the most recent Correspondence File Book. Because the issues are arranged by date, they consolidate the activities by all parties into a convenient form. Similarly, it would only take a few moments at the end of each month to flip back through the most recent Correspondence File Book to review all occurrences within the month. Outstanding and open items should jump out at the reviewer, who will then be well positioned for immediate follow-up. If one performs these reviews while standing by the copy machine, it takes only a moment to create a new file copy that will facilitate such follow-up by serving as an attachment to a follow-up correspondence.

Items to be filed in the Correspondence Book in reverse chronological order will include:

1. Duplicate copies of all letters to and from the Company and to any other party regarding the project in any way (copies of which are received by the Company). Be sure that each document is stamped or is otherwise clearly identified with the date of receipt.

2. Copies of all transmittals (both to and from the Company) regarding everything but shop drawings and other submittals for approval. Refer to the Transmittal Procedure and Use section of this Operations Manual for further instruction.

Note that the approval submittal transmittals are generally not included in the Correspondence File Books. This is because the duplicate copies of the approval submittal transmittals and related records will be kept in their own separate set of file books—also arranged in reverse chronological order, as described in following sections. With those transmittals, the record transmittal prepared by the respective architect or engineer will be attached to the copy of the Company transmittal for the same document. Refer to the Shop Drawing Summary Record Procedure section of this Operations Manual for further instruction.

In all cases, be certain that each item is clearly marked with the proper file instruction. For all Letters of Transmittals, refer to the appropriate section of this Operations Manual regarding Transmittal Procedure and Use. All Company correspondence should already contain a file instruction as described in Section 2.3.3. If any document or correspondence does not contain such file instruction, write the instruction at the bottom of the document before it is routed for filing. One method of file instruction that may be consistently used is as follows:

Example File Instruction

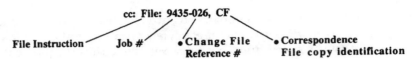

2.4.5 The Chronological File

Another supplementary filing system that often proves invaluable is the Chronological File (or chrono file)—one that files letters not by subject matter, nor by to whom the letter was sent, but by the date the correspondence was written.

Oftentimes it is easier to remember when a certain letter was written than whether or not it went to the architect or engineer or owner or government agency. All it takes is to make another copy of that letter or correspondence and file it *chronologically* instead of by subject matter.

This file becomes particularly useful when the project manager or the project engineer can remember roughly *when* a letter was written but, because the subject matter may have been filed in any one of a number of subject files, can't remember *where* it was filed. A quick flip through the chrono file during the time in question should allow for faster retrieval.

2.4.6 Clarification/Change Log

In a form that is consistent with the other file books prepared for the project, prepare a 3″ or 4″ hard-cover three-ring binder labeled as Clarification/Change Log in order to match the project set, and use tabbed inserts to separate the Log Book into four major divisions identified as follows:

- Clarifications/RFIs Log (Requests for Information)
- Change Order Summary
- Clarifications/RFIs
- Clarification Sketches

Insert the RFI Summary Log form as described in Section 4.4 of this Operations Manual in the first section of the Clarification/Change Log Book. The Change Order Summary Log form as provided in Section 4.9 of this Operations Manual is to be placed in the second section of the Log Book. The third and fourth sections of the Log Book are reserved as areas in which to file copies of the actual transmittals, clarification sketches, etc., that accompany any specific Request for Information (RFI) or clarification directive, along with their actual responses. All of these records are to be placed in the Log Book in reverse chronological order, as with the Correspondence File Books, and additional binders as needed for sections 3 and 4 of the original Log Book. The two Log Book sections (Clarifications/RFIs and Change Order Summary) should remain in the front portion of the Clarification/Change Log Book #1.

2.4.7   Subcontractor Summary and Telephone Log

The Subcontractor Summary and Telephone Log book may become one of the project manager's and project engineer's most effective tools with which to manage the army of subcontractors that will make up the construction force. It is intended for use in the central office (or where the central project engineering function will be consistently performed—at the jobsite at a larger project, for example). Its purpose is to consolidate all relevant subvendor information in a way that will display the status of all outstanding requirements, as well as provide a mechanism that will allow the project manager and project engineer to keep firm control over all Company-subvendor written and verbal communications. Its format will allow the accomplishment of these objectives in a way that will actually save time while increasing the project manager's and project engineer's control over the project.

Prepare the Subcontractor Summary and Telephone Log Book as either a 1″ hard-cover three-ring binder (for smaller projects), or a 2″ hard-cover three-ring binder (for larger projects). Again, prepare the binder in the color to match the project set, and label as "Subcontractor Summary." Insert into the binder one set of alphabetically tabbed dividers, which will eventually receive each Subcontractor Summary Form as described in Section 4.4 of this Operations Manual, followed by the Subcontractor/Supplier Reference Form (Section 4.5) and the Telephone Log Form (described in Section 4.4) for each respective subvendor as their agreements are confirmed and they are incorporated into the project. The manner in which the Log and record forms are to be used is specifically treated in those identified sections of this Operations Manual. At this point, it is sufficient to note that this Log Book will quickly assume the roles of several key management functions. Beginning as the project telephone book, the Subcontractor Summary and Telephone Log Book will serve as the focal point around which all subvendor communications will be controlled.

2.4.8   Jobsite Subcontractor Performance Summary and Telephone Log

In a manner similar to the Subcontractor Summary and Telephone Log described in Section 2.4.7 to be used by the Project Manager and the Project Engineer, the Jobsite Subcontractor Performance Summary and Telephone Log Book is prepared for use at the jobsite by the senior project superintendent. Depending on the size of the project, prepare either a 1″ or 2″ hard-cover three-ring binder (again, with color to match the project file set). Label the binder as "Jobsite Subcontractor Log." Insert one set of alphabetical dividers that will later receive the individual Subcontractor/Supplier Jobsite Performance Forms as described in Section 4.19 of this Operations Manual. Along with each Performance Form, insert copies of the same Subcontractor/Supplier Reference Form described in Section 4.5 of this Operations Manual and as included in the Subcontractor Summary and Telephone Log described in Section 2.4.7, along with the Telephone Log form as described in Section 4.4 for each respective subvendor as they are assigned to the project. Insert a starting set of 30 sheets of the Performance Form and 300 to 600 sheets of the Telephone Log Form. These starting sheets will be used as the project develops. Copies of the Reference Form will be forwarded by the Project Engineer to the jobsite superintendent for incorporation into this Log Book as they are received, completed by the Project Engineer from the respective subvendors. As noted, the use of the Jobsite Subcontractor Performance Summary and Telephone Log Book is treated in detail in those sections of the Operations Manual that describe the use of each form included.

2.4.9 Submittal Log

It is simply not important to record the manner in which submittals are processed from each subvendor, through the constructor's review, to the design professional and/or the owner for approval, back to the constructor for coordination, and ultimately returned to the subvendor for their use. This process can and should be orchestrated as a proactive management activity to compel all of these things to occur in their proper form and content, and within the specific requirements of the progress schedule. In other words, it is wholly insufficient to simply expect the submittal/approval process to occur just because the contract requires it to be so. The project manager and project engineer are to assume complete responsibility for the management and enforcement of the proper and timely completion of this most important project management function.

The Submittal Log Book will be the focal point about which management activity will be orchestrated. It will be the location at which the specification requirements for submittals will be cataloged, the dates of performance identified, and action by responsible parties summarized.

As Submittal Log systems have become computerized in recent years, the biggest problem that has crept into such management systems has been a dramatic loss of visibility. The nature of these records that include voluminous and tedious detail have been allowed to overwhelm the record system. While it is true that the information is almost always there, it is too often arranged in a manner that can make it difficult or impossible to use the data on a day-to-day basis in order to compel performance. Instead, these cumbersome computerized systems can too often be useful only in a de facto analysis in a claim situation. In all of these cases, the problems with records retrieval and use boil down to the form of reporting and presentation. The objective, therefore, in the development of the form and format for the Submittal Log information will be to allow information to be presented to the project staff in its most effective, useful form. Logged data can and should be arranged in a manner that will identify all necessary requirements, clarify relationships, present actual performance relative to required performance, and allow a mechanism for expediting. They should be arranged and presented to management in a way that restores visibility of the data, instead of burying the information in an unusable mass of technical complication. Again, the details of these procedures are presented in the respective sections of this Operations Manual. This section here only prepares the vehicle (Log Book) for its eventual use toward these objectives.

For use, then, by the project manager and project engineer at the central project engineering location, prepare either a 1″ or 2″ (depending on project size) hard-cover three-ring binder (color to match the project set), and label as Submittal Log. Insert the starting set of a minimum of 30 sheets of the Submittal Log Form as described in Section 4.9 of this Operations Manual. The Submittal Log Book is now ready to fill this important function.

2.5   Recovering a Letter Previously Mailed

Although this topic is not directly related to the construction and use of any Log Book or other project record document, it is related to the ongoing administration of the communications between and among offices. If it ever becomes necessary, it is useful to know that it is possible to recover a letter after it has been mailed in the U.S. Postal System. In order to accomplish this, request PS Form 1509 from any branch office of the U.S. Post Office. The success of the use of the form, of course, depends upon how soon the recovery effort is implemented after the document to be retrieved was originally mailed.

2.6   Field Labor Time Reporting

2.6.1   General

Accurate reporting of field labor is, of course, necessary in order to process the respective payroll correctly. The other objective of the procedures is to substantiate actual costs as they are being applied to the individual project components. This is important not only to allow management to compare actual to planned productivity—and to thereby evaluate job-cost performance—but if done correctly can be the mechanism that will allow management to prove actual or potential cost overruns that are not the responsibility of the Company and may be recoverable from another party by the terms of a contract.

Most ongoing construction operations have adequate payroll processing methods for purposes of employee compensation. Because of this, such systems are beyond the scope of this book and are not treated in detail here. Even the largest companies, however, can do a very poor job of differentiating between costs spent on work of the original contract and costs spent on the performance of changed or added work. These companies never really know where their money is going. In the worst cases, the cost to perform added or changed work can be unintentionally combined with the job-cost categories for the work of the original contract. In such cases, it becomes impossible to substantiate actual money spent on changed or added work—even after it becomes obvious that the changed or added work has been completed. Such a lack of precision in job-cost documentation not only impacts the way the project performance is reported on an accounting basis, but can adversely affect the way estimates are prepared and bids are submitted on future projects. And so because of this, any improper and imprecise payroll reporting system on one project can affect other projects as well. Finally, an imprecise job-cost reporting method that fails to differentiate between work performed on the original contract and that performed on changed or added work can place the Company at a serious disadvantage if it is attempting to substantiate its actual costs and damages in a claim situation.

For these reasons, it is necessary to have a field labor payroll reporting system that not only ensures precision in compensation to the individuals comprising the workforce, but also accurately assigns those values to the proper job-cost categories for both work of the original contract and for added or changed work interjected into the process after work had begun.

2.6.2   Field Payroll Report Form

The Field Payroll Report Form is broken into the specific activities that are performed on a given day by a particular employee. A numbering system will be assigned to the possible labor categories as determined by the project work estimates. If there is no current activity number for a new item of work being done, contact the project engineer to secure a new added number assignment.

The labor hours are reported against those activity numbers. The result will be a job-cost report that will identify the exact labor cost actually applied to the individual job components.

This kind of detail will provide valuable comparisons to the cost estimates for the respective activities. Actual costs will be tracked against the estimates as they occur, in order to allow identification of potential problems before opportunities for correction have passed.

Finally, the accurate detail of the records will provide an indisputable account of the actual cost of time and material, change order, or any matter in dispute. In each of these cases, absolute attention must be devoted to timeliness and accuracy of the information.

2.6.3   Procedure

1. Each hourly wage individual on the project staff is required to prepare and maintain the forms on a daily basis. *There are no exceptions. All* salaried field personnel are required to use the *Weekly Administrative Time Sheet* as described in Section 2.7 of the Manual.

2. Most field personnel are assigned to a single project. For those with any multi-project responsibility, a single form (or sets of forms) must be completed for *each individual project.* Never combine projects on the same form for any reason.

3. Treat change orders, potential change orders, work done under protest, and work that is or may be the subject of a claim as a separate project. The general project number with the change suffix will serve as the complete "Job Number."

2.6.3
Example Change File Identification

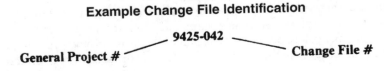

General Project #   9425-042   Change File #

4. The forms should be filled out throughout the day as the activities are changing, but in any case they *must* be filled out *every day* at an absolute minimum.

5. Use the *Project Administrative Activity List* included in Section 2.8 of the Manual as the source for the activity numbers to be used to identify the particular administrative activities.

6. To report activity information for specific field work (concrete formwork, rough carpentry, etc.), secure the general activity list from the central office as determined for the specific project.

   This is a listing of each trade activity to be performed by Company personnel that has been prepared from the original bid package estimate for the work category. Refer to Sections 4 and 8 of the Manual for details on its preparation.

7. All completed forms are to be returned to the central office each week on the designated day. Check with your Accounting Department for the requirement.

2.6.4 Sample Field Payroll Report Form—Completed Example (*page 2.40*)

2.6.5 Sample Field Payroll Report Form—Blank Form (*page 2.41*)

2.7 Administrative Time Reporting

2.7.1 General

Because Project Managers, Project Engineers, Project Executives, Estimators, and Administrative Assistants are salaried personnel, it is not necessary to report their labor hours for payroll. The purpose of properly reporting administrative time is to provide management with accurate information regarding:

- Where administrative overhead is spent
- Where the largest percentage of overhead is spent in terms of the project's life cycle
- What percentage of administrative overhead is spent in the field and in the central office
- What charges to apply to changes, delays, interruptions, accelerations, and reschedules
- What are the actual costs of the project's general conditions components

What is necessary is for the Company to be able to determine on a current basis the divisions of time spent by salaried personnel on the various planned and other activities. Reasons for this include:

2.6.4
Sample Field Payroll Report Form—Completed Example

FIELD PAYROLL REPORT

PROJECT: __PLAINVILLE LIBRARY__   NO: __9415__   Week Ending: __1 / 22 / 94__

EMPLOYEE	SUN		MON		TUE		WED		THUR		FRI		SAT		TOTAL HOURS
	Act No	Hrs	Act No	Hrs	Act No	Hrs	Act No	Hrs	Act No	Hrs	Act No	Hrs	Act No	Hrs	
No. **5467**			0151	4	0151	4	0151	3	0151	3					14
			0155	3	0155	4	0155	3			0155	4			14
Name:			0190	1			0190	2	0190	3	0190	4			10
ED FREDRICK									0210	2					2
	Total		Total	8	Total	8	Total	8	Total	8	Total	8	Total		40
No. **3320**			0180	2	0180	4	0180	4			0180	2	0810	2	14
			0110	2					0110	4	0110	2			8
Name:			1900	4	1900	4	1900	2	1900	2					12
M. LYDEN							1400	2							2
									1550	2	1550	4			6
	Total		Total	8	Total	8	Total	8	Total	8	Total	8	Total	2	42
No. **4141**			2200	4	2200	4	2200	4	2200	2	2200	3			17
			2210	4							2210	3			7
Name:			2100	2	2100	4			2100	2	2100	2			10
L. TEE					3550	2	3550	2							4
							3500	2	3500	4					6
	Total		Total	10	Total	10	Total	8	Total	8	Total	8	Total		44
No.															
Name:															
	Total		Total		Total		Total		Total		Total		Total		
No.															
Name:															
	Total		Total		Total		Total		Total		Total		Total		

1. For planning reasons, a knowledge of where time is really being spent on routine operations of the individuals is helpful for determining future time/cost requirements for particular assignments.

2. Of particular interest from that point will be the time soaked up by changes, problems, or on other activities that are otherwise not the direct responsibility of company individuals to perform directly (such as managing a subcontractor's subcontractor).

   In all these cases, it is crucial to maintain accurate records of *all* costs and efforts related to these changed conditions if the Company is to retain any ability to recoup these extra costs and eventually remain whole.

2.6.5
Sample Field Payroll Report Form—Blank Form

FIELD PAYROLL REPORT

PROJECT:_____ NO:_____   Week Ending: _____/_____/_____

EMPLOYEE	SUN		MON		TUE		WED		THUR		FRI		SAT		TOTAL HOURS
	Act No	Hrs	Act No	Hrs	Act No	Hrs	Act No	Hrs	Act No	Hrs	Act No	Hrs	Act No	Hrs	
No._____ Name: _____															
	Total ___		Total ___		Total ___		Total ___		Total ___		Total ___		Total ___		
No._____ Name: _____															
	Total ___		Total ___		Total ___		Total ___		Total ___		Total ___		Total ___		
No._____ Name: _____															
	Total ___		Total ___		Total ___		Total ___		Total ___		Total ___		Total ___		
No._____ Name: _____															
	Total ___		Total ___		Total ___		Total ___		Total ___		Total ___		Total ___		
No._____ Name: _____															
	Total ___		Total ___		Total ___		Total ___		Total ___		Total ___		Total ___		

3. Even in the cases of central office personnel such as project executives, estimators, accountants, and so on who divide their time simply by project and not
by specific activity within each project, the information will be extremely useful
if it becomes necessary to display the disproportionate amounts of time being
spent by these individuals in problem situations.

To treat these individuals' time reporting in a more simplified (but more
generalized) manner, their records can be maintained on a simplified project-
by-project basis and consolidated monthly.

2.7.2   Procedure

1. The Weekly Administrative Time Sheet is to be kept by *all* salaried jobsite personnel, and by any other individuals so designated by Company procedure to report their project time split into the individual activities.

    Site superintendents will generally have single-project responsibility. All offsite and central office positions will generally have multiproject responsibilities to varying degrees.

2. Throughout and at the end of each day, general supervisory people with multiproject responsibility and all central office staff (estimators, purchasing, etc.) list each project worked on at the left side of the form. An estimate of the amount of time spent on each activity within each project is included in the appropriate day column.

    Use the Project Administrative Activity List of Section 2.8 of the Manual as the starting listings of activity numbers to be assigned on the form. Any additional activities and number listings as may become necessary will be determined by project engineering as coordinated with project accounting per the requirements of Sections 4 and 8 of the Manual.

3. Treat change orders, potential change orders, work done under protest, and work that is or may be the subject of a claim as a *separate project*. The general project number with the change suffix will serve as the complete "Job Number."

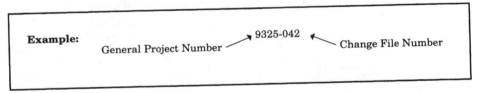

4. For those individuals who are authorized by general management to report their time simply divided by project, use the *Monthly Administrative Time Sheet*. Understand that, although completing this form requires less time and effort on the part of the individual, the respective information has accordingly been greatly generalized. Consider this carefully before the procedure is officially adopted.

5. The form must be turned in to central accounting at the designated day each week (or at the end of each month for the Monthly Form) as a condition for release of the individual payroll check.

2.7.3   Sample Weekly Administrative Time Sheet—Completed Example *(page 2.43)*

2.7.4   Sample Weekly Administrative Time Sheet—Blank Form *(page 2.44)*

2.7.3
Sample Weekly Administrative Time Sheet—Completed Example

WEEKLY ADMINISTRATIVE TIME SHEET

Form No. __1__

NAME _L. LEONARDO_   MONTH _JON_   19 _94_   WEEK NOS. _1 2 3 4_
POSITION _ADMIN. ASSISTANT_   REVIEWED BY _SC_   DATE _1·31·94_

(2) WEEK PERIOD (Sunday to Saturday)

PROJECT NAME	PROJ. NO.	S	M	T	W	T	F	S	S	M	T	W	T	F	S	TOTAL
FIRE HOUSE	9205		2	4	3	2	3			3	2	4	4	5		32
AJH CNTR.	9320		1	4	3	3	2			3	3	2	4	3		28
JEFFERSON SCH.	9330		3	0	1	2	1			1	2	1	0	0		11
COMM. SQUARE	8925		2	0	1	1	2			1	1	1	0	0		9

(2) WEEK PERIOD (Sunday to Saturday)

PROJECT NAME	PROJ. NO.	S	M	T	W	T	F	S	S	M	T	W	T	F	S	TOTAL
FIRE HOUSE	9205		3	2	2	4	4			2	3	4	3	3		30
AJH CNTR.	9320		3	2	2	1	4			2	4	4	3	3		30
JEFFERSON SCH.	9330		1	2	3	3	0			3	1	0	1	1		13
COMM. SQUARE	8925		1	2	1	0	0			1	0	0	1	1		7

SIGNATURE _L. Leonardo_

2.8   Project Administrative Activity List

2.8.1   General

The Project Administrative Activity List is to be used by all hourly and salaried personnel for their time reporting as described in Sections 2.6 and 2.7 of the Manual. If new categories become necessary to accurately describe all activities, secure new numbers and activity descriptions from the project engineer who will secure the information in accordance with the requirements of Section 4 of the Manual.

Because actual and potential changes are *treated as separate projects,* there is no need to make the distinction within the individual activity descriptions.

2.7.4
Sample Weekly Administrative Time Sheet—Blank Form

WEEKLY ADMINISTRATIVE TIME SHEET   Form No. _____

NAME _____	MONTH _____ 19 ____	WEEK NOS. __ __ __ __
POSITION _____	REVIEWED BY _____	DATE _____

(2) WEEK PERIOD (Sunday to Saturday)

PROJECT NAME	PROJ. NO.	S	M	T	W	T	F	S	S	M	T	W	T	F	S	TOTAL

(2) WEEK PERIOD (Sunday to Saturday)

PROJECT NAME	PROJ. NO.	S	M	T	W	T	F	S	S	M	T	W	T	F	S	TOTAL

SIGNATURE_____

The activities themselves will apply to the respective contract or changed work as so designated on each form.

2.8.2   Project Administrative Activity List

Activity Number	Activity Description
_____	Set up/maintain field office
_____	As-Built Documents
_____	Field Studies/Reporting
_____	Field Engineering & Layout

———————	Field coordination
———————	Off-site travel time
———————	Design coordination/clarification
———————	Finalize change designs
———————	Negotiate/finalize change order
———————	Winter conditions preparation
———————	Progress schedule generation
———————	Progress schedule updating and reporting
———————	Quantity survey estimating
———————	Cost Estimating
———————	Secure sub-bid proposals
———————	Estimate/proposal preparation
———————	Job Meeting attendance & documentation
———————	Special Meeting attendance & documentation
———————	Subcontractor submittal review/processing
———————	Company submittal review/processing
———————	Material expediting
———————	Organize/monitor/document backcharges
———————	Subcontractor payment review/approval
———————	Prepare/process general payment application
———————	Cost progress reviews/documentation
———————	Staff hiring/interviews
———————	Exit interviews
———————	Staff performance evaluations
———————	Proposal preparation
Activity Number	*Activity Description*
———————	Proposal resubmission/negotiation
———————	Progress photographs
———————	Field Reporting
———————	Prebid conferences
———————	Scope review conferences
———————	Value engineering
———————	Purchasing
———————	Safety reviews
———————	Accident investigation/reporting

2.9   Expenses and Reimbursements

2.9.1   General

It is the policy of the Company to reimburse employees for all normal and necessary expenses incurred in connection with company business. Good expense reporting and tight budget control are as important as efficient production.

2.9.2   Allowable Items

1. Fuel, oil, tolls, parking expenses directly related to company business. Use of your personal vehicle must be approved in advance by the Project Manager.
2. Lunch, or other entertainment for customers, design professionals, and owner representatives only if generally approved in advance by the Project Manager.
3. Small procurement for the project in amounts not exceeding $50.00. All small tools, supplies, etc., that have not been completely expended must be immediately turned over to the Company after their use is complete.
4. Material, tool, equipment, or supply procurement for the project in amounts exceeding $50.00 only if approved by the Project Manager in advance.
5. Maintenance items for company-issued equipment (batteries and film for cameras, etc.).
6. Overnight travel expenses only with the approval of the Project Executive.

2.9.3   Guidelines for Expense Reporting

1. Expense reports are to be submitted in neat form, complete, in ink.
2. Tape appropriate receipts in order to $8\frac{1}{2}'' \times 11''$ paper, and attach to the respective expense report.
3. Completed Expense Reports are to accompany the Weekly Time Sheet or Administrative Time Sheet (Sections 2.6 and 2.7 of the Manual).

2.9.4   Sample Expense Report—Blank Form (*page 2.47*)

2.9.4
Sample Expense Report—Blank Form

TO_____FROM_____

FOR PERIOD ENDING	SUN. CITY		MON. CITY		TUES. CITY		WED. CITY		THUR. CITY		FRI. CITY		SAT. CITY		TOTALS	
1 HOTEL MOTEL																1
2 BREAKFAST																2
3 LUNCH																3
4 DINNER																4
5 PLANE-RAIL BUS FARE																5
6 LOCAL TAXIS BUS FARE																6
7 AUTO EXPENSE REPAIR-TIRES SUPPLIES																7
8 GAS AND OIL																8
9 LUBRICATION AND WASH																9
10 GARAGE PARKING																10
11 PHONE TELEGRAMS																11
12 TIPS																12
13 TOLLS																13
14 ENTERTAINMENT																14
15																15
16																16
TOTALS																

MILEAGE RECORD		
END OF TRIP		REMARKS:
LESS — START		
MILES PER TRIP		

I HEREBY CERTIFY THAT THE ABOVE EXPENDITURES REPRESENT CASH SPENT FOR LEGITIMATE COMPANY BUSINESS ONLY AND INCLUDES NO ITEMS OF A PERSONAL NATURE.

SIGNED_____

DATE	REPAYMENT RECAP	AMOUNT	APPROVAL	CASHIERS MEMO
	ADVANCE RECEIVED			
	REIMBURSED			CHECK NO.
	TOTAL			
	EXPENSE FOR WEEK			DATE AMOUNT
	OVER OR SHORT			

General Contracts

3.1   Section Description

3.1.1   General

This section is intended to be used by all company personnel having any project management, project engineering, or site superintendence responsibilities. It should be read through in sequence, thoroughly understood, and referred to at any time. The issues regarding contracts, rights, and duties of the contracting parties, and specific application of contract language are the fundamental bases of the interaction between the owner, design professionals, and construction provider.

The section is generally divided into three areas: the first describes the various forms of project delivery methods, associated forms of agreement, and the general rights and duties of the parties; the second addresses concepts that apply to contracts in general; and the third devotes a large amount of material to the specific issues affecting the various forms of construction contracts.

Contemporary project management demands the ability to quickly and completely assess the total picture of each situation almost as it is occurring. From that position of complete information, we will all be in a better position to assess all options and to visualize each one through to its final conclusion—even if that conclusion must be arbitration or litigation. The better we can develop our ability to do this, the better we will be able to quickly determine appropriate responses for the moment, and for the time periods following. The details necessary to be able to efficiently string our action-items together will immediately become clear—right down to the correspondence distribution and file procedure.

3.1.2   The "Conduit" Principle

As a general contractor, construction manager, or other type of prime contractor, it is important to preserve the distinction as being the construction provider, and to avoid inadvertently or otherwise assuming any responsibility of either the design professionals and/or the owner vis-à-vis the contract. As the construction provider, it is our responsibility to coordinate the "work." It is not our responsibility to coordinate the design. It is not our responsibility to "interpret" the Contract Documents, to "approve" submittals, or through any other mechanism unnecessarily assume any liability that correctly belongs to the design professionals and the owner.

Lump sum general contracts, design-build contracts, and the various forms of construction management contracts all have major differences in the fundamental relationships between the parties. Additionally, they each contain numerous subtle differences in both their expressed relationships and in the manner in which those relationships are treated. However well-meaning we may be, as construction providers we continually get ourselves into many unnecessary problems. We do this simply because even though our formal contracting arrangements may be different to varying degrees, many projects continue to "feel" like most others in terms of their relationships and the way we tend to treat each other. On every project there is always an owner, design force, and construction force—each doing what pretty much looks like the same activities that are done on every other project. If we are not careful, our own management or staff can allow themselves to slip into a type of complacency with respect to our contracting relationships.

Our problems as construction providers in this regard distill to two levels:

1. We must continually remind ourselves that every contract is indeed different. It is almost too simple to be necessary, but in truth it is crucial that we always be aware of the specific language for the current situation on the particular project. We must constantly remind ourselves to check, verify, and then check again before proceeding on an item in the way that we've proceeded with similar items in the past. If we are not careful, it will otherwise be much too easy to dig a hole much deeper before we can begin climbing back out.

2. Many of us are often too eager to interpret the documents and to direct subcontractors and suppliers directly with respect to their required performance under the particular contract. In the interest of moving the project along, and with the intention of taking a "can-do," "take-the-bull-by-the-horns" action on what may seem on the surface to be a frivolous or opportunistic action on the part of a subcontractor, or in just trying to do what seems to be the right thing, we too often unintentionally assume the liability for these actions that more correctly belongs with the designers and/or the owner.

Read and reread the section of this Operations Manual entitled "The Pass-Through Clause." Understand clearly that each subtrade is bound to the owner to complete its work to the same extent that the prime contractor is responsible to the owner for the work of that particular subcontract. In turn, understand that the owner is bound back to those same subcontractors (through the prime contractor) in the same manner and to the same extent. Understand that we as construction providers should not interpret the documents (and thereby assume the responsibility for the interpretation)—it is not our "obligation," and it is not our "right" to interpret the contract documents either. Most typically, there is sure to be very detailed language placing responsibility for such contractual interpretation on either the design professionals or the owner.

Don't be too quick to jump into the middle of an issue between a subcontractor and the contract documents. Learn to keep the owner and design professionals squarely in the line of fire, and take the actions necessary to compel the decisions that you are entitled to in order to keep the contractual relationships intact.

Develop this perspective, and hold onto it dearly as you consider the discussions of this and other sections of this Operations Manual.

3.2 Contract Structures, Relationships, and the Contracting Parties

3.2.1 General

Much of the material in this Operations Manual is devoted to describing the manner in which the day-to-day operations of project management, project engineering, and site superintendence are carried out on behalf of the Company. For the most part, the language of the manual assumes a lump sum, general contracting form of project delivery, as well as a conventional set of relationships among the contracting parties.

Prior to the 1980s, the lump sum, general contracting form of project delivery was by far the most common. Throughout the decade of the 1980s, Construction Management with Guaranteed Maximum Price, Agency Construction Management, and other hybrid forms of construction management became very common. The primary motivator for the proliferation of construction management instead of lump sum general contracting essentially was the desire by owners to shorten the total design-construction process. The "construction management" system was therefore implemented in order to allow certain components of construction to proceed while other components were still on the drawing board.

Eventually, construction management fell out of favor in the eyes of many owners and design professionals—and even among some "construction managers." The primary reason for this disenchantment is based upon fundamental misunderstandings with respect to the relationships among the parties. Owners and design professionals seem to have been "expecting" one set of relationships—because every project "feels" like every other one. These same individuals, however, drafted construction agreements that said something very different. The result was a rash of projects that experienced extreme cost overruns and even delays that exceeded the time that otherwise might have been consumed in the more conventional design-bid-build lump sum general contracts.

For these reasons (and to be fair to the design-build community, and for other reasons as well), owners and design professionals began to look for yet another project delivery method that might accomplish the objective of early-start fast-tracked construction while regaining "control" over changes, claims, delays, and even the finger-pointing between contractors and designers. Attention was focused in many circles to reconsider the Design-Build method of project delivery—but sometimes with new twists.

And so as the industry moved from the 1980s into the 1990s, as many owners and design professionals became disenchanted with the various forms of Construction Management for various reasons, the Design-Build project delivery method began to achieve a new status as the delivery method of choice among many owners and design professionals. As the 1990s have closed, and as we now move into the new century of construction in America, Design-Build has been gaining prominence in many circles, while losing stature in others. All other forms of project delivery are being given renewed consideration. The wishful thinking of the 1980s and 1990s with respect to dreams of claims avoidance seems to have given way to a sober acceptance of the simple fact that the best way to avoid claims and other similar problems is to be clear in our forms of agreement, to manage each aspect competently, and above all, to be fair in our dealings. If we can accomplish all of these things, the form of project delivery will have much less influence on the degree of problems that we will experience on our projects.

In order for us as professionals to be the most effective in managing our projects, we must begin with a thorough understanding of the various forms of contracting arrangements and the specified relationships among the contracting parties. From there, we should develop an adequate working knowledge of the practical realities of these relationships, and of the ways that they are managed—and mismanaged.

This section of the Operations Manual, then, goes on to describe the major types of project delivery methods in their basic forms. Following that, the general rights and responsibilities of the parties are summarized. Together, the information will form a backdrop against which all construction management objectives, applications, and activities will be applied.

3.2.2 Contract Structures

There are several common forms of contract and project delivery arrangements that may be selected by an owner for a construction project, depending upon the individual circumstances and personal preferences. The basic forms of these arrangements are often modified in numerous ways. The actual contractual arrangement that we wind up working under might ultimately look very different from the "basic" format that may have been the basis for the particular contract. Often, however, we may still find that contracts are drafted with their familiar formats generally intact. In either case, the key is that the strict wording of each contract must be studied thoroughly and be completely understood in its specific detail. As those who work intimately with contracts should know, the modification of a single word can alter—and even reverse—the meaning of the entire document (consider the judicious placement of the word "not," for example, in almost any paragraph that you can imagine).

"Historical" relationships (the way in which things are "usually" done) may no longer apply (and may never apply again). Just because the roofer, for example, never cuts masonry reglets as part of its normal operations doesn't mean that it's not the roofer's responsibility to do precisely this work on this project. Or just because the mason contractor never provides the portion of the masonry anchor that is welded to the structural steel columns doesn't mean that it's not the mason contractor's responsibility to provide precisely this work on this project.

The plans, specifications, general conditions, special conditions, special provisions, working procedure, etc., must be read completely. The contract must be understood in its smallest detail. Assumptions and blind dependence upon past experiences have absolutely no place in contemporary construction contracting.

Finally, there is an important distinction to be made between contract relationships and contact relationships. This sounds almost too simple to require discussion. The truth, however, is that "contact" (communication, day-to-day dealings) too often leads to actions, inactions, and assumptions of responsibilities and liabilities due to misunderstandings—and inadvertent, incorrect assumptions—with respect to contract relationships and authorities.

The communication, or "contact," circles between the functional administrative divisions of every construction project include an owner, the design function, and a construction force. The communication format tends to occur (largely due to various practicalities) in nearly the same fashion and in the same manner project after project. The communication relationships tend to be preserved to very large degrees in a similar manner regardless of differences between technical contract structures.

Because of this, it is the first thicket in which the snare is laid—the huge potential for misunderstandings that eventually lead to cost increases, delays, claims, and other disruptions. At the site where all the activity is, the job "feels" like most other jobs. The traditional owner–architect–general contractor relationship becomes a difficult perspective and mindset to dissolve if some other contractual associations and relationships are specified. Even though the same individuals seem to be performing the same or similar activities, the differences lie in authorities, responsibilities, liabilities, and even timing.

So it is here in this general atmosphere of common forms of communication where a new game may begin to be played inadvertently by old rules. It is this environment in which problems, disputes, claims, and changes are nearly certain to become a way of life.

The most common structures of construction contracting arrangements can be classified as follows:

1. Cost Plus a Fee Contract

2. Cost Plus a Fee Contract with a Guaranteed Maximum Price (GMP)

3. Lump Sum Contract

4. Construction Management (CM)—Two types: At Risk and Agency (for Fee)

5. Design-Build

3.2.3 The Letter of Intent

The Letter of Intent is a temporary form of contract that generally is the precursor to one of the more formal types listed in this section. There are several reasons for an owner and a contractor to employ a Letter of Intent:

- An owner may be desirous of starting demolition in a recently vacated office space in anticipation of attracting a new tenant.

- An owner, receiving a verbal loan commitment from a lender, may be eager to commence construction, but reluctant to execute a formal contract until he or she has received written confirmation of the construction loan.

- When a project is "fast tracked," an owner may wish to proceed with a limited amount of work such as site clearing or limited authorization to procure items with long lead times and requiring shop drawings.

A Letter of Intent is very specific as to the exact scope of work to be executed; the method of determining the cost of the work, i.e., cost plus or lump sum; and the event that will terminate the agreement.

The key components of a Letter of Intent are:

1. A clear definition of the scope of work via an architect/engineering drawing or a narrative describing the exact nature of the work.

2. The cost of the work, i.e., lump sum or cost plus a fee or GMP, with a list of reimbursable costs and documentation to substantiate those costs.

3. A payment schedule.

4. A date when work is to commence and the date when work is to be completed and therefore cease.

5. If a subsequent construction contract is being considered, a statement that the scope of work and all associated costs will be credited to any associated scope and costs contained in the formal construction contract.

6. A termination clause setting either a time limit on the work or an event that would trigger termination.

It is important to clearly state the "costs" that will be included in the work. As in other forms of contracts where reimbursable costs are included, attaching an Exhibit to the Letter of Intent listing all such reimbursable costs will be helpful in overcoming any misunderstandings between the owner and contractor.

3.2.4 The Cost Plus a Fee Contract

Although this form of contract appears to be rather easy to comprehend, without an agreement on what constitutes *reimbursable* costs and what constitutes *nonreimbursable* costs, the Cost Plus a Fee contract can generate lots of misunderstandings. And quite often a Cost Plus a Fee contract is executed where the owner is given an "order of magnitude" for the final cost which, depending upon a rather defined scope, can be another area of misunderstanding.

The writer was involved in one such Cost Plus contract where the owner kept increasing the scope of the work and the $250,000 initial ballpark figure ballooned to $950,000. Even with constant weekly updates to apprise the owner of the added cost of replacing all light fixtures in a 40,000 square foot store and replacing all of the flooring, among other additions, they were still shocked by the final price. It is therefore very important to document any and all changes to an initial scope of work when a Cost Plus contract is employed.

The Cost Plus contract is frequently used to deal with emergencies. Damage from a fire, a flood, wind, or rain demands quick action, and an owner is likely to ask his or her contractor to just jump in and get the work done. There are instances other than emergencies when a Cost Plus contract will be considered, such as a commercial owner desirous of completing space for a new tenant with a move-in deadline or the need to begin construction of a limited scope on a small project prior to final completion of the plans and specifications.

To avoid misunderstandings over what constitutes a reimbursable or nonreimbursable cost, a good reference document is AIA Document A111, Cost of the Work Plus a Fee with a Negotiated Guaranteed Maximum Price. This contract form contains a list of costs, both reimbursable and nonreimbursable.

Reimbursable costs include:

1. Labor costs including burden.
2. Wages and salaries of the contractor's administrative and supervisory staff when stationed in the field with the owner's approval. If any non-field-based personnel are to be reimbursed, they should be listed with the proviso "with the owner's approval."
3. Taxes, insurance, employer contributions, assessments.
4. Subcontractor costs.
5. Cost of materials and equipment incorporated in the completed project.
6. Cost of other materials, temporary facilities, and related items fully consumed in the performance of the work.
7. Rental costs for temporary facilities, machinery, equipment, and hand tools *not customarily owned by construction workers* whether rented from the contractor or others.
8. Cost of removal of debris from the site.
9. Costs of document reproduction, faxes, telephones, postage, expedited delivery services, and reasonable petty-cash disbursements.
10. Travel expenses by the contractor while discharging duties connected with the work.
11. Cost of materials and equipment stored off-site—if approved in advance by the owner and accompanied by a bill of sale and storage and transit insurance.
12. Related portion of insurance and bond premiums.
13. Sales and use taxes.

14. Fees—assessments for building and related permits, agency inspections.

15. Fees for testing—laboratory and field.

16. Royalties—license fees for use of a particular design, process, or product.

17. Data processing costs related to the work.

18. Deposits lost for causes beyond contractor's control.

19. Legal, mediation, and arbitration costs *with the owner's prior approval.*

20. Expenses incurred by the contractor for temporary living allowances.

21. Cost to correct or repair damaged work provided that such work was not damaged due to negligence by the contractor or is judged nonconforming by the owner's design consultants.

Nonreimbursable costs include:

1. Salaries and other compensation of the contractor's personnel stationed at the contractor's principal office, *except as specifically provided for in the contract.*

2. Expenses of the contractor's principal office.

3. Overhead and general expenses *except as provided for in the contract.*

4. The contractor's capital expenses.

5. Costs due to negligence of the contractor.

6. In the case of a GMP contract, costs, other than approved change orders, which would cause the GMP to be increased.

7. A catchall—any costs *not specifically included* in the Costs to be Reimbursed.

This list can be added to or modified to meet individual project needs and should be attached to the construction contract as an Exhibit.

Several other attachments are helpful in identifying costs and avoiding misunderstanding as to what constitutes "cost":

- A complete hourly rate breakdown for each trade anticipated to work on the project. Section 3.2.5 is an example of carpenter rates for a union worker for both regular and premium time work.

- A list of management/estimating/purchasing personnel hourly rates (if applicable).

- A list of equipment proposed for use on the project with hourly rates, exclusive of the operator's hourly rate, which can be attached. Section 3.2.6 provides an example.

3.2.5 Labor Rate Breakdowns—Journeyman Carpenter—Regular Time and Premium Time (*page 3.12*)

3.2.5
Labor Rate Breakdowns—Journeyman Carpenter—Regular Time and Premium Time

BURDEN RATES FOR TRADES

Updated 09/01/06

JOURNEYMAN—STRAIGHT TIME

		MASSACHUSETTS BOSTON CARPENTERS 09/01/06	EASTERN CARPENTERS 09/01/06	BOSTON ZONE 1 LABORERS 06/01/06	OTHER ZONE 2 LABORERS 06/01/06
TRADE: CONTRACT DATE:					
HOURLY RATE:		$33.28	$27.83	$25.10	$23.50
EXPENSE:					
UNION BENEFIT:		$20.32	$20.08	$15.85	$14.10
INSURANCE:					
GL RATE per Hourly Wage:		2.8%	2.8%	2.8%	2.8%
GENERAL LIABILITY:		$0.93	$0.78	$0.70	$0.66
WC RATE per $100:		$4.59	$4.59	$10.99	$10.99
WORKER'S COMP:		$1.53	$1.28	$2.76	$2.58
OCIP:					
TAXES: FICA	0.0765	$2.55	$2.13	$1.92	$1.80
FUTA	0.008	$0.27	$0.22	$0.20	$0.19
SUTA	0.0877	$2.92	$2.44	$2.20	$2.06
SDI	0.0012	$0.04	$0.03	$0.03	$0.03
TOTAL BURDEN EXPENSE PER HOUR:		$28.55	$26.96	$23.66	$21.42
SUBTOTAL COST PER HOUR:	Other fringe (10%):	$61.83	$54.79	$48.76	$44.92
		$6.18	$5.48	$4.88	$4.49
TOTAL COST PER HOUR:		$68.01	$60.27	$53.64	$49.41
BURDEN PERCENTAGE Including Other Fringe:		104.37%	116.57%	113.71%	110.24%
Less: OCIP CREDIT:		($2.46)	($2.06)	($3.46)	($3.24)
BURDEN/MARKUP EXPENSE PER HOUR:		$32.27	$30.38	$25.08	$22.67
TOTAL COST PER HOUR:		$65.55	$58.21	$50.18	$46.17
OCIP JOBS BURDEN PERCENTAGE:		96.98%	109.18%	99.92%	96.45%

RATES SUMMARY:

NON-OCIP CHARGEABLE RATES:	$68.01	$60.27	$53.64	$49.41
OCIP CHARGEABLE RATES:	$65.55	$58.21	$50.18	$46.17

3.2.6 List of Equipment Proposed for Costs to be Reimbursed (*page 3.14*)

3.2.7 The Cost Plus a Fee Contract with a Guaranteed Maximum Price (GMP)

The GMP contract is frequently used for fast-tracked projects, allowing the owner to receive a maximum project cost while retaining the potential for cost savings. The GMP contract approach permits an owner to lock in a maximum project cost even before the contract documents—the plans and specifications—are 100% complete, quite often when the plans and specs are only 70% to 80% complete.

Because the GMP contract is generally executed prior to the remaining 20% to 30% of the design having been completed, some of the pitfalls in this contract format are:

- What did the owner expect the final design to include?

- What did the contractor assume would be included in the final design and therefore include as costs to cover the remaining portion of the design?

- What did the architect actually develop in that last portion of design work, and did it conform to the same concept the builder assumed?

Much the same caveats for the Cost Plus contract described previously apply to the GMP project, defining costs to be reimbursed and costs not to be reimbursed. However, another very important component of the GMP contract is a document developed to define the scope of work anticipated by the contractor when it receives the 100% complete drawings. This document or Exhibit is the Exclusion, Inclusion, or Contract Qualification Exhibit. Included in the body of the GMP contract or attached to it as an Exhibit, will be, in detail, what the contractor, in effect, has included in its scope of work independent of what may be shown in the final documents. This document will be reviewed with owner and design consultants to obtain a meeting of the minds before appending it to the contract for construction.

An experienced general contractor will have assumed certain architectural work; finishes; and coordination involving mechanical, electrical, and plumbing rough-ins and equipment connections in the as yet, not fully designed portion of the contract documents. It will base its price on these assumptions, which if left unsaid and unwritten, will most surely be cause for misunderstandings when the 100% complete documents are produced. For example, the contractor may have assumed medium quality in the offices and hallways, ceramic tile floors and painted walls in the men's and ladies' rooms on each floor only to find out that the final design included a high-end 54-ounce carpet and ceramic wall tile, wainscot height in each bathroom. Without a complete contractor Inclusion or Exclusion list, these upgrades may become a part of the contractor's obligation under the terms of the 100% complete contract documents.

3.2.6
List of Equipment Proposed for Costs to Be Reimbursed

PROJECT		SLIP#		DATE
** INCLUDED IN STANDARD RATE				
EQUIPMENT	**# OF ITEMS**	**COST PER DAY / HOUR**	**HOURS IN USE**	**COST PER ITEM**
1 GAS WELDER / CABLES		$ 26.00		$0.00
2 ELECT. WELDER / CABLES**		$ 7.00		$0.00
3 BOTTLES / GAUGES /				
HOSE / CART / TORCH**		$ 3.50		$0.00
4 CHAIN FALL / COMEALONG		$ 1.10		$0.00
5 T.C. / IMPACT GUN**		$ 8.00		$0.00
6 4" & 7" GRINDER**		$3.00		$0.00
7 EXT. / STEP LADDER**		$2.00		$0.00
8 STAGING / EA. SECTION		$ 4.00		$0.00
9 DECK SAW**		$ 8.00		$0.00
10 ELECT. LIFT 20' OR LESS		$10.00		$0.00
11 ELECT. LIFT OVER 20'		$23.00		$0.00
12 ADD FOR GAS LIFT NO FUEL		$20.00		$0.00
13 GENIE LIFT		$8.00		$0.00
14 HAMMER DRILL**		$4.25		$0.00
15 SCREW GUN / HAND DRILL**		$3.00		$0.00
16 AIR ARC GEAR**		$3.25		$0.00
17 COMPRESSOR		$25.00		$0.00
18 MAG. DRILL		$4.20		$0.00
19 BOOM LIFT 120-0'		$125.00		$0.00
20 BOOM LIFT 100-0' OR LESS		$100.00		$0.00
21 LN25 WIRE FEEDER**		$15.00		$0.00
22 PORTAPOWER**		$2.50		$0.00
23 GENERATOR				$0.00
24 STUD WELDER (ELECT)		$20.00		$0.00
25 STUD WELDER (DIESEL)		$50.00		$0.00
26 PLANKS		$0.58 PER DAY		$0.00
27				$0.00
28				$0.00
			SUBTOTAL	$0.00

CONSUMABLES	**COSTS PER**	**CHARGES**	**AMOUNT USED**	**COST PER ITEM**
1 OXYGEN (154 CL. EA. TANK)**	CL.	$0.25		$0.00
2 ACETYLENE (120 CF EA)**	CF.	$0.50		$0.00
3 PROPANE (20 LB EA TANK)**	LB.	$1.00		$0.00
4 WELDING WIRE**	LB.	$2.00		$0.00
5 GASOLINE / DIESEL FUELS	GAL.	$3.25		$0.00
6 OILS	QT	$2.25		$0.00
7 ABRASIVE WHEELS 4" & 7" **	EA.	$2.00		$0.00
8 GLASSES / LENSES**	EA.	$5.00		$0.00
9 GLOVES**	EA.	$8.00		$0.00
10 RESPIRATOR FILTERS**	EA.	$5.32		$0.00
11 CHOKERS (1/2" OR LESS)**	EA.	$28.00		$0.00
12 TIPS / BITS (1/2" OR LESS)**	EA.	$18.00		$0.00
13 SAFETY CABLE (FURNISH ONLY)	LF.	$0.35		$0.00
14				$0.00
			SUBTOTAL	$0.00

OTHER	**COSTS PER**	**CHARGES**	**AMOUNT USED**	**COST PER ITEM**
1				$0.00
2				$0.00
			SUBTOTAL	$0.00
			20% O.H & P.	$0.00
		GRAND TOTAL COST PER SLIP		$0.00

Section 3.2.8 is only one small portion of a much more detailed Contract Qualification Exhibit, this one pertaining to some renovations at a commercial urban project.

3.2.8 Qualification or Exclusion and Clarification Exhibit (*page 3.16*)

3.2.9 Lump Sum, General Contracting (GC) (*page 3.17*)

Lump Sum, General Contracting is what may be thought of as the "traditional" form of project delivery. In this structure, the construction force referred to as the "contractor" is brought into the arrangement sequentially as the third part of a clearly defined design-bid-build sequence.

In this arrangement, the design-bid-build sequence prescribes that the entire project will be completely designed during the "design phase." The completed design is then "bid" among various competing contractors. The contract is either awarded to the lowest prequalified bid (in the case of a public project) or negotiated between owner and contractor. The successful contractor is then responsible for providing a "complete" project—as defined in the set of Contract Documents—for a single, "lump sum."

In the lump sum general contracting structure, the owner contracts directly (and individually) with the contractor. The owner also contracts directly (and individually) with the architect. There is no direct contractual relationship between the architect and the prime, or general, contractor, but there are important operating practicalities that will impact the manner in which the entire contract is managed.

The architect's general responsibilities beyond the production of the design and the general contract documents are to administer the owner-contractor contract, functioning as the owner's agent. The prime, or general, contractor (the GC) enters into individual agreements and ostensibly is solely responsible for the complete performance of each of the individual subcontractors and suppliers (subvendors). The owner looks directly to the general contractor for satisfaction of performance, and accordingly, the general contractor's dealings with the subvendors are, in theory, intended to be transparent to the owner. The subvendors, in turn, look (at least initially) directly to the general contractor for resolution of difficulties—even if those difficulties originate in the owner's contract documents.

By subcontracting, the general contractor also shifts much of the legal responsibility for the performance of each portion of the work directly to the individual specialty subcontractors. Just as the owner and general contractor have agreed to a fixed amount to be paid to the contractor for the performance of the entire contract work in the lump sum agreement, so too does each agreement between the general contractor and each subcontractor. In both cases, however, the art of managing the general, or prime, contract will remain fundamentally based in the ability of the general contractor to compel and enforce performance on the parts of the

3.2.8
Qualification or Exclusion and Clarification Exhibit

(This happens to be for a renovation project in an urban area)

Exhibit J—Exclusion and Clarifications

The following items are excluded from our proposal:

1. Abatement of any hazardous materials is excluded with the exception of lead-based paint on the structural steel as required for the new installation.
2. We have excluded handling or shipment of any contaminated soils. Removal of soil and urban fill to an unlined site is included.
3. We have excluded any underground obstructions.
4. We have assumed that the owner will furnish all required temporary power.
5. We have included an Allowance for Winter Protection in the amount of $50,000 and will notify owner prior to reaching this amount.
6. Police protection, if required for any street closings, is excluded from this proposal.
7. We have included Temporary Protection from all areas other than the second floor from Cols A-1 to A-7 to B-1 to B-7, the third floor—same as second floor, and the fourth floor from Col C-6 to C-10 to D-6 to D-10.
8. Masonry restoration was not included except as needed for the new structural steel installation.
9. No spray fireproofing is included other than as required for the new structural steel.
10. We have assumed that all furniture, artwork, wall hangings, and equipment will be removed by the owner and stored out of the construction area.
11. We have included removal of finishes with the "Area of Work" only.
12. We have not included Specification Section 07199 since the limits of the area where work is to be included is unclear.
13. We have not included an overhead coiling door as specified in Specification Section 08334 inasmuch as none is shown.
14. We assume all signage and furniture is furnished and installed by others.
15. Fire protection is included in Rooms 235 and 248. This work is not shown on the drawings.
16. Builder's risk insurance premium costs are to be paid directly by the owner.
17. All building permit costs are excluded and are to be paid directly by the owner to appropriate city authority.

3.2.9
Lump Sum, General Contracting (GC)

owner and each subvendor—strictly as required by the contract documents. This is important not only as applied to what is included in the contract, but can be crucial in the manner in which it applies to those items that for one reason or another fail to be included in the contract (however they may have been "intended").

3.2.10 Construction Management (CM)

There are two basic types of construction management contracts:

- *Agency or CM for Fee.* The CM acts solely as the owner's agent, and the fee structure is based upon a percentage of the total project cost plus a list of reimbursable expenses. The owner issues contracts to subcontractors and purchase orders to vendors approved by the CM, and the owner pays subcontractors and vendors directly. The liability of the CM is limited to the "Standard of Care" provision similar to that of a design professional.

- *At risk.* The CM acts as the owner's representative and its fee and list of reimbursable expenses is included in the contract sum. The CM often deals directly with contractors and absorbs any cost overruns. The liability of the CM is similar to that of a general contractor with a lump sum or GMP contract.

3.2.11 Agency Construction Management (ACM) (*page 3.18*)

"Construction Management" can mean many different things to different people. In a very real sense, every contractor is and always has been a "construction manager," in that management regularly spends the bulk of its time orchestrating direct-hire labor, equipment, subcontractors, suppliers, and money (cash flow).

3.2.11
Agency Construction Management (ACM)

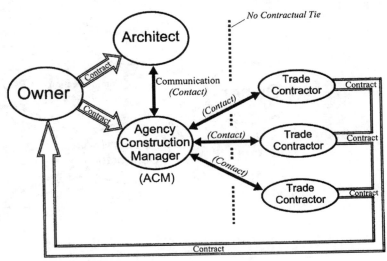

Management "coordinates" the work with other contractors and outside requirements, plans and schedules the work, communicates the requirements to all parties, and does whatever it can to compel performance on the part of each individual player. Finally, the "contractor-construction manager" clearly documents each activity and develops skills for effective problem resolution.

As the term is being used here, however, the functions and services (to the various extents as may be defined in the specific agency construction management agreement) are by design being carried out on behalf of the owner. An "agent" construction manager—the ACM—can be thought of essentially as an extension of the owner's staff, with the assigned responsibility to manage the owner's construction project. Somewhere along the line, the realities of the contractor-as-construction manager became apparent to construction owners. This thought soon became combined with the idea that if the owner were somehow able to manage the construction of his or her own project, the owner would also realize the benefits of the profit that otherwise would have been earned by a general contractor. As icing on the cake, the owner would also have the luxury of being able to directly control the "contractor" forces without otherwise assuming company-to-company liability for such direct intervention.

Refer to the preceding figure for illustration of the specific practical and contractual relationships in the Agency Construction Management arrangement. The agency construction manager (ACM) contracts directly with the owner to provide the distinct services in the consultant capacity. There is direct contact between the ACM and the individual trade contractors, much the same as in the general contracting sense, but there is no direct contract between the ACM and those trade contractors. There are usually not many "subcontractors" on ACM projects (if any at all) because the specialty contractors do not contract directly with the construction management firm. Instead, the specialty contractors are managed by the CM firm but formally contract individually with the owner. Because of this, they are generally referred to as "trade contractors."

Because of the lack of any contractual tie between the ACM firm and each of the trade contractors, the ACM usually bears no direct liability for the performance or nonperformance of those trade contractors. Each of the individual trade contractors is—in contractual effect—a separate prime contractor because of its direct contractual relationship with the owner. The ACM firm manages each of the trade contractors essentially just as it would if it were a general contractor or "at risk" construction manager, but the important difference lies in the lack of direct contractual responsibility for performance.

Advantages of this method of project delivery can begin during the design phase. Input of construction expertise (provided by the ACM firm) early in the process—preferably during the concept stage—may significantly improve construction efficiencies further down the road. Through such involvement, the owner might ultimately obtain the same function of the project while actually reducing facility cost and operating expenses. Better knowledge of local construction sequences, conditions, and labor markets, etc., for instance, may significantly reduce total time needed for the design-construction process—thereby saving everyone money. An owner may benefit from a contractor's superior working knowledge of sensible ways to format the various bid packages and apportion the work among the separate trade contractors. This kind of assistance can make it much easier to deal with the trades not only during the construction phase, but also through the management of callbacks and guarantee work. By being able to take a different look from a fresh perspective at the same "standard" specification sections and divisions and to impart professional judgment that can improve the effectiveness of the contract, direct cost savings, greater benefits for the same money, or both can be achieved.

Still another reason that owners have considered the ACM arrangement is to provide for the capability of "fast-tracking" the design and construction phases without losing control over the individual trade contract structures. In a fast-tracked environment, certain portions of the work can be bid, awarded, and begun at the site well before the entire facility design is complete. The foundation work can start while the structural steel design details are being finalized. Similarly, interior partitions can begin before the final wall finishes are selected. These individual bid packages are managed by the agency construction manager as their respective designs become finalized (or even "substantially" finalized). By being managed by the ACM on a fee basis, additional costs (caused to the trade contractors as a result of subsequent changes to previously completed designs) are passed back to the owner to pay for as change orders. Depending on the project, disposition of the owner, competency of both the design professionals and the ACM firm, and the manner of interaction among all of these companies, these types of problems can actually be minimized, and the benefits gained in the reduction in overall project time can more than offset any such increases in this type of direct cost.

3.2.12 Construction Management with Guaranteed Maximum Price (CM w/GMP)

This form of construction management is also known as a CM at Risk inasmuch as the CM will be guaranteeing a contract sum, whether it be a Lump Sum or a GMP-type contract.

Opponents of this form of CM argue that the construction manager is serving two masters—the owner and themselves since they have assumed a risk in guaranteeing the final contract sum. Critics state that the CM could weigh judgments on certain matters nonobjectively if these matters would impact its fee. While this may be true, a CM not acting strictly in the owner's risk will tarnish its reputation and make it more difficult to obtain new work.

Although this type of construction manager again performs the same functions, the CM now contracts directly with and is ultimately responsible for the performance of each respective subtrade. The contract and subtrade relationships in the CM w/GMP format are in fact identical to those shown in Section 3.2.9, Lump Sum, General Contracting. In practice, therefore, the construction manager in the Guaranteed Maximum Price context actually appears to become a general contractor—with the dubious distinction that it does so with a set of contract documents that all parties—the owner, design professionals, and construction manager—agree are substantially incomplete.

It is this fundamental nature of CM w/GMP contracts that is the genesis of most of the misunderstandings—and serious problems—that can and do occur much too often in this type of arrangement. Because of the words "Guaranteed" and "Maximum" into GMP designation, the owner is often under the misconception that little to nothing can occur contractually that will cause the cost of the project to exceed its budget. After all, the price is "guaranteed." Problems, however, are "guaranteed" to occur because the "Guaranteed" Maximum Price is prepared with the knowledge and acceptance by all parties that the contract documents are indeed substantially incomplete.

How, for example, could it be possible for a construction manager (or anyone else, for that matter) to prepare a "maximum" price for a "100%" completed project—when everyone is well aware that the design is not yet complete? It should be clear that the only way such a price can be prepared is for the construction manager to "reasonably" assume the manner in which the remaining (as of yet undesigned) portion of the project will eventually be completed. A long list of such assumptions becomes necessary in order to define the manner in which the construction management firm "anticipates" that the remaining design will be completed. And so the "Guarantee" of the Guaranteed Maximum Price is simply a guarantee of the total cost if the project is built precisely in the manner that the construction manager has anticipated that design would eventually be completed. Thus, each

Guaranteed Maximum Price proposal carries with it some mechanism by which the construction manager has explained to the owner and design professionals its list of assumptions and anticipations. It may be called something like "Basis of Estimate," "Qualifications and Assumptions," or something very similar, but in every case it is the specific list of the manner in which the construction manager has essentially "completed" that 40% or so of the design that has yet to be done as of the time the Guaranteed Maximum Price is prepared.

This is the area that becomes the fundamental point of misunderstandings and disagreements among the construction manager, owner, and design professionals. When the design professionals eventually complete the designs, any departure from the assumptions made by the construction manager in the complete Guaranteed Maximum Price portion of the agreement becomes a source of added costs—and possibly associated delays and disputes.

The American Institute of Architects (AIA) has a number of contracts specifically devoted to construction management.

The Construction Management Association of America (CMAA), in 2005, revised its basic contract format and offers the following CM contracts:

A-1: Owner and Construction Manager contract

A-2: Owner and Contractor contract where the CM represents the Owner's interest

A-3: General Conditions document, a companion to the base contract

A-4: Owner and Design Consultant contract

Defining the Responsibilities of a Construction Manager. A review of CMAA's A-1 document reveals the basic responsibilities of a construction manager and sums up the most common activities an owner expects of his or her CM:

Predesign Phase
- Prepare a Construction Management Plan (the Plan) taking into consideration the owner's schedule, budget, and general design requirements.
- Assist the owner in the selection of an architect.
- Assist the owner in the review and preparation of a contract with the architect.
- Assist the owner in conducting a Designer Orientation Session regarding the project scope, schedule, and budget.

Time Management
- Prepare a Master Schedule for the project.
- Prepare a Milestone Schedule for the Design Phase.

Cost Management
- Conduct a market survey to provide the owner and design team with current information about labor, materials, and equipment costs and availability.

- Prepare a Project and Construction Budget based upon the Construction Specifications Institute's (CSI) divisions of work and identify any contingencies for either design or construction, or both.
- Report to the owner and designer the estimated cost of various design and construction alternatives, recommendations to the budget, costs relating to efficiency, usable life, maintenance, energy, and operational costs of the building's components.

Management Information Systems (MIS)

- Develop a management information system between the owner, CM, and designers as well as other parties to the project.
- The MIS shall include procedures for reporting and communication during the design phase.

Project Management—Design Phase

- Provide the owner with potential bid packages.
- Conduct an initial Project Conference, and conduct progress meetings with the owner and designers to review the Plan, schedule, design phase schedule, budget, and MIS.
- Monitor the design to ensure compliance with the Plan, and coordinate the flow of information between the owner, designers, and others as required.
- Review the design documents, and make recommendations on constructability, scheduling, clarity, consistency, and coordination of the design documents.
- Facilitate the owner's review of the contract documents.
- Coordinate the transmittal of design documents to the various regulatory agencies.
- Revise the Master Schedule as required, and monitor the Design Phase Milestone Schedule.
- For cost management, prepare an estimate for each submittal of the design development drawings and facilitate decisions by the owner when changes are required to the program. Provide value engineering recommendations to the owner.
- Prepare and distribute schedule maintenance reports and project cost reports that compare actual and estimated costs.

Procurement Phase

- Prequalify bidders, and assist the owner in preparing and placing notices to solicit bids.
- Expedite bid documents to all bidders.
- Conduct a prebid conference and coordinate procedures to disseminate questions and answers resulting from that conference.
- Assist the owner in opening and evaluating bids, and make recommendations regarding the acceptance or rejection of bids.
- Assist the owner in preparing construction contract documents.

Construction Phase

- Conduct a preconstruction conference in consultation with the owner and designers, and verify that the contractor has provided all permits, bonds, and insurance as required.
- Provide an on-site management team to implement contract administration as an agent of the owner.
- Establish and implement procedures for reviewing and processing RFIs, RFCs, shop drawings, samples, proposals for substitutions, change orders, payment applications, and logs.
- Review RFIs, RFCs, shop drawings, and other submittals and make recommendations before passing them on to the appropriate party, that is, architect, engineer, inspector, or owner.
- Conduct project site meetings, and prepare meeting minutes and distribute to all concerned parties.
- Review the contents of all change orders, whether generated by the owner, designers, general contractor, or subcontractor, and make the necessary recommendations to the owner.
- Establish a program to monitor the quality of the project.
- Require the contractor to prepare a Safety Program and submit it for review and approval or comment.
- Render to the owner decisions involving any disputes arising between the contractor and owner.
- Receive all Operation and Maintenance manuals and warranties, and deliver to the owner and designer.
- Determine when the project has reached substantial completion, in conjunction with the designers, and prepare a list of incomplete work or work that does not conform to the contract documents.
- After consulting with the designers, determine when the project has attained final completion.

3.2.13 Design-Build

Design-Build is basically a different project-delivery system that requires the formation of a design-build team to provide an owner with both design and construction in one contract, such as lump sum, GMP, and CM, as well as some specific new forms specifically designed for the design-build process. Because Design-Build involves a somewhat new concept, some institutional changes, and additional legal and licensing considerations, Section 7, Design-Build Project Administration, is devoted solely to describing this process.

3.2.14 AIA Integrated Project Delivery and the ConsensusDOCS® Contracts

In an August 2012 survey conducted by the Construction Management Association of America, the types of contract format most prevalent in the vertical construction market were found to be as follows:

Design-Bid-Build (DBB)	60%
Construction Management at Risk (CMAR)	25%
Design-Build (DB)	15%
Integrated Project Delivery (IPD)	less than 1%

This last contract format, IPD, was one of a series of revised contracts released by the American Institute of Architects (AIA) in the period 2007–2009. An entirely new category of contracts, ConsensusDOCS®, also evolved around this time; these contracts were written and endorsed by a coalition of thirty-eight leading construction industry organizations, including the Associated General Contractors of America (AGC).

These contracts stress collaboration among all parties to the design and construction process, and invite the owner, architect, general contractor, and key subcontractors to work together, sharing their expertise and knowledge, to develop project goals beneficial to all. Although a complete reading of each of these IPD contracts is necessary to obtain the full concept of Integrated Project Development, a brief look at several of these AIA and ConsensusDOCS® contracts will provide project executives, project managers, and project superintendents with the basics of this collaborative movement.

The Six Principles of Integrated Project Delivery

1. Involve all participants in project development and execution as soon as possible in the programming process.
2. Identify the prime technologies to be incorporated into the project—Building Information Modeling (BIM), Sustainability, Green Criteria, Schedule, Economic Performance—across the project's life.
3. Develop detailed budgets early on to allow for more rapid assessment of design decisions. This budget or cost assessment is made available to all key parties for review and comment.
4. Develop performance goals, including a method for measuring and monitoring performance.
5. Develop project incentives dealing with costs, schedule, and quality; determine how these incentives will be distributed to each party, and obtain approval of each party in accepting risks and rewards.
6. Link the Preliminary Schedule, which is developed early on, to the project's development model and adjust the schedule accordingly.

Let's now look at selected key provisions of these Integrated Project Development documents.

- AIA Document A195-2008™: "Standard Form of Agreement Between Owner and Contractor for Integrated Project Delivery IPD)"

- AIA Document A295-2008™: "General Conditions of the Contract for Integrated Project Delivery"
- Document C191-2009™: "Standard Form Multi-Party Agreement for Integrated Project Delivery"

AIA Documents B195™ and A195™. These contracts between Owner and Architect (B195) and Owner and Contractor (A195) establish an Integrated Project Delivery system. AIA A195-2008™ provides the business terms and conditions for the owner-contractor agreement, whereas AIA B295-2008™ provides the General Conditions for the IPD, setting forth the contractor's duties and obligations for each of the phases of the project, along with the duties and obligations of the owner and the architect. AIA A195 is based upon a Guaranteed Maximum Price agreement.

The following diagram reflects the contract relationship document between A295, A195, and B195 (contract between Owner and Architect for an IPD project).

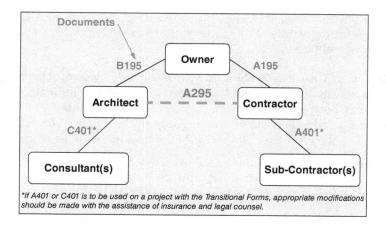

Other articles in AIA C195™ provide for incentive compensation for a portion of the differences between Actual and Target Costs. All parties can participate in establishing Goal Achievement Compensation, determining the amount and the method of compensation when all project goals have been achieved.

The following excerpts from AIA Document A295™, AIA Document A191™, and AIA Document C191™ are reproduced with permission of the American Institute of Architects, 1735 New York Avenue, N.W., Washington, D.C. 20006.

AIA Document A295™. AIA A295™ defines the IPD approach, which is reflected in many of its several articles. Let's start with Article 1—General Provisions:

Article 1.1—Purpose—The Owner, Architect and Contractor have agreed to plan, design and construct the Project in a collaborative environment following the principles of Integrated Project Delivery and to Utilize *Building Information Modeling* [italics added] to maximize the use of their knowledge, skills and services for the benefit of the Project. The Architect and Contractor will deliver the Project in the following phases, which may overlap: Conceptualization, Criteria, Design, Detailed Design, Implementation Documents, Construction and Closeout.

Article 1.3.13—Integrated Project Delivery: Integrated Project Delivery [is an] approach that integrates people, systems, business structures and practices into a process that collaboratively harnesses the talents and insights of all participants to reduce waste and organize efficiency through all phases of design, fabrication and construction.

Article 4.2.3—The Contractor shall provide estimating services throughout the design of the Project … provide estimates of the total cost to the Owner to construct all elements of the Project designed or specified by the Architect and shall include contractor's general conditions costs, overhead and profit (Contractor's Estimate). The Contractors' Estimate shall not include the compensation of the architect, the cost of land, rights-of-way, financing, contingencies for changes in the Work, or other costs that are the responsibilities of the Owner. The Contractor's Estimates shall increase in detail and refinement as the Architect progresses with the preparation of the Criteria Design, Detailed Design and Implementation Documents.

AIA Document C191™-2009 Standard Form Multi-Party Agreement for Integrated Project Delivery. This document, as the excerpts that follow indicate, is one possible agreement between the Owner, Architect and Contractor. The relationships among all parties are shown in the diagram below and the following articles provide the flavor of the collaboration the documents seek to create.

Article 1.1.2—The Parties intend to establish a Target Cost for the Project and to amend this Agreement to incorporate the Target Cost. To the extent that the Actual Cost is less than the Target Cost, the Parties shall share in any savings realized in accordance with the terms of this Agreement. To the extent the Parties have agreed to Project Goals, they shall be set forth in Target Criteria Amendment and the Parties shall be compensated for achieving the Project Goals as specified in this Agreement.

Article 2 of Document C191™, entitled "Management of the Project," creates a Project Executive Team. This team, comprised of representatives of the owner, architect, contractor, and "additional parties, if any," is designated to plan and manage the project, but not on a day-to-day basis. This function is relegated to a Project Management Team, composed of owner, architect, and contractor representatives who provide the same roles as they would in a typical construction project: the owner's rep acting as a bridge between architect and contractor by approving payments, reviewing changes in scope, etc.; the architect providing architectural services commensurate with that profession; and the contractor performing the "Contractor's Work." These three representatives, once again, do not supervise each other's employees.

Article 4 includes some provisions unique to the collaborative nature of this multi-party agreement:

> Article 4.2.4—Compensation for Labor Costs When Actual Costs Exceed the Target Cost.
>
> When Actual Costs for the Project exceed the Target Cost, as adjusted under the Contract Documents, the Owner's obligation to reimburse the other Parties' Labor Costs shall be as follows:
>
> [1] The Owner shall reimburse the Parties for all Labor Costs in accordance with Section 4.2
> [2] The Owner shall not be required to reimburse the other Parties for any further Labor Costs incurred.
> [3] Other (identify)

3.2.15 Comparison of Selected Articles from AIA A201™ and ConsensusDOCS®

A comparison of certain articles from AIA A201™ and ConsensusDOCS® reveals some basic differences and similarities.

The ConsensusDOCS® Concept. The ConsensusDOCS® contract approach is founded, much like the AIA concept, on a collaborative approach among owner, design consultants, general contractor, and key subcontractors. Selected provisions of ConsensusDOCs®300 illustrate this "working together" environment.

ConsensusDOCS® 300—Tri-Party Collaborative Project Delivery Agreement

> Article 3—Collaborative Principles—This article creates the Collaborative Project Delivery (CPD) Team consisting of the Owner, Design and Constructor representatives who sign Joining Agreements that provides authority to a Management Group (fully described in Article 4).

Article 3 includes an option to limit the designer's and contractor's liability (including the cost of damages, negligence, and breach of contract arising out of the agreement) in an amount not to exceed that which has been established by the CPD Team, unless such claims are reimbursed by the insurance policies carried by the corresponding parties.

> Article 4—Reference to management by the group established in Article 3 is elaborated on here. This management group, consisting of representatives from the owner, the design consultants, and the contractor, will conduct regular meetings to review the project's status and issues affecting progress or delays and conflicts that may result in potential claims.

Article 6—This article deals with development of design and collaborative pre-construction services. This provision dictates that the design services, working collaboratively with the CPD Team, will develop design, quality, costs and schedules jointly. The article encourages, but does not require, the utilization of Building Information Modeling (BIM); however, ConsensusDOCS® 301 is an addendum that adds the BIM requirement. Provisions in DOCs 300 include development of "Target Value" produced by the constructor, key trade contractors, vendors, and design consultants who are also charged with continually seeking savings by identifying options to improve constructability and functionality.

Article 7—Project Planning, in this article, includes several processes that affect planning; one of them is "pull planning," a schedule approach that looks at preceding activities that are not to be started sooner than is necessary to allow for the continuous performance of subsequent activities. The CPD Team is to develop a Make-Ready Look Ahead Plan and schedule weekly meetings to review and evaluate the work performed as well as the work planned for the coming week.

Article 8—The Construction Budget is addressed in this article, which requires the budgeting process to be an ongoing activity. The updating of costs as design develops commences with a preliminary cost model, proceeds to a schematic cost model, a design development model, and finally a construction document cost model.

Article 12—Trade contractors, subcontractors, suppliers, and design consultants are discussed in Article 12; this article aims to bring trade contractors and suppliers into the design process at an early stage so they can provide their experience and expertise in this key stage of project development. These trade contractors are also required to participate in coordination activities to ensure that conflicts and routing problems are eliminated, or at least dramatically reduced.

Article 23—This section is entitled "Dispute Resolution." Direct discussions among members of the CPD Team and the Management Group are required to resolve disputes or disagreements early on. Dispute mitigation allows for two options: selection of a Project Neutral or a Dispute Resolution Board (DRB) to settle disagreements. Mediation would follow if these two approaches fail; if mediation doesn't work, binding arbitration or litigation would be the final approach.

Many of the concepts relating to collaboration—that is, working together toward a common goal—set forth in this ConsensusDOCS® contract, as well as those in the AIA Integrated Project Delivery contracts, could be introduced into any standard owner-contractor contracts. The Project Manager and Project Superintendent could introduce several of the collaborative-type provisions into their standard operating procedures even without formal contract provisions, since they stress the need for and advantages of working together as a team.

ConsensusDOCS® 301—Building Information Modeling (BIM) Addendum addresses the implementation of this multidimensional design concept. Although it is a required attachment to an IPD-type contract, it can be attached to other contract formats as well.

A Comparison Chart of AIA 2007 Contract Documents and ConsensusDOCS® Contracts

ConsensusDOCS
Construction Contracts Built by Consensus

Document Comparison Chart

2007 AIA Number	ConsensusDOCS Number	ConsensusDOCS Document Name
A101 (+ A201)	200	Owner/Constructor Agreement & General Conditions (Lump Sum)
A102		Owner/Constructor Agreement & General Conditions, Short Form
A103	235	(Cost of Work)
A107 (+ A201)	205	Owner/Constructor Agreement, Short Form (Lump Sum)
A132-2009 (+ A201)	801	Owner/Construction Manager Agreement & General Conditions (CM is Owner's Agent)
A133-2009 (+ A201)	500	Owner/Construction Manager Agreement & General Conditions (GMP, Preconstruction Option) (CM-at-Risk)
A134-2009 (+ A201)	510	Owner/Construction Manager Agreement & General Conditions (Cost of Work, Preconstruction Option)
A141–2004	410	Owner/Design-Builder Agreement & General Conditions (Cost Plus w/ GMP)
	415	Owner/Design-Builder Agreement & General Conditions (Lump Sum)
A305-1986	221	Constructor's Statement of Qualifications
A310-2010	262	Bid Bond
	260	Performance Bond
A312-2010	261	Payment Bond
A401	750	Constructor/Subcontractor Agreement
A441–2008	450	Design-Builder/Subcontractor Agreement
A701-1997	270	Instructions to Bidders on Private Work
B101, B103	240	Owner/Design Professional Agreement & General Conditions
B104, B105	245	Owner/Design Professional Agreement & General Conditions, Short Form
B143–2004	420	Design-Builder/Design Professional Agreement
B214	300	Tri-Party Integrated Project Delivery (IPD) Agreement
B305-1993	222	Design Professional's Statement of Qualifications
C106	200.2	Electronic Communications Protocol Addendum
C191-2009 (+ A295)	300	Tri-Part Collaborative Project Delivery Agreement
D503	310	Green Building Construction Addendum
E201	200.2	Electronic Communications Protocol Addendum
E202	301	Building Information Modeling (BIM) Addendum
G701-2001	202	Change Order
G701/CMa-1992	525	Change Order/Construction Manager Fee Adjustment

www.ConsensusDOCS.org

2007 AIA Number	ConsensusDOCS Number	ConsensusDOCS Document Name
	291	Application for Payment (GMP)
G702-1992	292	Application for Payment (Lump Sum)
G703-1992	293	Schedule of Values
G704-2000	280	Certificate of Substantial Completion
G704/DB–2004	481	Certificate of Substantial Completion for Design-Build
G714	203	Interim Directed Change
G716-2004	204	Request for Information

3.3 Responsibilities of the Contracting Parties

3.3.1 Section Description

Each agreement to provide construction services necessarily deals with the management of the relationships among the owner, design professionals, and the construction provider. Although, as we have seen, it is certainly important to maintain a clear understanding of the specific contractual relationships being considered for a particular project, there are general principles that govern the manner in which the parties to construction agreements should treat each other. The first step in any evaluation of contract performance responsibilities should be, of course, to review the specific contract documents in order to confirm just what the particularities of the relationships for the specific project are. Any confusion, gray areas, or language that appears to be inconsistent or otherwise different from what you might expect should be clarified as early as possible.

In the most basic contracting model, all documents and money flow from the owner to the contractor, and the product (the project) moves back from the contractor to the owner. The basic system is elaborated upon to include an architect and/or engineer. The design function and its various representations contractually become extensions of the owner as far as the contractor is concerned. The "contact" issues discussed earlier in this section of this Operations Manual must be kept in perspective with respect to the "contract" realities.

In summary, the contractor has the right to expect:

1. To have all project documents be completed and be correct to the point that they can be depended upon with confidence.

2. To be dealt with fairly.

3. To be paid on time for acceptable performance.

In turn, the owner has the right to expect:

1. To get what is specified and paid for.

2. To get it at the appropriate level of quality.

3. To get it on time.

If motivations are clear and aligned in the proper direction, intentions are clearly understood, all documents are produced without flaws, all communication is professional and timely, and everyone approaches each issue fairly and equitably, everything just might go as planned. If any of these items are to any degree less than perfect, project success will depend squarely upon the contractor's ability to move construction forward while resolving issues on a reasonably timely basis.

3.3.2   General Responsibilities of the Owner

Many types of disclaimers of owner responsibilities and warranties—particularly of the "express" variety—are frequently found riddled throughout today's construction contract. While it is true that these disclaimers must be clearly understood if they do exist in print, it is equally true that some types of attempted evasions of responsibilities may often be overly ambitious in their attempt to shift the owner's responsibilities onto the contractor. In some cases, efforts to create iron-clad disclaimers possibly can backfire, apply intermittently, or mean exactly what they say. Each one should therefore be scrutinized very closely and thoroughly understood.

When not barred from a specific contract, the law imposes several warranties, duties, and responsibilities on the owner—whether or not they are specifically highlighted directly in the contract language. These general responsibilities include the following:

1. In the case of public contracting, using sound discretion in evaluating the qualifications of apparent low bidders.

2. In the case of public contracting, preserving the integrity of the public bidding and contract award systems.

3. Funding the work, including all changes.

4. Providing all surveys describing physical characteristics on the site (unless such surveys are clearly part of services to be provided by either the contractor, construction manager, or design-build firm).

5. Securing all necessary easements and authorizations.

6. Warranting the adequacy of the plans and specifications.

7. Warranting the suitability of furnished materials.

8. Disclosing of superior knowledge.

9. Implementing prompt action on clarifications and changes.

10. Acting within time periods that are within the reasonable contemplation of the progress of the schedule.

11. Providing "final" interpretation of the contract documents.

12. Cooperating.

13. Ultimately assuming responsibility for the design professionals.

(1) Preserving the Integrity of the Public Bidding and Contract Award Systems.   Public bidding requirements are normally set forth very clearly and specifically. They can often be extensive and in some cases comprise an extreme amount of detail. Beyond the price proposal itself, bid bonds or other bid security will be required, various affidavits and statements may be included, and the bid itself may be expanded to detail specific components with unit prices. The bid forms may require that certain subbidders be named, along with their prices. All bid documents need to be delivered in complete and proper form by a particular day and time.

The bidder who completes the forms and requirements to the last detail as required in the stated bid procedure seals off at least those portions of the bid from further consideration and/or competition. Those bidders who do not complete a particular section in a very real sense maintain the option to keep shopping around for the particular item. At the extreme, if improper security for performance has been submitted, such a bidder may not even be qualified to perform the work. If this type of bidder is ultimately awarded the contract, it is patently unfair to the other bidders who may have been completely responsive to all of the bidding requirements. Accordingly, if the nonresponsiveness in bidding is allowed by the public agency to continue, all bids in the system would soon degenerate. Left unchecked, the public bidding system itself would be left in ruins.

After the requirements of strict conformance to the document forms of submission have been met, the fundamental nature of public bidding requires that all bids be submitted precisely on time or before. 2:00 p.m. means 2:00 p.m.— and not 2:05 p.m. In the words of the Practicing Law Institute:

> "If late bids could be accepted, the integrity of the competitive bidding system would be violated because the late bidder would have the opportunity to obtain superior knowledge and, hence, there would be multiple opportunities for abuse."

In addition, the late bidder would have actually had more time to prepare his or her bid in the first place, thereby compounding the unfair advantage.

Acceptance of late bids at public openings is becoming increasingly rare, but it still occurs. When nonresponsive bids are brought to the attention of the other bidders, it is important that all responsive bidders protest immediately and compel the public agency to do the right thing by rejecting the nonresponsive bid.

(2) Funding the Work, Including Changes.   Many contractors would list this item as the owner's first requirement. In the case of a public bid, if the funding source has not been properly arranged, it is, at best, unfair to those contractors who took the time and expended the energy (and money) to prepare and submit a bid for the project. In any event, when the project is under way, the owner's primary responsibility as far as the contractor is concerned is to secure funds necessary to allow payments to occur strictly as prescribed in the contract. These parameters include the following:

a) The timing of payables.

   Payments to the contractor must be made after receipt of each invoice within the time periods specified, or within the requirements of appropriate laws (the contract language itself may determine which takes precedence). If the contractor has met all procedural requirements, the owner will become responsible for subsequent delay in the work and possibly for consequential effects that may include such items as interest on late payments.

b) Conformance to rates and amounts corresponding to actual job progress.

If particular quantities of materials can be substantiated to have been delivered to the jobsite and/or put in place in an acceptable manner considering all project requirements, the owner must pay for that quantity at the prescribed rate. Any alteration by the owner for convenience may create a hardship for the prime contractor and affected subcontractors, for which they may be entitled to compensation.

c) Contingency for changes.

Change orders are a normal part of the construction process. Because of the fact that changes occur on virtually every construction project, the owner has an obligation to provide financially for their eventuality. Responsible project funding will incorporate some additional percentage (often ranging between 5% and 10%) of the project based bid value that is to be set aside in order to be available to accommodate legitimate changes as they occur. Owners who fail to provide an adequate funding contingency for such changes are setting their projects up to experience major difficulties. In such projects that are inadequately funded, even if an owner's representative agrees with a contractor with respect to the need to increase the Contract Sum because of needed change orders, the change may be rejected—or at least significantly delayed—simply because the funds are not in place to deal with the issue properly. Although it is true that the denial of an otherwise appropriate change order for this reason can be considered to be "bad faith," it can occur much more often than it should.

(3) Provide All Surveys Describing the Physical Characteristics of the Site. The contractor cannot be responsible for the correct layout and prosecution of the work unless the information upon which its activities are founded are fundamentally correct. Generally, it is the owner's responsibility to provide all complete and accurate relevant data as it is or may become necessary for correct installation of the work. Certain aspects of this information include the following:

a) Proper Establishment of Property Lines and Contract Limit Lines.

The first and the most basic of these kinds of information is the establishment of property lines and the Contract Limit Lines within which the contractor has the right (and obligation) to operate. Again, it is much more common than it should be that components of the work otherwise required as part of a particular contract are indicated to be completed outside of precisely indicated Contract Limit Lines. In such cases, even if the work still remains within property lines, the Contract Limit Lines formally define those boundaries within which the contractor must operate.

b) Site Material Composition.

If significant excavation or other subsurface work is included in the particular project, boring data or other test data describing the subsurface material composition that will have to be dealt with as part of the contract are in the final analysis fundamental prerequisites for completing the work of the contract. While it is very true that there has been a compelling movement in construction contracting to provide such subsurface data in construction contracts as "informational," instead of defining such data as specifically necessary for determination of scope and cost, the fact remains that such information still has a clear purpose. More and more construction owners are working very hard to shift the risk for unknown (or undisclosed) subsurface conditions from themselves directly to the contractor through complicated, confusing, and often very cryptic language in the contract documents with respect to subsurface information. This subject is so important that every construction professional should become intimately familiar with the specific issue and be prepared to deal with its nuances in every situation.

c) Baselines and Benchmarks.

Normally, the contractor is responsible to lay out the work, as a typical specification clause states: "...from two (2) baselines and a benchmark furnished by the owner."

Under normal circumstances, the baselines and benchmark must be physically located on the site, and the owner should assume complete responsibility for and warranty their correctness. The project is then normally laid out by the contractor relative to these baselines and benchmark provided by the owner. Any subsequent errors in the project layout that may be transmitted from incorrect baselines or benchmark information should therefore remain the responsibility of the owner.

d) Utility Locations.

The owner is normally responsible for providing accurate locations of all existing utilities. Locations of telephone lines and so on may be necessary in order to tie into new building services, or may be necessary only to prevent accidental interruption. Correct sanitary and storm line locations and elevations are critical to the design of their underground systems' correct drainage and tie-ins. Invert elevations are also necessary in order to allow proper estimates of the amount of excavation and backfill for the respective utilities. If, for example, the invert elevations indicated on the drawings as 48.0′ are actually 43.0′, the additional 5′ of excavation depth may now require shoring or greater trench width—all substantially increased costs. If this error is due to an improper representation on the drawings of the subject utility locations, the owner in most cases will be likely to be responsible for associated increases in cost to complete the work under these changed conditions.

e) The Effects of Adjacent Properties.

The relationship of adjacent properties may be significant to the construction phase at a project site. An important piece of information that may not be made available in the contract documents and would not be readily apparent in a prebid site investigation by the contractor, for example, may be the drainage characteristics of a surrounding piece of property. If, for example, surface water from several acres of land that is adjacent to the construction site drains into a swale that creates an active water course through the site that lasts several days every time it rains, this fact may not be considered to be something that should have been "reasonably anticipated" by the contractor. If so, the responsibility for resulting downtime, additional drainage and dewatering operations that may be required, and rework of the affected areas can rest with the owner. This can be the case if such a water course had not been disclosed to bidders and is not apparent or otherwise discernible in any reasonable prebid site investigation.

(4) Securing and Paying for Easements and Authorizations. The contractor must be provided with the physical facility in which to complete its work. Barring any specific contractual language to the contrary, the contractor may have the right to assume that the entire footprint within the Contract Limit Lines is available at any time during the period of the contract as necessary to allow the performance of the work as determined by the contractor's "means and methods." If the contract contains some other unusual restriction, such as sequential access to different areas of the site, such information should be clearly included in the contract documents. It should be sufficient in detail in order to allow the contractor to plan its work both in the bidding phase and throughout the construction phase. To the extent that such language is included, it remains the owner's obligation to comply with its part of the bargain as well. Specific considerations in this regard include the following:

a) Site Access.

The contractor must have access to the site in order to perform its operations. Even if such access is limited or restricted, as, for example, may be the case when working within a military complex, a security area of a manufacturing facility, or some similar environment, the peculiarities must be made known to the contractor prior to bid. The consideration of these types of issues with respect to probable lost time and attendant costs must be allowed. If these or any other types of access restrictions have not been made clear prior to the commencement of the work and are not apparent in the reasonable prebid site investigation, the contractor cannot be expected to carry the extra burden without additional compensation.

Some reasonable facility should be provided for workers to park, enter into, and leave the site. If not, specific conditions that address these types of matters should be provided in the bid documents in order to allow contractors to provide appropriate pricing for unusual circumstances. If it is either expressed or implied in the contract documents or in the reasonable prebid site inspection that space is available for material storage (to allow billing the payment for such materials before they are actually installed in the project) and that space suddenly becomes unavailable, the contractor may be faced with another surprise that will prevent invoicing for materials that now need to be stored off-site. Again, if this is not a problem that should have been anticipated by the contractor during the bidding process, the contractor may not be required to bear the costs of material storage, multiple handling, carrying charges, and other problems resulting from late payment for off-site stored materials. These conditions should be studied carefully in order to allow the contractor to take appropriate steps toward equitable adjustment.

The point of this section is simply that if any considerations other than those that can normally be expected to be encountered during the life of the project (considering the type of project) exist, the owner has a specific obligation to disclose them. If the conditions have been changed, the responsibility usually rests with whomever caused the changed condition.

b) Agency Approvals.

The Environmental Protection Agency (EPA), the Occupational Safety and Health Administration (OSHA), the Inland Wetlands Commission, and any number of federal and local agencies may have some level of jurisdiction over certain areas of the facility design and construction. The approval of these agencies of the facility, or some component thereof, may be necessary either at the design development stage or at some time prior to the achievement of the Certificate of Occupancy. It is most often the owner's responsibility to secure these types of approvals prior to the start of construction. In the absence of any noted exception, the contractor usually will have the right to expect that all of these loose ends have been taken care of.

(5) Warranting the Adequacy of the Plans and Specifications. The Spearin Doctrine states that the owner warrants the adequacy of the plan and specifications, and therefore bears the responsibility for any defects in them [(United States v. Spearin, 248 U.S. 132, 136, 39 S. Ct. 59 (1918)]. Defects in the plans and specifications can exist in many forms but can be thought of as boiling down into the two distinct categories of product and time. Most defective specifications problems involve the accuracy of the technical specifications. Such issues can involve mistakes in the way information is communicated (errors), the failure to provide some necessary information (omissions), the failure to coordinate various components of the design (errors), or inconsistencies and incompatibilities between and among design components (errors). Time, then, most often becomes a consequential consideration that is related to the inability of the contractor (or anyone else for that matter) to provide for the (changed) scope of the contract within the amount of time originally contemplated for the performance of the work.

Accordingly, the owner warrants that:

1. The facility and its various components can, in fact, be built as designed.

2. All pieces of the design will fit together in the way indicated.

3. It is physically possible to accomplish all this by employing reasonable means (unless a more elaborate procedure is clearly spelled out in the contract).

4. All the components necessary to complete the project are available within time frames necessary to allow completion of the project within the contract time specified.

From this fundamental warranty flows the responsibility for accurate description of the divisions of responsibilities between the different parts of the work. It is normally the owner's responsibility to ensure that the plans and specifications are clear, complete, comprehensive, and comprehensible (unambiguous). This is referred to as the "Four Cs of Contracts." Responsibilities for the individual pieces of work must be clearly spelled out—without duplication, and without holes.

As noted, the second fundamental owner warranty involves adequate time within which to complete the work of the contract. If the contract allows, for example, 320 working days to complete a project, the owner warrants that the project can in fact be built by reasonable means within those 320 working days. The bidder will normally have a right to expect that by employing reasonable methods and care, he or she will complete the project by or before the total amount of contract time has been used up. The designers (representatives of the owner, for which the owner is responsible) have had every opportunity to confirm that all specified materials and equipment can in fact be purchased and delivered within the constraints imposed by a logical construction sequence that will allow completion of the project within the prescribed period of time. The contractor has the right to depend upon this information. So, for example, if it is discovered that by employing reasonable schedule logic the project time will be overrun, the owner may hold title to the additional time needed to complete the project (and/or for the costs associated with acceleration of the construction that becomes necessary to meet the original completion date—by employing unusual or excessive efforts).

(6) Warranting the Suitability of Furnished Materials.   In projects where the owner provides material and/or furnishes equipment to the contractor for use in the work or for incorporation into the project, the law provides that a warranty that these items will be suitable for their intended purpose exists [Thompson Ramo Woolridge, Inc. v. United States, 175 Ct. Cl. 527, 361 F.2d 222 (1966)]. Beyond ensuring the fundamental compatibility of materials, the owner is also responsible for the timing and coordination of the respective items in the same manner and to the same extent as every other subcontractor and supplier to the project. Shop drawings and other coordination information must be submitted and distributed correctly and in a timely manner, and deliveries of owner-furnished materials and equipment must be made within the current requirements of the contractor's progress schedule. Because of this, the owner would become responsible for delays to the project—and associated costs—caused by delays in providing this information and/or materials and equipment.

(7) Disclosing Superior Knowledge.   The owner has an absolute duty to disclose to the contractor "superior" knowledge of any item that it may directly or indirectly have that is related to the work, where that knowledge is either unknown to or has not otherwise been available to the contractor. Considering the "boring data" example discussed in item #3 above, assume that the boring and soils information did in fact exist for the interior portions of construction, but that it was either intentionally or inadvertently left out of the contract. Such an omission can be the result of a simple error, or may be the result of a "nothing-to-lose" attitude that the contractor might somehow absorb the cost of working through the rock (undisclosed condition) once it is finally encountered. An owner with a "bad faith" motivation might consider that such information can either (properly) be provided in the contract, thereby guaranteeing added costs, or gambles with the chance that the contractor may or may not later make an issue out of it. Even though this type of attitude may actually border on fraud, it is unfortunately one plausible scenario. Likewise, where the owner's superior knowledge of a fact, such as the unavailability or inadequacy of the specified material, would lead to reduced costs, improved efficiency, or simply the earlier exposure of the problem, the owner has the obligation to so advise the contractor.

(8) Taking Prompt Action on Clarifications and Changes. Good construction contracts recognize the need for and importance of speed with respect to issuing clarifications and reviewing/approving Change Orders. This is true whether or not these documents are produced by and for public bureaucracies or for private development contracts. Although not as often openly admitted, this recognition is an acknowledgement that change orders do, in fact, interfere with and disrupt the orderly sequence of the work. They accordingly must be resolved as quickly as possible in order to minimize their total direct and consequential impacts on construction.

In the case of large public agencies, this recognition may manifest itself as a prescribed set of unique procedures that in effect transcends all other standard practices as they relate to normal project correspondence. The following passage, taken from a state Bureau of Public Works "Working Procedure" section of the General Conditions, is an example of this exception procedure:

> "To expedite change orders during the course of construction, proposals are to be submitted directly to the Construction Supervisor (in the numbers of copies requested) with a copy to the Chief of the Construction Section (in duplicate), the District Construction Supervisor, and the Architect."

(9) Providing "Final" Interpretation of the Documents. Although the architect normally researches and prepares recommendations for technical matters relating to design, when involving cost and/or time, these corrections, changes, and interpretations are communicated to the owner, who issues the final position on the matter in the case of public projects. In the case of private development projects, greater amounts of authority may be invested into the design professionals. Varying amounts of "informal" authority may be accorded to architect, other design professional, or even some other member of the owner's organization, but actual authority should be clearly defined in the contract. Because no contractual tie normally exists between an architect and a contractor, however, the ultimate responsibility for the architect's performance in these situations continues to rest with the owner.

It is true that certain contract structures incorporating strict, unmodified provisions of certain documents such as the AIA Document A201 General Conditions of the Contract for Construction prepared by the American Institute of Architects may include language either empowering or requiring (depending upon your perspective) the architect to "interpret" some or all questions and issues relating to the contract documents, and may also provide that such interpretation would be "final." In most public contracts, though, the "right"— and obligation—to interpret the contract is more specifically bestowed directly upon the owner.

(10) Working Cooperatively.   Just as the owner's contract documents are likely to include an express requirement on the part of the contractor to "cooperate" with separate contractors and other entities, the owner has a corresponding duty to cooperate with the contractor to the best of his or her abilities. This cooperation includes an obligation to avoid impeding, hindering, obstructing, or otherwise interfering with the work of the contractor. It is as simple as it sounds and often boils down to the personalities involved.

On the one hand, the contractor should be held responsible for assuming at least a certain impact from normally anticipated effects that should reasonably be anticipated in the given contracting environment. The bureaucracy of any public agency, for example, must be anticipated by the contractor to some extent. In the case of public projects, it would not be considered reasonable for a contractor to claim that it should not be responsible for costs associated with dealing with the normal operations of an existing bureaucracy. The owner, on the other hand, is responsible for ensuring that its own procedures and response times do not exceed those limits expressed in the contract or implied by industry standards.

(11) Assuming Ultimate Responsibility for the Design Professionals.   The architect and his or her engineers have many duties and responsibilities. Although there may be few "express" warranties in the contract as far as the owner responsibilities go, many implied warranties of performance exist. Professional liabilities are assumed for the activities of these people and companies. The continuing contact and dialogue between the contractor and the architect that is typical on nearly every project can sometimes confuse direct-line responsibilities if the specific language of the particular contract is not clearly understood and enforced by all parties. A review of the contract structures described in the early part of this section reaffirms that unless the architect coincidentally assumes some other formal role in addition to his or her design responsibilities, no formal contractual relationship exists between the contractor and the architect. The architect remains an agent of the owner. The owner therefore ultimately assumes the liability for the architect's performance, or lack thereof.

3.3.3   General Responsibilities of the Architect

As the owner's agent, the architect is responsible for the technical design within the context of the owner's responsibility (refer to point #9 of the preceding section).

The responsibilities outlined in this section can, therefore, be accurately thought of as an extension of the owner's responsibilities when considered in the contractual sense.

The details of many of the architect's responsibilities, such as review and approval of shop drawings, may be expressed clearly in the contract. In the absence of such clear definition, however, the items noted here will generally apply.

By engaging an independent design professional, the owner makes an attempt to secure the best possible design given the respective parameters of the particular project. At the same time, an attempt may also be made to shift complete responsibility for design onto the architect. This is normal and expected in conventional owner-architect relationships. In order to preserve these intentions, the owner then clearly and completely conveys all project objectives and necessities to the architect for consideration in the complete design. Participation in the actual design work should then be avoided by the owner as much as possible, if the owner wishes to be sure that the design liability remains solely with the design professionals, at least to the greatest extent allowed by current laws.

As far as the contractor is concerned, all of these considerations remain strictly between the owner and designer. The contractor needs only to consider them both as a single entity and to think of itself as dealing simply with the "owner."

Given these considerations, and in the absence of specific and enforceable language to the contrary, the architect is generally responsible to the owner for:

1. Production and coordination of the plans and specifications.

2. Technical accuracy of all documents prepared by the design professional.

3. Compliance with all applicable codes.

4. Interpretation of the contract documents.

5. Submittal review and approval.

6. Prompt, timely response.

7. Evaluation of the work.

8. Diligence, skill, and good judgment.

(1) Production and Coordination of the Plans and Specifications.   It is the architect's first responsibility to indicate clearly and completely all items of work in sufficient detail on the plans and/or in the specifications in order to accurately describe the complete project with sufficient particularity. All this has to be accomplished comprehensively, in a coordinated manner, and without overlap.

Too often, however, attempts may be made in the contract to impose upon the contractor at least some degree of the design professional's responsibility for the completeness and/or correctness of the plans and specifications. Such efforts may include incorporation of "exculpatory" clauses that are designed to shift the risk—or the burden—of one-party (architect) onto another party (the contractor). An example is language such as:

> "The plans and specifications are complementary. The contractor is responsible to provide wall work shown on the plans, whether or not adequately described in the specifications, and all work described in the specifications, whether or not specifically indicated on the plans, as if called for by both."

In the final analysis, however, the ultimate responsibility to adequately describe each building component does rest squarely with the designer, however uncomfortable the assumption of the liability may seem to them. That adequate description of each building component, depending upon the particular project delivery method contracted for, may include both the technical specification in the physical design itself, along with the clear identification of the individual responsibility (specifications section/division and/or particular subvendor).

In theory, in lump sum general contracting and other fundamental types of project delivery methods, the contractor may very well be entitled to assign individual subcontracts for each specification section strictly "per plans and specifications," with complete confidence that when the procedure is completed the entire project will have been covered—once, and without overlap.

All necessary pieces for the complete project should be accounted for through that process. The responsibility for each may thereby be correctly specified and assigned without misunderstandings or disputes among the various subcontracted bid packages.

(2) Technical Accuracy of All Documents Prepared by the Design Professional. It is the responsibility of the architect to be sure that the design is complete and correct, and that this accuracy is preserved throughout the production of the plans and specifications. The contract documents must be technically accurate in every respect. If, for example, a 5-hp motor is specified for a particular piece of equipment, the designer warrants that a 5-hp motor is available for the designed application. In addition, if that particular 5-hp motor requires 220 V of three-phase power, it is again the designer's responsibility to be sure that the provision is included in the contract, along with the properly assigned responsibility for the final equipment connections. As an additional example, if a specified roof insulation R-factor is necessary and the thickness of the insulation is clearly indicated in the plans, there is a very good probability that the contractor should not be held responsible for any increases in roof blocking that may become necessary due to increases in the roof insulation thickness that may occur because the indicated insulation thickness did not measure up to the specified thermal performance. If the architect wants a blue finish on certain equipment, blue should be normally available. If a particular boiler or other equipment is specified, that equipment should fit between the walls of the boiler and equipment rooms.

It is not normally the contractor's responsibility to confirm these design coordination issues before the specified boiler, the 5-hp motor, or the roof insulation are placed on order. It is important to understand these distinctions clearly in order to defend against the chronic abuse that often surrounds the word "coordination."

(3) Compliance with All Applicable Codes.   The architect is responsible to be sure that the specific design as it is assembled and integrated into the project complies with all fire, safety, and all other applicable building codes in every respect. If the door between two spaces needs to bear a fire rating, for example, it is up to the architect to indicate in the contract the precise rating that the code actually dictates. In this example, it is not enough to leave the contractor with oversimplified instruction to "provide all doors in accordance with the fire code." Similarly, unless specific engineering activities are incorporated into the respective subcontract, such as may be the case in the Fire Protection Systems specification section, it is again the designer who must specify pipe sizes, etc., as those technical requirements relate to appropriate code restrictions. From that point, the designer is completely within his or her rights to require installation and workmanship in accordance with applicable codes and standards. Lacking any clear specific design requirement in the particular contract, however, the designer should not require the determination by the contractor of any component of the design itself.

(4) Interpretation of the Contract Documents.   Depending upon the exact contractual relationships, the duties of the architect to interpret the documents for both clarifications and dispute resolution vary widely. The first determination of contractual interpretation creates an early friction point. This is often the case because an initial response to many types of questions raised by contractors due to apparently incomplete designs is often related to the design professional's "intent" in the preparation of the design. The reality, however, is simply that it is not the "intent" that had been priced by the contractor in the original bid, but the specific indication in the contract language.

In any event, the different possible relationships between owner and the architect may leave a range from no authority to absolute authority vested in the architect for interpretation and "final" decision on all matters relating to design and construction. Most often, however, the architect's role is to review conflicts, proposed changes, and the like, and to submit specific recommendations to the owner for the owner's ultimate decision and action.

The AIA A201 Document—General Conditions of the Contract for Construction, prior to the updated 2007 edition, stipulated that "The Architect shall be the interpreter of the requirements of the Contract Documents and the judge of the performance thereunder by both the owner and the Contractor." As the new wave of collaborative construction contracts were issued by the AIA and other organizations in 2007 and 2009, this atmosphere of working together was incorporated into Article 4 of AIA A201 (2007 edition), which states:

Interpretations and decisions of the Architect will be consistent with the intent of, and reasonably inferable from, the Contract Documents and will be in writing or in the form of drawings. When making such interpretations and decisions, the Architect will endeavor to secure faithful performance by both Owner and Contractor, will not show partiality to either and will not be liable for results of interpretation or decisions rendered in good faith.

In every case, know what the contract exactly says with respect to ultimate authority for interpretation of the contract. Know what your rights are and what your position is, and do not allow authority to be assumed by those who do not rightfully possess it.

(5) Submittal Review and Approval.   Review and approval of shop drawings and other submittals for the various reasons specified in your contract are ongoing activities that are most likely to be the most time-consuming of all the architect's activities, particularly during the first third of the construction phase of the project. It is for this reason, combined with the fact that there exists such great potential for abuse, that this subject justifies special attention.

Assuming proper and timely submissions by the contractor, the architect is normally responsible to receive and act upon each approval submission in a manner that should be precisely described within the body of the contract itself. Lacking that precise description, architectural action should be conducted within the parameters customary in the trade.

Before continuing, it is important to note that much too often contractors—and very often their subcontractors—fail to comply with specific submittal requirements that may in fact be very clearly detailed in the contract. Subcontractors fail to provide complete, clear, and coordinated information, and inexperienced or otherwise diverted members of the general contractor's staff can "rubber-stamp" submittals and ship them off for review by the design professionals without having properly performed their own review and coordination. These types of inappropriate actions taken (or avoided) by contractors on submittals can justifiably upset the design professional, cause unnecessary delays in the submittal/approval process, and even give the other contracting parties the idea that the contractor is not fulfilling its own duties and obligations. To the extent that this may be true, the design professional may be justified in its criticisms of the contractor, and in its belief that it may be entitled to compensation for the extra time needed to process the inappropriate submittals. As contractors, then, it is important as a very practical matter that all subvendors are compelled to comply with submittal requirements in every respect—right down to the number of copies specified—and the work of each separate bid package should be thoroughly coordinated with all others and with the contract itself. If all of these things are accomplished consistently, the contractor will have every right to expect the design professionals to fulfill their own obligations completely and on time.

Considering the above, the responsibilities of the design professionals with respect to review and approving submittals generally include the following:

(A) Conformance to Requirements.   This principle requires a precise, detailed review by the design professional of every significant component of the submittal in order to confirm that the respective item proposed by the contractor (or its subvendor) meets all design and performance criteria as originally specified in the contract. This process is implemented in order to confirm that the product meets the stated requirements and that the product is not simply submitted as someone would now wish it to be. "Standard colors" does not mean "special colors," and "_" insulated glass does not mean "1" insulated glass." Obvious efforts to embellish the contract should be easy to spot (because the contractor has performed its own coordination activities). It is important for contractors to exercise the discipline necessary and to check all review comments that may have been made by each design professional involved in the process in order to determine if it has any potential to impact cost or time of performance.

(B) Provide Missing Design Information. In many instances, complete preparation of the submittal by the contractor or its subvendor may not be possible because design information is lacking. Dimensions sufficient to make precise location calculations may not, for example, be available. The precision of detailed, large-scale shop drawings may expose conflicts that may require significant redesign, or that may require nothing more than the designer's decision as to which possible alternative may be preferred.

While it may be true that the contractor is normally held "responsible for dimensions," this is normally not so when insufficient information was originally included in the contract. If original dimension information is lacking, the architect may become "responsible for dimensions."

(6) Prompt, Timely Response. No matter how narrow or broad the scope of responsibilities of the design professionals is as designated in the contract, the architect (and his or her subconsultants) are obligated to perform their work in time frames that are compatible with needs of the project schedule. The American Institute of Architects AIA Document A 201, General Conditions of the Contract for Construction, provides that the architect has an express obligation to take all actions "with reasonable promptness so as not to cause a delay in the work." Even if "reasonable promptness" is not specifically defined in terms of the number of days considered to be acceptable, the durations of the review/approval activities by the architect and its subconsultants may be established by customary trade practice, by confirmation or clarification at early job meetings, or by calling attention to specific requirements in the appropriate correspondence as individual situations may dictate.

Common activities that the architect will normally be required to perform at such "reasonably prompt" rates include:

- Review and approval of shop drawings and other submittals.
- Review and recommend or approve Change Orders.
- Prepared Change Order designs.
- Approve Applications for Payments.

- Issue documentation (meeting minutes, transmittals, etc.).
- Conduct site inspections and investigations.
- Perform, monitor, and process tests and other performance evaluations.
- Prepare clarifications.
- Respond to all ongoing operating questions.

The architect is responsible to properly and completely respond to both usual and unusual situations within either the stated or implied time constraints. If such time constraints are consistently being stretched by the design professional, that person or organization may risk bearing the responsibility for delays and other consequential damages that may result.

(7) Evaluation of the Work.   The architect has a responsibility to satisfy himself or herself that the work is being performed in accordance with the design, as well as with applicable workmanship standards. The architect may not necessarily be responsible to be intimately familiar with every nut and bolt of construction as the work is progressing. Fundamentally, it is a responsibility of the respective subcontractors or trade contractors to install their portions of the work correctly in the first place—the first time. Ignoring the site until a problem finally comes up, however, is an unacceptable extreme that should not be tolerated.

The activities that make up this ongoing evaluation of the work as determined specifically in most contracts can include the following:

(A) Inspection.   The architect is normally responsible for making regular visits to the jobsite in order to familiarize himself or herself generally with the progress and the quality of the work as it is being performed. It is not enough to simply review the progress photographs and try to get a picture of the job progress through the correspondence. The architect must be on the site in order to confirm that the work is progressing along the lines expected, and the architect must do so as the work is progressing (not long after). It will probably not be considered reasonable, for example, for the architect to wait until all the brick is up before it is determined that the color of the mortar is not close enough to the sample in order to be considered acceptable.

Finally, the architect should be available on a frequent, periodic basis, and at any other time required in order to provide answers to questions and to resolve minor conflicts as quickly as possible, without having to cause the contractor to hunt unreasonably and for a long period of time in order to secure simple answers.

(B) Testing.   This function or responsibility may actually be split between the architect and owner, or be incorporated within the contract in some other manner. Most often, construction contracts provide that inspections and testing of the work as it is being performed should be an activity to be performed and paid for by some owner's agent as opposed to the contractor in order to avoid conflict of interest. Regardless of who performs the testing and inspection responsibilities, the results of such efforts are normally subject to review and approval by the architect. In many cases, the architect should therefore conduct those activities to the extent of its responsibility completely and on time.

(C) **Evaluation.** The architect is responsible to either determine entirely or to confirm the owner's evaluation of the "fitness" of the work, along with the associated dollar value. Specific activities included in the process may be:

- Determine/confirm/verify the amount of work put in place by the contractor, along with corresponding payment value.

- Confirm acceptable material quality and workmanship standards.

- Reject work that does not conform to the contract.

- Determine the dates of substantial completion and final completion.

- Issue stop-work orders in cases in which defective and nonconforming work is being allowed by the contractor to proceed.

(8) Diligence, Skill, and Good Judgment. It is unusual to find many "express" warranties of architects (and their subconsultants) in most construction agreements. These design professionals do, however, implicitly warrant that they have exercised diligence, competence, skill, and good judgment throughout the design process and contract preparation. Moreover, they are considered to have performed all of their activities in accordance with the professional standards of the community in which the work is to be constructed. With the exception of the most obvious errors (patent errors), the contractor has the right to assume that the information provided in the contract documents is complete and sufficient in order to allow an accurate estimate, and to allow the contractor to construct the work as designed. More subtle differences that become apparent as the work progresses are then the responsibility of the party that causes or contributes to the deficiency.

3.3.4 General Responsibilities of the Contractor

As one observer put it some time ago, a contractor is a gambler who never gets to cut, shuffle, or deal. It is the general, or prime, contractor who will likely have the multitude of "express" responsibilities to the contract riddled throughout the documents. In the worst cases, this type of language can be pervasive and applied almost in a shotgun pattern designed to drop an (unwary) elephant. The manner in which the various contractor responsibilities and warranties are catalogued, labeled, defined, recatalogued, and repeated for emphasis is most often in very sharp contrast to the way in which the corresponding responsibilities and obligations of the owner and the design professionals have been left to be "implied." The contractor may be left to look to outside authorities—such as applicable laws and the increasingly nebulous idea of "trade custom"—in order to search for guidance that will specifically define such responsibilities.

In efforts to insulate themselves from liabilities and extra costs that may be born out of defects in the contract or in its administration, the drafters of construction contracts can combine layer upon layer of phrases, clauses, references, standards, boilerplate, exculpatory language, and the like in a scramble to shield themselves from every conceivable issue. This process can continue to the point at which the front-end documents of construction contracts have finally become fatter than the technical specifications themselves.

Ironically, it is an overly enthusiastic abundance of complicated dissertations on the contractor's responsibilities that more often actually increase the likelihood of confusion and conflict. Such efforts that create such multiple layers of excruciatingly complex language often carry with them a significant risk of conflicting statements and even "requirements" that may be contrary to public policy and be therefore unenforceable. This whole idea, however, often does not seem to be taken seriously by the drafters of our construction agreements. The greater the number of statements made, the more detailed and complicated the statements are, the greater the number of specification interfaces, the greater probability of contradiction—and therefore ambiguities—every step of the way. It's simple mathematics and probability. A specification that is too fat as far as the front-end (General Conditions and other general requirements) is concerned can have more problems than a specification that might be considered by designers to be "too thin." It seems that some time ago, the principle of "saying it clearly and saying it once" has given way to "saying it over and over again in different ways and in different places."

In any event, don't be shaken by a seemingly overwhelming collection of confusing and exculpatory language. This is not to say that it should be lightly considered. Every word must be read and thoroughly understood. Many inclusions may in fact be very serious and require specific and focused attention. If there is any doubt regarding the applicability, legitimacy, or enforceability of any provision included in your contract, consult with your superiors and possibly with a competent construction attorney. Confirm exactly where you stand with respect to the written word before proceeding.

Besides the strict technical requirements of the work itself, then, the responsibilities of the contractor will generally fall within these categories:

1. Duty to Inquire.
2. Reasonable Review.
3. Plan and Schedule the Work.
4. Lay Out the Work.
5. Supervise, Direct, and Install the Work.
6. Adequate Workmanship.
7. Correction of Patent Errors.
8. Coordination of All Parts of the Work.
9. Review, Submit, and Coordinate Shop Drawings.
10. Properly Process All Contract Payments.

11. Provide Adequate Insurance.

12. Adherence to Safety Standards.

13. Warranty of Clear Title.

1. Duty to Inquire. The bid documents usually require bidders to bring any questions regarding inconsistencies, conflicts, or ambiguities to the attention of the owner and the design professionals prior to the bid. Such requirements may become an elaborate, strained effort intended to shift responsibility onto the contractor for a fundamental rule of contract interpretation: that ambiguities are resolved against the party who drafted the documents. In its practical application, owners and design professionals have often attempted to hide behind the contractor's "duty to inquire" in even very subtle and hidden confusions and ambiguities in the contract language. In the final analysis, however, it is the obviousness or significance of an ambiguity that may place a duty upon the contractor to inquire about work to request clarification regarding such ambiguity. This may be true whether or not the specific responsibility to do so is clearly called out in the contract language. Certain legal cases have determined, however, that no such duty to inquire was determined to be any obligation of the contractor when the discrepancy in the contract documents was subtle in nature or hidden from a reasonable review.

This is the theory that is generally being applied by the courts, but in arbitration (where the arbitrator is more free to consider an issue in any manner in which he or she sees fit) the foregoing distinction applies probably even more so.

2. Reasonable Review. The contractor is responsible for "reasonable" review of the contract—but it is not responsible for complete search of documents to be conducted with the intent of determining flaws that may or may not be contained within those documents. The contractor is a businessperson performing many diverse and distinct functions every day and is not customarily involved in the preparation, review, or coordination of design work. While the contractor's daily activities are being performed, one of these activities—bid preparation—is commonly being performed within just a few weeks immediately prior to the respective bid. In sharp contrast are the owner, architect, engineers, agencies, and whoever else had in many cases months or years in which to prepare and coordinate the contract documents. They've had enough time (and opportunity) to determine for themselves how the respective portions of the work will be divided so as to ensure that the sum of the parts will in fact equal a whole (completed project). If all that conscientious and time-consuming effort still results in a flawed set of documents, how can it be considered to be reasonable to expect the contractor to discover such defects during the short time immediately prior to a bid?

Because the owner warrants the adequacy of the plans and specifications to the contractor, the contractor should have the right to depend upon those documents. The contractor is therefore normally entitled to assume that the documents are correct [(with the exception of dealing with glaring, patent, and obvious errors as discussed in item (1.) above]. The drafters of the contract had every opportunity to be sure that the contract was correct in every respect before asking any businessperson to sign off to it after a comparatively short review period. It is this concept that underscores the rule of contract interpretation—that ambiguities are to be construed against the drafter.

3. Plan and Schedule the Work.   There can very often be any number of ways in which to complete a particular project. The original bid may likely anticipate one reasonable sequence of construction that is apparently achievable, given the technical requirements of the contract and the stated time constraints. Assuming proper design workability, the contractor is normally responsible to anticipate the various components that will eventually make up the whole project in terms of sequence of construction. The interrelationships of each of these components should be determined, and a project plan should be completed as soon as possible. Timetables are added and a calendar is imposed upon the plan in order to transform the plan into a schedule.

If properly prepared, the resulting catalog of items, dates, times, and inter-dependencies will represent a complete method in which to build a project—all as can be reasonably construed from the information presented in the contract documents. It follows, then, that if any item had been misrepresented or confused in the documents, the resulting effect on the progress and schedule may belong to the party who created the original flaw. Such effects may manifest themselves in the form of interruptions, resequences, accelerations, or delays.

It is important to recognize that schedules will change. Courts have begun to more consistently recognize this fact of construction life. Contemporary legal thought regarding construction schedules is that for a schedule to remain valid it must be periodically updated. All significant effects of all parties must be considered; none must be emphasized more than can be justified, and none must be left out for convenience. This consistent recognition of the need for periodic schedule updates is also a recognition of the fact that the details of construction are likely to change periodically as well.

A final thought is that if a schedule is to commit the owner, architect, subcontractors, and suppliers with any level of effectiveness, the updated information must be communicated to and acknowledged by those respective parties in a timely manner, just as with any other proper documentation. Too many contractors seem to take the approach that their construction schedules should be prepared in a vacuum—and even with some degree of secrecy. Schedules are prepared and dictated to subvendors without having allowed evaluation or input by those subvendors. This issue is treated in more detail in the section in this Operation Manual dealing with Project Schedules. For purposes of this section, however, it is important to recognize that this type of closed-door treatment of progress schedules is most often a shortsighted approach that can hurt the entire effect much more than it might help. If, instead, the input, acknowledgement, and even approval of the owner, architect, and subvendors is solicited and achieved, such effort may go a long way in eventually proving the reasonableness of the baseline schedule and of the agreement of all parties to perform according to it. This approach can in contrast be a very powerful force that should not be underestimated.

4. Lay Out the Work. The contractor is normally responsible to physically lay out the work from the baseline and benchmark information provided by the owner, and then to proceed to lay out the remainder of the project. If errors permeate the work as a result of the owner's error in the beginning data, the contractor might not be held responsible for consequential effects (assuming that its own layouts had been conducted at least to the level of the professional standard to be expected for that activity). This may be true unless it is a type of error that should have been discovered by a competent contractor who is complying with its own professional standard practice.

5. Supervise, Direct, and Install the Work. The first responsibility for supervision of the work is to enforce participation in and compliance with the construction schedule. Progress must be consistent with the commitments and arrangements established throughout the planning and scheduling of the contractor's activities. This participation is orchestrated beginning with the overall game plan of job sequence, to the placement of field offices, material staging areas, and equipment control, through policing temporary protection and facilities, to the physical installation of the specific job components, and implementation of all close-out activities.

As the work is being installed, it is the methods, techniques, sequences, and procedures that are and remain the responsibility of the contractor. These are specifically different from ownership of the design details' workability, as described in the section outlining the architect's responsibilities. In other words, the contractor begins with a legitimate assumption that the job components will in fact fit together (the workability). It becomes the contractor's assignment from that point to assemble those components in a logical, efficient sequence of activities.

6. Adequate Workmanship. A good specification will incorporate precise tolerances within which the respective work is to be installed, if it is to be considered "acceptable." Clear descriptions of these parameters will accordingly eliminate any necessity for subjective evaluations. Such clear, nonsubjective descriptions may end arguments before they have an opportunity to blow themselves out of proportion. The quality-control specification for the placement of a concrete floor, for example, may require the finished floor surface to simply "be level," or it may require the surface to "be level within 1/8″ in 10′ in any direction."

In the first case, evaluation of performance will be very subjective indeed with respect to the meaning of the word "level." Does "level" mean "flat"? Can any degree of "waviness" exist in the "level" slab?

In the second case, there is no question regarding the acceptability of the floor as it relates to its "degree" of "level." All that is needed is to simply measure it. The slab either meets the 1/8″ requirement or it does not. The question of acceptability, in other words, is reduced to arithmetic. It is clearly quantifiable and is not at all subjected to opinion. This type of specification is in marked contrast to one that would have required the same floor slab to merely have been "level." In the first example, the debate over "level" may last long after the building is occupied.

Lacking a precise technical definition, the next area in which to look for a description of an acceptable level of quality are any industry standard specifications that may have been incorporated into the contract documents by reference. Some standards are more applicable than others. Assuming a precise technical reference, such as requiring a particular weld to "conform to the requirements of AWS (American Welding Society) Code for Welding in Building Construction AWS D1.1-82," the contractor will still be responsible for the specific, measurable requirements as if they had been directly included in specification language originally.

In the event that no technical specification or reference standard is included in the contract documents, it may be highly unlikely that there would not be some boilerplate expressly requiring the contractor to install the work in accordance with applicable building codes as they relate to that portion of the work. If beyond this the specification was so grossly incomplete as to leave out even these most basic requirements, the law may impose other restrictions. In this regard, the contractor implicitly warrants that the work will be performed in accordance with standard levels of workmanship and within the descriptions of accepted trade practice as defined in the community in which the work was performed. This idea of conformance to community standards cannot, however, be used to excuse the contractor from performing a more stringent requirement if such requirement is unambiguously described in the contract.

Finally, specific procedures outlined in manufacturers' instructions will in most cases need to be followed precisely if the warranties for those items are to remain intact. If the detailed manufacturers' information conflicts with the stated contract requirements, the owner and the design professionals should be the ones to make the decision regarding the direction to take. The decision to move away from procedures recommended by a particular product manufacturer is very likely to carry with it definite liabilities. These decisions are therefore left with other parties, if at all possible. Barring this type of conflict, the manufacturer's directions will generally take precedence over other noted instructions.

7. Correction of Patent Errors.   In most cases, contractors do not bear responsibility for the designer's or owner's errors in the contract documents. This is true unless the errors are so obvious or glaring (patent) that a competent contracting professional should undeniably have discovered them through a reasonable review. Therefore, where it can be demonstrated that an error can be considered to be patent, the contractor has a clear duty to bring such an error to the attention of the owner prior to bidding—or at the very least immediately upon its discovery.

For example, a Mason contractor's experience will likely be that most (if not all) cavity walls that he or she has ever constructed have had some kind of through-wall flashing integral to the design. If the design detail on a particular project shows no through-wall flashing of any kind in any cavity wall, the defect in the design may be considered to be so obvious to an experienced (reasonably competent) masonry contractor that it will be considered to be incumbent upon that contractor to bring the design flaw to the attention of the owner. It remains, of course, that the definition of what qualifies as "obvious," "glaring," or "patent" is the fertile gray area that may itself defy easy resolution. In such a situation, the discussion of trade custom and usage may be most relevant.

8. Coordination of All Parts of the Work.   The word "coordinate" can too often be used in an attempt to stretch the contractor's responsibility with respect to the contract documents. In its worst case, the word can be used to abuse. Dictionaries define the verb "coordinate" as "to arrange in order" and "to harmonize in common action or effort." In as many dictionaries as I have consulted, I have not been able to locate the definition of "to find the mistakes made by others and to correct them at no additional cost." All too often, this, however, seems to be the definition that some owners would have if the contractor would allow it to be.

The coordination activities that a general contractor is responsible to perform as they relate to the construction process amount to:

1. Assembling pertinent information from all sources that may possibly affect a certain portion of the work.

2. Correlating that information with the respective specific job requirements as indicated in the contract.

3. Distributing relevant information to those requiring it in time to incorporate the information in their own work.

All of these activities must be accomplished while meeting the constraints imposed by the current progress schedule. The general, or prime, contractor is in a real sense nothing more than a conduit in this regard. "Coordination" becomes an ongoing mechanism by which defects in the contract documents may be exposed in time to avoid expensive impact on the project. "Coordination" is not the vehicle by which contractors become responsible for mistakes in the design. Do not let it become one.

9. Review, Submit, and Coordinate Shop Drawings.   Technically, this activity of shop drawing review, submission for approval, and distribution should be included in item (8) above, "coordinate," but is given special treatment here because as a contractor activity it is one of the most common and time-consuming. It is also one that affects virtually every component of every project.

Shop drawings submitted for approval must clearly and comprehensively describe the specific product(s) that they represent without ambiguity. They must include the exact information regarding how those respective products interact with the other parts of the work. Where differences exist between the item being submitted for approval and that which is specified, such differences should be highlighted in a manner that will allow their specific consideration without the risk of oversight. Failure by a contractor to note or otherwise highlight such differences will not likely relieve such a contractor from the responsibility for those differences, even if the submission bears the architect's formal approval.

The contractor is responsible to coordinate all dimensions insofar as necessary to install the work properly relative to the surrounding construction. The contractor is not normally required to invent dimensions, providing dimensions originally missing from the design. That is the architect's responsibility. If a structural steel shop drawing detailer, for example, takes upon himself or herself to decide upon the column location as might seem to be logically correct but not necessarily as a result of calculation or extrapolation of other data on the plans, such a determination must be called to the attention of the designer for that designer's verification and approval. Otherwise, the detailer, and therefore the contractor, may likely bear the full responsibility for how well (or how poorly) that "missing" dimension fits within architectural, plumbing, ductwork, finish work, or other considerations.

Finally, the confirmed and approved submittal information must be forwarded to all who might require it in time to incorporate it without interrupting its other activities. If an approved pump, for example, is mounted on a beam and requires 208-V three-phase power, the information must be distributed to all affected subcontractors in time to allow them to complete their shop drawings for the affected items. The rule to remember is "when in doubt, send it."

10. Properly Process All Contract Payments. The contractor is normally responsible for payment of all costs directly related to the physical completion of the work (once having been paid for such work by the owner). These include the cost of labor, materials, equipment, tools, machinery, transportation, and sales and consumer use taxes. In addition (in a straightforward lump sum general contract arrangement), the responsibility for all costs to provide for all items necessary and incidental to the work, such as temporary heat, light, utilities, etc., is also likely to be included. This description is as it is on many construction projects. Know, however, what the specific requirements are on the particular project, and manage them accordingly.

Beyond the absolute responsibility, most construction agreements normally require that the contractor make payments to its subvendors in the manner and to the same extent that payments for the work of those subvendors has been made by the owner to the prime contractor.

The pay-when-paid clause, long a staple in many contracts between the general contractor and subcontractors on both private and public work, has been challenged in recent years. In some states, the clause has been declared null and void. Thus, it's best to check the state statutes before withholding payment from a subcontractor, if and when this situation arises in the state in which you are currently working.

The Federal Miller Act requires contractors to post a payment bond on federal construction projects, but this requirement also has "pay-when-paid" implications.

In a 2001 case, *Walton Technology v. Westar Engineering*, the subcontractor sued the contractor under the Miller Act, requesting payment even though the general contractor had not been paid. The federal appeals court ruled that the surety was responsible despite the contractor's "pay when paid" clause, so the subcontractor was able to reclaim the amount owed by the general contractor by utilizing this "call on the bond."

11. Provide Adequate Insurance.   As a standard article in construction contracts, the contractor will be required to obtain predetermined amounts of insurance coverage prior to allowing any work to begin at the site. Such insurance coverages will normally include:

1. Worker's compensation insurance.
2. Fire and extended coverage insurance against fire and other risks normally included in standard coverage endorsements.
3. Public liability and property damage insurance.
4. Contractor's protective public liability insurance covering the operations of subcontractors.
5. Automobile insurance for owned or hired vehicles.

12. Adherence to Safety Standards.   The contractor is responsible for regularly, at all times, and without the need for any specific notice to take all necessary precautions for the safety of all individuals on the construction site. Safeguards for the protection of both the workers and the public as may be required by job conditions, progress of the work, and applicable authorities such as OSHA may include warning signs, barricades, scaffolding, lights, fire extinguishers, proper ladders and walkways, and similar items. Workers are to wear hard hats and proper clothing and use correct safety equipment (harnesses, etc.) in appropriate situations. Contractors should be thoroughly familiar with the requirements of an adequate safety and loss control program and treat the whole issue of safety as its highest priority.

13. Warranty of Clear Title.   This contractor responsibility is most often in the form of an express warranty located within the General Conditions or other similar section of the contract documents. The guarantee of clear title is the owner's assurance that the project is free of liens or other encumbrances on all materials, labor, and equipment incorporated into the work. A typical clause in such contracts may read as follows:

> "The contractor represents and warrants that the title to all work, materials, and equipment for which payment shall have been made to the contractor shall vest in the owner upon said payment, free and clear of all liens, encumbrances and adverse interests of any kind whatsoever; however, the contractor shall remain liable for damage and loss to said work, materials, and equipment, until such time as the Project is fully completed and accepted by the owner, and final payment is made pursuant to the contract."

3.3.5 The Pass-Through Principle

In the interest of clarity and simplicity, discussions in the preceding sections were generally confined to the owner, architect, and contractor, without many specific references to subconsultants or subcontractors. This approach was taken intentionally, and within the context of an idea that is referred to as the "pass-through" principle. The "pass-through" principle is elaborated upon in other sections of this Operations Manual. Simply stated here, the pass-through idea means that the subconsultants are generally responsible to the architect (primary design consultant) in the same way that the architect is responsible to the owner (and vice versa). Similarly, each subcontractor is generally responsible to the general (prime) contractor or construction manager in the same manner and to the same extent that the prime contractor or construction manager is responsible to the owner for the work of that particular subcontractor (and vice versa). This issue is clarified at this point in order to confirm that consideration of these important entities has not been neglected but has been included within the parameters of the relationships as summarized here.

In addition, it is often difficult to read and write about many construction topics without encountering words like "reasonable" or "acceptable" or other such words that may appear on the surface to clarify levels of quality and performance expected. In practical application, however, exactly the opposite is most often the case with such words. It is important to realize that whenever any adjective is used to describe some measure of acceptability that is not specifically definable and/or quantifiable (such as "flat" or "hard," as opposed to "level within 1/8″ in 10′ in any direction" or "95% compaction"), a flag should pop up in your mind that yet another gray area is sure to develop into muddy water if the subjective criterion is not clarified immediately and definitively. The word "reasonable" has been and will regrettably continue to cause serious problems in all types of construction contracts.

3.4 General Principles of Contracts as Applied to Construction

3.4.1 General

There are several basic rules of contract interpretation that generally apply to all contracts, regardless of the industry. These principles form the basis for the initial application of the law in deciding how conflicts will be resolved. Even if these ideas were absent in the form of strict rules, their principles are useful to apply to determine the practical structures of various situations and options. They mesh closely with the next section that will apply additional rules directly to construction contracts.

The descriptions are in short outline form to describe the principles in the most direct manner. They have been selected for inclusion here because of their common application to our industry. Discussions have been kept direct so they can be immediately understood and applied; discussion of theoretical reasonings are accordingly kept to a minimum.

Finally, it is important to be aware that discussions of "Contract" principles include the entire contract, including the plans, specifications, and all other contract documents as described in Section 2.4.2 of this Operations Manual.

3.4.2 Reasonable Expectations

The first principle of contract law is to protect the reasonable expectations of the parties. The first step in doing this is to be able to confirm what those expectations are (or what they should have been)—to determine *whose* meaning is the right one.

In an effort to reduce subjectivity in the process, the courts have sought to develop an objective standard. The standard of a "reasonable expectation" has accordingly been described as the meaning that would be attached by normally intelligent people competent in their profession, with complete knowledge of all related facts.

Application Rules. The following ideas will aid in the application of the principle as described above:

1. The ordinary meaning of language is given to words unless circumstances show that a different meaning is applicable.
2. Technical terms and works of art are given their technical meaning unless the context indicates a different meaning.
3. A writing is interpreted as a whole, and all writings forming a part of it are interpreted together.
4. All circumstances are taken into consideration.
5. *If the conduct of the parties defined a particular interpretation, that meaning is adopted.*
6. Specific terms are given greater weight than is general language.
7. Separately negotiated or added terms are given greater weight than standardized terms (boilerplate) or terms that are not specifically negotiated.

3.4.3 Ambiguities Resolved against the Drafter

The law assumes that those drafting the contract (contract documents) will provide for their own interests and that they have reason to be aware of uncertainties. Not only should the contract drafters have had a clear idea of their intent, but they have had every opportunity to be sure that those intentions are clearly enough defined so as to allow reasonable, competent people to understand and accommodate them. The owner and design professionals typically may have had a year or more to be sure that the contract documents are clear and complete. Their attorneys have presumably had every opportunity to express each requirement clearly. It is unreasonable to expect a contractor to find flaws in the short time prior to bid.

3.4.4   Right to Choose the Interpretation

If anything is subject to more than one reasonable interpretation, the contractor has the right to choose the interpretation. The reasons are generally the same as those described for resolving ambiguities. If such a reasonable interpretation is available, there is no duty to continue to search for other "reasonable interpretations." All that is necessary is that your interpretation be reasonable; it does not even have to be the "most" reasonable.

3.4.5   Trade Custom

The Uniform Commercial Code defines "usage of trade" custom as:

> ...any practice or method of dealing having such regularity of observance in a place, vocation, or trade as to justify an expectation that it will be observed with respect to the transaction in question. (U.C.C. Sec. 1-205(2))

A 2×4 is not $2'' \times 4''$, but $1\frac{1}{2}'' \times 3\frac{1}{2}''$. An $8''$ CMU is $7\frac{5}{8}''$. Customs always seem unusual at first, but through familiarity eventually make sense. If the designer referred to "a 2×4," you could rely on a dimension of $1\frac{1}{2}'' \times 3\frac{1}{2}''$. If, however, that designer referred to "$2'' \times 4''$ blocking," be careful. Does the "Ambiguities" rule apply?

To establish trade custom as the definition of meaning, it must be shown that the custom is followed with absolute regularity. It is not sufficient that it is "usually" done in a certain manner; it must be a usage that is observed in virtually all cases in the area in which the contract is being performed.

Finally, trade custom cannot be relied on as an excuse if the requirement is clear and subject to one reasonable interpretation. When the roofing contractor says, "I never cut reglets," read the specification. He may be cutting reglets on *this* job.

3.5   Key Principles of General Contracts

3.5.1   General

This section continues the principles of Section 3.2, applying the additional ideas directly to construction contracts. The descriptions are again in short outline form to describe the principles in the most direct manner, and have been selected for inclusion here because of their common application. Unique to our industry, these contract rights and requirements are very consistent throughout the majority of contemporary construction contracts. Discussions are direct so they can be immediately understood and applied; discussions of theoretical reasonings are again kept to a minimum.

3.5.2   "General Scope of Work"

The scope of work that must be completed in order to fulfill contractual obligations must be clearly and completely defined. Beyond that explicitly described the scope will be interpreted to include work that is *plainly and obviously* necessary for a complete installation. Strained interpretations that attempt to add work that is not specifically shown must be subjected to the most intense scrutiny. Consider the entire discussion of Section 3.2. From there, apply "Intent" vs. "Indication" (Section 3.5.3), "Reasonable Review" (Section 3.5.4), and "Performance" and "Procedure" Specifications (Section 3.5.6).

3.5.3 "Intent" versus "Indication"

In a disagreement or dispute over scope of work, architects or engineers may try to apply what they *meant* to say if they cannot find anything in the documents that precisely indicates the item in question. Typically, some general conditions clause containing language to the effect that "It is the *intent* . . . to include all items necessary for proper execution and completion of the work," may be relied upon by the owner.

The key to the applicability (or lack) of design intent as a means to fill gaps left in the design goes to whether the work could be "reasonably inferrable." Design intent will apply only if the gap being bridged is so obvious that a professional contractor would not normally overlook it. Refer to "Trade Custom" (Section 3.4.5) and "Reasonable Review" (Section 3.5.4) for related discussion.

Applicability Test. To assess the applicability of design "intent" to force a contractor to complete work not specifically indicated, answer the following questions:

1. Can the remaining work shown be completed without the extra work in question?

2. Is there more than one way to complete the extra work?

3. Is the extra work not usually encountered by the type of trade now being considered to construct it?

An affirmative answer to any of these questions will throw into question the applicability of "intent." For "intent" to apply, it will essentially need to satisfy the same test as that for trade practice (Section 3.4.5); that is, the work in question must be required with *absolute regularity* in all such cases.

3.5.4 "Reasonable Review"

Contractors are responsible for reviewing all bid and contract documents as they relate to their work. What can be seen, or *reasonably inferred,* gets priced. The contractor *is* responsible for the disclosure of *patent errors* and the contractor is normally required to conduct a prebid site inspection.

While it is certainly desirable to be able to discover mistakes, inconsistencies, or other problems with the contract in time to minimize their effects on the project, it is not the contractor's *responsibility* to do so. The contractor is *not* required to perform a complete search of documents seeking out hidden flaws in the contract, making subsurface explorations, or undertaking any other extreme investigations. That should have been done by the owner and/or the design professionals prior to the bid.

3.5.5   Disclosure of Patent Errors

Beyond the responsibility for "Reasonable Review" as described in Section 3.5.4, the contractor may be held responsible for disclosing *patent errors*. These are mistakes so obvious or glaring (patent) that any competent contracting professional should have observed them through a reasonable review.

> **Example:**   A masonry contractor's experience will be that masons *always* incorporate flashing and ties in *every* cavity wall they ever build. If the details on *this* project, however, show no flashing or ties, the defect may be considered to be so obvious that the masonry contractor should have at least raised the question.

It is interesting to note that the more "obvious" the error, the more in question the designer's competence becomes.

> **Defense:**   One argument for being paid extra for the work in question is simply that if the design mistake had been observed prior to bid, the cost for the additional work would have been added to the bid at that point. The greater risk, however, is that of consequential damage resulting from the work's being constructed with the error, and the necessity of removing incorrect work and rebuilding.

3.5.6   "Performance" and "Procedure" Specifications

A performance specification describes the ultimate *function* to be achieved, leaving the means and method up to the contractor.

> **Example:**   Provide foundation perimeter insulation having a k = .20 @ 75 degrees F, and a minimum compressive strength of 20 lb/ft.

A procedure specification describes the material to be used and its physical relationship to the surrounding construction. It should detail the qualities, properties, composition, and assembly.

> **Example:**   Provide extruded closed-cell polystyrene board as manufactured by XYZ Corp., 2″ thick × 24″ wide.

Application.   There is nothing inherently "wrong" with either type of specification. Performance specs, for example, may shift responsibility for the design onto the contractor, but they also increase the contractor's options and control.

Performance specifications may be used more often in descriptions of mechanical and electrical equipment, but they will even be used for things like concrete (minimum compressive strength of 3000 psi @ 28 days).

Procedure specifications must be correct in every respect in order to avoid problems. If it is properly followed, the risk of its design success will remain with the designer.

If a procedure *and* a performance specification are provided for the same item, the contractor is fulfilling its obligation by providing the exact material described. There is no further duty to confirm that the described material also meets the performance description—that was up to the designer in the first place.

3.5.7 Change Clause

The Change Clause authorizes the owner to alter the work if the change in question falls within the general scope of the original agreement. If it does not, there is no inherent "right" of the owner to change the work.

Basic Elements. Most Change Clauses will incorporate similar provisions. Typically, they include:

1. Adjustments to the contract may only be effected by a change order.

2. The change order must be in writing, signed by both parties.

3. The change order must specify adjustments to both the contract price *and* the net effect on the project time.

4. The change order will be for work, that is, within the scope of the original contract.

5. No changed work is to be performed without a properly executed change order (except in the case where the contractor must act in an emergency to prevent injury or property damage).

When asked to perform changed work without a finalized change order, if an AIA contract document is in force, the owner should be directed to the article dealing with Change Orders and that section of the article relating to the Construction Change Directive (CCD), whose purpose is to allow change order work to proceed when there is either no time to assemble a cost or lack of agreement on cost. Section 8.7 in the chapter on change order work spells out the details of initiating and proceeding with change-order "extra work" via the CCD.

If a contract other than an AIA Owner/Contractor document is in force, present this CCD approach and obtain agreement to proceed with the extra work following those principles.

3.5.8 Pass-Through Clause

The Pass-Through Clause (sometimes referred to as the Conduit or Flow-Down Clause) incorporates into each subcontract by reference all the rights and responsibilities of the prime contract to the owner as they relate to the work of the respective subcontract.

Example: The general conditions may make "the contractor" responsible to provide "all scaffolding, hoisting equipment, etc., as may be necessary to perform the work."

The Pass-Through narrows this to mean the plumbing subcontractor provides all these items as necessary to complete the plumbing work, the plaster sub provides these items as necessary for the plaster work, and so on.

The authority of the Pass-Through Clause comes from each individual specification section that references the general conditions, supplementary general conditions, etc., to be incorporated into the requirements of the respective technical specification section. It thereby brings all those requirements to bear directly on the work.

3.5.9   Dispute Clause

The Dispute Clause provides the specific procedure for resolution of serious problems. It may detail a progressive series of steps (such as appealing to higher authorities) or may simply describe the ultimate option (... arbitration ...).

Know what your Dispute Clause says, and always follow its instruction precisely. Don't make your problem worse by taking some action to resolve an issue that is not technically correct.

3.5.10   Authority (Formal/Constructive)

Descriptions of formal authority are spread throughout the general provisions of the contract. Beyond this, correct authorities are often confused because of familiarities with past relationships, or by the *constructive actions* of the parties.

Use extreme caution when determining where the formal authority lies for the situation at hand. The contract may "clearly" describe that the "commissioner" is the only one who can resolve the problem, but you may be able to show that those kinds of issues (changes of lesser cost, for example) were in fact "resolved" by someone else before, and that you've even been paid for it. You therefore had every right to rely on the *actual relationship* that was previously experienced that *constructively altered* the contract procedure.

Don't let others unnecessarily use boilerplate remarks to hide behind "authority." Use the actual job history to define the actual relationships.

3.5.11   Correlation of Contract Documents

Neither party to a contract is supposed to take advantage of any obvious error. Bad contracts do not live up to any technical or moral obligation to guide proper, responsible decisions. They cop out, trying to get the best of all worlds while covering up mistakes. They'll say something like "... in the case of a conflict, the more expensive detail will apply."

Better general contracts, on the other hand, will attempt to provide some mechanism to more definitely determine which kind of information takes precedence in the case of a conflict. [The Pass-Through Clause (see Section 3.3.5 of the Manual) ensures that the application will flow down to each respective "Plans and Specs" subcontract.] This kind of treatment will also be found in most federal and many state construction contracts. If there is no language to guide you, Company Policy will follow the typical rules as stated below to resolve conflicts between the plans, specifications, and other contract provisions:

1. Amendments and addenda take precedence over the specifications.

2. The specifications take precedence over the plans.

3. Stated dimensions take precedence over scaled dimensions.

4. Large-scale details take precedence over smaller-scale ones.

5. Schedules take precedence over other data given on the plans.

In addition to these rules that hopefully will be clearly stated in the contract, correct legal consideration provides that specific terms, conditions, and requirements are given greater weight than is given general language designed to cover a multitude of circumstances (boilerplate). It is recognized that the specific language was designed with the intention to accommodate a unique circumstance. It is therefore reasonable to assume that it so considers the precise conditions. Supplementary General Conditions will therefore take precedence over the General Conditions.

3.5.12 Force Majeure

Force majeure is a term given to delays that are the fault of neither party to the contract. They include those forces that are either unforeseeable, or otherwise beyond the control of either party, such as strikes, severe weather, and acts of God. These delays usually permit the contractor to file for and be granted an appropriate extension of contract time, but will not allow either party to be compensated for the time or expenses. The owner correspondingly cannot seek actual or liquidated damages.

3.5.13 Impossibility and Impracticability

It's been said (courts have held) that there is no excuse for nonperformance due to misfortune, accident, or misadventure. This blanket approach, however, has many times been found to be too harsh. Exceptions to the rule have accordingly been explained in terms of constructive or implied conditions.

The doctrine of impossibility of performance is still evolving. One modern application has been determined as follows:

> A thing is impossible in legal contemplation when it is not practicable; and a thing is impracticable when it can only be done at an excessive and unreasonable cost.

The doctrine describes the gray area dealt with by courts in attempting to be responsive to practices in which the community's interest in having the contract strictly enforced is outweighed by the commercial senselessness of requiring performance.

Application. When the issue is raised, the condition of performance must be shown to be required by some changed circumstance in a process that involved the following steps:

1. A "contingency"—something unexpected—must have occurred.
2. The risk of the unexpected occurrence must not have been allocated to either party by agreement or "custom."
3. The occurrence of the contingency must have rendered the performance commercially impractical.

3.5.14 Termination

The remarks here apply to agreements between the contractor and owner and those between the contractor and subcontractor.

For any termination to be appropriate, performance must be so poor as to be considered a "material" breach (as opposed to a "minor" or "immaterial") breach. The victim of a material breach may regard the contract as having been abandoned or terminated. The victim of a minor breach may sue for damages, but must continue its performance under the contract.

Because of the enormous significance of any move to terminate, the determination of what constitutes a material breach versus a minor one becomes *the* question.

A termination clause does not *require* a party to terminate the contract if there is a clear right to do so; it is merely an option. The party considering such a termination could instead allow the offending party to complete the contract, and sue for damages related directly to the potential cause for termination.

3.5.15 Notice

In a modern general contract, there will be literally dozens of references to some kind of "notice" requirement as related to almost anything. Most critical will be those relating to changes and claims, but they can apply to many other items.

It is not necessarily the date of the first formal correspondence that qualifies as the effective notification date. Notification can most often be considered to have been achieved if there has been an understanding in the mind of the recipient. The purpose of "written" notification is simply to remove doubt as to when this had been achieved. Notification, therefore, can sometimes be said to have occurred in many types of communications, including:

- Telephone or other conversations
- Meetings
- Other letters
- Shop Drawings
- Schedule Updates

It is not a good feeling to have lost a legitimate issue on technical notification grounds; don't put yourself or the Company into that position. When in doubt, get *some* kind of notification confirmed.

Such notification for a specific kind of issue (change order, for example) will be required to be in writing, and delivered within a precise number of days of the issue's "occurrence." It is important to realize, however, that it can be argued that the operative date beginning the notification period will be from the point at which it was first realized that there will definitely be an effect on the contract time and/or cost—not the point at which it was simply realized that an event occurred.

Be aware of your notice requirements in your specific agreements and comply with them in all cases. In so doing:

1. *Always establish the earliest possible legitimate date.* If the issue was discussed at a job meeting or in a phone conversation, refer back to it in the first part of the notification. "Confirming our conversation of (date)," will be enough.

2. *Notify everyone who might possibly require it.*   Do this either directly or by copy.

3. *Specifically name the individual(s) involved in the prior notification.*   "Confirming my conversation with your office," is too weak. Give names.

4. *Get into the habit of confirming all potentially significant discussions immediately.*   Any piece of information that has any potential to affect the project should be recorded in a manner that corresponds to the level of potential effect. Even if a seemingly insignificant item begins as a plain note in a file, it can form the basis of a later formal notification if one becomes necessary.

5. *Be concise, but be clear.*   Refer to Section 2.3.2, *"Rules of Effective Project Correspondence,"* and apply them in all your communications.

As a practical matter, then, the reason for "notice" with respect to changes or potential changes on a construction project is to provide the party being notified with the opportunity to determine for itself whether or not a change to the contract must proceed, and if so, the manner in which it will proceed. Timely notice may provide such a party with opportunity to mitigate such changes or to otherwise impact the manner in which a change will proceed. Under these conditions, it is commonly recognized that if a party is not given timely notice, it may be stripped of opportunity to deal with the change, and its rights may thereby be prejudiced.

In the above example, the case of delay to a project caused by rain, however, notice simply affords the owner the opportunity to satisfy itself as to the fact of the rain delay and to create any related documentation that it deems appropriate. In this condition, notice does not afford the owner any opportunity to alter the effect of the rain delay on the project, or otherwise mitigate such a change. Because of this, the owner's rights with respect to such rain delays may not be prejudiced in any manner if notice is not deemed proper for any reason. If proper notice were indeed lacking, the only issue to be resolved would be agreement between the owner and contractor on the actual condition of the rain event itself.

3.5.16 Proprietary Specifications

General. A proprietary specification is one that limits competition. On private contracts, owners have the right to specify exactly what they want if they're willing to pay for it. Although there are still formal and informal procedures to consider "equals" and "substitutions" (see Section 3.8), there is no obligation on the part of the owner to provide for fair competition among competing products.

On public contracts, however, the owner *does* have such a responsibility. Specifications are supposed to be written with the clear intention to encourage competition, and allow as many vendors as reasonably possible access to the project's "market." To this end, each product description should name at least three "acceptable" manufacturers and/or add the words "or equal" to the list of named sources. In the case where this language treatment has not been complied with, public policy, state statute, or federal regulations may provide the needed basis to allow competition.

Application. Although the specification may appear on the surface to comply with the requirements by naming alternative sources and by using the words "or equal," the difficulty may next lie in the fact that except in the most simple product descriptions (like a steel stud) it is rare to find two products that are made precisely the same, or have the same list of technical specifications.

The problem is compounded with the complexity of those product descriptions. A Douglas fir stud is a Douglas fir stud, but carpets will have two dozen or so technical criteria, with no two carpets having the same list. The problem may be even worse in the case, for example, of mechanical equipment.

Read Section 3.8 on "equals." If your proposed product is being rejected because it does not precisely match up to a long list of technical items, the specification may be considered to be unnecessarily restrictive so as to illegally limit competition, or the problem may simply be the designer's placement of an inappropriately large amount of weight on some technical criteria that do not go to the essence of the product needed. Consider again the definition of *equal*: "The recognized equivalent in substance, form, and function. . . ."

3.6   General Contract Bonds and Insurance

3.6.1   General

A bond is the guarantee of one party for the performance of another. Construction bonding is essentially a three-party contract between the principal (the contractor), the obligee (the owner), and the surety.

In a real sense, the first purpose of a bond is to identify the actual ability of a contractor to get one. This ability tends to separate unqualified contractors out of the process. Before any surety guarantees the performance of any general contractor, it will have conducted the most intrusive and detailed investigations of the contractor's financial strength to carry the type of work contemplated, and of its management ability to deal with all factors of production. The company principals will personally have had to demonstrate absolute commitment to the surety for its actual performance by providing large amounts of financial security to back its promises of performance and competence. Nowhere else is it more true that a bonded contractor has "put his money where his mouth is."

Finally, it is important to note that every bonded contractor will have a profound, sober understanding that if it ever needs to rely on a surety to cover performance, it will be the one and only time. The contractor not only risks the security provided for the bond, but will in all likelihood be unable to get another bond. Except in the most unusual circumstances, the contractor will effectively be on the way out of business—or at least out of its ability to bid bonded work.

There are many types of bonds. As related to general construction contracts, the prevalent ones are bid, payment, and performance bonds.

3.6.2   Bid Bonds

A bid bond is an assurance to the owner that if selected, the contractor will actually proceed with the contract at the bid price. If the contractor does not, the bid bond becomes payable to the owner as compensation for damages sustained.

Application Issues

1. *The bid bond may not stipulate that the surety providing the bond will also provide subsequent payment and performance bonds.* Such language protects the surety against having to provide such bonds for questionably bid contracts that would have been awarded. For this reason, many bid procedures also require a surety letter of intent to provide payment and performance bonds if the contractor is awarded the project.

2. *Bid mistakes may be justification for not being held responsible for a bad bid.* Refusal of a successful bidder to go ahead with the contract is usually because of a significant error in the bid. If the contractor can offer "clear and convincing" proof of such an error, there is a chance that both the contractor and surety will be relieved of the liability. Rules of such a mistake are tough, and include the ideas that the mistake:

   - Must be one of fact, and not one of law or judgment.
   - Must be of such grave consequences that enforcement of the contract would be unconscionable.
   - Must relate to a material feature of the contract.
   - Must not have come about through the contractor's culpable negligence.
   - Must not prejudice the owner except to loss of bargained-for performance.

   It should be clear that if the contractor's mistake was this bad, the surety will not jump at the chance to guarantee the contractor's next bid.

3. *A material change to the contract will void the principal's obligation to the bid.* If a material change in the terms and conditions of the contract that could not be reasonably anticipated when the bid was submitted occurred after the bid, the principal may be entitled to withdraw its bid.

   Examples of material changes may include:

   - Delay in awarding the contract beyond the time specified in the invitation to bid
   - Changes in the method of construction from those specified in the bid documents
   - Changes in the materials specified
   - Massive "corrections" of documents that otherwise would have resulted in significant change to the bid price

3.6.3 Performance Bonds

The performance bond protects the owner from the contractor's failure to complete the contract in accordance with the contract documents by indicating that a financially responsible party stands behind the contractor to the limit of the penal amount of the bond. Most performance bonds will accordingly require such a guarantee for 100 percent of the amount of the contract price. The entity protected is usually only the named obligee (the owner), with such protection not usually extending to third parties.

The Miller Act of 1969 requires performance bonds on all government projects. For other projects of significant size, it is much more than just a good idea for the owner.

The performance bond, however:

1. *Does not provide any guarantee that the contract work will be completed as specified for the contract price.*

2. *Limits the surety's liability to a specific dollar amount—the cost to complete the work.* This will not normally include consequential damages.

3. *Can be discharged if the owner permits a cardinal change in the contract.* A cardinal change is one that fundamentally alters the scope of contract performance.

4. *May be discharged if the owner violates contract terms that are prejudicial to the surety.* Failing to provide builder's risk insurance, for example, may be such a violation.

3.6.4 Payment Bonds

Labor and Material Payment Bonds protect those who have supplied material and labor to a project, first because there may be no lien rights against public properties. The bonds also protect owners from liens or other claims made against the property on nonpublic projects after completion of the work, and after final payment has been made to the contractor.

The Miller Act provides direct protection by way of Payment Bonds for *those who have a direct contractual relationship with the prime contractor or with a subcontractor.* Accordingly, those who furnish material or services to sub-sub's materialmen or third-tier subcontractors would not be covered by the Act.

The bonded contractor must therefore police all payments made through the chain of subcontractors and primary suppliers to be certain that all payments have in fact been made to all intended suppliers and materialmen. If such payments are misapplied, the bonded contractor may be required to pay twice for the same work.

Typically covered items include:

1. Materials incorporated into the work, delivered to the jobsite, or furnished pursuant to the contract documents even if not delivered to the site.

2. Labor performed at the jobsite, and labor performed in fabricating materials off site pursuant to the contract.

3. Freight and transportation costs.

4. Equipment rental and repair costs.

5. Fuel and maintenance items.

6. Insurance premiums.

7. Unpaid withholding taxes (on federal projects).

8. Union pension and welfare benefits.

9. Some categories of legal interest and attorney's fees.

Items usually not covered include:

1. Bank loans (even if the loans were used to pay project costs).

2. Claims of liability or damage arising out of performance of the contract.

3. Withholding taxes on state or private contracts.

Note that in the determination of items to be covered, it has been upheld in many states that a supplier only need to "reasonably believe" that its products were shipped to your jobsite. If a subcontractor, for example, picked up materials and noted your job as their destination on the receipt, you may have a problem.

Likewise, some subcontractors may arrange to have material shipped and even delivered to your job because the job is nontaxable, then take the materials to another project. If you allow such a thing to go undetected, you may be ultimately held responsible if there is a payment problem involving those materials.

3.6.5 Insurance

Insurance, as opposed to bonds that provide *guarantees*, is a *loss sharing mechanism* that guards the policy holder from damages that may be incurred in the future.

Standard contract insurance requirements are usually limited to the following:

- *Commercial general liability (CGL)*. A policy that provides third-party coverage to operations and premises either owned by or under the control of the contractor. The CGL policy provides both bodily injury and property damage liability coverage.

- *Contractor's professional liability insurance (CPL)*. A policy that provides payment on the part of the contractor for damages caused by bodily injury or property damage, or a combination of both, when caused by the insured.

Builder's risk insurance is generally excluded from the contractor's basic insurance requirements, and if requested by the owner, is treated as an addition to the scope of the contract, leading to an increase in the contract sum. This type of insurance is also known as *course of construction* insurance since it provides coverage for loss or damage to the structure during the course of construction. There are two basic forms of builder's risk insurance:

- *All Risk*. A policy that covers all risk except that specifically excluded

- *Named Peril*. A policy that covers only specific risks spelled out in that policy

3.6.6 Worker's Compensation Insurance

Nowhere does the maxim "Safety pays" have more meaning than in the insurance premiums computed by the state government agency administering its worker's compensation insurance program. A poor accident rate by either a subcontractor or a general contractor's workers will cause worker's compensation rates to increase where they will remain at a higher level for three years subsequent to the year in which the poor accident record was incurred. After that three-year period, a better safety experience will cause the rates to be reduced.

Worker's compensation insurance premiums are established by applying the following formula:

$$WCIP = EMR \times \text{Manual rate} \times \text{Payroll units}$$

where   WCIP = worker's compensation insurance premiums
             EMR = experience modification rate (the multiplier established by the contractor's accident experience based upon benefits paid to employees who have filed claims)
    Manual rate = rate structure for various trade crafts, ironworkers, carpenters, laborers (each type is classified as a "family," and each family has a four-digit number corresponding to claims experience of that group)
    Payroll units = number obtained by dividing contractor's annual direct labor costs by 100

A poor accident record resulting in a higher worker's compensation rate can, in effect, increase the contractor's or subcontractor's overhead by a point or two, a key element in remaining competitive in a tight market. And since it will take three years to erase this bad record, a general contractor or subcontractor will be saddled with this penalty for many years.

3.6.7 Subcontractor Default Insurance

Subcontractor default insurance protects the general contractor from some losses incurred when a subcontractor defaults on its project. This insurance is frequently used when bonded subcontractors are required but a particular subcontractor cannot obtain a bond.

The default insurance includes costs to complete the unfinished portion of the subcontractor's work in case of default, but there is usually a deductible or copay provision attached to the policy and a cap on the amount of the liability.

3.6.8 Controlled Insurance Programs (CIP)

Controlled insurance programs can either be contractor controlled (CCIP) or owner controlled (OCIP), but in either case it is basically a wraparound policy that allows the purchasing power of one large insurance policy instead of multiple smaller ones to effect a lower premium.

If the insurance is contractor controlled, the GC will provide the insurance for its own crews as well as for its subcontractors and receive a credit from the subcontractors for the amount they would have spent on their individual policies, which, in theory, should total to much less than the sum of those individual policies. Owner-controlled insurance programs will require the general contractor, along with all of its subcontractors, to provide credits to the owner who will administer the insurance program.

Type of Coverage	Conventional Approach	CIP Approach
Worker's compensation	Each GC and Sub furnish	Held for all parties*
CGL	Owner, GC, and Sub furnish	Held for all parties
Builder's risk	Owner or GC furnish	Held for all parties
Auto liability	Each GC and Sub furnish	Same as conventional

*OCIPs and CCIPs are not generally cost effective for small projects, but on large ones, say, over $50 million, significant savings can be accrued since the insured can present one large account rather than a series of smaller ones from many different companies.

3.7   Shop Drawing "Approval"

3.7.1   General

The concept of shop drawings, or "approval submittals," is constantly being tried—by design professionals and subcontractors and materialmen—both sides of the coin. Motivations are rooted in everything from simply trying to avoid the time and effort necessary to do the job right to direct and strained efforts to avoid responsibility and liability.

On the design side, chronic problems exist such as:

1. The use of "creative" words and language on shop drawing stamps
2. Actual failure of the design professional to check for compliance with design intent
3. Direct attempts simply to avoid having to spend the time necessary for proper review and other action that the construction force may be entitled to (particularly on equals and substitutions)
4. Failure to take action in a reasonable amount of time
5. Taking inappropriate full or partial rejection positions
6. Refusing to accept the consequences of his or her "approval," regardless of what the stamp might say

On the side of those who prepare, submit, or pass-through submissions of shop drawings, problems include:

1. Failure to comply with contractual requirements regarding the manner, scale, amount, and type of information presented in the submittal
2. Failure to submit such documents in time to avoid delay in the work

3. Failure to coordinate the work with contiguous work

4. Attempts to avoid responsibility for properly accommodating field dimensions or other field conditions

5. Failure to provide sufficient number of copies for proper distribution

6. Failure to properly review third-tier submittals for compliance with the contract

7. Failure to properly review third-tier submittals for compliance with the primary subcontract

8. Failure to clarify responsibilities between primary vendors or subcontractors and their subs or suppliers

9. Inappropriate treatment of "equals" and "substitutions"

10. Failure to properly distribute information to all those who might need it—in time to avoid conflict

Actual procedures for securing, reviewing, and processing all approval submittals are treated in great detail in Section 4, "Project Engineering." This section will focus on contractual responsibilities and liabilities in the submittal process, and is closely related to the discussion of equals and substitutions in Section 3.8.

3.7.2 "Approval" Abuse

The most common first-line approach architects and engineers use to this end has been to change the "approval" statements and to add strained, exculpatory language to their shop drawing approval stamps:

1. The word "approved" has been totally eliminated from their vocabulary. Words like "No Exceptions Taken," "Examined," "Reviewed," and (my favorite) "Not Rejected" are used in attempts to somehow reduce their liability.

2. Paragraphs describing what they are and are not doing make the stamp larger than the paper that was stamped.

In most instances, these efforts have produced no beneficial effects, and in many instances, they have had decidedly adverse effects:

1. Courts have clarified that the use of words other than "approved" does *not* bar a liability. They acknowledge that:
   a. Only the person with the "big picture" could carry the responsibility for design safety and integrity.
   b. The contractor relies on the architect's written authorization to proceed with the work at all. Whatever the actual word may be, it is considered as an express statement authorizing the work to proceed as described in

the submittal. It cannot be construed to mean "it's OK to fabricate and install the work this way for now, but I may reject it later."

2. The inclusion of detailed language may actually be an express statement by the designer that he or she is actually *not* performing the reviews or taking the action that is required by the owner/architect agreement, the owner contractor agreement, or even the general conditions of the contract. The designer may therefore actually be setting himself or herself up for a case of gross negligence.

3.7.3 General Contractor Liability

Under prevailing standard general contracts, both the general contractor and the design professionals share a responsibility for review and "approval" of submittals. It has been arranged in this way in an effort to increase the likelihood that shop drawing errors will be detected and corrected.

Contractors' review/approval responsibilities principally are to:

1. Check for conformance to the contract documents.

2. Highlight deviations from those documents, either because of necessity (conflict, impossibility, etc.) or other reason.

3. Coordinate the work with contiguous work.

Unless there is extremely unusual language in the general contract, it is *not* the contractors' *responsibility* to find mistakes or otherwise correct the contract documents themselves.

3.7.4 Appropriate Contractor Action

The specific shop drawing review procedures for both subcontractor and in-house submittals are detailed in Section 4, "Project Engineering." The action described here is to control any abuses attempted by design professionals to avoid approval liability.

Discussion of related actions regarding equals and substitutions is incorporated in Section 3.8.

When confronted with any action on your submittals that is other than "approved," be aware that:

1. The owner/architect agreement is almost certain to contain language requiring that the architect *approve* submittals. There won't be anything requiring the architect to "not reject."

2. The owner/contractor agreement will instruct the contractor that "the architect will review and *approve*" your shop drawings. You have an absolute right to rely on such statements.

Take the following steps to correct, or at least clarify, the meaning of language and reliance of the construction force on it to allow construction to proceed:

1. Advise the Owner and design professionals of their responsibility to *approve,* per the owner/contractor agreement.

2. The designers will likely refuse. They may even admit openly that it is their professional liability insurance company that is preventing them from using the word "approved."

3. Point out the facts of the owner/architect agreement and the owner/contractor agreement as described above. Unfortunately, this will still in all likelihood not induce the design professionals to change.

4. Point out the apparent effort to avoid approval responsibility as described in the contract, and the practical problem of not being able to release anything for construction until the Owner clarifies just what "words" will so allow the work to proceed.

5. If the Owner and design professionals agree to change the language, don't wait for it. Send the Sample Letter #1 to the Owner Clarifying Shop Drawing "Approval" of Section 3.7.5, thereby:

   - Confirming the decision
   - Making the action retroactive, in order to avoid having to resubmit everything that has been acted on to that point

3.7.5   Sample Letter #1 to the Owner Clarifying Shop Drawing "Approval" (*page 3.75*)

The Sample Letter #1 to the Owner Clarifying Shop Drawing "Approval" accommodates the ideas of Section 3.7.4 for a condition in which the owner agrees to alter the language to be used on the shop drawing stamps. Specifically it:

1. Confirms the Owner decision to require the design professionals to correct the language on the shop drawing approval stamps.

2. Defines the words that had been used so far as meaning "approved," thereby eliminating the necessity of resubmitting previous submittals with the otherwise inappropriate language on the stamps.

3.7.6   Sample Letter #2 to the Owner Clarifying Shop Drawing "Approval" (*page 3.76*)

The Sample Letter #2 to the Owner Clarifying Shop Drawing "Approval" follows through on the suggestions on Section 3.7.4 in the face of a condition where the owner "agrees as a practical matter" with the purpose of the words on the shop

3.7.5
Sample Letter #1 to the Owner Clarifying Shop Drawing "Approval"

Letterhead

(Date)

To: (Owner)

RE: (Project)
           (Company Project #)

SUBJ: Shop Drawing Approval

Mr. (Ms.) (   ):

This letter is to confirm our discussion of (date):

1.   The design professionals have been using shop drawing approval stamps with the words "No Exceptions Taken" in lieu of "Approved," and "Note Markings" in lieu of "Approved as Noted"; contrary to the requirements of General Conditions Article (   ).

2.   You confirmed today that the stamps will be corrected to reflect the proper use of the word "Approved" and "Approved as Noted"; the other language will be dropped.

3.   In order to avoid the necessity of having to resubmit and redistribute all submittals that bear the words "No Exceptions Taken" or "Note Markings," you confirmed that on those submittals the words "No Exceptions Taken" are construed to mean "Approved," and the words "Note Markings" are construed to mean "Approved as Noted."

4.   In order to allow the work of those previous submittals to proceed without further interruptions, please confirm this letter by signing it in the space provided and returning it to my attention.

Agreed to By: _____   Date: _____

Very truly yours,

COMPANY

Project Manager

cc: Architect
        File: Submittal Approval
                 CF

3.7.6
Sample Letter #2 to the Owner Clarifying Shop Drawing "Approval"

Letterhead

(Date)

To: (Owner)

RE: (Project)
           (Company Project #)

SUBJ: Shop Drawing Approval

Mr. (Ms.) (   ):

This letter is to confirm our discussion of (date):

1.   The design professionals have been using shop drawing approval stamps with the words "No Exceptions Taken" in lieu of "Approved," and "Note Markings" in lieu of "Approved as Noted"; contrary to the requirements of General Conditions Article (   ). General Conditions Article (   ), however, states that the architect will approve shop drawings. We therefore require that the design professionals correct their stamps to properly reflect such approvals. Note that the owner is as much entitled to this security from the designer as the construction force is, by virtue of its owner/architect Agreement.

2.   Although you agree that the constructive purpose of the architect's action on such submittals is to release the item for use in the project, you have refused to require the design professionals to actually correct their stamps' language to meet the requirements of the contract.

3.   Accordingly, in order to avoid delay to the project, the words "No Exceptions Taken" will be construed per your direction specifically to mean "Approved, and "Note Markings" will be construed to specifically mean "Approved as Noted"; all strictly within the context of the owner thereby expressly directing the work to proceed as indicated in the respective submittals. These operative definitions must be relied upon in order to allow any work to proceed. Please confirm your agreement by signing this letter in the space provided and returning it to my attention.

Agreed to By: _____   Date: _____

Very truly yours,

COMPANY

Project Manager

cc: Architect
        File: Submittal Approval, CF

drawing stamp, but does not ultimately force the design professional to change it. Specifically, it:

1. Defines those words that will be construed by the construction force to actually *mean* "approved," regardless of what the word actually is, along with whatever qualifications that may have been part of the discussion.

2. Notifies the Owner that the construction force is relying on those words specifically as the owner direction to proceed with the work as it is described in the submittal.

3.8   Equals and Substitutions

3.8.1   General

The concepts of "equals" and "substitutions" are very often misapplied by owners and designers. Unless corrected, they almost always automatically place "equal" products and submissions immediately into the "substitution" category in order to force the contractor either to forget using the product (reject the submission) or to secure some kind of credit or other "reimbursement," if they decide that they can actually live with the product.

Reasons for rejection range from an honest objection to the product's being proposed for some otherwise understandable problem to the idea that the designer's office simply doesn't want to spend time reviewing any products that it had not specified.

3.8.2   Typical Contract Treatment

Many general contracts contain language, either in the general conditions or in the bid documents themselves, requiring that *substitutions* be submitted within a certain number of days after contract award, or after the bid, or even with the bid. This will be the first line of attack on the submittal for any product that was not named in the specification.

Because of the realities of project buyout logistics, it is usually improbable that all bid items will be bought within those usually very short time periods and that it will be practically possible to secure all project submittals within such aggressive time frames. The other hurdle will be the treatment of the submittal itself, whether or not it satisfies any "substitution" time requirement. The designer is simply going to want what was specified, for good or bad reasons. Be prepared, therefore, for an uphill battle. Consider Section 3.5.16, Proprietary Specifications, for related discussion.

Better contracts will go on to define exactly what products will be considered "equal," and what will be considered a "substitution." Know in every case precisely what the contract language is, and deal with it directly. In the absence of such clear definitions, the following are the operative ideals:

Equal. The recognized equivalent in substance, form, and function, considering quality, workmanship, economy of operation, durability, and suitability for the purpose intended, and not constituting a change in the work.

Example: If a 2 × 6 16-ga. structural steel stud manufactured by ABC Corp. is specified, a 2 × 6 16-ga. structural steel stud manufactured by XYZ Corp. should be considered an *equal*.

Substitution. A replacement for the specified material, device, or equipment which is sufficiently different in substance and function to be considered a change in the work.

Example: If a 2 × 6 16-ga. structural steel stud manufactured by ABC Corp. is specified, a 2 × 6 Douglas fir stud should be considered a substitution.

3.8.3   Application

Submissions for product substitutions and equals from subcontractors and suppliers will be accepted by the company for submission to the project designers only after it is confirmed that they have first met all relevant requirements of the individual subcontract or purchase order. If they have not, it is important to understand that processing such a submittal to the designers will likely expose the Company to otherwise unnecessary coordination, late deliveries, and even possible product liability. These considerations must, however, be weighed against the right of the design professional or owner to interpret the documents. You may after all be forced to pass through the submittal for their decision anyway. If this turns out to be the case, get it out and back as quickly as possible.

After specifically considering these references and concerns, proceed as described below for those submissions so confirmed to be appropriate.

Substitutions. Double-check that the submission is in fact a substitution. Don't assume too easily. Is it really a *change* in the work? Legitimate reasons to consider a real substitution include:

1. The material described in the specification is no longer available.
2. It can be reasonably shown that the material described in the specifications is inadequate, inappropriate, or will not otherwise serve its intended purpose for whatever reason. Remember—substitutions do *not* necessarily mean a credit change order.
3. Material deliveries or other problems related to the specified material or equipment or its supplier will cause unacceptable delivery or erection schedules or will otherwise affect contiguous construction.
4. There is significant price and/or construction advantage with a proposed substitution to warrant the effort necessary for its approval, considering the strong possibility of having to offer an acceptable credit.

Upon such reconfirmation, comply explicitly with any/all stated submission requirements for substitutions, considering:

■ Time of submission
■ Number of copies

- Form, manner, and content of all submittal information
- Requirements for comparative information on "specified" items

For both equals and substitutions, there is usually some requirement to include a complete submission of the originally specified item, in order to allow the reviewer to conduct a side-by-side comparison of all product features and information. This is usually the third place where a subcontractor's submission will be deficient (first: it will be late; second: there will only be one copy).

If such requirements are so explicitly stated, there is just no excuse for an improper submission. Don't give anyone such a simple excuse for rejecting the submittal.

Exception: Submissions requiring preparation of detailed, individually drafted drawings, or other type of submission that will require elaborate, expensive preparation will just not be available from those (originally specified) vendors—at least not without considerable expense. In this case, point out the practical problem, and submit on the product itself, complying with all submissions requirements.

After all these requirements have been met, follow the procedures described in Section 4.10, "Submittal Requirements and Procedures," to expedite the actual submission.

Equals. The procedure for treatment of submittals is similar to that for substitutions. The differences lie in the need or right of the GC or subvendor to make such a change, and the final treatment of the submittal by the contract. True "equals" stand a much stronger chance of seeing the light of day:

1. Review, as with substitutions, the respective subcontract or purchase order to confirm that there's a need for any action other than insisting that the subvendor provide specified material. Upon such reconfirmation, comply explicitly with any/all stated submission requirements for substitutions, considering time, quantities, and form.

2. Process the submittal in accordance with the procedures described in Section 4.10, "Submittal Requirements and Procedures."

3. Consider if the specification can be considered proprietary (refer to Section 3.5.16). If so, consider calling attention to this idea in the submission transmittal.

4. Be prepared to have your submission treated as a substitution, with the result that the first designer reaction to it will be flat rejection on inappropriate grounds. Consider ways to head off this time-wasting posturing.

3.8.4 Perspective

Throughout the substitution or equal process, it is important that the general contractor maintain appropriate perspective. Remember the "Conduit" Principle of Section 3.1.2. In one real sense, the GC is the custodian of each subcontractor's rights relative to the owner's general contract, and as such, must take all actions regarding its submissions as responsibly as possible. On the other hand, that subcontractor is signed to the same responsibilities as you are (refer to the Pass-Through Clause of Section 3.5.8), and the GC in the final analysis has no "right" to interpret the documents (refer to Section 3.7).

When an owner's (designer's) rejection of an equal or substitution is received, it is important that you transmit that action clearly, completely, and immediately to the vendor affected, also transferring the designer's direct order to provide that which is specified. If you maintain this perspective, you will:

1. Keep the subvendor responsible for the product, for its effect on contiguous construction, and for its delay that it subjected the project to—if the designer is right.

2. Have kept all liabilities for extra cost and extra time that may ultimately be determined to have not been the fault of the subvendor or the owner—if the designer is wrong.

3.9   Responsibility to "Coordinate": Use and Abuse

3.9.1   General

As a general contractor or construction manager, your responsibility is to coordinate the *work*. It is not to invent information and not to coordinate the design. When a problem surfaces, it has become common practice to criticize the contractor for failure to coordinate. The irony is that much more often it is actually the process of coordination that discovered other problems with the design in the first place. The word is used to abuse, and the contractor must be sensitive to such efforts to distort responsibility.

3.9.2   Operative Definition

Coordinating the work does *not* mean inventing information. It means:

- Securing relevant information from those responsible for generating it
- Fitting that information into the project requirements and confirming its appropriateness
- Communicating that information to those responsible for finally approving its incorporation into the project
- Distributing that information to all those requiring it, in order to allow them to complete their work at all those points where it interfaces with the information in appropriate time

In other words, it means finding, assembling, and distributing information. It is *not* making up information, or fixing the mistakes in the specification when they're finally discovered. That's the designer's job.

Example.   Two subcontractors both take exception to a certain item of work. After your own review, you conclude that each subcontractor actually reasonably inferred from their respective specification sections that the work would be done by the other sub-contractor. The owner will therefore likely argue:

1. The work is specified *twice*; give me a credit, and
2. The GC should have "coordinated" the item; you determine who's going to do it, and just get it done.

The appropriate response will be based on ideas related to those principles outlined above. You will demonstrate that:

1. You *did* coordinate; that's what discovered the problem in the first place.

2. Each subtrade reasonably interpreted the specification to determine that the work was not included in their competitive bid (refer to Section 3.5.3, "Intent" vs. "Indication" and Section 3.5.4, "Reasonable Review"). The work was not, in fact, incorporated twice. It was not incorporated at all. A change order *is* therefore appropriate, but at an increase, not decrease, in cost.

Refer to Section 3.9.4, Sample Letter to Owner Regarding Lack of Design Coordination, as a further example of proper treatment.

3.9.3 Coordination Drawing Guidelines

The contract specifications generally provide the general contractor with information regarding the requirement to prepare and submit coordination drawings for approval by the architect. There are further instructions to the contractor to notify the architect of any problems, conflicts, etc., that occur during this process and provide suggestions to correct any conflicts.

There are generic coordination drawing protocols that can act as guidelines for this process.

Definition of coordination. Coordination drawings are reproducible drawings showing work with horizontal and vertical dimensions required to avoid interference with structural, framing MEP rough-ins, fire protection, elevators, and other thru floor and wall penetrations.

What the contractor is to provide. The contractor will usually provide ¼" scale drawings showing elevation, sections, to include:

1. Structural framing members—size and location
2. Partitions
3. Column lines
4. Ceiling heights

These drawings will show the installation of all different components, products, and materials fabricated and furnished by others, to be installed within the walls, chases, in and above the ceilings, and within mechanical spaces and electrical spaces.

The contractor is to allow for maximum accessibility for the required maintenance, service, and repair of these components in the walls and above the ceilings. The contractor is to work out all "tight" conditions prior to the installation of those components.

The drawings. The drawings should include the size and locations of sleeves, core drill areas, and block-outs as well as items embedded in concrete walls, floors, and beams.

Contractor suggestions. Any contractor suggestions of modifications to walls, chases, and ceiling heights to allow for proper installation of all components must be submitted to the architect/engineer for approval. Modifications to system configurations, for example, redesign of duct sizes, fittings, and piping layouts, are to be submitted to the architect/engineer for approval. If these modifications to walls, chases, ceilings, or the components required to fit in these spaces result in increased costs to the contractor, the architect/engineer must be notified accordingly.

When conflicts occur. In the event of conflicts involving the layout of work, there are some priorities that can be implemented to assist in resolving the conflicts. In order of precedence they are:

1. The structure and partitions have the highest priority.
2. Equipment locations and access.
3. Ceiling system and light fixture (recessed) installation.
4. Gravity drain lines.
5. High-pressure ductwork and associated devices.
6. Large pipe mains, valves, and associated devices.
7. Pneumatic tube and material conveying systems.
8. Low-pressure ductwork, diffusers, grilles, and registers.
9. Fire protection piping, devices, sprinkler heads.
10. Small piping, tubing, electrical conduit.
11. Sleevs through partitions.
12. Access panels.

3.9.4   Sample Letter to Owner Regarding Lack of Design Coordination *(page 3.83)*

The Sample Letter to the Owner Regarding Lack of Design Coordination follows through on the ideas of Section 3.7.2 in the face of having discovered an error or conflict in the contract documents and having been accused of failing to "coordinate." Specifically, it:

1. Describes the conflict in as concise terms as possible.

2. Indicates directly how each side of the conflict was interpreted, and that it was reasonable to so interpret.

3. Clarifies that it is the GC's responsibility to coordinate the *work*—not to coordinate the design.

3.9.4
Sample Letter to Owner Regarding Lack of Design Coordination

Letterhead

(Date)

To: (Owner)   CERTIFIED MAIL
                                      RETURN RECEIPT REQUESTED

RE: (Project)
           (Company Project #)

SUBJ: Design Coordination

Mr. (Ms.) (   ):

We agree that it is our responsibility to coordinate the work. It is not, however, our responsibility to coordinate the documents.

Per my letter of (date), the work (brief description) is actually described in both Sections (   ) and (   ) to be provided in the "other" section. Accordingly, neither subcontractor carried the cost of the item.

After reviewing the specification sections you reference, it is apparent that both subcontractors were justified in their views. Having made such reasonable interpretations, it is neither the subcontractors' nor the general contractor's responsibility to complete a search of documents to discover such mistakes in them; that is the designer's responsibility.

Moreover, it is the owner's responsibility to properly interpret the documents per General Conditions Article (   ). If you feel that the work has, in fact, been properly specified, it is your responsibility to provide the precise specification section. If you cannot, or for any reason will not, we will have no basis to direct the subcontractor to perform its "plans and specifications" work.

Accordingly, please confirm that a change order will be processed to cover the cost of this work.

Very truly yours,

COMPANY

Project Manager

cc: Architect
       File: Change File (   )
                CF

4. Confirms that you *did* coordinate; that's what disclosed the designer's failure to coordinate the design.

5. Points out that the design flaws are not your problem and that you're entitled to a change order to cover the extra cost.

3.10   The Schedule of Values

3.10.1   General

This section discusses principles of the Schedule of Values and appropriate application in terms of identifying price structures to be used throughout the payment processing procedures and controlling would-be abuses during change pricing. The actual procedures for assembling, correlating, and communicating all this information for both the subcontractors and in the general process to the owner is described in detail in Section 4.

Refer also to Section 4.8, "Subcontractor Schedule of Values," for important related discussion.

3.10.2   Principles

The Schedule of Values is for partial payment invoicing processing *only:*

1. It is *not* a schedule of contract unit prices.

2. It is *not* to be used as any basis for adjustment in the contract sum.

> **Example:**   A schedule of values typically will include breakdowns of quantities and unit prices. If it is eventually determined that actual quantities for an item finally total to a number less than that shown on the schedule, the owner cannot adjust the line-item price downward. That kind of treatment is only appropriate where items are clearly part of specific unit price, allowance, or alternate contract price qualifications.

3.10.3   Level of Detail

The schedule should be broken down into at least a sufficient amount of detail that will allow clear demonstration of the actual completed work. Many individuals resist providing great amounts of detail in the schedule of values. The fact, however, is simply that the greater the level of detail in the schedule, the easier it will be to consistently substantiate your billings.

Prior approval of the detailed schedule of values removes to that degree an amount of subjectivity inversely correlated with that level of detail. If, for example, a detailed schedule is approved at the beginning of the project, simply having the percent complete of the individual line items approved in the regular payment procedure gets their values approved automatically.

In contrast, an oversimplified schedule will have fewer, but larger, values for each general item. The larger the respective value, the more difficult it will be to justify in terms of percent complete in a given period. The final irony is that in order to actually justify billings on a large lump sum item, you'll have to make up a detailed breakdown anyway. This will be an unofficial, unapproved breakdown that will then be subjected to intense scrutiny and debate.

Avoid the problem. Have your schedule of values in proper level of detail approved at the project's onset, and thereby keep your billing reviews smooth and without major incident.

3.11   Requisitions for Payment and Contract Retainage

3.11.1   General

This section discusses contractual considerations of the requisitions for payment and retainage. Refer to Section 4 for additional procedures for development, submission, approval, and management processes.

Refer also to Section 4.8, Subcontractor Schedule of Values, for important related discussion.

3.11.2   Maintenance of Billing Accuracy

By contract, statute, or public policy, the general contractor will be obligated to release funds to subcontractors in the same manner and extent that those funds have been released to the general contractor by the owner on behalf of that subcontractor. To put it another way, if the owner approves 80% payment to the general contractor for concrete, the general contractor must then release 80% to the concrete subcontractor.

It should be clear to any prime contractor, then, that if the concrete subcontractor, for example, is not really 80% complete, it is *very* risky to allow yourself to overbill, intentionally or unintentionally, for the item; legally the extra money will only have to be turned over to the subcontractor, who will not be entitled to it. Add to this the idea that if the general contractor is bonded and the subcontractor is not, it will be the *general contractor, not* the subcontractor, who will be guaranteeing completion on the work to the owner. This is probably the strongest argument against allowing any subcontractor to be overpaid.

It is therefore crucial that the subcontractor's schedule of values be comprehensive, in sufficient detail, and equitable—with particular scrutiny given to the real value of the *last* 25% or so of the subcontract item. All eyes must be continually focused on the real value of *remaining work to be completed*. Otherwise, if you intentionally or inadvertently get overpaid on a subcontractor's behalf, you will be setting yourself up to be responsible to overpay the subcontractor, and thereby increase the risk to the Company of exposure to incomplete work.

3.11.3 Correlation of Subcontractor Schedule of Values with the General Schedule of Values

It is too common for subcontractor schedules of values to the general contractor to be too loosely correlated with the general schedule of values to the owner. This may be in large part because the general schedule of values is one of the first project documents to be expedited to the owner. Therefore it is prepared and submitted before every subcontractor's schedule of values is received or approved, possibly even before major bid packages are finalized.

If this is allowed to happen, an inordinate amount of subjectivity will be introduced into the process that will make correlating the amount due subcontractors with those amounts actually received on their behalf from the owner quite a challenge.

Avoid the arguments. Establish a procedure in the general schedule of values submission to the owner in which the values of major bid packages that have not yet been resolved are left as single lump sums in the first submission. When the respective schedules of values are received and finalized for the subcontractors, the lump sums on the general requisitions will be supplemented with this detail that will remain in that form for the duration of the project. Done in this way, early billings for the first items of work in process will not be delayed because of a late submission of the general schedule of values that was waiting for subcontractor breakdowns, and all eventual billings for each subcontractor will correlate on a one-to-one line-item basis with the general schedule of payments to the GC.

3.11.4 GC/Prime Contractor Retainage

The amount to be withheld from the general contractor as retainage from each payment is normally specified in the general conditions. Understand, however, that federal and state statutes dictate public policy regarding the actual amounts allowed to be withheld on public projects. If your contract, for example, states that 10% will be withheld, statute, or even the regulations of the funding source may require, for example, that only 5% be withheld. The regulations determining the procedure may be different even among agencies of the same state.

If the retainage amount on your contract seems excessive, check with the relevant laws for a possible source of relief.

3.11.5 Limits on Subcontractor Retainage

There is likely to be some statutory provision in each state that is intended to regulate the maximum amount of retainage that can be held by a general contractor on a subcontractor. Many states, for example, may provide that the general contractor can only withhold the same percentage as is being held on the GC by the owner. Even where it cannot be shown that there is a clear statement in existence, astute subcontractors may attempt to apply the Pass-Through

Clause (see Section 3.5.8) to make the general percentage value apply. They will accordingly argue, for example, that if the owner is holding 5% on the GC, the GC may not withhold 10% on the subcontractor.

In these cases, it is important to realize that there may be circumstances that materially alter the justification for such an interpretation.

> **Example:** If the general contractor has 100% payment and performance bonds to the owner, and the subcontractor has no such bonds to the general contractor, the clean pass-through relationship has been materially altered. The additional retainage may be considered an appropriate amount of needed security for faithful performance to help make up for the lack of the subcontractor's bonds.

3.12 Liquidated Damages

3.12.1 General

The liquidated damages provision is often misunderstood in terms of its real potential impact on the project. Most times, it is perceived to be a heavy weight suspended from a string over the contractor's head that could snap at any moment. The reality is usually nowhere near this severity, and the truth on a particular project may wind up actually being a comfortable limitation on the owner's ability to make trouble.

It is crucial that you develop a complete perspective on the liquidated damages realities on each individual project both at the project's start and whenever there is trouble potentially associated with a nonexcusable (or alleged nonexcusable) delay. Have a clear idea on both the owner's rights and abilities and those of each subcontractors' pass-through (see Section 3.5.8) responsibilities.

3.12.2 Definition

It is recognized that:

1. If the owner is delayed in its ability to occupy the project and use it for its intended purpose, it will suffer damage.

2. It is extremely difficult, time-consuming (and therefore very expensive), and very subjective to define the actual extent of those damages, and to distill those values to a daily assessment to be directly assigned to any delay.

The "liquidated damages" provision of the contract is therefore provided to simplify the whole thing: if such a delay occurs, damages on the part of the owner will be justified. It is nothing more than a simple stipulation of both contracting parties as to the value of daily damages to be assigned *if the owner is actually delayed through the fault of the contractor*. It is there *only* to avoid the argument of determining the *value* of damages.

Example #1: A road project is delayed a month. Is the public "damaged" because of its inability to use the road? If so, how much is the damage worth? Is the public agency administering the contract damaged? Why? After all, isn't its funding tied to local politics?

Example #2: A commercial client is delayed in occupying its space. Lost rent may be easy to calculate, but in this market would they really have rented the space at all? If it is a manufacturing, sales, or distribution organization, what is the value of not being able to ship orders for that time period?

Liquidated Damages does not *justify* the concept of damages, does not make them in any way automatic, and does not in any way make it any easier for the owner to prove or actually collect damages. It is an arithmetic convention designed to avoid only the argument of defining the actual value of any damages.

3.12.3 Concepts and Clarifications

Several concepts and clarifications are necessary in order to properly apply the principles of this section:

1. *"Liquidated Damages" is not a "Penalty."* The term "Liquidated Damages" is used first specifically to avoid the use of the word "penalty." It has, for example, long since been interpreted by the courts that if a contractor is working under threat of a "penalty" for failing to complete the project by a prescribed date, then the contractor would be correspondingly entitled to a "reward" to the same extent if the contract is completed *prior* to that date. What's good for the goose is good for the gander. The use of the word "penalty" is therefore avoided in order to preclude the contractor's ability to collect any such "reward."

2. *"Liquidated Damages" is not an automatic cost that occurs after the original contract end date has passed.* Liquidated Damages cannot be assessed against the contractor for delays incurred that fall within the scope of any excusable delay clause—for delays demonstrated to be not the fault of the construction force. The owner is just as responsible to prove its case of justification for any damages in the first place.

3. *The presence of a Liquidated Damages provision may in some cases actually improve your position to negotiate or settle problems.* Because the amount set forth in the contract as the value of damages has the effect of *limiting* the owner's ability to assign more than that amount, it immediately controls the size of threats and restricts posturing and legal maneuvering designed to puff up the size of claims or counterclaims by the owner. You know exactly the maximum size of any potential owner claim—the owner can't so easily quantify yours.

3.12.4 Technical Defenses and Considerations

1. *Apportionment.* If *both* the contractor and the owner have contributed to the cause of the total delay, the contract may provide that the responsibility be

"apportioned" between the contractor and the owner. Liquidated Damages will in this way be assessed against the contractor for its portion of the delay, and the contractor would be entitled to prove its damages for the owner's portion of the delay.

If the contract does not provide for such apportionment, some courts have held that, by contributing to the delay, the owner totally waives any claim to Liquidated Damages. Other courts, however, have held that the contractor is only entitled to a "credit" for that part of the delay caused by the owner— another way of saying "apportionment."

2. *Application in Contract Abandonment.* If a contractor breaches a contract by totally abandoning performance, some courts have determined that the Liquidated Damages provisions may not apply. These courts distinguish between a breach by delay and a breach by abandonment, finding that the parties intended to apply the Liquidated Damages only to delays when the contractor completes the work.

3. *Application after Substantial Completion.* Courts generally refuse to allow Liquidated Damages if the contractor's performance under the contract has been *substantially completed* by the date specified in the contract and only punchlist or detail work remains. Substantial Completion has generally been defined in a manner consistent with that used by the American Institute of Architects as:

> ...the date certified by the architect when construction is sufficiently complete in accordance with the contract documents, so the Owner can occupy or utilize the Work or designated portion thereof for the use for which it is intended.
>
> It is therefore generally recognized that when the owner has possession or use of the property, an assessment of Liquidated Damages for trivial details in the work would amount to a penalty.

4. *Application in Cases of Partial Occupancy.* If you've built or renovated a school, for example, and 80% of the facility had been accepted and occupied by the date stipulated in the contract, is the owner justified in assessing 100% of the value of Liquidated Damages?

In these types of cases, it has been successfully argued that by its definition, the concept of Liquidated Damages can only apply when the entire project is delayed. In the case of partial occupancy, therefore, is it inappropriate to assign even a portion of Liquidated Damages?

Be extremely careful on this one. If the food service equipment, for example, is the item of incomplete work, and if you successfully argue that Liquidated Damages does not apply, you may be opening yourself up for an assessment of *actual* damages. The school, for example, now has to arrange catering, and it has a food service staff doing little or nothing for which they still have to be paid. You may be better off negotiating a percentage of Liquidated Damages if it comes to that.

3.13   Guarantees and Warranties

3.13.1   Definition of Terms

The words "Guarantee" and "Warranty" have different meanings, and they should not be construed as being interchangeable.

Although the two terms are often characterized as being the same, in strict legal usage, the terms are different. A *warranty* has been defined as an absolute liability on the part of the warrantor, and the contract is void unless it is strictly and literally performed. A *guarantee* is a promise not imposing any primary liability on the guarantor, but binding him or her to be answerable to the default of another.

In other words:

A *warranty:*

- Binds a party to the terms of its contract
- Implies present or future liability
- Has no time limitation on the liability if "unrestricted"
- Applies usually to products and their qualities

A *guarantee:*

- Binds a third party to the terms of another's contract (*Example:* Performance Bond)
- Implies present or past liability
- Has a stated time limit usually
- Applies to indebtedness or performance of one's duties

3.13.2   Date of Beginning Coverage

General contracts usually require guarantees and warranties to begin from some stage of contract completion at which the project can be used for its intended purpose, that is, "substantial completion." Be sure that all warranties list the term "substantial completion" as well as the corresponding actual date as the start of the warranty period.

Refer to the respective specification section, general conditions of the particular contract, and other subcontract provisions to finalize complete requirements and other necessary, specific language.

Most subcontractor and supplier guarantees and warranties will be required to be countersigned by the general contractor prior to delivery to the owner. Verify all such requirements.

3.13.3   Express versus Implied Warranties

1. *Express.*   Many products carry a standard written (express) warranty that:
   a. Is usually in force for a prescribed period (often one year) from date of *shipment.*
   b. Does not cover damages in shipment or abuse by other trades before acceptance.

It can be difficult to get a distant product manufacturer to provide for all the specific components of a required warranty after the order has been placed. Even if the subtrade responsible will positively make up the difference in such warranties, these supplements may not be accepted by the owner.

Be sure that the complete product warranty provisions are part of the purchase agreement in the first place—either directly to that vendor or passed through the affected subcontractor.

2. *Implied.* There is usually an implied warranty in every transaction or contract, unless the documents clearly state otherwise, that requires the product or goods to be reasonably fit and sufficient for the purpose for which they are to be used.

Example: When a product is supplied in response to a performance specification (see Section 3.5.6), there is an implied warranty that the product will perform as required for that purpose.

Example: When an owner requires a contractor to build in accordance with plans and specifications provided or prepared by an architect or engineer, the owner impliedly warrants that those plans and specifications will be sufficient to achieve the desired results, that the project *can* be built as designed, and that all specified materials *can* be ordered and delivered within the time necessary.

If you cannot get your specified materials in time because they were not available through no fault of yours, the *designer* will have violated *his or her* implied warranty of availability of such products within the time required to complete the contract.

3.13.4 Submission Dates

1. The subcontract or purchase order is the first document that requires a respective subvendor to provide guarantees and warranties in accordance with the requirements of the contract documents. The Sample Letter to Subcontractors Regarding Submittal Requirements (Section 4.10.3) is the next early warning reminder.

2. The *acceptable* date of substantial completion must be confirmed with the owner. If this is not done in advance, you will run the serious risk of spending all that time and energy securing guaranties and warranties with unacceptable dates. They'll all have to be redone.

3. When the date of substantial completion has been so confirmed, all responsible vendors must be notified that:

   a. Immediate submission of the document(s) is required, *in proper form.*
   b. Final payment cannot be released without such proper, timely submission.
   c. Any damages resulting from delays in final payment to the general contractor directly related to the failure to provide such proper, timely submissions will be charged accordingly.

Refer to the Sample Subtrade Guarantee/Warranty Notification Letter of Section 3.13.6 to follow through on the procedure.

3.13.5   Form

1. If the contract documents are complete, they will have either the specific language and form for each respective guarantee and warranty delineated or a direct reference to the correct document. In such cases, it is a straightforward matter to define the requirements and enforce compliance. Beyond that, any requirements beyond those in the project documents that have been added by way of subcontract or purchase order must also be confirmed.

2. In the absence of a specifically required form, request one from the owner that will be acceptable. Review such a form for adherence to the contract without embellishment.

3. It is common, but not automatic, that subtrade guarantees and warranties are to be countersigned by the general contractor. Do not countersign if it is not specifically required.

3.13.6   Sample Subtrade Guarantee/Warranty Notification Letter *(page 3.93)*

The Sample Subtrade Guarantee/Warranty Notification Letter is designed to compel all responsible parties to submit what is usually one of the last requirements of their contracts. It is a form letter that follows through on the suggestions of this Section 3.13 by:

1. Requiring timely submission, in proper form.

2. Advising of the date that the guarantee or warrantee is to take effect.

3. Advising that payments will be affected until the proper submission is received.

4. Putting the party on notice that problems directly or indirectly resulting from delay in the proper submission will be charged.

Have a supply of letters on hand to be filled out and sent by the project manager or project engineer immediately after receipt of an official correspondence from the owner confirming the date of substantial completion.

Refer also to Section 4.19, Securing Subcontractor/Supplier Guarantees and Warranties, for important related discussion.

3.13.7   Guarantee Period Checklist *(page 3.94)*

The submission of all guarantees and warranties to the owner as required by the contract documents will be necessary prior to completion of the project. The submission of these documents may also be necessary in order to comply

3.13.6
Sample Subtrade Guarantee/Warranty Notification Letter

+---+
| |
| **Letterhead** |
| |
+---+

Date: _____

To: _____   Confirmation of Fax:
       _____   Fax #: _____
       _____

ATTN: _____

RE: Project:_____
           Company Project #: _____

SUBJ: Guarantees / Warranties: Section(s) _____

Mr. (Ms.) _____ :

The Substantial Completion Date of the Project is _____, 20 _____. As you are aware, this is the date on which your guarantees and/or warranties as required in the subject specification section(s) take effect.

Submission of your guarantees and/or warranties *in their specific, proper, and complete form* is required by _____, 20 _____ in order to allow the current payment to be processed.

Failure to provide your correct guarantees and/or warranties by the above date will delay the submission of the general requisition to the owner that includes retainage reduction and final payment to the entire construction force.

Please don't allow yourself to be in the position of being responsible for delay in such payment: submit your proper documents as requested. Your prompt compliance is crucial.

Please call if you require further clarification.

Very truly yours,

COMPANY

Project Manager

cc: Accounts Payable
       File: Final Payment
                Sub-Vendor File:_____
                CF

3.13.7
Guarantee Period Checklist

Specification Section	Bid Package	Subvendor	Guarantee Item	Guarantee Expiration Date	Date Delivered (In Correct Form)

with the contractual definition of "substantial completion" and can therefore be very significant in terms of preserving the ability to stop the calculation of the Contract Time at its earliest possible point.

Because of this, consideration of the guarantees and warranties can begin with the formal need to submit such guarantees and warranties to the owner simply as part of the contract requirement. From that simplistic view, however, the additional reality is that it is simply in the prime contractor's (or construction manager's) best interest to be sure that all guarantees and warranties as required by the contract documents are in fact provided by each subcontractor and supplier who by those contract documents has been required to provide these guarantees of continued performance to—and through—the prime contractor or construction manager. In other words, any failure by the prime contractor or construction manager to secure such guarantees or warranties will not in any way relieve the prime contractor or construction manager of the identical responsibility of continued performance with respect to the owner. Such a failure may, however, strip the prime contractor or construction manager of the corresponding ability to compel and enforce the same continued performance on the part of the subcontractor or supplier who had originally been responsible for it.

And so the project manager's and project engineer's first objective in organizing, securing, verifying the proper form and content, distributing, and enforcing all guarantees and warranties as required by the contract documents is to protect his or her own company first. From that point, timely compliance with the requirements of the contract documents will be a matter of course.

One method that will facilitate the organization and management of all project guarantees and warranties is the timely use of the Guarantee Period Checklist in Section 3.13.7.

The effort should have begun at the very beginning of the project, during the preparation of the Subcontractor/Supplier Reference Form (described in Section 4.5 of this Operations Manual), the Subcontract and Purchase Order Distribution Procedure (described in Section 4.7), the Subcontractor Schedule of Values (described in Section 4.8), and Submittal Requirements and Procedures (described in Section 4.10). At that time, the entire contract should have been studied methodically during the initial effort conducted to identify and consolidate every submittal necessary to be provided by every subvendor. During the preparation of each subvendor bid package requirement and submittal package, any guarantees and warranties called for should have been identified.

In addition to that effort, the Guarantee Period Checklist of this section should also be used—beginning at the very start of the project. As each individual specification section is reviewed, any guarantee or warranty requirement that is identified as part of that effort should be listed in the Guarantee Period Checklist by specification section and vendor responsible. Once so identified, the time and manner in which each respective subvendor has or has not complied with the requirement can be individually dealt with throughout the project during the management and use of the Subcontractor/Supplier Reference Form. Results of that effort with respect to guarantees and warranties will also be posted on the Guarantee Period Checklist.

Subsequent to its original preparation, the Guarantee Period Checklist should be sent to the appropriate design professional and/or owner for review and confirmation. When transmitted to these parties, state that the Checklist is the result of having reviewed the complete set of contract documents in an effort to consolidate each requirement. Request review and confirmation of the content of the Checklist. Advise that the finally approved Checklist will be used as the basis for enforcement of submission of each document from a respective subvendor, and for delivery of all guarantees and warranties to the owner—specifically in compliance with any and all stated guarantee/warranty requirements of the contract. Refer to Section 3.13.8, Sample Letter to Owner Transmitting the Guarantee Period Checklist, for assistance in following through on the recommendations of this section.

As the project nears substantial completion, the Guarantee Period Checklist is reviewed in order to determine any outstanding documents and is used as the focal point for organizing all efforts to enforce all guarantee and warranty submission requirements.

3.13.8   Sample Letter to Owner Transmitting the Guarantee Period Checklist (*page 3.97*)

The Sample Letter to Owner Transmitting the Guarantee Period Checklist that immediately follows accommodates the recommendations of Section 3.13.7 above. Specifically it:

- States that the Checklist is the result of having reviewed the complete set of contract documents in an effort to consolidate each requirement.

- Requests review and confirmation of the content of the Checklist.

- Advises that the finally approved Checklist will be used as the basis for enforcement of submission of each document from a respective subvendor, and for delivery of all guarantees and warranties to the owner—specifically in compliance with any and all stated guarantee/warranty requirements of the contract.

3.13.8
Sample Letter to Owner Transmitting the Guarantee Period Checklist

---

Letterhead

(Date)

To: (Owner)

RE: (Project)
         (Company Project #)

SUBJ: Guarantee Period Checklist

Mr. (Ms.) (   ):

Attached is the (company name) Guarantee Period Checklist that has been prepared for use on the project. The Checklist consolidates each requirement provided in the contract documents for the submission of guarantees and warranties, and is intended to represent the complete list of all such requirements.

Please review the content of the checklist in detail in order to confirm that no requirements have been omitted. If any items need to be added or modified in any way, please advise in writing by or before (insert date). If the Checklist is complete as it is, we will appreciate your written confirmation of same as well.

When we have received your written remarks, we will use the Guarantee Period Checklist for management and enforcement of all guarantee and warranty submittal requirements. The documents referenced therein will then become the basis of all submittal requirements of the contract, including those necessary for substantial completion.

Thank you for your consideration.

Very truly yours,

COMPANY

Project Manager

cc: Architect
         (Official Project Distribution)
         File: Guarantee Period Checklist
                  CF

3.13.9 **Sample Notification of Pre-Punchlist Requirement**

Contractor's Letterhead

Addressed to the Subcontractor

Attention: Subcontractor's Project Manager or Owner

Re: Project Name and Number

Subject: Pre-Punchlist Requirement

Dear Mr. (Ms.) (   ),

Prior to your demobilization from the jobsite, we will inspect your work and prepare a punchlist that will be transmitted to you on/about (date). We will expect this work to be completed no later than (date) and prior to your full demobilization.

This punchlist, prepared by our project superintendent in conjunction with your on-site supervisor, is in addition to the architect/engineer's formal document that will be submitted to your office when received from (architect/engineer's name).

With best regards,

Project Manager

Project Engineering

4.1   Section Description

Project engineering is where the administration of construction is done. In a real sense, it is the how-to of building the project on paper—before it is built in the field.

The fit of the project engineering function within the organization is outlined in Section 1.5. Here we provide the step-by-step activities that are necessary to actually fulfill the Project Engineer's complex responsibilities. Sample full-size forms, checklists, flowcharts, word-for-word letters, and specific instructions are provided at each step of the way.

This section is very closely coordinated with:

- Section 5, Site Superintendence
- Section 6, Safety and Loss Control
- Section 10, Progress Schedules and Funds Analysis

Although this section includes those specific activities that fall within the restricted definition of the project engineering function, actual practice will necessarily dictate large amounts of overlap of individual responsibilities across these disciplines.

As this (or any other) section of this Operations Manual is being reviewed, it is most important to appreciate the lack of any distinction that should be made between systems designed to be used manually and systems—or components of systems—that are organized electronically.

It seems that a major problem we often get ourselves into with the implementation of computerized management systems is that we allow certain systems to almost "take over." In too many cases, managers and professionals can allow such electronic systems to dictate the manner in which information should be produced and records should be managed, instead of the other way around. We as managers must continually remind ourselves that it is we who must determine the forms, format, and procedures that we must use to manage and control our information, and we should be able to format electronic systems to carry out their functions as we have designed them. If any electronic information and document management system cannot be modified and formatted in the way that we determine to be necessary to conduct our own business, then it's time to find another system.

In the final analysis, electronic information and record management systems are most useful in their ability to deal with large volumes of information. In the use of any such system, we must be vigilant to guard against the implementation of "boilerplate" formats that, although they are assured to address the volume of documents generated on every project, may do so in a manner that will remove visibility. Too many of the systems in their "standard" formats out of the box create file and records retrieval methods that can hide the location of important information just as easily as they can keep it readily available.

Because of this irony, one of the most important functions that we as managers can enforce is the assurance that every information generation and management system is designed for the convenience of our managers first, and is designed with the objective of facilitating—and not making more difficult—the complex job of information management. Accordingly, this section (and most other sections of this Operations Manual) appears to dwell on manual methods of information management. It does this in recognition of the fact that even as we march steadfastly into the information age, the armies of smaller and mid-sized construction companies have limited staff who are charged with responsibility for all of the wide ranges of management functions as outlined throughout this Operations Manual. In many cases, it is simply recognized that the manual completion of a form or form letter can be very time-efficient, as opposed to using nearly the same amount of effort to prepare the same information for eventual inclusion in any electronic-management system. And so the final irony with many electronic-management systems is that the effort to "feed the monster" can for some functions actually double the effort necessary to complete the task manually.

4.1.1   Responsibilities of the Project Engineer

The specific assignments of the actual project staff will vary greatly from organization to organization and from project to project. The actual activities detailed in this section may, in practice, be divided among several people and even to a large extent redundant to the activities of other individuals—spread across the responsibilities of the Project Manager, Project Engineer, Site Superintendents, Scheduler, and any other staff enlisted to support these functions.

Refer to Section 1.5 for a summary overview of the project engineering responsibilities and to the table of contents of this section for specific related activity lists.

4.2   Trade Contracts and Subcontracts

4.2.1   General

Major and subtle distinctions exist between the various forms of agreement among the individual parties providing the particular goods and services that will make up a complete project. This Operations Manual does not treat the details of the particular terms and conditions that should be part of each such agreement, but does recognize that project management, project engineering, and site superintendence functions necessarily deal with such detail each day.

The distinction among the various types of agreements begins with the differences between suppliers and subcontractors. Although material suppliers may or may not provide labor off site to fabricate or assemble all or a portion of the materials or products eventually delivered to the project, these companies do not provide any labor on the jobsite to direct or install those items being provided to the project. Because of this, most of the issues surrounding such non-site-provided labor, such as the need for worker's compensation insurance, material staging areas, etc., do not apply to material purchase orders. Subcontractors, on the other hand, can therefore be thought of as any individuals or companies who do provide any amount of labor at the jobsite. It is this fundamental relationship that kicks into gear the myriad of specialized terms and conditions that deal specifically with such on-site labor issues.

The next distinction is to be made between the terms "subcontractor" and "sub-subcontractor." The term "subcontractor" is customarily reserved for those individuals or companies who contract directly with a general, or prime, contractor or the "at risk" construction manager. The subcontracts can therefore be thought of as a separate "tier" in the contracting arrangement flowing first from the owner to the general or prime contractor (the prime contract), and then down to the subcontractor (the subcontract). This arrangement very often continues to flow in the same direction in the form of additional agreements between the (first-tier) subcontractor and other companies that provide portions of the respective subcontract to the first-tier subcontractor. These companies are referred to as sub-subcontractors, or lower-tier subcontractors.

Finally, the distinction is made between the terms "subcontractor" and "trade contractor." As noted above, a subcontractor contracts directly with the prime contractor, who in turn contracts directly with the owner. A trade contractor, however, does not contract directly with the prime contractor. Instead, a trade contractor is actually an individual contractor who contracts directly with the owner. Confusion in this arrangement can arise because most trade contractors are in fact the same specialty contractors who are normally considered to be subcontractors in conventional lump sum general contracting arrangements. At this level, making a clear distinction between subcontractors and trade contractors may seem to be splitting hairs. In truth, however, this is a very important distinction to appreciate in order to ensure that contractually defined responsibilities remain intact and all manner of communication is conducted appropriately.

It is therefore crucial for all levels of project management and site superintendence to understand these distinctions clearly, to know by rote the company's standard terms and conditions in each purchase order, subcontract, and trade contract form and the reasons behind the inclusion of each term and condition. From there, it is important to be aware of those terms and conditions in each instance that are negotiable, those that are chiseled in stone, and those that may fall somewhere in between.

Use the Trade Contract/Subcontract Checklist in the following section as a guide to help ensure that all the important components of your trade contracts and subcontracts have been directly accommodated in some manner.

4.2.2 Trade Contract/Subcontract Checklist (*page 4.8*)

Prior to commencing negotiations with a Subcontractor, the following procedures should be followed:

- Review the contract with the Owner to determine the following:
  - Are there any unit prices that must be negotiated with the Subcontractor to comply with those committed to the Owner?
  - Are there any Allowance items that should be incorporated into a specific subcontract agreement?

4.2.2
Trade Contract/Subcontract Checklist

The following checklist is to be used as a guide in the preparation or finalization of a trade contract or subcontract. It is important to understand that laws vary across state lines, and trade practices may differ depending upon the geographic location of your business and/or the jobsite. Accordingly, before your specific contract form, procedure, etc., is determined, the advice of a competent construction attorney should be sought.

The items that follow are not intended to represent a conclusive list of all items that should be considered. Rather, they identify key issues that may be addressed and resolved at the time that a subcontract or trade contract is prepared—rather than after related problems arise.

1. **PROJECT/BID IDENTIFICATION**
    a. Project Identification:
        i.   Project Name: _____
        ii.   Address: _____
        iii.   Job #'s: Owner: _____ Architect: _____ Company: _____

    b. Bid Package:
        i.   Number: _____
        ii.   Description: _____

    c. Invitation to Bid Required: Yes: _____ No: _____   Date: _____

    d. Prebid Conference Required: Yes: _____ No: _____   Date: _____

    e. Specific Proposal Form Required: Yes: _____ No: _____   Ready By: _____

    f. Bid Due Date: _____

2. **TRADE CONTRACT / SUBCONTRACT**
    a. Project Identification:
        i.   Project Name: _____
        ii.   Location: _____
        iii.   Job #'s: Company: _____   Trade/Subcontractor: _____

    b. Project Owner: _____

    c. Project Architect: _____

    d. General Contractor/Construction Manager: _____

    e. Trade/Subcontractor:
        i.   Name: _____
        ii.   Address: _____
        iii.   Contacts: Phone: _____ Fax: _____ e-mail: _____

4.2.2
Trade Contract/Subcontract Checklist *(Continued)*

f. Scope of Work:

    i.   Description: _____

    ii.   Specification Section(s): _____

    iii.   Complete "Plans & Specs":   Yes: _____   No: _____

    iv.   Labor & Material: Yes: _____   No: _____

    v.   Design-Build Responsibility: Yes: _____   No: _____

    vi.   Equipment, Scaffolding, Hoisting included: Yes: _____   No: _____

    vii.   Storage & Protection included: Yes: _____   No: _____

    viii. Receipt/Handling/Storage/Protection of
           Material purchased by others: Yes: _____   No: _____

g. Shop Drawings or other submittals required: Yes: _____   No: _____

h. Trade/Subcontract Price:

    i.   Value: _____

    ii.   Lump Sum Prices: _____

    iii.   Unit Prices: _____

    iv.   Combination: _____

i. Alternate Prices:

    i.   _____

    ii.   _____

    iii.   _____

    iv.   _____

j. Buildings and Payments:

    i.   Deposits required: Yes: _____   No: _____

    ii.   Schedule of Values required: Yes: _____   No: _____

    iii.   Requisition submission dates, schedules: Yes: _____   No: _____

    iv.   Retainage (_____%) required :   Yes: _____   No: _____

4.2.2
Trade Contract/Subcontract Checklist *(Continued)*

k. Schedule:

    i.   Will give Notice to Proceed: Yes: _____   No: _____

    ii.   Person providing schedule information: _____

    iii.   Milestone Schedule Dates:

         (a) Submittal Preparation & Delivery: Date: _____

         (b) Item: _____   Date: _____

         (c) Item: _____   Date: _____

         (d) Item: _____   Date: _____

         (e) Item: _____   Date: _____

         (f) Item: _____   Date: _____

l. Insurance:

    i.   Worker's Compensation: Yes: _____   No: _____   Amount: _____

    ii.   General Liability: Yes: _____   No: _____   Amount: _____

    iii.   Special requirements: Yes: _____   No: _____   Amount: _____

m. Standards of Work:

    i.   Industry practice: Yes: _____   No: _____

    ii.   Applicable building codes: Yes: _____   No: _____

    iii.   OSHA: Yes: _____   No: _____

    iv.   Other requirements: Yes: _____   No: _____

n. Taxes, fringe benefits:

    i.   Responsibility: _____

o. Minimum wage standards:

    i.   Davis-Bacon   Yes: _____   No: _____

    ii.   Other: Yes: _____   No: _____

p. Changes:

    i.   Written change orders required: Yes: _____   No: _____

    ii.   Person authorizing changes: _____

q. Cleanup:

    i.   Responsibility to remove debris to dumpster/central jobsite location: _____

    ii.   Responsibility to remove debris from jobsite: _____

    iii.   Ability to backcharge for unacceptable performance: _____

4.2.2
Trade Contract/Subcontract Checklist *(Continued)*

r. Permits/Fees/Deposits:

    i.   Responsibility: _____

s. Equal Opportunity:

    i.   Mandatory requirement: Yes: _____   No: _____

    ii.   Good-faith effort required: Yes: _____   No: _____

t. Lien Waivers and Certifications of Payment:

    i.   Required: Each Payment: _____

                                  Final Payment: _____

u. Waivers of Claims:

    i.   Required: Each Payment: _____

                                  Final Payment: _____

v. Indemnification:

    i.   Required: Yes: _____   No: _____

w. Termination:

    i.   "Breach" = due cause: Yes: _____   No: _____

    ii.   Notice required

x. Trade/Subcontract Assignment:

    i.   Assignable: Yes: _____   No: _____

y. Dispute Resolution:

    i.   Arbitration: Yes: _____   No: _____   Rules: _____

    ii.   Mediation: Yes: _____   No: _____   Rules: _____

    iii.   Other: Yes: _____   No: _____   Rules: _____

z. Authorized contract execution:

    i.   Corporate resolution required: Yes: _____   No: _____

    ii.   Positive identification required: Yes: _____   No: _____

    iii.   Power of attorney   Yes: _____   No: _____

- Are there any restrictions on times when work can start or stop or restrictions on working on Saturday or Sunday? Are there noise abatement issues?
- Are there Alternates in the contract with the Owner who should be included in a subcontract agreement, and are there time restraints on how long the "contract" price will apply, and if so when the price expires?
- Are there restrictions on the amount of overhead and profit the Subcontractor can apply to change orders, and are there restrictions on sub-subcontract percentages?
- Are there any other pass-through provisions that should be reviewed with the Subcontractor prior to contract award?
- Do any warranties exceed the normal one-year period, e.g., for air-conditioner compressors, insulated glass panels, some types of flooring materials?

- Review the plans and specifications to determine whether a scope based upon "plans and specs" will be inclusive enough for award of a subcontract.
- Determine if it would be beneficial for the General Contractor to expand the scope of the Subcontractor's work to include other sections of the specifications. For example, should the window installer also caulk its work rather than having the General Contractor award a separate caulking contract, thereby establishing responsibility for complete weather integrity with one subcontractor rather than two?
- Determine if there are any special requirements not covered by the appropriate plans and specifications that the General Contractor wishes to have the Subcontractor include in its scope of work.

There are a number of basic questions to ask a Subcontractor during the initial interview even before proceeding to a checklist containing specific scope questions:

- Has the Subcontractor received the latest specifications and drawings? A prepared list of the current "contract" drawings will be helpful so that the Subcontractor can agree that it has received and acknowledged the drawings that will be incorporated into a subcontract agreement.
- Have all addenda been received and accepted?
- If the project is tax exempt, does the Subcontractor have the correct deductions in its proposal?
- Is the Subcontractor aware of the project schedule and when it will be required to commence work?
- Has the Subcontractor taken any exception to the plans and specifications? At times a subcontractor may have included an "or equal" material or equipment in its price, assuming acceptance by the Architect/Engineer. If so, it is obliged to divulge this information because the Architect/Engineer may not agree that the substitution is equal to the specified product.

4.2.3 Review the Contract Documents Prior to Negotiating Subcontract Agreements

Prior to scheduling a meeting to negotiate a subcontract agreement, a careful review of the contract documents should be conducted. A review of the contract with the Owner and a review of the specific section(s) of the specifications relating to the subcontract agreement and the appropriate contract drawings may reveal other requirements that will affect those negotiations.

Review the contract with Owner:

- Are there applicable unit prices in the contract that must be accepted by the subcontractor?
- Are there any terms and conditions that vary from the normally accepted ones? For example, are there limits on allowable overhead and profit for change order work, limits on working hours, or noise abatement restrictions?
- Are there Alternates containing subcontract work that are to be accepted by the appropriate Subcontractor?
- In particular, on some government contract work, a provision requiring payment to Subcontractors within a specified period from the time the General Contractor receives payment from that agency, in effect, abrogates the typical "pay when paid" clause in many subcontract agreements.

Review the contract specifications:

- In the specification section relating to the specific work to be awarded to the Subcontractor, is there a "Related Sections" referring to other specification sections that may also contain work to be folded into the subcontract agreement?
- Is the term "contractor" fully defined? Some specification sections use the term "contractor" applied to either the General Contractor or Subcontractor. When responsibility for certain tasks is "the responsibility of the contractor," defining this term becomes important.
- The question of furnishing "temporary light and power" is often "lost" in the specification section, missing from either Division 0 or 1 and also from the Electrical section. Clarification as to whether the electrical Subcontractor or the General Contractor must assume responsibility for the installation and removal of temporary lighting can become an important cost issue.
- Again in the Electrical specification section, terms must reflect the customs, fees, and permitting requirements of the local utility company. (This is particularly true if the MEP engineering was performed by an out-of-state firm not conversant with the rules and regulations of the utility companies in the project's geographic area.)
- Particularly in the mechanical and plumbing trades, are local labor practices and customs different from those in the specifications? Does the plumbing Subcontractor in this geographic area customarily perform its own trenching and backfilling as dictated in the specifications, or does this trade historically not have that responsibility, leaving it to the General Contractor to trench and backfill?

Review the contract drawings:

- As previously discussed, the contract drawings may contain some inconsistencies and omissions, and a careful review of the appropriate drawings may uncover some "gaps" in scope that can be filled by negotiating minor missing items into the subcontract agreement. It is not wise to refer to the scope of work merely as "per plans and specs" without a careful review of those plans and specifications.
- Are there "related" items of work on other drawings that might also be incorporated into this Subcontractor's scope of work? As an example, the mechanical specifications may require color-coding of some piping systems; however, this is not indicated on the drawings nor included in the Painting specification section but can only be found in the MEP specification sections.
- Is there merit in combining items of work with one Subcontractor? For example, combining masonry restoration work with the application of a water-repellent coating, normally installed by a painting contractor, places the waterproof integrity with one Subcontractor. A window installer caulking its work also places responsibility for water infiltration in one Subcontractor's domain.

4.2.4 Subcontractor Negotiation Forms (*page 4.15*)

Along with the sample Trade Contract/Subcontract Checklist shown in Section 4.2.2, the Subcontract Negotiation Forms that follow are helpful when these interviews are conducted. Although the initial preparation of a whole series of customized forms may require considerable work, once produced, the forms can become an integral part of the company's Subcontractor negotiation process. A description of each of the presented forms follows:

4.2.4(a) Subcontractor Form—Concrete, Cast-in-Place. Includes items typically required for this work and includes quantity checks so that one Subcontractor's scope of work can be compared with that of the competition, furthering assurance that all bidders have interpreted the drawings in the same way, more or less. This form also requires the interviewee to sign the form attesting to its contents and refuting some Subcontractor's future claim that they were misinterpreted during the interview process.

4.2.4(b) Subcontractor Form—Masonry. Provides an inclusive list of materials and operations that combine plans and specifications and "best practices" that apply to masonry work.

4.2.4(c) Subcontractor Form—Miscellaneous Metals. Requires the project manager or project engineer to review *all* drawings to ensure that some of these miscellaneous metal items that can appear on the civil (site), architectural, structural, and MEP drawings have been incorporated into the scope of work presented to the Subcontractor.

4.2.4(d) Blank Subcontractor Negotiation Form for Other Trade Applications. A blank form that can be used to create the matrix for the preparation of a whole array of other Subcontractor Interview Forms.

4.2.4
Subcontractor Negotiation Forms

SUBCONTRACTOR NEGOTIATION FORM   Page 1 of 2

PROJECT

TRADE	SPECIFIED SECTIONS	DATE
Concrete, Cast-in-place		

SUBCONTRACTOR	REPRESENTED BY	(AREA CODE) TELEPHONE NO.

BASE BID AMOUNT	ADDENDUM NO.	

ALTERNATES	UNIT PRICES	UNIT PRICES
(1)	(1)	(6)
(2)	(2)	(7)
(3)	(3)	(8)
(4)	(4)	(9)
(5)	(5)	(10)
SALES TAX	INSURANCE	

SCOPE OF WORK Including but not limited to the following:

This form must be completely filled out and signed by the subcontractor and general contractor's representative to provide a record of this negotiation meeting.

ITEM	YES	NO	EXPLANATION AND/OR COMMENTS
1. Quantity check: Total Concrete:_____ cy Foundation concrete: _____ cy Slab on grade concrete:_____ cy Slab(s) on deck concrete:_____cy Misc. concrete:_____ cy			
2. Quantity check: Site concrete:_____cy Mech/Electrical concrete_____ cy			
3. Reinforcing steel: Rebar tonnage:_____ lbs Welded wire mesh_____ sf Caps on all vertical bars, accessories, chairs, etc.			
4. Concrete mix design per specifications: Stone Lightweight			
5. Admixtures per specifications			
6. Special aggregates, cement color			
7. Slump as specified- no exceptions taken			
8. Forms - specify type for structural concrete. Liners for architectural concrete			
9. Chamfer strips and rustications			

	FINAL AGREED AMOUNT
GENERAL CONTRACTOR'S REPRESENTATIVE SIGNATURE	SUBCONTRACTOR'S REPRESENTATIVE SIGNATURE

ORIGINAL

4.2.4
Subcontractor Negotiation Forms *(Continued)*

CONTINUATION SHEET

SUBCONTRACTOR NEGOTIATION FORM   Page 2 of 2

ITEM	YES	NO	EXPLANATION AND/OR COMMENTS
10. Scaffolding, pumps, hoisting and lifting as required			
11. Provide layout and install box-outs, sleeves (by others), framed openings, embeds (by others) as required.			
12. Mock-ups as required			
13. Dowels for other trades			
14. Shoring-installation and removal only when written approval is received from engineer			
15. Additional work: _____% overhead, _____ % profit			
16. Schedule of Values to be submitted for approval by general contractor			
17. Subcontractor to provide temporary heat and winter protection for their work, as required.			
18. Subcontractor to provide line, grade, and all engineering required for the placement of their work.			
19. Contractor to coordinate and receive approval of placement of field trailer and lay-down areas for their work.			
20. All debris, excess materials to be removed from site.			
21. Subcontractor has received, read, and accepts subcontract agreement. Also it has received, read, and accepts all specification provisions including General and Special Conditions.			
22. Subcontractor to sign subcontract agreement, if awarded within ____ days of receipt, and cannot commence work on site until a fully executed contract is received by them.			
Detail sheets, when applicable are attached hereto and become a part of this negotiation session when initialed by subcontractor and (company).			

4.2.4
Subcontractor Negotiation Forms *(Continued)*

SUBCONTRACTOR NEGOTIATION FORM

Page 1 of 2

PROJECT

TRADE	SPECIFIED SECTIONS	DATE
Masonry		

SUBCONTRACTOR	REPRESENTED BY	(AREA CODE) TELEPHONE NO.
BASE BID AMOUNT	ADDENDUM NO.	

ALTERNATES	UNIT PRICES	UNIT PRICES
(1)	(1)	(6)
(2)	(2)	(7)
(3)	(3)	(8)
(4)	(4)	(9)
(5)	(5)	(10)
SALES TAX	INSURANCE	

SCOPE OF WORK Including but not limited to the following:

This form is to be completely filled out and signed by the subcontractor and the general contractor's representative to record the negotiation meeting.

ITEM	YES	NO	EXPLANATION AND/OR COMMENTS
1. Furnish samples, sample panel, and/or mock-up per specifications.			
2. Build in work of other trades as required.			
3. Scaffolding and hoisting for own work. Scaffold planks to be cleaned of excess at end of each workday.			
4. Layout and engineering for own work			
5. Bracing and protecting of all work			
6. Bond pattern, joint width and type (tooled, raked, flush, etc.) per specifications			
7. Mortar type, color (integral, not added) and mixes as specified			
8. Furnish and install all masonry reinforcement and anchoring devices in strict accordance with the specifications.			
9. Face brick-size, allowance, manufacturer			

FINAL AGREED AMOUNT

GENERAL CONTRACTOR'S REPRESENTATIVE SIGNATURE

SUBCONTRACTOR'S REPRESENTATIVE SIGNATURE

ORIGINAL

4.2.4
Subcontractor Negotiation Forms *(Continued)*

CONTINUATION SHEET

SUBCONTRACTOR NEGOTIATION FORM   Page 2 of 2

ITEM	YES	NO	EXPLANATION AND/OR COMMENTS
10. Special shapes (list)			
11. CMU –types (list)			
12. Sills, copings, decorative pieces- furnish and install			
13. Control joints, expansion joints; furnish- install.			
14. Grout fill of CMU and bond beams to include furnishing and installation of all rebars			
15. Weeps, pea gravel cavity wall drainage to be kept free of all debris, excess mortar			
16. Furnish and install all through-wall flashings and flashing materials supplied by others.			
17. Lintels - furnish, receive, unload and install all precast lintels. Install steel lintels supplied by others.			
18. Install rigid insulation by others in cavity walls as required.			
19. Reglets as required			
20. Non–acid clean-down, additional wash downs required to eliminate efflorescence within 45 days after completion			
21. Debris to be placed in dumpster as directed by general contractor			
22. Site Masonry (list)			
23. Subcontractor has received, read, and accepts general contractor's subcontract agreement.			
24. Subcontractor has received, read and accepts general contractor's safety program.			

4.2.4
Subcontractor Negotiation Forms *(Continued)*

SUBCONTRACTOR NEGOTIATION FORM   Page 1 of 2

PROJECT

TRADE	SPECIFIED SECTIONS	DATE
Miscellaneous Metals		

SUBCONTRACTOR	REPRESENTED BY	(AREA CODE) TELEPHONE NO.
BASE BID AMOUNT	ADDENDUM NO.	

ALTERNATES	UNIT PRICES	UNIT PRICES
(1)	(1)	(6)
(2)	(2)	(7)
(3)	(3)	(8)
(4)	(4)	(9)
(5)	(5)	(10)
SALES TAX	INSURANCE	

SCOPE OF WORK Including but not limited to the following:

This form must be completely filled out and signed by the subcontractor and general contractor's representative to provide a record of this negotiation meeting.

ITEM	YES	NO	EXPLANATION AND/OR COMMENTS
1. All miscellaneous metal items are to be furnished and installed, unless otherwise noted.			
2. Galvanized as required/ Shop primed as required			
3. Bituminous coating as required			
4. All accessories such as bolts, screws, clips, anchors, etc. to fabricate and install work			
5. Furnish inserts/anchoring devices set into concrete/ masonry for attachment to miscellaneous metal items			
6. Furnish inserts and/or anchoring devices for other trades as specified			
7. Shop assemble to greatest extent possible			
8. Field measurements as required			
9. Coordinate delivery, setting drawings, diagrams, templates, instructions, and directions for inserts, anchors, bolts or anchorages set by others			

FINAL AGREED AMOUNT

GENERAL CONTRACTOR'S SIGNATURE

SUBCONTRACTOR'S REPRESENTATIVE SIGNATURE

ORIGINAL

4.2.4
Subcontractor Negotiation Forms *(Continued)*

CONTINUATION SHEET

SUBCONTRACTOR NEGOTIATION FORM   Page 2 of 2

ITEM	YES	NO	EXPLANATION AND/OR COMMENTS
10. Subcontractor has accepted GC's safety program and all OSHA safety regulations			
11. Subcontractor to provide all temporary power unless general contractor agrees to provide			
12. If lay-down area is required, provide drawing for general contractor's approval			
13. Specified manufacturers or "Or Equal" only (burden of proof on Subcontractor)			
14. Insurance certificates to be furnished to meet limits as set forth in the contract specifications			
15. Subcontractor has received, read, and accepts terms/conditions of subcontractor agreement			
16. A payment and performance bond will be forthcoming, if requested			
17. For additional work _____% overhead and _____% profit will apply			
18. Subcontractor has received all dwgs, including A,M,E,P and includes all related work			
19. Subcontractor has received GC's base line schedule and accepts portion for their trade			
20. All items requiring factory finish to be in strict accordance with specification requirements			
21. Submit shop drawing schedule and delivery of each item upon receipt of approval			
22. Acknowledge attached detail list(s) to be appended to this form			

4.2.4
Subcontractor Negotiation Forms *(Continued)*

SUBCONTRACTOR NEGOTIATION FORM

Page 1 of 2

PROJECT

TRADE	SPECIFIED SECTIONS	DATE
SUBCONTRACTOR	REPRESENTED BY	(AREA CODE) TELEPHONE NO.
BASE BID AMOUNT	ADDENDUM NO.	
ALTERNATES	UNIT PRICES	UNIT PRICES
(1)	(1)	(6)
(2)	(2)	(7)
(3)	(3)	(8)
(4)	(4)	(9)
(5) SALES TAX	(5) INSURANCE	(10)

SCOPE OF WORK Including but not limited to the following:

This form must be completely filled out and signed by the subcontractor and the general contractor's representative to provide a record of this negotiation meeting.

ITEM	YES	NO	EXPLANATION AND/OR COMMENTS
1.			
2.			
3.			
4.			
5.			
6.			
7.			
8.			
9.			

FINAL AGREED AMOUNT

GENERAL CONTRACTOR"S REPRESENTATIVE SIGNATURE

SUBCONTRACTOR'S REPRESENTATIVE SIGNATURE

ORIGINAL

4.2.4
Subcontractor Negotiation Forms *(Continued)*

CONTINUATION SHEET

SUBCONTRACTOR NEGOTIATION FORM   Page 2 of 2

ITEM	YES	NO	EXPLANATION AND/OR COMMENTS
10.			
11.			
12.			
13.			
14.			
15.			
16.			
17.			
18.			
19.			
20.			
21.			
22.			

4.3   Project Files

4.3.1   General

The establishment of the project files is treated in detail in Section 2.4, Files and File Management. It is included in that section in order to allow those with responsibility for other portions of this Manual to be aware of the complete file structure system and to coordinate their activities with the rest of the organization.

4.3.2   File Responsibilities of the Project Engineer

In referring to Section 2.4, the Project Engineer will be the one responsible to prepare and maintain:

- The plans, specifications, and changes (2.4.3)
- The General Project File (2.4.4)
- The Clarification/Change Log (2.4.6)
- The Subcontractor Summary and Telephone Log (2.4.7)

In this way, each of the documents generated in this section will be properly consolidated with the complete project record.

4.3.3   Duplicate Files—Home and Field Offices

Certain files will be duplicated between the home office and the field, modified as indicated hereafter. Those include:

1. *Plans, Specifications, and Changes* (as described in Section 2.4.3). Both the field and office sets must be maintained and posted with all changes, modifications, clarifications, etc., on a current basis, and must be precise duplicates of each other.

2. *The General Project File.* The General Project File at the home office will be complete in every respect, as detailed in Section 2.4.4. The duplicate file in the field will be set up and maintained with the following modifications:

    a. Documents relating to solicitation, bid package evaluations or reviews, and subcontract negotiations will not be distributed to the field.
    b. Field copies of final, executed purchase orders, and subcontracts will have prices omitted for security reasons. This information will be available to the field staff as necessary, but no hard copy is to be left on-site.
    c. Assuming that shop drawing review by the project engineering function is being conducted at the home-office (off-site), copies of submittals for approval will not be sent to the field during the approval cycle unless unique circumstances require an expediting effort conducted from or with the field. This would be a special circumstance to be managed with the utmost caution.

Instead, the field will receive copies of all submittal transmittals so that all are aware of the current status of particular approvals. Only copies of "approved," "approved as noted," or equivalent (see Section 3.7, Shop Drawing "Approval") that are actually to be used for construction are to be distributed to the field file.

Beyond these qualifications, all other project correspondence and documents are to be distributed to the field on a current basis, noted with the same file instruction as for the home-office.

4.4   Subcontractor Summary and Telephone Log

4.4.1   General

Immediately after the award of a contract, prepare a 1-inch three-ring binder labeled *Subcontractor Summary*. On small to midsized projects, or projects that otherwise will not have relatively large numbers of approval submittals, this binder could be combined with the Submittal Log of Section 4.9.2.

This binder will be the location in which the summaries of each Subcontractor will be noted and performance compliances (and noncompliances) consolidated daily. It will provide at a glance the complete status on a current basis of each Vendor's actions and, more importantly, their specific deficiencies.

4.4.2   Purpose

A project will have dozens of vendors. Each vendor will have dozens of individual requirements, including, for example:

1. Submissions required by the project documents, such as:
   - Vendor approval by the Owner
   - Complete submission of shop drawings and samples
   - Periodic submission of certified payroll reports
   - Delivery of lien waivers for itself and for its vendors
2. Submissions required by its subcontractors, such as:
   - Delivery of insurances in proper form
   - Certified list of sub-subvendors
3. Adherence to specific subcontract requirements, such as:
   - Confirmation of schedule requirements
4. Responses to or confirmation of receipt of Company forms and special correspondence sent periodically, such as:
   - Submittal Requirements Letter of Section 4.10
   - Backcharge Notifications of Section 4.17
   - Construction Schedule transmittals and notices

These add up to hundreds of time-sensitive issues and requirements that must be policed by the Project Engineer on a current basis for each project to ensure that each vendor is continually being brought into compliance.

The purpose of the Subcontractor Summary and the Telephone Log is therefore to:

1. Provide a vehicle that will streamline the Project Engineer's review of each Subcontractor's compliance in terms of both performance and all the various submissions and compliances of their subcontracts.

2. Provide the space for a phone log that will consolidate very powerful information both for daily subcontract management situations and for backup in a dispute.

Together, the two components provide a quick but complete history on the total performance of the party. This will permit exerting constant pressure on the process to close all those cracks through which things can fall too easily. You'll be able to substantiate to the minute:

- Chronically necessary expediting efforts on your part
- Chronically missed dates or other performance failures
- Specific reasons why payments should be withheld, with all the backup
- Direct responsibilities for delays and interferences
- Substantiation of all backcharges, along with all those chances given to the other party to avoid them

4.4.3 Preparation

1. Arrange alphabetical tabs in the 1-inch binder.

2. A Subcontractor Summary Form (see Section 4.4.6) is inserted alphabetically for each Subcontractor and Supplier.

3. Place three or four pages of the vendor Telephone Log Form (see Section 4.4.8) immediately behind each Subcontractor Summary Form. Add to the supply during the project if it becomes necessary.

4. As other vendor summary documents become available (such as the Subcontractor/Supplier Reference Form of Section 4.5), place the completed form directly behind the Subcontractor Summary Form.

5. The front of the binder is reserved for telephone logs for the Owner, the Architect, and other design professionals.

4.4.4 Procedure and Use

1. Use the standard form. Add any specific requirements unique to the project (such as special approvals), and photocopy a supply of those *project* forms.

2. Add any requirements specific to a particular Subcontractor or Supplier (a bond or special insurances, for example) to that vendor's form. Do this *as* each subcontract or purchase order is being prepared.

3. *The most important idea is to develop immediately the habit of always having the binder open in front of you.* This aspect is crucial to using the power of this simple procedure to its fullest potential.

4. Use the binder as your regular telephone and address directory. In this way you'll automatically have it open to the vendor as the call is being made, or as the form letter is being prepared.

5. When you are phoned by any vendor, have the binder open to that vendor's summary *before you pick up the phone.*

6. Log all information *as it occurs.* Since the binder is already open, it takes less than a moment, for example, to note the date of the summary as the Subcontractor/Supplier Delivery Requirements Form Letter of Section 4.11.3 is prepared.

7. Enter *all* information, especially when nothing has happened. If, for example, all you get for your effort is "he's not in; he'll call you back," note those remarks along with who gave them, date, and time. A list of six of these in a row, for example, goes a long way in demonstrating the performance failures of the other party.

8. Using the telephone log in this way will consolidate all information by date *per vendor,* making the information immediately clear, available, and usable. This is in sharp contrast to including these kinds of notes, for example, on the Daily Field Reports. Records kept in that way will require hunting through the entire set every time a comprehensive summary of an effort is needed for a vendor. It's just too much work, which will never get done, or done properly.

9. Subtly make everyone aware of these records. Having the records is the first step to getting the upper hand in every situation every time. If the other parties are *aware* of the existence of such precise, detailed records, they'll learn fast that it is in their best interest to avoid games. When they see your ability to greatly simplify the usually extreme complexity of project relationships, they'll take another look, and decide that it's best not to fool with *this* relationship.

   Some ways to accomplish this include:

   a. Whenever a Subcontractor phones to question, to complain, or for any other reason:
   (1) Review that Subcontractor's performance summary before you pick up the phone.
   (2) Before the party gets a chance to say what his or her problem is, fire off *every* piece of paper that he or she has failed to deliver properly. After that, advise of all those dates (and times) that you've been trying to reach him or her about *your* problem. Then go through *your* issues. When this process has been completed, *then* you can ask the caller what *his* or *her* problem is.

b. In any formal or informal meeting:
  (1) Open the binder whenever a relevant issue is raised.
  (2) Let everyone *see* you're making notes in their file *as the discussion is occurring.* This action gives profound legitimacy at a later date to *your* information as being the *right* information. It will make everyone think before they speak, and force them to consider their actions (or inactions) very seriously.

10. If more elaborate conversation or an impromptu meeting has taken place, consider using, in addition to your notes, the Meeting/Conversation Record Form of Section 4.13.12.

4.4.5   Sample Subcontractor Summary Form—Completed Example
(*page 4.28*)

The sample Subcontractor Summary Form that follows is the form to be used to follow through on the procedure described in this section. A completed form is provided as an example of its proper use. Section 4.4.6 provides the blank form to be photocopied for actual use.

4.4.6   Sample Subcontractor Summary Form—Blank Form
(*page 4.29*)

4.4.7   Sample Telephone Log Form—Completed Example
(*page 4.30*)

The sample Telephone Log Form that follows is the form to be used to follow through on the procedure described in this section. A completed form is provided as an example of its proper use. Section 4.4.8 provides the blank form to be photocopied for actual use.

4.4.8   Sample Telephone Log Form—Blank Form
(*page 4.31*)

4.5   Subcontractor/Supplier Reference Form

4.5.1   General

The Subcontractor/Supplier Reference Form is to be sent to each Subcontractor and Supplier along with their respective subcontracts or purchase orders. The vendor will be required at that time to complete the form and return it immediately to project engineering's attention.

4.4.5
Sample Subcontractor Summary Form—Completed Example

PROJECT _TRADESMAN SCHOOL_   SUB _FOUNDATIONS CORP._   SECTION(S) _03200_
OWNER No. _BI-Q-277 (B)_   _292 AMITE BLVD_   _03300_
COMPANY No. _9900_   _COLCHESTER, CT 06920_ PHONE _203·624·1820_
SUB No. _—_   NAME _JERRY GRANT_   FAX _203·624·1841_

1.	Prefile Sub	_X_ No ____ Yes; #____		
2.	Owner Selected Sub	_x_ No ____ Yes		
3.	Subcontractor Approval Required	____ No _x_ Yes;	Date Submitted _1-7-94_	
			Date Approved _2-4-94_	
4.	Subcontract	Sent _2-7-94_	Received _2-19-94_	
5.	Plans & Specifications	Sent _1-8-94_		
6.	Information Request Form	Sent _2-7-94_	Received _2-19-94_	
7.	Schedule of Values Letter	Sent _2-7-94_	SOV Rec'd _2-19-94_	
8.	Submittal Requirements Letter	Sent _2-7-94_		
9.	Preliminary Construction Schedule	Sent _2-7-94_	Comments Rec'd ____	
10.	Final Construction Schedule	Sent ____		
11.	"Copy" Notice	Sent ____		
12.	Required Material Deliveries:	Item: _NONE_	Date: ____	

13. Scheduled Installations/Erections   Item _FOOTINGS_   Duration: _25 W.D_
FOUNDATIONS   _30 W.D._
PIERS   _10 W.D._

Correspondent: _JERRY GRANT_
Remarks/Qualifications: _FOUNDATIONS START (10) W.D. AFTER FOOTING START._
PIERS PLACED W/IN SAME TIME FRAME AS FOUNDATIONS

14. Materials Received: ____ Ahead of Schedule: Date: ____
         ____ On Schedule
         ____ Behind Schedule : Date: ____

Remarks: _SUBCONTRACT FOR LABOR - ONLY_

15.	Guaranties/Warranties	Requested____	Received____
16.	As-Built Documents	Requested____	Received____
17.	O&M Manuals	Requested____	Received____
18.	All Closeout Documents	Requested____	Received____
19.	Lien Waivers	Requested____	Received____
20.	Final Release Form	Requested____	Received____

21. Backcharge Summary

Date	Description	Amount
____	____	____
____	____	____
____	____	____

22. Conversations, Notes, & Remarks (Continue on additional sheets as necessary)

Date	Time	Description
_2-11-94_	_2:00 PM_	_JERRY ADVISED AMC IN TELECON THAT EXECUTED_
		SUBCONTRACT WILL BE RETURNED IMMEDIATELY &
		3-15-94 START LOOKS GOOD
_3-1-94_	_9:00 AM_	_TELECON; JERRY ADVISED PAUL D. THAT 3-15 MAY MOVE TO 3-16_

4.4.6
Sample Subcontractor Summary Form—Blank Form

PROJECT_____   SUB_____   SECTION(S)_____
OWNER No._____   _____
COMPANY No._____   _____   PHONE _____
SUB No._____   NAME_____   FAX   _____

1.	Prefile Sub	_____ No _____ Yes; #_____	
2.	Owner Selected Sub	_____ No _____Yes	
3.	Subcontractor Approval Required	_____ No _____ Yes;	Date Submitted_____
			Date Approved _____
4.	Subcontract	Sent _____	Received _____
5.	Plans & Specifications	Sent _____	
6.	Information Request Form	Sent _____	Received _____
7.	Schedule of Values Letter	Sent _____	SOV Rec'd _____
8.	Submittal Requirements Letter	Sent _____	
9.	Preliminary Construction Schedule	Sent _____	Comments Rec'd _____
10.	Final Construction Schedule	Sent _____	
11.	"Copy" Notice	Sent _____	
12.	Required Material Deliveries:	Item:_____	Date:_____
		_____	_____
		_____	_____
13.	Scheduled Installations/Erections	Item _____	Duration:_____
		_____	_____
		_____	_____
		_____	_____

Correspondent:_____
Remarks/Qualifications:_____
_____
_____

14.	Materials Received:	_____ Ahead of Schedule:	Date:_____
		_____ On Schedule	
		_____ Behind Schedule :	Date:_____
	Remarks:_____		

_____
_____

15.	Guaranties/Warranties	Requested_____	Received_____
16.	As-Built Documents	Requested_____	Received_____
17.	O&M Manuals	Requested_____	Received_____
18.	All Closeout Documents	Requested_____	Received_____
19.	Lien Waivers	Requested_____	Received_____
20.	Final Release Form	Requested_____	Received_____
21.	Backcharge Summary		

Date	Description	Amount
_____	_____	_____
_____	_____	_____
_____	_____	_____
_____	_____	_____

22. Conversations, Notes, & Remarks (Continue on additional sheets as necessary)

Date	Time	Description
_____	_____	_____
_____	_____	_____
_____	_____	_____
_____	_____	_____
_____	_____	_____

4.4.7
Sample Telephone Log Form—Completed Example

RECORD OF TELEPHONE CONVERSATIONS		Project: SOUTHINGTON STA. # 9890
Date	**Time**	**Discussion**
4.14.94	9:00 AM	M. JEFF CALLED PAUL GABRIEL (WESCON) TO REVIEW ITEMS 7 & 12 ON THEIR SUBMITTED SCHED. OF VALUES. PAUL OUT 'TILL NOON; WILL RETURN CALL.
4.15.94	8:30 am	M. JEFF CALLED PAUL BACK RE: 4.14.94 CALL; NOT IN YET. WILL CALL BACK
	1:10 pm	M. JEFF CALLED PAUL AGAIN; STILL NOT IN
	2:40 pm	PAUL GABRIEL CALLED BACK; M. JEFF REVIEWED THE S.O.V.; PAUL WILL RESUBMIT TODAY.
4.18.94	11:05 am	PAUL ADVISED AMC IN TELECON THAT ALL SHOP DWGS. WILL BE DELIVERED TO JOBSITE ON 4.22.94
4.21.94	10:30 am	M. JEFF CALLED PAUL GABRIEL TO CONFIRM SHOP DWG. DELIVERY. PAUL SAID "THEY'RE COMING BACK FROM THE PRINTER ON 4.22; WILL BE DELIVERED 4.23.94

4.4.8
Sample Telephone Log Form—Blank Form

RECORD OF TELEPHONE CONVERSATIONS   Project: _____ # _____

Date	Time	Discussion

4.5.2   Purpose

The purpose of the form is to greatly facilitate present and future communications. The form will:

1. Inform the vendor of the Company project identification and job number that is to be used on all correspondence.

2. Solicit and confirm the names and phone numbers of all contact persons within that vendor's organization who will be directly involved with the project in the various capacities.

3. Confirm the names, addresses, and phone numbers of all sub-subcontractors and Subsuppliers. This information will:
   a. Correlate the presence of subvendors with those indicated in the subcontract and purchase order process.
   b. Identify those vendors who may be considered for joint payments.
   c. Allow initiation of contact directly with the subvendors, which will greatly facilitate later material expediting efforts (see Section 4.11).

4. Update the Company file on the respective Subcontractor or supplier.

5. Determine or confirm the relative and absolute amounts of work that the subvendor is actually performing with its own forces and those that are being sub-sublet.
   a. The identification of all such sublet sources early on (before they become "issues") makes confirmation of all material scope, delivery, and cost information directly with the sources a matter of simple phone calls.

4.5.3   Procedure and Distribution

1. Send the reference form together with the subcontract or purchase order attached to the Subcontract/Purchase Order Transmittal Form Letter of Section 4.7.

2. Enforce its return *with the subcontract or purchase order.*

3. Upon receipt, immediately contact the vendor to complete any relevant information that is apparently missing.
   a. The most obvious omission will likely be the home and cell car phone numbers of key production people. The nature of this business is that if they're deserving of your order, you're deserving of the ability to contact these individuals in emergency or expediting situations. Do whatever is necessary to get these numbers *now*—well before you will need them.

4. Distribute the *completed* form to:
   a. The subvendor's general file
   b. The Subcontractor Summary Form of Section 4.4
   c. The Site Superintendent

4.6   Transmittal Form Letter Procedure and Use

4.6.1   General

The Transmittal Form Letter is one of the most common types of correspondence used throughout the industry. Its purpose is to document the movement of information and/or materials. Each time a shop drawing is transferred from the custody of one party to that of another, for example, it should be done by way of a transmittal.

Most individuals, however, do not use the form properly. In the rush to fill it out and get it over with:

- Individual names are neglected (personal accountability).
- File references get omitted.
- Reasons for the transmittal or required action are not clearly indicated.
- The method of delivery is left out.
- Important remarks are abbreviated or otherwise left unclear.
- Distribution of the item or of the transmittal information itself is not properly performed.

4.6.2   Proper Use of the Transmittal Form Letter

Do not let your transmittal become a low-value, time-wasting exercise. Follow these few rules as a routine application that will leave complete and comprehensive records where and when you need them. Use the Transmittal Form Letter when sending all shop drawings, samples, copies of letters, or almost anything else. In each case, be sure to:

1. *Send it to a specific person's attention.* If you don't know who will be handling the response, find out.

2. *Always note the complete file reference.* Refer to Section 2.3, Correspondence, and Section 2.4, Files and File Management. Indicate the designated file instruction on the Transmittal Form Letter in the same way as on any other correspondence.

3. *Indicate definite action on the letter.* The boxes on the form are there to be checked. Use them. If a document is attached, note it. If it has been sent under separate cover, indicate it properly as such.

4. *Indicate the method of delivery.* This can be a very important piece of information, particularly in cases where time or delay in response becomes an issue. It might make all the difference in the world, for example, whether the document had been faxed, mailed, sent by express delivery, carrier pigeon, or had been hand delivered.

5. *Use document identification numbers if there are any.* Describe the item(s) being sent concisely, *but in a manner that identifies them positively and distinguishes*

them from other items. A description of "Duct Drawings," for example, means absolutely nothing. Use instead the same description that is on the drawing's title block.

6. *Request definite action.* Again, the boxes are there to be checked off. Do not assume that the architect "knows" that you are expecting his or her office to "approve" the submission. If you need a response that has not been conveniently categorized for you, don't settle for what's there. Take the time to be certain that you indicate positively what action you need.

7. *Use the "remarks" area for short but important qualifications and notifications.* This is a legitimate project record document. The recipients are responsible for all information contained in it. Don't overlook opportunities for repeated action requests and other notifications.

8. *Indicate the correct distribution of the letter and its attachments.* Refer to Section 2.3.3, Correspondence Distribution, for specific instructions in this regard.

4.6.3 Sample Letter of Transmittal—Completed Example
(page 4.35)

A completed example of the sample Letter of Transmittal is provided here to display the approach presented in this section. Section 4.6.4 provides a blank form to be copied for actual use.

4.6.4 Sample Letter of Transmittal—Blank Form *(page 4.36)*

4.7 Subcontract and Purchase Order Distribution Procedure

4.7.1 General

The descriptions of the subcontract and purchase orders themselves, along with a discussion of relevant clauses and their applications, are specifically developed elsewhere.

This section describes the execution and distribution procedure to be followed, after a complete agreement has been reached, that will:

- Expedite the process.
- Preclude the possibility of unauthorized modifications.
- Distribute the documents to all those who need them in order to properly administer the project.
- Preserve the security of originally executed documents.

4.6.3
Sample Letter of Transmittal—Completed Example

LETTER OF TRANSMITTAL

To: STRAIT STEEL CORP.
      456 SUMMER ST.
      STAMFORD, CT 06362

Attn: JOHN STRAIT

Date: OCT 1, 1994
Project: BRIDGETOWN TREATMENT PLANT
Project #: 9660
Subject: ANCHOR BOLTS

TRANSMITTED: _X_ Attached ___Under Separate Cover   Via: _X_ Mail ___Express Mail ___Fax: #_____
are the following:

_X_ Shop Drawing(s)   ___ Contract Drawing(s)   ___ Specification(s)
___ Letter(s)   ___ Coordination Drawing(s) ___ Change Order(s)
___ Sample(s)   ___ Message (See Remarks)   ___ _____

Quant	Date	ID No.	Description
2	9.19.94	AB-1	ANCHOR BOLT LAYOUT & DETAILS

If items are not received as listed, please notify us immediately.

_____ For Approval   _____ For Your Use   _____ As Requested   _____ For Fabrication
_____ Approved   _X_ Approved As Noted   _____ Rejected   _____ Revise & Resubmit

Remarks: PLEASE CONFIRM BY 10·5·94 THAT ALL ANCHOR BOLTS & CAGES
WILL BE ON-SITE BY 10·15·94

cc: JOBSITE W/ ATT.
       BENT STEEL ERECTORS W/ATT.

File: STRUCT. STEEL W/ATT
        _____
        CF

COMPANY

David Jeffko
D. JEFFKO, PROJECT ENGINEER

4.6.4
Sample Letter of Transmittal—Blank Form

LETTER OF TRANSMITTAL

To: _____   Date: _____
        _____   Project: _____
                                                                             Project #: _____
Attn: _____   Subject: _____

TRANSMITTED: ___ Attached   ___Under Separate Cover   Via: ___Mail ___Express Mail ___Fax: # _____
are the following:
___ Shop Drawing(s)   ___ Contract Drawing(s)   ___ Specification(s)
___ Letter(s)   ___ Coordination Drawing(s)   ___ Change Order(s)
___ Sample(s)   ___ Message (See Remarks)   ___ _____

Quant	Date	ID No.	Description

If items are not received as listed, please notify us immediately.

_____ For Approval   _____ For Your Use   _____ As Requested   _____ For Fabrication
_____ Approved   _____ Approved As Noted   _____ Rejected   _____ Revise & Resubmit
_____ _____ _____ _____ _____ _____ _____ _____

Remarks: _____
_____
_____
_____
_____
_____
_____
_____
_____
_____
_____
_____
_____

cc: _____   COMPANY
      _____
      _____
File: _____   _____
      _____   _____
      CF

4.7.2 Procedure

1. Five copies of the subcontract or purchase order are prepared.

2. The Subcontract/Purchase Order Transmittal Form Letter of Section 4.7.3 is completed as appropriate.

   a. The first item is checked noting that (three) *unsigned* copies of the subcontract or purchase order are attached, and indicates that *all three* are to be executed and returned to the Company for final execution. This may appear to be a simple detail, but in practice it will stop any unauthorized attempts to modify the agreement in any way. If, for example, the documents were sent out *with* signatures, there would be no control over anyone wishing to modify the signed document. While it is true that it may be argued later on technical grounds that provisions so modified needed to be initialed by both parties, a presigned document sets the stage for significant arguments and problems at the jobsite.

   b. Other beginning requirements, such as delivery of the certificate of insurance and of the proper submittals, are also highlighted.

3. Copies of the Transmittal Form Letter and the subcontract or purchase order documents are distributed to:

   a. The respective vendor project file (to maintain copies of the agreement as actually tendered)

   b. The Correspondence File (CF) as backup

4. The three copies of the executed subcontract or purchase order and the complete Certificate of Insurance *in proper form and coverage amounts* must then be *delivered* by the Subcontractor or supplier to the Company *before any work is allowed to proceed at the site*. Under *no* circumstances will exception to this requirement be allowed.

5. If the subcontract or purchase order forms have been modified in any way when they are returned, the modifications must be dealt with immediately in a manner consistent with the specific contract approval authorities of the Company.

6. The three copies of the final documents are executed by the Company. The distribution will be made as follows:

   a. One original is returned to the vendor using another Subcontract/Purchase Order Transmittal Form Letter of Section 4.7.3, completing the lower portion of the form letter as appropriate.

   b. One original is distributed to the fireproof vault at Accounts Payable that houses all contracts.

   c. One original is distributed to the respective vendor file in the General Project File as set up per direction of Section 2.4.4, with duplicate copy to the Correspondence File.

7. One additional copy of the subcontract or purchase order is made deleting all prices. This modified copy will be distributed to the jobsite for use in confirming the actual scope of the vendor's responsibilities.

4.7.3 Sample Subcontract/Purchase Order Transmittal Form Letter—Completed Example *(page 4.39)*

The Sample Subcontract/Purchase Order Transmittal Form Letter included here is self-explanatory. It will be used as described in Section 4.7.2 to send copies of unexecuted documents for execution, copies of executed documents, and notification of other requirements. Check off the appropriate box for the distribution process being performed.

4.7.4 Sample Subcontract/Purchase Order Transmittal Form Letter—Blank Form *(page 4.40)*

4.7.5 Subcontract/Purchase Order Distribution Flowchart *(page 4.41)*

The flowchart displayed here presents the document flow process described in Section 4.7.2 for clarification.

4.7.6 Telephone and E-Mail Quotes

Although this Operations Manual does not deal directly with the subject of estimating or purchasing, both of these functions are necessarily intimately related to nearly every activity performed by project managers and project engineers. After the major purchasing effort has been accomplished, it is very common for various managers and staff members either on- or off-site to purchase miscellaneous and incidental items directly. Depending upon the specific company organization, these activities may or may not be conducted by or through a centralized purchasing department.

Whether or not conducted on the jobsite, by project management or project engineering staff members, or by specialized purchasing personnel, the nature of the contracting business dictates that many purchase/award communications between the prime contractor and the various subvendors are done on an expedited basis over the telephone and fax machine or by e-mail. While there is nothing inherently wrong with this process, the speed with which negotiations are conducted and the manner in which communications are performed by non-purchasing professionals creates its own special set of issues.

Today's telephone is capable of sending and receiving voice mail and text messaging, and wherever "telephone" is used in this Section, assume that the term also refers to the various "smart" phones available on the market today—regardless of which communication function is being used on any particular machine.

Beyond the normal issues associated with soliciting, negotiating, finalizing, and enforcing purchase agreements, several basic factors involving the special situation of the telephone or e-mail are too often misunderstood and mismanaged by project engineering, estimating, and bidding personnel.

4.7.3
Sample Subcontract/Purchase Order Transmittal Form Letter—Completed Example

---

	Letterhead

To: TRIANGLE MASONRY CORP.   Date: 3.28.94
   555 TERMINAL RD.
   BRITTON, CT 06555

ATTN: JOHN TRIANGLE

RE: Project: ENGINE CO. #4
      Company Project #: 9440

SUBJ: Subcontract (Purchase Order) Agreement

Mr. (Ms.) TRIANGLE   :

Please find attached:

__X__ Three copies of the Subcontract Agreement (Purchase Order). Please sign and return *all three copies* for final execution.
_____ One executed copy of the Subcontract Agreement (Purchase Order). Retain for your records.
__X__ The Subcontract/Supplier Reference Form. Complete as appropriate and return to my attention.
_____ Three copies of the Full Lien Waiver Forms to be executed and returned with the Subcontract or Purchase Order Agreement.
__X__ ADD'L COPY OF COMPLETE PLANS & SPECS

Please:

__X__ Deliver all appropriate Certificates of Insurance in correct coverage amounts. Be sure to include *all* requirements for additional insured.
__X__ Deliver __9__ copies of all required approval and coordination submittals to:
      __X__ This Office   _____ The Jobsite
__X__ Confirm that your current delivery and erection schedules meet current Project Schedule Requirements as coordinated with ANTHONY BARTA   , Project Superintendent.

_____
_____

Very truly yours,

COMPANY

George Raymond

GEORGE RAYMOND
Project Manager

cc: A/P, Jobsite
      File: Subcontracts, CF
              Subvendor File: MASONRY

4.7.4
Sample Subcontract/Purchase Order Transmittal Form Letter—Blank Form

---

Letterhead	

To: _____   Date: _____

_____

_____

ATTN: _____

RE: Project:_____
            Company Project #: _____

SUBJ: Subcontract (Purchase Order) Agreement

Mr. (Ms.) _____:

Please find attached:

____   Three copies of the Subcontract Agreement (Purchase Order). Please sign and return *all three copies* for final execution.
____   One executed copy of the Subcontract Agreement (Purchase Order). Retain for your records.
____   The Subcontract/Supplier Reference Form. Complete as appropriate and return to my attention.
____   Three copies of the Full Lien Waiver Forms to be executed and returned with the Subcontract or Purchase Order Agreement.

____   _____

____   _____

Please:

____   Deliver all appropriate Certificates of Insurance in correct coverage amounts. Be sure to include *all* requirements for additional insured.
____   Deliver _____ copies of all required approval and coordination submittals to:
        ____ This Office   ____ The Jobsite
____   Confirm that your current delivery and erection schedules meet current Project Schedule Requirements as coordinated with _____, Project Superintendent.

____   _____

____   _____

Very truly yours,

COMPANY

Project Manager   cc: A/P, Jobsite
                                                    File: Subcontracts, CF
                                                            Subvendor File:_____

4.7.5
Subcontract/Purchase Order Distribution Flowchart

(3) Copies of unexecuted
Subcontract or Purchase
Order w/ Transmittal Form
Letter
· (1) Copy to Vendor File
· (1) Copy to Corresp. File

To →

Subcontractor
or
Supplier

Return

All (3) copies of
Subcontract or Purchase
Order w/vendor's
execution

Company
Final
Execution

DISTRIBUTION

Return to
Subvendor:

(1) Original

Company
Fireproof
Vault:
(1) Original

Project
Subvendor
File:
(1) Original
(1) Copy to CF

Field:

(1) Copy with
all prices
deleted

Telephone or e-mail quotes occur very often during the prebid and bid phases of a project, but also continue throughout the life of a project. Such telephone quotes are normally the basis of an eventual written purchase agreement. Because of this, if the communication avoids directly addressing special terms and conditions that may not be normally found in "standard" forms of agreement, such a lack of complete understanding is guaranteed to lead to problems in achieving a final, executed agreement. In addition, a small but significant number of special conditions exist precisely because the quote was delivered over the telephone (or fax).

Assuming there is no issue regarding the establishment of a contract based on the e-mail/telephone quote, the prime contractor must be certain that all critical terms and conditions are established as well. Much too often, too little attention is paid to memorializing the complete and total agreement. Despite being used to very large degrees, purchase orders are often surprisingly brief, leaving out critically important conditions. In contrast, some purchase orders can contain too many terms in the fine print that unduly favor the supplier. Many suppliers, for example, will fax a quote with their own standard term imprinted thereon requiring "net 30" terms of payment, whereas both of you "fully intend" a purchase agreement that will reflect the terms of the prime contract (which itself will require a pay-when-paid, plus some value of retention). Once again, the best way to prevent problems resulting from these kinds of conditions is to have a clear understanding of the issue and have the necessary contract procedures in place from the very beginning.

The Uniform Commercial Code may come to the aid of contractors in this instance. In some cases, the UCC can be flexible with respect to determining the specific terms of the contract, as long as a quantity of goods under the contract is correctly stated. What, then, is the significance of this in the special case of the telephone bid?

Generally, it is reasonable to begin with the presumption that bidders wishing to provide materials have at least reviewed the relevant project plans and specifications. By then submitting a bid "per specifications," the material vendor will be bound to provide the items specified and in the amounts specified. That assumption, however, does the prime contractor no good when the key issues of delivery times and payment terms arise. Even if the issues aren't specifically treated in the project specifications, they are sure to be detailed from the owner's point of view and not the point of view of the prime contractor or construction manager. Even though the prime contractor will certainly rely on the pass-through principle (refer to that section of the Operations Manual for details), it is best if the prime contractor takes the time to be sure that its own requirements in these crucial areas are directly incorporated into the purchase order language itself.

What, then, would be the best way for the prime contractor to bind the material supplier to the contractor's standard purchase order terms, including specific delivery schedule(s), payment terms, and any other special conditions? If the bid is taken over the telephone or by e-mail and there is little or no time for a proper written confirmation, it is important that a minimum standard procedure be applied to the telephone quote process.

When recording information from a vendor being given as a quotation over the telephone, you must do more than simply confirm the basic vital information such as date of order, material description, material quantity, and material price. Once having recorded the basic information in some manner of writing or directly into a computer terminal, and having reviewed and verified such basic information, the work does not stop there. The person taking the telephone quote should also read to the vendor a standardized statement that includes the specific terms of your purchase order, and also confirm unique terms such as material delivery date(s). Once having read the standard and special terms and conditions, the purchaser should then specifically confirm that the vendor still wishes to go ahead with the deal under such circumstances. If the order is then placed, it will be that much more difficult for the vendor to claim that it was not aware and does not agree to the additional terms and conditions, no matter how one-sided they might be.

Accordingly, contractor personnel taking a bid over the telephone should have a form available that does more than merely record the basic information specifically pertinent to the particular bid. At the bottom of that form should be printed a standard statement to bidders that the contractor's personnel can read to a subbidder over the telephone as part of the process of confirming the correctness of the bid information that has been written down. Such a statement need not be elaborate, but it should inform the bidder that the bidder will be bound by the desired delivery and payment schedules and by other conditions of the contractor's standard purchase order.

Ideally, the contractor's form should also have a space at the end of the document for the contractor's personnel to sign or initial, confirming that this statement has in fact been read over the telephone to the individual indicated on the form. It is important to understand, however, that the inclusion of this type of record can backfire if the company is not careful to police its use consistently. The failure to sign or initial in the space provided, for example, could later be used as evidence that the statement had not been read (and therefore the vendor had not agreed to the particular terms and conditions).

Immediately upon finalization of the telephone quotation, a copy of the complete form, including the standard statement confirmed to have been read to the vendor, should be sent by fax, with hard copy by mail. This procedure should be part of the written confirmation process that establishes the existence of the contract as of that immediate point in time. The standard form may also have a box to be checked and/or initialed that indicates that the complete standard company purchase order will be sent as follow-up to the telephone quote and will contain the terms and conditions reflected therein.

The confirmation process described above will usually provide the contractor with a significant advantage in the event of any later dispute over the terms of the contract. Generally, such records are then reviewed as powerful confirmation of the facts because they had been prepared contemporaneously with the bid. Such documentation will therefore most often constitute powerful evidence that the particular vendor actually knew of the terms and conditions and did agree to them. Once this advantage has been achieved, however, the contractor must be careful to preserve the relationship during the process of exchanging supplementary documents at a later date and finalizing the formal purchase order.

Once such written confirmation of the telephone quote, including confirmation of the contractor's standard and special terms and conditions, is sent, the vendor receiving such confirmation may respond in different ways. In some cases, the vendor may respond with specific objections or with counter proposals that modify or supplement the heretofore agreed-upon terms and conditions. A more common method often taken by vendors is simply to ignore the contractor's written confirmation and to send back their own standard order form with their own fine print. Such a response may be made because there is genuine disagreement with some term or condition, or it may actually be nothing more than the particular vendor's regular operating procedure. Whatever the reason for this volleyball game, this type of situation can create special problems if it is not anticipated and controlled early.

If the vendor's response contains terms and conditions that alter or add to the contractor's terms, it may wish to condition the acceptance of the order upon agreement with these new terms. If this occurs, a dispute with respect to this contract exists. If this happens, the effort taken as described above to confirm the complete understanding and agreement with specific standard and special terms and conditions may hopefully act as a deterrent to this type of vendor maneuver. As such, it may be satisfactory evidence for the contractor if cooler heads ultimately do not prevail and the issue is litigated.

One way for the contractor to work to minimize such situations is to include a statement in its standard form (which had been read as part of the telephone quote process) that no variation from these terms will be accepted. Unfortunately, the simple inclusion of such a statement in the written and spoken terms cannot be solely relied upon. If the contractor is faced with the "volleyball game" of terms and conditions as described above, the written objection to anything inappropriate must be sent to the offending subvendor immediately. Such an objection should also reaffirm the contractor's original statement that no change in its terms and conditions will be allowed. Hopefully, this time of action will confirm that there should then be no debate over which terms are part of the contract.

4.7.7 Telephone Quote Checklist *(page 4.45)*

The Telephone Quote Checklist that follows is a summary of the items identified in Section 4.7.6, Telephone and E-Mail Quotes. Use it to prepare your own standard operating procedure with respect to telephone quotes.

In summary, then, the telephone quote procedure used by contractor personnel who are taking quotations over the telephone in order to secure materials is as follows:

1. Verify all information specific to the particular purchase order.

2. Read a statement of the company's standard terms and conditions, as well as any special terms and conditions particular to the project, and verify same.

3. Immediately follow up in complete detail.

4.7.7
Telephone Quote Checklist

1. Verify all specific information

   a. Specific material:
      i.   "As specified?"
      ii. "As equal?"
      iii. "As a substitution?"

   b. Price:
      i.   Lump sum?
      ii. Unit price?
      iii. Applicable taxes?
      iv. Other?

   c. Price Incentives:
      i.   Discounts for combined or consolidated orders?
      ii. Other conditions for reduced costs?

   d. Submittals:
      i.   Per plans and specifications (form and content)?
      ii. Number of copies?
      iii. Date of submittal delivery (In total? In parts?)

   e. Material Delivery(ies):
      i.   Time after receipt of approved submittals?
      ii. Single shipment or broken up?
      iii. Shop inspection availability (Required? Advisable?)

   f. Acceptance of General Contract Terms and Conditions:
      i.   No Exceptions?
      ii. Acceptable Exceptions?
      iii. Unacceptable Exceptions?

   g. Payment terms:
      i.   Pay-when-paid?
      ii. Pay-if-paid?
      iii. Retainage to be withheld? If so, what percentage?
      iv. "Net" payment terms?
      v.   Copayment agreement necessary?

2. Read statement of company standard terms and conditions

   a. Inform the bidder that:

      i.   It will be bound by the desired delivery and payment schedules,

      ii. All other terms and conditions of the standard purchase order are required,

      iii. No variations from these terms in conditions will be accepted, and

      iv. The Company considers the agreement to be final as of this date, and is relying upon it.

3. Have an area at the end of the form for company personnel to initial, confirming that the statement was read over the telephone to the specific individual indicated on the form. (Note that the failure to initial this area can later be used as proof that the statement was *not* read.)

4.7.7
Telephone Quote Checklist *(Continued)*

4. Include a statement that the formal purchase order including all Company standard terms and conditions as well as special terms and conditions particular to the project will be prepared based upon the information herein and forwarded to the vendor for final execution.

5. Immediately fax and mail the completed and initialed form to the subvendor, and request immediate confirmation by signing and returning a copy of the form.

6. Note that generally, if the Telephone Quote form is not returned by the vendor within 10 days or if it is otherwise not responded to, the contract is likely to be considered enforceable. (Check with local laws.) If, however, the vendor responds with modified terms, the contract may be confirmed to exist but may be subject to further negotiation.

7. If a vendor responds to the written confirmation of the Telephone Quote with either modifications to terms, additions to terms, or even its own Purchase Order form, it must be properly objected to in writing. The objection must include a remark that reaffirms the original statement that "no changes in terms will be allowed."

4.8   Subcontractor Schedule of Values

4.8.1   General

There are a number of requirements that materially affect the Subcontractor's proper Schedule of Values, including:

- Time of submission
- Level of detail
- Authority to approve or reject its composition
- Amount of work performed by the Subcontractor's own forces
- Amount of work performed by sub-subcontractors
- Value of materials provided
- Value, source, and composition of equipment provided
- Value of final tie-in and system operation activities
- Value of all project close-out efforts
- Value of guarantees, warranties, instruction, manuals, etc.

These considerations necessarily vary greatly between bid packages for both practical and firm contractual reasons. A framing contractor, for example, will not have the same "final tie-in" or "close-out efforts" as the HVAC contractor. Its level of necessary detail will also be much less. That same framing contractor, for example, is not likely to have long-distance suppliers of expensive, sensitive equipment that has elaborate guarantee requirements.

The requirements of each bid package must therefore be considered individually and have *all* considerations applied to it in order to confirm the final determination of all its particular requirements.

Refer to Section 3.10, The Schedule of Values, and Section 3.11, Requisitions for Payment and Contract Retainage, for important related discussion.

4.8.2 Time of Submission

It is crucial that the Subcontractor's Schedule of Values be delivered to the Company in the absolute shortest period of time. The general Schedule of Values of the Company to the Owner is one of the earliest documents to be expedited if the first payments are to be processed to all vendors on time. Further, every subvendor's schedule must correlate directly with the manner in which its prices have been represented to the Owner, or there will be major scope/value payment disputes between the Company and its subvendors in every payment cycle. These considerations cannot be stressed enough.

Refer to Sections 4.7 and 4.8 regarding subcontracts. There you will find instructions for indicating the subvendor's Schedule of Values directly on the subcontract or purchase order form itself for those more simply defined bid packages.

Use the principles of this section and of Sections 3.10 and 3.11 to properly determine that subvendor's schedule, and whenever the opportunity is available to do so, have the entire matter put to rest at the time of subcontract or purchase order execution.

If the timing of the bid package or other considerations do not allow for the immediate finalization of the completed, detailed subvendor's Schedule of Values, follow the directions of the later portions of this section to expedite the complete process.

4.8.3 Correlation with the General Schedule of Values

Section 3.11.3 discusses this consideration from the General Schedule of Values perspective. The most common dilemma is that often the General Schedule of Values must be submitted to the Owner long before the Schedules of Values of major or complex bid packages are finalized between the General Contractor and the respective subvendors. An approach that can help to deal with the problem, thereby avoiding continuing payment issues with both the Owner and the subvendors, follows:

1. Prepare the General Schedule of Values to the level of available detail with respect to the work of all technical bid packages.

2. For any bid package that will be submitting requisitions for work in the first requisition period (such as site work or concrete) there is no option. Those bid package Schedules of Values *must* be completely resolved between the Company and the respective Subcontractors by the time the General Schedule of Values is submitted.

3. Many (or at least several) bid packages for work that will not be performed during the first payment period may not yet have had their detailed breakdowns completely resolved as of the time of the first submission of the General Schedule of Values.

     For these cases include a lump sum for the entire bid package. This will allow the Owner to approve the relative amount of the bid package in the approval of the General Schedule, while it will allow you flexibility to provide ultimately a breakdown and level of detail that directly correlate to the Subvendor's Schedule of Values for its billing throughout the remainder of the project.

     Review this procedure completely with the Owner and design professionals before proceeding. Note in those discussions that billings against such lump sums will not occur until the appropriate breakdowns have been submitted and approved. Having the total approach properly explained is usually sufficient to confirm the procedure.

4. When the Subvendor's Schedule of Values is finalized in accordance with the recommendations of this section, immediately submit it for approval by the Owner in the same procedure as the original General Schedule of Values. Upon approval, include the additional detail on the General Schedule for use as such throughout the remainder of the project.

4.8.4   Level of Detail

Section 3.10.3 describes these considerations from the perspective of the General Schedule of Values. Here note first that most Subcontractors have a tendency to resist providing sufficient levels of detail in their Schedules of Values until forced to do so.

The procedure to determine the appropriate level of detail is very straightforward. In its simplest idea, it boils down to the following:

1. Visualize the manner and sequence in which the work will be accomplished physically.

2. Visualize the determination that will become necessary in order to demonstrate the level of completion of a particular component, and to provide some facility to the Owner's representatives to verify the bill.

    a. Is a single item titled *Branch Wiring* sufficient, for example, to determine the actual amount of work completed without going through financial gymnastics each period, or will it be necessary to detail the individual levels of completeness in various areas of the project?

    b. Will *cubic feet of concrete* be enough to properly determine the correct relative amount of reinforcing steel throughout the footings, walls, columns, beams, and so on, or will you wind up in a "discussion" every time the Owner will want to release payment for steel on a prorated basis while your actual payables to the steel vendor are heavy in a certain area?

3. Separate Material and Labor Prices
   a. Do this first as a practical matter in helping to improve the detail that will facilitate the other considerations of this section.
   b. Second, it may turn out to be one of the "unwritten rules" of the Owner to pay only for materials after they have been installed if there is no approved breakdown to the contrary.

      This practice can have the potential for disaster if, for example, you or your Subcontractor is on a 30-day payment cycle and the material is delivered on the 5th of the month, but won't be installed until the 20th of the following month.

      Directly avoid the problem by having all material identified separately, billable per the express terms of the Owner/Contractor agreement.
4. Determine and/or confirm appropriate unit price values as a method of at least generally confirming the relative correctness of the billing items.
5. Give careful and detailed attention to all project close-out activities. Do not overlook the values of such items as:
   a. System tie-ins and balancing
   b. Hookups or other connection with work of other trades
   c. Submission of guarantees, warranties, maintenance and operating manuals, as-built drawings, and other special documentation
   d. Special instruction to be provided to Owner maintenance personnel
   e. "Incidentals," such as labeling, key plans and charts, and color-coding of components
6. Give special attention to confirming the complete, real values of the remaining 25% to 30% of the work.
   a. If left to its own process, it will be a sure bet that by the time the work has been approved to be "75% complete," there would not be a Contractor on the planet—including the one who is responsible—who could actually complete the work for the remaining 25% of money available.

      In the worst of these situations the subvendor can leave or be terminated at a late point of completion, leaving everyone with the major problem of completing the work for the lesser amount of available money remaining. In the best situations it will be like pulling teeth to have that subvendor who stays with the job finish its odds and ends on time, complete its punchlist, and provide its documentation. *Keep appropriate values assigned to all items, and the work stands a much greater chance of getting done without major incident.*

4.8.5 Sample Subvendor Schedule of Values/Requisition for Payment Form

The sample Subvendor Schedule of Values/Requisition for Payment Form included here is to be sent to each vendor for its use in properly representing its Schedules of Values and subsequent applications for payment.

In cases where the Schedule of Values has been finalized by subcontract execution, the line-item listing will correlate directly with that indicated in the subcontract or purchase order. In cases where the Subvendor's Schedule of Values is still to be submitted and approved, it will become the form of submission.

In either case, once the Subvendor's Schedule of Values is finally approved, it will become the periodic requisition form for that bid package for the life of the project. Extras, credits, and backcharges will be added as new line items as the project progresses.

4.8.5.1   Sample Subvendor Schedule of Values/Requisition for Payment Form—Completed Example #1 *(page 4.51)*

4.8.5.2   Sample Subvendor Schedule of Values/Requisition for Payment Form—Completed Example #2 *(page 4.52)*

4.8.5.3   Sample Subvendor Schedule of Values/Requisition for Payment Form—Completed Example #3 *(page 4.53)*

4.8.5.4   Sample Subvendor Schedule of Values/Requisition for Payment Form—Blank Form *(page 4.54)*

4.8.6   Sample Subvendor Schedule of Values/Requisition for Payment Form Letter *(page 4.55)*

The sample Subvendor Schedule of Values/Requisition for Payment Form Letter that follows is used to:

1. Transmit the Schedule of Values/Requisition for Payment Forms.
2. Advise of the manner of preparation of the subvendor's Schedule of Values.
3. Notify of the required submission date.
4. Identify the approved form as the future requisition form.
5. "Remind" the subvendor of its contractual requirement to submit its periodic requisition on time in order to avoid problems.

Use this procedure to expedite all subvendor Schedules of Values that, for whatever reason, have not been able to be finalized as of the contract execution date.

4.8.7   Sample Subvendor Schedule of Values Approval/Rejection Form Letter *(page 4.56)*

The sample Schedule of Values Approval/Rejection Form Letter that follows is used to return the subvendor's Schedule of Values submission, directly indicating that either:

■ It is approved as it is, and is accordingly to be used as the basis for all applications for payment, or

■ It is not approved, along with an appropriate modification that will allow its approval.

4.8.5.1
Sample Subvendor Schedule of Values/Requisition for Payment Form—Completed Example

SCHEDULE OF VALUES / REQUISITION FOR PAYMENT

Co: HEAT & AC, INC.   Application No.:   Bid Package: HVAC
Proj.: FIREHOUSE ADD'N   Date:
No.: 9424
Period: Page 1 of 1

	SCHEDULE OF VALUES				REQUISITION FOR PAYMENT				
Item #	Description	Quant	Unit Price	Total	Prev. Billed	This Application	Total Complete To Date	% Comp	Balance To Complete
1	SUBMITTALS	L.S.	L.S.	2,000.⁻					
2.	STL. DUCT MAINS	230 LF	20.⁻	4,600.⁻					
3.	STL. DUCT BRANCHES	570 LF	15.⁻	8,550.⁻					
4.	DUCT INSULATION	800 LF	4.⁻	3,200.⁻					
5.	FAN-COIL UNITS (MAT)	10 UN	900.⁻	9,000.⁻					
6.	FAN-COIL UNITS (LAB)	10 UN	300.⁻	3,000.⁻					
7.	BOILER (MAT)	1 UN	3,500.⁻	3,500.⁻					
8.	BOILER (LAB)	1 UN	L.S.	2,000.⁻					
9.	CHILLER (MAT)	1 UN	3,000.⁻	3,000.⁻					
10.	CHILLER (LAB)	1 UN	L.S.	2,000.⁻					
11.	TEMP. CONTROLS	L.S.	—	2,600.⁻					
12.	SYSTEM BALANCING	L.S.	—	3,200.⁻					
13.	AS-BUILTS / O&M MAN.	L.S.	—	2,500.⁻					
14.	GUARANTIES /WARRANT.	L.S.	—	2,000.⁻					
	TOTALS			$51,150.⁻					

4.8.5.2
Sample Subvendor Schedule of Values/Requisition for Payment
Form—Completed Example #2

SCHEDULE OF VALUES / REQUISITION FOR PAYMENT

Co: HEAT & AC, INC.   Application No.: 1   Bid Package: HVAC
Proj.:   FIREHOUSE ADD'N   Date: APRIL 30, 1994
No.:   9424
Period: APRIL, 1994   Page   1 of   1

	SCHEDULE OF VALUES					REQUISITION FOR PAYMENT				
Item #	Description	Quant	Unit Price	Total		Prev. Billed	This Application	Total Complete To Date	% Comp	Balance To Complete
1.	SUBMITTALS	L.S.	—	2,000.-			2,000.-	2,000.-	100	0.0
2.	STL. DUCT MAINS	230 LF	20.-	4,600.-			3,680.-	3,680.-	80	920.-
3.	STL. DUCT BRANCHES	570 LF	15.-	8,550.-			4,275.-	4,275	50	4,275.-
4.	DUCT INSULATION	800 LF	4.-	3,200.-					0	3,200.-
5.	FAN-COIL UNITS (MAT)	10 UN	900.-	9,000.-					0	9,000.-
6.	FAN-COIL UNITS (LAB)	10 UN	300.-	3,000.-					0	3,000.-
7.	BOILER (MAT)	1 UN	3,500.-	3,500.-					0	3,500.-
8.	BOILER (LAB)	1 UN	L.S.	2,000.-					0	2,000.-
9.	CHILLER (MAT)	1 UN	5,000.-	3,000.-					0	3,000.-
10.	CHILLER (LAB)	1 UN	L.S.-	2,000.-					0	2,000.-
11.	TEMP. CONTROLS	L.S.	—	2,600.-					0	2,600.-
12.	SYSTEM BALANCING	L.S.	—	3,200.-					0	3,200.-
13.	AS-BUILTS / O&M MAN.	L.S.	—	2,500.-					0	2,500.-
14.	GUARANTIES/WARRANT.	L.S.	—	2,000.-					0	2,000.-
	TOTALS			$51,150.-			9,955.-	9,955	19.46	41,195.-

4.8.5.3
Sample Subvendor Schedule of Values/Requisition for Payment
Form—Completed Example #3

SCHEDULE OF VALUES / REQUISITION FOR PAYMENT

Co: HEAT & AC, INC.   Application No.: 2   Bid Package: HVAC
Proj.: FIREHOUSE ADDITION   Date: MAY 31, 1994
No.: 9242
Period: MAY 1994   Page 1 of 1

Item #	Description	Quant	Unit Price	Total	Prev. Billed	This Application	Total Complete To Date	% Comp	Balance To Complete
1.	SUBMITTALS	L.S.	–	2,000.-	2,000.-	0.-	2,000.-	100	0.-
2.	STL. DUCT MAINS	230 LF	20.-	4,600.-	3,680.-	920.-	4,600.-	100	0.-
3.	STL. DUCT BRANCHES	570 LF	15.-	8,550.-	4,275.-	3,420.-	7,695.-	90	855.-
4.	DUCT INSULATION	800 LF	4.-	3,200.-		1,600.-	1,600.-	50	1,600.-
5.	FAN COIL UNITS (M)	10 UN	900.-	9,000.-				0	9,000.-
6.	FAN COIL UNITS (L)	10 UN	300.-	3,000.-				0	3,000.-
7.	BOILER (MAT)	1 UN	3,500.-	3,500.-		3,500.-	3,500.-	100	0.-
8.	BOILER (LOB)	1 UN	L.S.	2,000.-		1,000.-	1,000.-	50	1,000.-
9.	CHILLER (MAT)	1 UN	3,000.-	3,000.-		3,000.-	3,000.-	100	0.-
10.	CHILLER (MAT)	1 UN	L.S.	2,000.-				0	2,000.-
11.	TEMP. CONTROLS	L.S.	–	2,600.-				0	2,600.-
12.	SYSTEM BALANCING	L.S.	–	3,200.-				0	3,200.-
13.	AS-BUILTS / O&M MAN.	L.S.	–	2,500.-				0	2,500.-
14.	GUARANTIES/WARRANT.	L.S.	–	2,000.-				0	2,000.-
15.	CO#1 · CHANGES TO FAN-COIL UNITS	–	–	3,000.-				0	3,000.-
16.	CO#2· DUCT CHANGES	–	–	2,000.-		1,800.-	1,800.-	90	200.-
TOTALS				56,150.-	9,955.-	15,240.-	25,195.-	44.8?	30,955.-

4.8.5.4

Sample Subvendor Schedule of Values/Requisition for Payment Form—Blank Form

SCHEDULE OF VALUES / REQUISITION FOR PAYMENT

Co: Application No.:   Bid Package:
Proj.:   Date:
No.:
Period: Page   of

| | SCHEDULE OF VALUES | | | | | REQUISITION FOR PAYMENT | | | |
Item #	Description	Quant	Unit Price	Total	Prev. Billed	This Application	Total Complete To Date	% Comp	Balance To Complete
	TOTALS								

4.8.6
Sample Subvendor Schedule of Values/Requisition for Payment Form Letter

Letterhead

Date: _____

To : _____
     _____
     _____

ATTN: _____

RE: Project:_____
        Project #:_____

SUBJ: Schedule of Values/Requisition for Payment

Mr. (Ms.)_____:

Attached are the Schedule of Values/Requisition for Payment Forms, to be used as follows:

1.   Prepare & Submit Your Schedule of Values

        On the left side of the form, separate your total contract into individual line items that accurately reflect the manner in which material and equipment will be delivered to the jobsite, and the way in which the work will be performed. Provide sufficient level of detail that will allow the Owner, Architect, and this office to properly evaluate your periodic requisitions for payment. Also include appropriate categories of your general conditions and project close-out items.

        Submit your Schedule of Values to this office by _____, 20_____ for approval.

2.   Requisition for Payment

        When the Schedule of Values is approved, it becomes your Requisition for Payment Form.

        By the 25th of each month, submit your Requisition on the same form by indicating on the right side of the form the amounts completed for each line item. Compliance with this procedure is mandatory if your payments are to be processed for the respective period.

Please call if you have any questions.

Very truly yours,

COMPANY

_____
Project Manager

                                          cc: File:_____, CF
                                              A/P

4.8.7
Sample Subvendor Schedule of Values Approval/Rejection Form Letter

```
┌─────────────────────────────────────────────────────────────┐
│                                                               │
│                         Letterhead                            │
│                                                               │
└─────────────────────────────────────────────────────────────┘
```

Date: _____

To: _____
      _____
      _____

ATTN: _____

RE: Project:_____
          Company Project #: _____

SUBJ: Schedule of Values

Mr. (Ms.) _____:

Attached is:

____   Your Schedule of Values approved as submitted. Please use this form on all periodic Requisitions
        for Payment that are to be submitted to this office by the 25th of each month.

____   Your Schedule of Values marked to suggest more appropriate breakdowns, and/or with questions
        regarding the submission. Please revise and resubmit as soon as possible in order to avoid delay in
        processing your requisitions.

Please call if you have any questions.

Very truly yours,

COMPANY

_____
Project Manager

cc: A/P w/att.
      File: Subvendor File:_____
               Subcontracts, CF

Use it to expedite processing of problem submissions, and to confirm the correct paper trail and chronology for those cases where payment or payment timing to subvendors is affected.

4.9   Shop Drawing and Submittal Management

4.9.1   Operating Objectives of the Project Engineer

Shop drawings and other submittals for approval are required in virtually every bid package. It is the responsibility of the Project Engineer to:

1. Identify all submittals required to be processed for each bid package.

2. Coordinate the schedule of submittals with the Purchase and Award Schedule in order to ensure that all submittal and review processes are conducted within the time requirements of the progress schedule.

3. Police the performance of each subvendor to enforce compliance with all document requirements with respect to form, content, and time.

4. Review each subvendor's submittals for compliance with the contract documents and with its own subcontract or purchase order, considering scope of work and coordination with contiguous work.

5. Confirm that each submittal is in the proper form and includes the correct number of copies as required for all submission and distribution requirements with respect to both specified items and any special requirements for the submission of anything to be considered an equal or a substitution (refer to Section 3.8 for important related discussion).

6. Keep all submittals moving smoothly through each designer's office to ensure that everyone takes appropriate action within the "reasonable period of time" required in the specifications.

7. Keep the design professionals' actions appropriate.

8. Be prepared to take quick, decisive action in the face of incorrect, late, or otherwise inappropriate submittal review action on the part of the Architect and Engineers.

9. Distribute copies of all submittals and other appropriate information to everyone who will need them—in good time as dictated by the progress schedule.

For important directly related discussion refer to:

- Section 2.4, Files and File Management
- Section 3.5.6, "Performance" and "Procedure" Specifications
- Section 3.5.16, Proprietary Specifications
- Section 3.7, Shop Drawing "Approval"
- Section 3.8, Equals and Substitutions

4.9.2 Submittal Log

The Submittal Log is a collection of forms to be used on each project to catalog each document's flow through the entire approval and distribution process. It is the place in which to record each submittal's respective characteristics along with the resulting action by each party and all distribution information.

It provides at a glance:

- The listing of all submittals as required (this will serve as an important checklist to prevent overlooking any requirement)
- The chronological order of submissions
- The required and actual dates of each submission
- The date by which designer action is needed
- The actual Owner or designer action and dates
- The required and actual distribution

The log will be an important document control tool that will be key to expediting the complete and comprehensive submittal control effort.

4.9.3 Submittal Log Procedure

1. Place a supply of forms in a 1-inch three-ring loose-leaf binder. Depending upon the size of the project, it may be reasonable to combine this log with the Subcontractor Summary and Telephone Log of Section 4.4 in the same binder.

2. Arrange the forms by specification section or by bid package number, identified in the same manner as the General Project File described in Section 2.4.4.
   a. This correlates the Submittal Log directly with the complete project file.
   b. It provides a research method based on the specification as a second way to research subcontractor information (the first being the alphabetical method of the Subcontractor Summary).

3. Research each individual specification section for each bid package. List all specific submissions required on the respective vendor's log form.

4. Determine any other submittal not yet specifically requested that you may require for proper coordination of the work. Add to each vendor's list as the shop drawing approval process for all subcontractors proceeds. The process itself will often identify such necessary or desired submittals.

5. Insert the name of each Subcontractor who will be requiring a copy of the approved submittal for coordination in one of the headings "_____ COPY."
   a. For example, if you know (or have been told) that the Concrete Subcontractor will need the Steel Subcontractor's anchor bolt layout, insert the name of the Concrete Subcontractor in one of the columns in the Steel Subcontractor's log form. Place a check mark in the corner of the Concrete Subcontractor's distribution box adjacent to the anchor

bolt submittal line. This will remind you of the distribution requirement when you are finally processing the submittal. Upon receipt of the approved anchor bolt drawing from the Architect, note in the same box the date when the drawing was finally distributed to the Concrete Subcontractor.

6. As you receive information requests or distribution requirements for submittals during the overall shop drawing process, include those names in the respective distribution sections.

   a. For example, if you receive an early request or correspondence from the Concrete Subcontractor requesting anchor bolt locations for light poles that are being provided by the Electrical Subcontractor, immediately turn to the Electrical Subcontractor's Submittal Log Form. Insert the name of the Concrete Subcontractor at "_____ COPY" at the top of a distribution column and place a check next to the submittal item for the light poles. This will provide an automatic important reminder to send the submittal to the Concrete Subcontractor when it finally goes through the system.

7. Insert the dates by which the respective submittals are required. Get the information from your construction schedule and/or your purchase/award schedule. Use "early finish dates" if your scheduling system identifies such information.

8. Have the Submittal Log for the respective vendor in front of you *while* you are writing the transmittal either to the design professional or back to the vendor. After completing the transmittal, *immediately* transfer all relevant information to the Submittal Log. If you don't do it at that moment, it will never get done.

4.9.4 Sample Submittal Log Form—Description of Terms

The procedure for the use of the Submittal Log Form is described in Section 4.9.3. Included here are the definitions of the terms included on the actual forms provided in the following sections.

1. *COPIES.* The number of copies required to be submitted for approval that will allow *complete* subsequent distribution. Insert the actual number received.

2. *DESCRIPTION.* Title of the drawing being submitted.

3. *COMPANY DRAWING NO.* If required by the specification or by project management, submissions may be stamped to correlate with the applicable specification section or bid package, or they may simply be identified in their chronological order of submission. Both approaches can also be combined in a single numbering system as, for example:

   - Section 07210
   - 07210-2 Second submission within that section

   If no particular numbering system is used, leave the box blank.

4. *SUB DRAWING NO.*   Insert the actual shop drawing number indicated in its title box whenever there is one that positively identifies the document.

5. *REQ'D BY.*   Insert the date by which submission of the construction schedule is required. Use the "early submission date" if your schedule or purchase/award systems will provide such information.

6. *APPR REQ'D BY.*   Insert the date by which the architect approval is required. As above, use the "early submission date" if your schedule or purchase/award systems will provide such information.

7. *FIRST SUBMISSION.*   Insert the date when the respective submission is initially transmitted from the Company to the Architect for approval. The date must correspond with the date on the transmittal form used.

8. *RETURNED FROM ARCHITECT.*   Insert the date when the respective submission is *received* back from the Architect after review or action.

9. *SECOND SUBMISSION.*   Insert the date of any resubmission of an item requiring correction and resubmittal, either for final approval or for the record. Refer back to *FIRST SUBMISSION.*

10. *ACTION/ACTION CODE.*   Identify the action taken by the Architect on the respective submittal; for example, A = approved, AN = approved as noted, and so on. Add actions and codes as the specific project requires; for example, R = rejected, NM = note markings.

11. *FILE COPY.*   Insert the date that a copy of the respective submission is filed into your own General File (refer to Sections 2.4 and 4.3). Every submission with any action or inaction is to be kept in the General Project File.

12. *JOB COPY.*   Insert the date when a copy of the respective submission is sent to the project Superintendent for use in construction of the work. Distribute to the field *only* submissions actually to be used for construction.

13. _____ *COPY.*   Insert the date when the copies of the respective submittals are transmitted to other Subcontractors and suppliers requiring the submittal for coordination of their own work. Note the name of the Subcontractor or supplier at the _____ COPY heading.

4.9.5   Sample Submittal Log Form—Completed Example (*page 4.61*)

A completed example of the sample submittal Log Form follows through on the procedures described in Sections 4.9.3 and 4.9.4. Section 4.9.6 provides a blank form to be photocopied for actual use.

4.9.6   Sample Submittal Log Form—Blank Form (*page 4.62*)

4.9.5
Sample Submittal Log Form—Completed Example

SUBMITTAL SUMMARY RECORD

Project: PLAINVILLE LIBRARY   Sub: QUALITY STEEL CORP.   Phone: (203) 555-6450   Sect(s): OS100

Proj. #: 9415   914 LONG HILL RD.   Fax: (203) 555-9450   OS200

Owner #: A-244   GREENVILLE, CT 06555   Contact(s): JOHN IRONHEAD   OS300

ACTION CODE:

A	Approved	ANR	Approved as Noted; Rev & Resub.	NX	No Exceptions Taken
AN	Approved as Noted	NA	Not Approved	NM	Note Markings
FYU	For Your Use	R	Rejected	NMR	Note Markings; Rev & Resub.

DOCUMENT		DOCUMENT #		1st SUBMISSION			2nd SUBMISSION			DISTRIBUTION		
Quant	Description	Company	Sub	Sent	Ret.	Action	Sent	Ret.	Action	File	Site	(COL.) ( )
9	A.B. #1	OS100·1	A.B-1	1/4/94	1/24/94	NMR	2/15/94			1/24/94	1/24/94	1/24/94
9	A.B. #2	OS100·2	A.B-2	1/4/94		NM						
9	ERECTION DWGS	OS100·3	E-1	1/11/94		NM						
9	DETAILS	OS100·4	S-1	1/11/94		NX						
		OS100·5	S-2			NX						
		OS100·6	S-3			NM	2/15/94			1/24/94	1/24/94	1/24/94
		OS100·7	S-4			NMR						
		OS100·8	S-5			NM						

Sample Submittal Log Form—Blank Form

SUBMITTAL SUMMARY RECORD

Project: _____ Sub : _____ Phone: _____ Sect(s): _____

Proj. #: _____ Fax: _____

Owner #: _____ Contact(s): _____

ACTION CODE:

A	Approved	**ANR**	Approved as Noted; Rev & Resub.	**NX**	No Exceptions Taken
AN	Approved as Noted	**NA**	Not Approved	**NM**	Note Markings
FYU	For Your Use	**R**	Rejected	**NMR**	Note Markings; Rev & Resub.

DOCUMENT		DOCUMENT #		1st SUBMISSION			2nd SUBMISSION			DISTRIBUTION		
Quant	Description	Company	Sub	Sent	Ret.	Action	Sent	Ret.	Action	File	Site ( ) ( )

4.10   Submittal Requirements and Procedures

4.10.1   Action Responsibility

The General Contractor's interest in reviewing submittals for approval is first to confirm compliance of its vendors and subvendors with its own agreements. Furthermore, the General Contractor and the Architect share the responsibility for shop drawing review from the project liability perspective.

The General Contractor receives, reviews, and "approves" each submittal and forwards it to the Architect and the Engineers for their approval to incorporate the item(s) into the work.

The individual responsibilities of each party to the process must be made clear in order to:

1. Expedite the process.
2. Enforce compliance with all subcontract or purchase order requirements.
3. Keep relationships between vendors and subvendors clear.
4. Keep all responsibilities intact.
5. Ensure that all documents are complete in all respects.
6. Force prompt, correct action on the part of all parties.
7. Distribute (coordinate) all information.

Section 3.7 reviewed important approval responsibilities of the design professionals, and Section 4.9 developed the file system for shop drawing and submittal management. This section will provide the system to:

1. Expedite the submissions from the subcontractors and suppliers.
2. Review each submittal for compliance with the subcontract/purchase order agreement and with the contract documents.
3. Keep all subcontractor and supplier responsibilities with respect to the material, installation, and submittal requirements themselves where they belong.

4.10.2   Submittal Responsibility

The shop drawing preparation and submittal procedure is normally detailed in the contract documents. Specific requirements should include size of the documents, number of copies, information required, and even some distribution requirements.

The Pass-Through Clause (see Section 3.5.8) and the opening remarks of every technical specification section will tie the Subcontractor or supplier to the general and supplementary general conditions, which may include the overall requirements. Beyond that, the individual specification sections themselves may have specific lists of items to be submitted.

The first actual notification of specific submittal requirements should be in the subcontract or purchase order itself. The specific *date* of complete compliance should be specified therein. It is, however, in project management's best inter-est to continually offer assistance to a reasonable extent. Such assistance, how-ever, should be informational only.

If any material or equipment is being submitted as "equal" or as a "substitu-tion," do not take these categories and the contractual meaning of these words for granted. There may actually be worlds of differences in each submittal process. There may even be differences in the Owner's procedure as well. Know how the contract specifically defines these terms, along with the specific procedure for their treatment. Refer to Section 3.8, Equals and Substitutions, for more in this regard.

4.10.3 Sample Letter to Subcontractors Regarding Submittal Requirements *(page 4.65)*

The sample Letter to Subcontractors Regarding Submittal Requirements that follows will first help you to increase your own familiarity with the technical requirements and highlight them for your Subcontractors. It will make it easier for your Subcontractors to be aware of the requirements, and of the fact that the Company takes them very seriously. The letter will later serve as a convenient checklist for follow-up.

The letter is to be filled out and sent to each Subcontractor and supplier on or about the time that the respective subcontract or purchase order is sent. It will thereby:

1. Call attention to all required submittals.

2. Serve as first notice that the party is absolutely expected to comply with all requirements.

3. Serve as additional notice (besides the subcontract itself) that the party is responsible for all information contained in the plans and specifications.

4. Advise of where the party can get all relevant information (plans and specs) in order to avoid any possible claim that they don't have them available.

5. Serve as notice that all approved submittals of contiguous work required by the Subcontractor and supplier for coordination of their submittals are avail-able for inspection at the jobsite (your failure to have directly sent that Subcontractor its own copy is therefore no excuse).

6. Require that your job number be included to help with your own file procedures.

In order to use this letter:

1. Review the Owner's specification requirements for all submittals as they are included in the respective subcontract or purchase order. Indicate them in the letter.

2. Review the subcontract or purchase order and its standard and special con-ditions. Indicate these special requirements in the area provided in the letter.

3. Send the letter, making sure to send copies to the Site Superintendent and other relevant project personnel.

4.10.3
Sample Letter to Subcontractors Regarding Submittal Requirements

---

Letterhead

Date: _____

To :   _____
          _____
          _____

ATTN: _____

RE: Project:_____
          Project #:_____

SUBJ: Submittal Requirements
          Section(s)_____

Mr. (Ms.)_____:

The contract requires your submission of the following items. Note that this list is for your convenience only; the omission here of any item does not relieve you of the requirement.

Please submit:

___	Certificate of Insurance	___	Installation Instructions
___	Performance Bond	___	Delivery Time After Approval
___	Labor & Material Payment Bond	___	Erection/Installation Time
___	Certified Payroll Reports	___	Tests/Inspection Reports
___	Payroll Ledgers	___	Guarantees/Warranties
___	Shop Drawings:_____Copies	___	Full Lien Waiver Forms
___	Erection Drawings	___	As-Built Plans & Specs
___	Product Specifications	___	_____
___	Employer EEO Reports	___	_____
___	MSDS Sheets	___	_____
___	Samples	___	_____

Please be aware of the proper form, content, and time requirements for each submission, and comply in every respect. Contact me immediately if you have any questions.

Thank you for your cooperation.

Very truly yours,

COMPANY

_____
Project Manager   cc: File:_____, CF

4.10.4   Submittal Review Checklist *(page 4.67)*

It is project management's responsibility to review the shop drawings, product data, and samples to confirm their compliance with the criteria set forth in Section 4.10.3. The actual process performed by those individuals responsible should become a matter of detailed routine.

Using a checklist each time will help you to:

- Avoid overlooking important considerations
- Make certain that all approval submissions meet all requirements
- Force prompt, correct action on the part of the design professionals
- Coordinate each in a timely manner with all those who need the information
- Follow up on resubmissions and noncompliances

Beyond this, it is a helpful practice during your reviews to use markers of different colors than the design professionals (typically red). You might even suggest to the designers that they use different colors among themselves. This will make it very easy later on to confirm the origination of any remark.

4.10.5   Sample Form Letter to Subcontractors Regarding Shop Drawing Resubmittal Requirements *(page 4.68)*

The sample Form Letter to Subcontractors Regarding Shop Drawing Resubmittal Requirements that follows is to be used to follow up on those submittals that for whatever reason have been returned to the Subcontractor or supplier for correction and resubmittal for final approval. It is an effort to compel an unresponsive Subcontractor or supplier to devote proper attention to *timely* compliance. Specifically:

1. It "reminds" the recipient to devote more attention to the project. It makes the recipient aware of the critical nature of his or her prompt action, and that you're not about to relax until the process is completed.

2. It documents your efforts to coordinate the work. These kinds of records will become extremely important in the event that untimely or inappropriate action on the part of the subvendor results in slipping material deliveries. It will also provide support for backcharges or other more serious claims.

3. It notifies the party of potential backcharges that may result from inattention to the project.

To use the letter effectively, review the Submittal Log of Section 4.9 periodically. Immediately send out the letter to Subcontractors and suppliers who are delinquent in their responses to the resubmission requirements. Sending "second-request" and "third-request" copies of the original letters in case of continuing lack of response prompts more drastic action.

4.10.4
Submittal Review Checklist

SUBMISSION REQUIREMENTS

All approval submissions contain:
1. Project Title & Job Number ____
2. Contract Identification ____
3. Date of Submission (or Revision) ____
4. Dates of Previous Submissions ____
5. Names of contractor, supplier and/or mfgr. ____
6. Identification of all products with specification section numbers ____
7. Field dimensions - clearly identified as such ____
8. Relation to adjacent and/or critical features of the work ____
9. References to applicable standard specs ____
10. *Clear* identifications of deviations from the Contract Documents ____
11. All other pertinent information as may be required by the specifications or the Company, such as:
    - Model Numbers ____
    - Performance Characteristics ____
    - Dimensions & Clearances ____
    - Wiring or Piping Diagrams ____
12. Manufacturer's standard drawings include:
    - Modifications to delete information not applicable to this project ____
    - Supplemental information specifically applicable to this project ____
13. Check the specifications for add'l requirements ____

SUBMITTAL REVIEW PROCEDURE

1. Ensure that subcontractors and suppliers submit materials promptly ____
2. Determine and verify:
    - That the sub has incorporated and will guarantee all field dimensions ____
    - All field conditions and construction criteria have been accommodated ____
    - That the product either complies with the specification requirements in every respect, or that every deviation has been properly identified, and includes its respective complete explanation/justification ____
3. Coordinate each submittal with both field and contract document requirements ____
4. Research and confirm all "justifications" for any deviation from the contract documents. Do this *before* submitting the documents for approval ____
5. Determine if a credit or addition to the contract is in order, based upon any changes in the submission ____
6. Determine if any backcharges to any other subcontractor or supplier are in order as a result of changes required by this item ____
7. Determine that the submission is timely, and that the material conforms to required deliveries ____

8. Positively identify by responsibility all "Not By Subcontractor" or "By Others" kinds of remarks. Correct as necessary *before* submission to the architect for approval. ____
9. Compare all resubmissions with the file copy of the previous submission. Confirm that all required corrections have been made. ____

DISTRIBUTION

1. Upon receipt of submittals bearing the stamp indicating architect action, distribute copies to:
    - Jobsite File ("For Construction" documents only)
    - Record Documents File ____
    - Other affected subcontractors and suppliers ____
    - The supplier or fabricator ____
    - Anyone else who may need the information in order to coordinate the work properly ____

FOLLOW UP

1. Monitor the time that it takes for the approval process.
    - Be sure that the architect is giving proper, timely attention ____
    - Be sure that all delays and other inappropriate actions are duly noted in the correspondence ____
2. Be certain that the design professionals:
    - Include all information required of them by way of questions in the submittals ____
    - Do not overstep their authority ____
    - Do not overstep their professional capacities ____
    - Do not add work without regard for the established change order procedure ____
    - Include only meaningful action that will allow proper completion of the submittal ____
    - Affix the *accepted* stamp and initial/sign it ____
    - Clearly indicate any requirements for resubmittal, or approval of the submittal ____
3. Upon distribution of the submittal back to its originator:
    - Reconfirm the delivery schedule(s) ____
    - Confirm that the submission is being returned in good time for the subcontractor or supplier to meet its own requirements ____
    - Note any significant information for the next construction schedule update ____
    - Begin any actions that may be necessary to resolve problems that may have been exposed by the review process ____
    - Begin any necessary change order procedure ____
4. Be sure that the Submittal Log form is used and maintained *as each part of the process is completed*. Complete the respective log entry

4.10.5
Sample Form Letter to Subcontractors Regarding Shop Drawing
Resubmittal Requirements

Letterhead

Date: _____

To: _____
_____
_____

ATTN: _____

RE: Project:_____
           Company Project #: _____

SUBJ: Shop Drawing Resubmission Requirements
           Section(s) _____

Mr. (Ms.) _____:

Copies of your shop drawings *requiring immediate correction and resubmission for approval* have been returned to you on the dates listed below. Immediate resubmission is required in order to avoid additional delays and related costs resulting from untimely action.

Drawing No.	Date Returned to Your Office	Description
_____	_____	_____
_____	_____	_____
_____	_____	_____
_____	_____	_____
_____	_____	_____
_____	_____	_____
_____	_____	_____

Please contact me immediately to confirm the date(s) of your resubmission(s). Time is of the essence.

Thank you for your cooperation.

Very truly yours,

COMPANY

_____
Project Manager

cc: Jobsite
File: Vendor File_____
        CF

4.10.6   Reproductions of Submissions—Subvendor Responsibility

The proper quantity of submittal documents is specified in the subcontract, the purchase order, or elsewhere in the contract documents. On a public project, for example, it will be common to require sufficient numbers of submittals that will provide for final distribution to:

- The Architect (home and field?)
- The Owner's administrative and field offices
- The Company's home and field offices
- The Company's general file
- The Subcontractor's or supplier's home and field offices

Beyond this, any engineer or consultant will need one or two copies, and all Subcontractors needing the submittal for coordination of their own work will need an office and a field copy.

It is easy to see, for example, that it may be necessary to have up to fifteen copies of the HVAC duct shop drawing. Even the simplest submissions may need upward of ten copies for only basic contract distribution.

Too many Subcontractors and suppliers, however, take it upon themselves to submit only a few copies of each submittal—obviously because of the inconvenience and cost. If this abuse is allowed, it will interfere with the otherwise smooth flow of document distribution, and will greatly add to the cost of reproductions for the Company.

It is project management's responsibility to enforce each submittal requirement. The most effective way to this enforcement is to assign a cost to the service.

Because the objective is to enforce the Subcontractors' and suppliers' compliance and *not* to go into the copy business, the charges assigned to the effort should intentionally be very high. They should raise a little anger and shock the offending party into compliance. It should become quickly evident to them that it is much less of a problem to just fulfill their responsibilities than to argue over large backcharges.

4.10.7   Sample Reproduction Backcharge Notice
Form Letter *(page 4.70)*

The sample Reproduction Backcharge Notice Form Letter follows through on the ideas of Section 4.10.6. Specifically, it:

1. Highlights the contractual requirements for proper, complete submissions.
2. Notifies of the intention to levy backcharges and associated costs.

After having sent out this letter, be prepared for an immediate and abrupt phone call. Stay calm and simply advise the party that failure to provide the correct submission is truly an interference. The party can avoid these charges by simply delivering what the party agreed to.

4.10.7
Sample Reproduction Backcharge Notice Form Letter

Letterhead

Date: _____

To : _____
  _____
  _____

ATTN: _____

RE: Project:_____
    Project #:_____

SUBJ: Failure to Provide Proper Content of Submissions

Mr. (Ms.)_____:

Per your contract, it is your responsibility to be sure that the proper number of copies for all required submittals be sent on time to be processed. Failure to comply interferes with the expedient flow of documents, and results in significant additional production costs.

Be advised that all copies made by this office of any of your submissions because of insufficient quantities necessary for proper distribution will be charged to your account as follows:

    8-1/2" x 11" . . . . $ 1.00 Each Leaf
    8-1/2" x 14" . . . . $ 2.00 Each Leaf
    11"   x 17" . . . . $ 3.00 Each Leaf

Prints:
    18"   x 24" . . . . $ 5.00 Each Leaf
    24"   x 36" . . . . $ 8.00 Each Leaf
    Larger   . . . . $ 10.00 Each Leaf

Before final payment is released, your account will be adjusted to reflect the number of copies charged.

Very truly yours,

COMPANY

_____
Project Manager

cc: Jobsite
    Vendor File:_____
    Backcharge File
    CF

4.11 Subcontractor Delivery Requirements

4.11.1 General

After the approval process has been completed, focus must be shifted to actually securing the materials within acceptable time frames. This section reviews techniques to expedite the delivery of materials and enforce compliance with all Subcontractor and supplier delivery requirements.

4.11.2 Expediting Subcontractor/Supplier Fabrication and Delivery Schedules

Begin with the attitude that it is *always* necessary to expedite *everything*. The construction business is founded on the squeaky-wheel concept. If your wheel isn't squeaking, you can be sure that all the other wheels with which your Subcontractors are working are. If you keep *your* wheels squeaking, you'll get *your* material on time.

This section identifies considerations and activities that may be applied in various circumstances to help avoid delivery problems in the first place, or to help bring problem deliveries back on line. They are not to be considered to be in any particular order, but as a menu from which to select appropriate actions.

1. *Identify the true source.*   Begin with the subcontract itself to require the disclosure of all actual sources of all materials and components. Continue with the Subcontractor/Supplier Reference Form of Section 4.5.

2. *Establish contact with the direct source as early as possible.*   Know early whom to call to directly verify all information handed to you by your first-tier Subcontractor. Develop the relationship as soon as possible.
   a. Explain that if vendors want to be sure that they get paid properly, it is in their best interest to communicate directly with you:
      (1) *You are* the one with the payment bond.
      (2) *You are* the one who will be processing payments, with the power to determine joint-payment structures.
   b. In most cases your immediate result should be that the reliability of your delivery information will improve dramatically. You'll also be able to use this direct information as a gauge to help determine the reliability of the delivery information for other items that you'll be getting from the same first-tier Subcontractor.

3. *Confirm the relationship history of the subsupplier with the first-tier Subcontractor.* You may, for example, discover that there really is no relationship to speak of, and the supplier is really very happy to be talking with you.

4. *Understand the specific payment terms between the Subcontractor and its supplier.*   Too often they are dramatically different than those between you

and your Subcontractor. Even though it is not technically your problem, these cases can have a tendency to introduce additional conflicting forces between you and your Subcontractor.

    a. One consideration may be to offer subsuppliers a copayment agreement in return for allowing them a pay-when-paid. This might go a long way in relieving pressure during normal production cycles.

5. *Consider visiting the fabrication location.*

    a. When you're told "the parts are on the shop floor," consider responding with "great; I'll be there this afternoon—you can show them to me." You'll thereby either truly satisfy yourself that the information is real, or you will have exposed an all too common "check-is-in-the-mail" syndrome of this industry. At the very least, you'll get your vendors to consider their answers to you more carefully.

    b. Even when all information is absolutely real, just the act of visiting a remote location to see your items communicates your genuine concern for the efforts and hard work on the part of that vendor to get you your things on time. Many times it may need nothing more than this appreciation to give just the added impetus that keeps *your* material fabrication moving along. Without any doubt, shop visits have been *the* most powerful and consistently successful of all expediting efforts.

6. *Point out that if the item is delivered by the end of the month, it will go on that month's requisition to the Owner.* In other words, if the item comes a week later, everyone is going to have to wait an additional 30 days to even submit the invoice.

7. *Identify whether the shipment is arranged direct or by common carrier.* Many shipments either are waiting for a full truck or will be consolidated by common carrier. It is often a reasonable upcharge to get the materials on their own direct truck.

8. *Consider accelerated payments.* If, for example, the supplier is on a 30-day or even a pay-when-paid payment term, consider offering a c.o.d. or other advanced form of payment. Don't necessarily give it away, however. Even though it will speed up delivery, it may also be justification for a further price discount.

9. *Consider advising the vendor that you're considering changing the item.* If the delay is the fault of the vendor, consider doing what you can to develop the idea that it might lose the order altogether. Do this with extreme caution. Make it believable, and *actually be ready to do it,* or you might actually be digging a deeper hole if the vendor drops *you.*

Throughout these efforts, those that become necessary as a result of the primary Subcontractor's failure will be chargeable to that Subcontractor under the acceleration and backcharge provisions of the subcontract agreement. Refer to Section 4.17, Backcharges, for more in this regard.

4.11.3   Sample Subcontractor/Supplier Delivery Requirements Form Letter *(page 4.74)*

The sample Subcontractor/Supplier Delivery Requirements Form Letter is designed to provide a record of proper notification of required deliveries. It should become a matter of habit to return the individual submission of the shop drawing to the Subcontractor after approval action to indicate expected delivery and request positive confirmation (see Section 4.6, Transmittal Form Letter Procedure and Use). If this had been done, the Delivery Requirements Letter will become at least the *second* such notification.

If acknowledgment is not received by the date indicated in the letter, stamp a copy of the letter with "Second Request" and indicate on the second letter that failure to respond is interfering with project coordination.

4.12   Request for Information

4.12.1   General

The Request for Information (RFI) will actually become either a request for information or a request for confirmation of a clarification. It is a procedure to be used in cases where it is necessary to:

1. Confirm your interpretation of a detail or other understanding as to the way a component of the work should proceed.

2. Secure the written direction or other clarification from the appropriate party that is necessary in order to allow the work to proceed.

The Request for Information becomes a crucial element of the complete project record. Learn to express your thoughts on it, and your project documentation will improve dramatically.

4.12.2   Use of the RFI Form Letter

1. The RFI is to be used in all cases for information clarification involving the design professionals, Subcontractors, suppliers, and Owner representatives. Use the chronological order of RFI initiation as the RFI #.

2. Summarize the issue in Section 1 of the RFI. Be specific; refer to plan details, specification section numbers, and so on. If it is a confirmation of a conversation or other direction, specifically state it. Name names; include dates.

3. Do the best you can to actually confirm the answer to the question first. This, of course, will greatly speed up your ability to proceed with the work. The form will then be used as confirmation.

4. If you are unable to confirm the answer *now,* send the form with the written statement of the issue to the responsible party with Section 3 checked. Use fax, hand delivery, or Express Mail if necessary. Do whatever it takes to get the form into the hands of the other party immediately. Stress the urgency of a quick but complete response.

4.11.3
Sample Subcontractor/Supplier Delivery Requirements Form Letter

---

Letterhead

Date: _____

To : _____
_____
_____

ATTN: _____

RE: Project:_____
Project #:_____

SUBJ: Delivery Requirements
Section(s):_____

Mr. (Ms.)_____:

Copies of "Approved" or "Approved as Noted" Shop Drawings were returned to you on the dates listed below.

Per the current construction schedule, delivery of each item is required on the respective dates indicated in order to avoid delays and related costs resulting from nondelivery.

Drawing No.	Description	Date Returned to Your Office	Required Delivery Date of Material
_____	_____	_____	_____
_____	_____	_____	_____
_____	_____	_____	_____
_____	_____	_____	_____
_____	_____	_____	_____
_____	_____	_____	_____
_____	_____	_____	_____
_____	_____	_____	_____
_____	_____	_____	_____
_____	_____	_____	_____

Please confirm your ability to meet these delivery dates by signing this letter where indicated below and returning it to my attention. If any item will not be delivered by the required date, submit your complete explanation. Your complete response is required by _____, 20_____.

Very truly yours,

COMPANY

_____
Project Manager

Delivery Dates Confirmed:

By:_____Date:_____

**4.12.3   Sample Request for Information (RFI) Form
Letter—Completed Example (*page 4.76*)**

**4.12.4   Sample Request for Information (RFI) Form
Letter—Blank Form (*page 4.77*)**

4.12.5   Use of the RFI Tracking Log (*page 4.78*)

The RFI Tracking Log can be kept in its own section of the binder being used for the Submittal Log of Section 4.9, adding additional log forms as necessary.

1. Log the information *as it occurs*.
2. Indicate the chronological order of the RFI as the RFI #.
3. Complete all "initiation" information as of the RFI preparation:

   TO   The recipient.

   DATE   Date of the RFI.

   CONFIRM   Check if the RFI is confirming a clarification or direction.

   ACTION REQUEST   Check if the RFI is requesting direction.

4. Periodically (often) review the log to identify outstanding information or outstanding written acknowledgments of your confirmations.
5. Routinely send "Second Request" copies of the original RFI.
6. Complete the ACTION REC'D information as it is received and keep the log current at all times.

4.13   Project Meetings

4.13.1   General

Project meetings are generally divided between regular periodic job meetings and special meetings held to address specific unique problems or circumstances. In either case they should be treated the same in terms of their arrangement, participants, rules of conduct, and documentation.

The discussions of this section will apply to all meetings, in particular regular job meetings. Additional considerations applying to special meetings are developed in Section 4.13.14.

4.13.2   Meeting Purpose

Job and special meetings are there to *solve problems*—not just to rehash the same items that you discussed last week. These meetings are critically important to the quick and complete resolution of every item affecting the project—if they are *managed* and not avoided. They are not simply there to record history, but to force action, pinpoint accountability, and support your actions.

4.12.3
Sample Request for Information (RFI) Form Letter—Completed Example

---

Letterhead

Date: JULY 22, 1994   REQUEST FOR INFORMATION: RFI #: 16

To: STATE ARCHITECTURAL UNIT   Project: BURLINGTON POLICE TWR
     165 BURECRATIC BLVD,   Project #: 9550
     CITY, STATE 00000
                                      _____ Hand Delivered   X Mailed
ATTN: JOHN ARCH   X Faxed: 666-9999

Subject: Section (s): 03200 CONCRETE FOUNDATION LAYOUT
             Specification/Plan References: S-1, A-4

1.   Problem/Information Requested:
     DWG S-1 DET. 6 REQUIRES THE FROST WALL AT COL. LINE 3
     TO BE 14" THICK.

     5/A-4 REQUIRES THE SAME WALL TO BE 12" THICK (IN ORDER
     TO WORK PROPERLY WITH THE BRICK SHELF)

                              Information Requested By: MARK LEONARDO

2.   Response Confirmation:
     YOU CONFIRMED IN OUR 9:40 AM TELECON TODAY THAT
     DETAIL 5/A-4 WILL BE FOLLOWED
     (6/S-1 WILL BE CORRECTED ON THE AS-BUILT DRAWINGS)

                              Approved By:_____ Date:_____

3.   Clarification/Action; please respond by or before _____, 19____ in order to
     minimize delay or interference with the ability to proceed with the work:

                              By:_____ Date:_____

cc: Superintendent
     OWNER ON-SITE REP
     RFI #16
     File: CONC_____, CF

4.12.4
Sample Request for Information (RFI) Form Letter—Blank Form

Letterhead

Date: _____   **REQUEST FOR INFORMATION: RFI #:___**

To: _____   Project: _____
_____   Project #: _____
_____

ATTN: _____   _____ Hand Delivered   _____ Mailed
_____ Faxed: _____

Subject: Section (s): _____
Specification/Plan References: _____
_____

1.   Problem/Information Requested: _____
_____
_____
_____
_____
_____
_____

Information Requested By:_____

2.   Response Confirmation: _____
_____
_____
_____
_____
_____
_____

Approved By:_____   Date:_____

3.   Clarification/Action; please respond by or before _____, 20_____ in order to
minimize delay or interference with the ability to proceed with the work:
_____
_____
_____
_____
_____
_____

By:_____   Date:_____

cc: Superintendent
_____
_____

File:_____, CF

Use of the RFI Tracking Log

RFI TRACKING LOG

Project: _____

No: _____

| RFI # | SUBJECT | INITIATION | | | | ACTION RECEIVED | | REMARKS |
		TO	DATE	CONFIRM	ACTION REQUEST	Complete ?	DATE	

4.13.3 Day and Time

Mondays and Fridays will be sparsely attended. Even if you can manage participation, attention spans will be divided, and you'll never be able to maintain proper consistency from meeting to meeting without constant bird-dogging of all required participants.

Tuesdays or Thursdays are best. Whenever the rest of the world schedules anything, it is sure to be on Wednesday, creating the inevitable conflict. Wednesdays do, however, remain much better than Mondays or Fridays.

Insist on morning meetings, starting between 10:00 and 11:00 A.M. This will:

- Allow a few precious minutes for last minute preparations
- Give people a chance to get to the meeting without having to fight morning or noon hour traffic
- Catch attendees before the rest of the day is allowed to interfere with their schedules, thereby improving meeting attendance
- Leave the rest of the day to act on critical issues before they're allowed to cool
- Improve your chances of catching someone else "in" if you need to phone or visit them today to resolve something
- Leave time to force others to change *their* plans for the rest of the day in order to resolve a current problem
- Keep everyone working toward getting through the agenda and finishing the meeting

The closer you get to lunch, the quicker things seem to get resolved. In marked contrast, meetings held after lunch are conducted in slow motion. They're sure to be sparsely attended, move at half-speed, and close with most of the original agenda still intact. You can't have a full head and a full stomach at the same time—one of them has to be empty.

4.13.4 Location

Always meet at the site. Don't waste time arguing over whose project it is, or get caught in the one-upmanship game of showing would-be authority by fighting over meeting location. Heads of bureaucratic government agencies, for example, have a habit of trying to arrange for all regular and special meetings at their own offices or at some other location convenient for them.

The jobsite is where the issues live. Get out of the field office, walk to the problem, point at it. Misunderstandings will clear up in seconds, and you'll move closer to resolution.

4.13.5 Participants

The absolute minimum will be the Owner's representative(s), the Architect, construction field personnel (Superintendent, Project Manager), and construction administration (Project Manager, Project Engineer). It unfortunately is getting to be too common on small and midsized projects for the mechanical, electrical, structural, or other special engineers not to attend, usually because the Architect retains these consultants on an hourly basis during the construction phase of the project. That is *their* problem, not yours. If you need or want the engineer present for any valid reason (including expediency), get him or her there.

The arrangement may otherwise work if you have no engineering problem (?!). If you do, however, their lack of attendance is your guarantee that a *minimum* of a week will be added to any related design resolution. If your agenda has design-related problems, insist on these individuals' attendance. If they then do not attend, you will have every right to call foul and highlight the potential schedule impact resulting from the extra delay in resolution.

4.13.6 Subcontractor Participation

Job meeting attendance is *mandatory* for all Subcontractors who:

- Are about to work
- Are working
- Just "finished" working
- Have any potential for being involved in any agenda item

Too many times, Subcontractors are allowed to avoid meetings because of the realization that they will have to sit there for an hour (maybe two) until their item comes up. Don't allow it. If a Subcontractor affected by any discussion is not present:

1. Issue resolution is delayed.
2. Extreme amounts of effort will be necessary to coordinate subsequent discussions between multiple parties.
3. *You* will remain responsible for the timely coordination of all information to those who are not there.
4. Even if you conduct proper coordination and distribution of relevant information, this adds significant time and effort.

In contrast, the time inconvenience of the individual Subcontractors is a small price to pay when compared with the fact that:

1. The issues were coordinated and resolved in minutes with all affected people present.
2. There is no time lapse in your coordination, or with your ability to immediately mobilize any subsequent action.

Refer to Sections 4.13.7 and 4.13.8 for help in inducing such attendance.

4.13.7   Sample Letter to Subcontractors Regarding Mandatory Job Meeting Attendance *(page 4.82)*

The sample Letter to Subcontractors Regarding Mandatory Job Meeting Attendance that follows is designed to induce Subcontractor participation in your meetings. Specifically, it notifies all project Subcontractors that:

1. It is mandatory that anyone performing or about to perform *any* work on the site is absolutely required to participate in all job meetings during that period.
2. Meetings will be held on the dates scheduled and will start *on time.* Attention is expected to be given to these requirements.
3. It is each Subcontractor's responsibility to be aware of all information as it relates to its work, and to make all efforts necessary to ensure proper coordination.
4. Each Subcontractor is absolutely responsible for all information contained in the job meeting minutes. This includes completeness, accuracy of description, noted commitments, and timetables.

4.13.8   Sample Letter to Subcontractors Regarding Lack of Job Meeting Attendance *(page 4.83)*

The sample Letter to Subcontractors Regarding Lack of Job Meeting Attendance is designed to deal with a particular Subcontractor who does not properly attend your meetings. Specifically, it confirms your conversation with that Subcontractor who, despite your coordination efforts, has failed to attend your meetings.

1. It emphasizes that the lack of attention is creating unnecessary interferences and inconveniences.
2. It states that interferences, delays, and additional costs resulting from the lack of attention will be entirely that Subcontractor's responsibility.
3. It reiterates that it continues to be the *Subcontractor's* responsibility to be aware of all project requirements as determined in those meetings, and to comply with them in every respect in a timely manner.
4. It "reminds" the Subcontractor of the next job meeting.
5. By copy of the letter it makes certain that the field representative's boss is aware of the absence and your feelings toward it. If that doesn't get some reaction, then you probably have other problems with that company.

4.13.9   Meeting Action Rules

1. Schedule meetings as *you* need them to be.
    a. During the job start-up, shop drawings are flying, construction and bid package coordination efforts are being compressed into small time frames, and resulting questions are multiplying.

4.13.7
Sample Letter to Subcontractors Regarding Mandatory Job Meeting Attendance

---

Letterhead

(Date)

To: (List all Project Subcontractors)

RE: (Project)
           (Company Project #)

SUBJ: Mandatory Job Meeting Attendance

Mr. (Ms.) (   ):

Your contract requires your participation in the regular job meetings. These meetings will begin on (Date) and will be held on alternating (insert day of the week). This schedule may be adjusted from time to time; it is your responsibility to be aware of the current job meeting schedule.

Each subcontractor is required to attend every job meeting prior to, during, and immediately after the work of the subcontract is being performed. This includes the submittal preparation/submission stage.

Please note that your attendance is *mandatory*. Your failure to attend will result in the need for excessive efforts by others to coordinate their work with yours. You will be held responsible for all information contained in the meetings, including timetables, commitments, and determinations of responsibility as set forth.

Thank you for your cooperation.

Very truly yours,

COMPANY

Project Manager

cc: Jobsite
       File: Meetings
               CF

4.13.8
Sample Letter to Subcontractors Regarding Lack of Job
Meeting Attendance

Letterhead

(Date)

To: (Subcontractor failing to
             attend a specific meeting)

RE: (Project)
             (Company Project #)

SUBJ: Lack of Job Meeting Attendance

Mr. (Ms.) (   ):

Per our conversation this date, your failure to attend today's job meeting as required is interfering with job coordination and completion. As you know, it continues to be your responsibility to be aware of all project requirements and to accommodate them completely and in a timely manner. Please be advised that you will be held responsible for all interferences, delays, and added costs resulting from this lack of attention.

The next job meeting will be held on (insert day and date) promptly at (insert time).

Very truly yours,

COMPANY

Project Manager

cc: Jobsite
        File: Vendor File:_____
                 Meetings
                 CF

      b. Insist on weekly meetings during these aggressive periods. Too often those responsible for acting on a job meeting item will not look at the item until the day before (or the morning of) the next job meeting. Deadlines seem to be the earliest date that you can hope for any action. Get these deadlines (the next meeting) as close together as possible.

      c. As the project settles into a pace, it's up to you if you think it will be OK to "relax" into biweekly meetings. If, however, you get any hint that too much time is spent between issue identification and resolution, immediately get back to the weekly schedule.

2. *Always* start on time—regardless of who is late and of how many times that person has been late.

      a. One or two times having to sneak into an ongoing meeting will usually solve the problem. If it does not, chronic offenders should be confronted at the meeting. Let those who do get there on time know that you appreciate their efforts, and that *none* of you appreciate the lack of consideration being demonstrated.

      b. If for some reason you cannot begin (the Owner's representative is driving in from another city, and the design professional refuses to start . . .), consider stating at the actual start of the meeting that you now need an adjustment of the agenda to hit the important topics because *you* must leave on time. *Get and keep control over the meeting. Let everyone know that your time is valuable, and that they need to learn to respect that. You've managed to keep your commitments; you have a right to expect others to do the same.*

3. Enforce mandatory attendance.

      a. Do not tolerate absence or neglect. When an expected attendee is missing, it disrupts the agenda and loses time. When the agenda item comes up, phone the person right in the middle of the meeting. If a speakerphone is available, put him or her on it. Preface your conversation with "we expected you here, but since you're not, we've got you on the speaker, so . . ." and move right into the issue. A mild reprimand for their lack of consideration (a strong one for repeated offenders) is definitely appropriate. Be matter-of-fact and businesslike. These people will be caught at least a little off-guard, and hopefully be embarrassed enough to be sure they won't put themselves through it again.

4. End each item with a *resolution*. If the issue itself is not finalized, end with a determination of a specific action to be made by a particular individual (by name) by a certain date. Nail it down.

5. Ongoing, comprehensive records must be kept in a way that keeps:

      a. The project record clear and correct

      b. Everything on the front burner

      c. Everyone accountable to and for his or her actions (and inactions)

Refer to Section 4.13.10, Meeting Minutes, for related action rules.

4.13.10   Meeting Minutes

All meetings must at least be recorded, but that in itself is not nearly enough. If the meeting is conducted properly:

- Each item is given its relevant place on the agenda (see Section 4.13.13).
- The *timing* of each item can be controlled.
- Each issue has either been completely resolved or ongoing items have had definite steps determined for resolution, along with naming those responsible and confirming action timetables.

The minutes themselves will:

1. Organize the agenda.
2. Establish the method of identification and correlation to the rest of the project record.
3. Keep everything *visible*—and organized so that unresolved items automatically get more irritating the longer they go unresolved.
4. Nail down accountability, that is, keep people directly identified and *personally responsible* for specific actions.
5. Display the cause and effect of timely and untimely actions.
6. Fulfill important notification responsibilities under the contract (see Section 3.5.15).

The following guidelines should be followed in recording the minutes:

1. *Use a standard form.*   Whether on a word processor or kept on standard forms made for the purpose, a standard layout should:
   - Display the project identification, meeting number, date, time, location, participants, distribution, and other requirements unique to the project
   - Prompt the recording of all relevant information while reducing the risk of oversight
   - Get everyone used to the information display

   This will improve the understanding of the issue identification and correlation features of the minutes and minimize the risk of someone overlooking his or her responsibilities or otherwise claiming ignorance.

2. *Identify each meeting numerically.*   If they are regular job meetings, identify each meeting in numerical sequence. If it is a special meeting, call it Meeting #1.

3. *Assign each item its own number that will never change.*   The third issue raised at job meeting #4, for example, will be identified as "4.3."
   - If you are conducting weekly job meeting #8, for example, and you find yourself still considering item 4.3 under "Old Business," you automatically know that the issue is four weeks old.

■ If you are a supervisor who does not attend the regular job meetings, simply reviewing any meeting minutes and comparing the item numbers with the meeting number will give you an instant (and sometimes painfully clear) indication of the way things seem to be going.

4. *Use a title for each item, and keep titles consistent each time the item is mentioned.* This will clarify the subject, speed research, and facilitate correlation with other topics.

5. *Include all appropriate references in the job title.* If it is the subject of a change order, bulletin number, etc., keep these identification numbers in the description.

6. *Be concise but complete.* Use outline format whenever possible.

7. *Require definite action.* Never leave any issue without the specific step-by-step program identified that will resolve it. Assign people and times to each step.

8. *Name names.* Do not say "the Owner will respond by. . . ." Instead, say "Mr. Dunn stated. . . ." Let everyone see their names in lights—it will be harder to make excuses.

9. *Insist on the precise accuracy of all statements as they are recorded.* If you are not keeping the minutes yourself, keep precise notes. If the minutes either represent an issue inaccurately or omit relevant discussion, highlight the correction at the next meeting. If it is not a regular meeting, immediately distribute your written correction.

10. *Include a "verification requirement" of all information contained in the minutes.* The minutes are an important job record that will be used to substantiate every cause and effect issue. If its accuracy is questioned, its usefulness will be compromised. At worst it may pervert the record in a manner that will hurt you and the Company.

11. Include as part of the standard form on every meeting record a statement to the effect that anyone noting any error or omission in the document is to notify the writer by or before the next meeting, or by a particular date, and that failure to do so constitutes acceptance of all information contained as it is represented. This will end many later arguments regarding the legitimacy of particular remarks.

4.13.11 Sample Job Meeting Minutes Form—Pages 1 and 2 *(page 4.87)*

The Job Meeting Minutes Form that follows is arranged to accommodate the requirements that must be provided for at each meeting. In addition, the form accommodates the objectives of this section by:

1. Providing appropriate areas to prompt the inclusion of all relevant job meeting identification and distribution information.

4.13.11
Sample Job Meeting Minutes Form—Pages 1 and 2

Project:
Proj. #:

Job Meeting No.:
Date:
Location:

Page 1 of 2

ATTENDING:		DISTRIBUTION	
Name	Company	Name	Company
		Attendees:	

NOTICE TO ATTENDEES AND MINUTES RECIPIENTS:
If any of the following items are incomplete or incorrect in any way, please notify the writer.
Failure to advise of such corrections by or before the next job meeting constitutes acceptance
of all information contained herein as represented.

SUBJECT	ACTION REQUIRED	
	By	Date

4.13.11
Sample Job Meeting Minutes Form—Pages 1 and 2 *(Continued)*

Project:
Proj. #:

Job Meeting No.:
Date:
Location:

SUBJECT	ACTION REQUIRED	
	By	Date

2. Encouraging documentation in accordance with the "outline form" recommendation.

3. Providing a convenient area for highlighting important necessary action, including person responsible and date required by.

4. Including the important notification to correct any errors or omissions in the record of noted discussions.

4.13.12 Sample Meeting/Conversation Record Form *(page 4.90)*

The Meeting/Conversation Record Form that follows is to be used where:

1. A verbal exchange has been more elaborate than should be included as only a note in the Telephone Log of Section 4.4, but not as elaborate as a formal meeting.

2. It is necessary to confirm instructions, notification to proceed with work in a certain way, or other distribution of information.

3. It is important to confirm the appropriate record of discussion to the participants and to others, to properly document the actual project record with regard to the subject items, and to verify the legitimacy of your record of the exchange.

Accordingly, the main difference between using this form and simply elaborating on your own notes in the Telephone Log is the distribution and confirmation of the information.

4.13.13 Meeting Agendas

1. *Concept.* If you control the agenda, you will control the timing and content of all discussions. You will also control what will and, perhaps more importantly, what will not be said. Agendas can:
   - Clarify or hide real objectives
   - Force quick decisions
   - Restrict discussions or permit digressions
   - Establish firm timetables or allow open ends
   - Guide actions directly toward problem resolution in the shortest possible time

   Regular job meetings have prearranged agendas in the form of "old business" and a fairly standard format for new items. Even there, however, the agendas of individual items can still be controlled.

2. *Action rules.* The action rules that follow can be conducted as clear, distinct steps, as they appear in the case of larger, or special, issues. Even for simple issues, however, each step is still part of the process when approaching every issue at every regular or special meeting, no matter how formally or informally applied.

4.13.12
Sample Meeting/Conversation Record Form

Record of:

_____ Meeting   Date: _____   Project: _____
_____ Telephone Conversation   Time: ____:____ AM / PM   Proj. #: _____

Subject: _____
_____

Present / Calling: Distribution:

_____   _____
_____   _____
_____   _____
_____   _____
_____   _____

Discussion:

_____
_____
_____
_____
_____
_____
_____
_____
_____
_____
_____
_____
_____
_____
_____
_____
_____
_____
_____
_____
_____
_____
_____
_____
_____

Prepared By: **NOTICE:**
                                    Failure to notify the writer of any necessary corrections to this
_____   Record constitutes acceptance of all information as it is repre-
Date: _____   sented.

a. *Confirm attendance prior to any meeting.* If specific individuals are required to resolve any issue, make certain they are present (see also Sections 4.13.6 through 4.13.8 and 4.13.14).

b. *Think through the entire problem resolution process.* List each step, along with the corresponding people necessary to settle it. Catalog the specific actions necessary by each individual, along with the early and latest acceptable dates for their actions. This catalog will become your agenda.

c. *Be sure that you are thoroughly prepared to discuss every issue and subissue.*
   - Have complete information.
   - Have your presentation package finalized.
   - List every conceivable option, along with the corresponding answer.
   - Practice; know all the bases and be able to cover them spontaneously.

d. *Prior to the meeting, notify all expected participants of your specific agenda.* This will remove excuses on their parts that they do not have appropriate people available, have not secured required information, or are not otherwise prepared to discuss *and resolve* the issues.

e. *Include the expectation of problem resolution in your agenda notification.*

f. *Be aware of other people's efforts to control or divert your agenda.* Be prepared to force corrections.

g. *Have good reasons for forcing compliance with your prearranged agenda.* Know and be believable with your own excuses for *not* discussing items that are not on the planned list:
   - Key or affected people are not available.
   - There is limited time available to cover the prearranged items; the new items must be tabled for another time.
   - Use "I don't know" if you need to.
   - You have every right to be unprepared for "surprises" (whether you're prepared or not. . .).
   - Do not discuss any item unless you are fully prepared to do so.

4.13.14 Sample Letter Confirming a Special Meeting *(page 4.92)*

The sample Letter Confirming a Special Meeting is an example of a confirming letter designed to follow through on the recommendations of the previous sections. Specifically, it:

1. Confirms the meeting parameters of date, place, and time.

2. Establishes the complete meeting agenda.

3. Lists all those expected to attend.

4. Notifies all expected attendees by copy of the letter.

4.13.14
Sample Letter Confirming a Special Meeting

Letterhead

(Date)

To:

RE: (Project)
            (Company Project #)

SUBJ: Specific Issue(s)

Mr. (Ms.) (   ):

Confirming our conversation today, a meeting will be held at my office on (Date) at (Time) to resolve the subject issue(s). The specific agenda is as follows:

1.   (Primary Issue #1)
        a.   (Sub Issue #1.1)

2.   (Primary Issue #2)
        a.   (Sub Issue #2.1)
        b.   (Sub Issue #2.2)

As we discussed, please be sure that your Mr. (Name), Ms. (Name), and any other people necessary to completely resolve the issue(s) are present.

Thank you for your consideration.

Very truly yours,

COMPANY

Project Manager

cc: (List all those named in the letter)
        (List all those definitely or potentially affected)
        Jobsite
        File: Vendor File:_____
                CF

4.14   Preparing for Project Close-Outs at Project Start-Up

A successful project completion begins at the project's inception. A well-planned process for project close-out will benefit all parties—the General Contractor and its Subcontractors who will receive final payment and retainage quickly, and, most importantly, the Owner who will retain the same positive opinion of the contractor that the Project Manager has worked diligently to create during construction.

The Owner's first impression of the construction team will come at the opening phase of construction and the first Owner/design consultant/Project Manager meeting. The tone of the project will be set at this time, and it is critical to set the right one.

The Project Manager must familiarize himself or herself with the terms and conditions of the contract with the Owner and the contract documents—the plans and specifications, especially those that pertain to project start-up and project close-out procedures. These special provisions need to be disseminated to the field crew, Subcontractors, and vendors.

4.14.1   Starting off on the Right Foot—A Review of the Contract with the Owner

The contract with the Owner will contain several provisions relating to performance and the flow-through process that binds Subcontractors to the same provisions that apply to the General Contractor. These flow-through or pass-through provisions will be addressed at the first project meeting with the Owner so that any required clarifications can be communicated to the appropriate Subcontractors and vendors. This list will include:

- Establishing the date when requisitions are to be submitted and procedures for submitting "pencil" copies for review before official submission.
- Reviewing restrictions on allowable overhead and profit for the General Contractor, Subcontractors, and second- and third-tier Subcontractors.
- Determining if the contract contains unit prices that need to be passed on to Subcontractors during buyouts.
- Noting any allowances and Alternates, and with respect to the latter, determining if there is a time frame in which the Alternate must be selected.
- If a contingency is included in the contract, determining if there are restrictions on items that can be charged to this account.
- Determining if the contract contains a liquidated damage clause, and if so, what actions the contractor must take to negate the imposition of these damages.
- If the General Contractor is to submit a list of Subcontractors to the owner for approval, determining how much time the Owner has to approve or disapprove the list.

- Determining if the Owner has a responsibility to appoint a representative, and if so, what the time frame is for such an appointment and what duties this "Owner's rep" will have.

- Determine if there are restrictions on the hours of construction, noise abatement concerns, and traffic patterns to follow with respect to site access and egress.

- Deciding whether there are any other provisions in the contract that ought to be "passed through" to Subcontractors.

4.14.2   A Review of the Specifications

A review of the specifications ought to produce three checklists, one for the Project Manager, one for the Project Superintendent, and one for the Subcontractors. Each checklist ought to highlight those provisions that pertain to its corresponding party's contract administration. For example,

1. A list of required mock-ups.

2. Requirement for timely submission of proposed change orders.

3. Requirement for timely submission of shop drawings, number of copies, format.

4. Requirement for timely submission of Requests for Information (RFIs).

5. Coordinated drawing preparation, review, and approval process.

Inspections and testing should include:

- Soils testing
- Concrete testing (cast-in-place) and off-site production of precast, if applicable
- Infiltration and exfiltration tests of underground storm sewer
- Hydraulic and pressure testing of underground plumbing and fire protection lines
- Mill reports for structural steel
- Weld, bolt-up steel connections [if tension control (TC) bolts are not required], shear stud testing
- Mortar cube testing for masonry work
- HVAC ducts pressure integrity
- Above-ground plumbing and fire protection installations
- Acoustical and/or insulation batt installations
- Flashing and roofing items of work
- Various substrates or framing before being encapsulated or enclosed

Record drawings should include:

- Record of all changes, either those made due to field conditions or those representing approved scope changes
- Inclusion of any accepted Alternates
- Dimensional changes to correct any "contract" dimensions
- Elevations related to site work, line and grade, rim elevations of manholes and catch basins
- Inverts and pitch of underground storm and sanitary lines
- Floor-to-floor heights and floor-to-ceiling elevations
- Structural changes as evidenced by approved shop drawings
- Location of concealed items, plumbing valves, HVAC access panels, and filter locations
- Fire dampers and adjustable HVAC flow dampers
- Heat tracings when enclosed or encapsulated

Testing/adjusting/balancing (TAB) should include:

- Testing procedures to determine and verify the design performance of equipment
- Recording of adjustment to achieve the required fluid flow and/or cubic feet per minute (CFM) air patterns at the terminal units
- Providing evidence of balancing the proportional flows within the plumbing and HVAC distribution systems to meet the design criteria

4.14.3   Punchlist and Warranty Items *(page 4.96)*

The Project Manager, in the absence of a formal company policy relating to punchlist work, should establish a procedure to deal with the creation and resolution of Subcontractor punchlist items, and possibly state a goal of a Zero Tolerance Policy that could include the following procedures:

- Perform an inspection of each trade's work that will be conducted jointly by the Project Superintendent and the Subcontractor's foreman. This procedure could be written into the subcontract agreement, stating that this is in addition to, and prior to, the issuance of a formal punchlist by the Owner. All pre-punchlist work is to be completed prior to the Subcontractor's final demobilization.
- Establish a method whereby incomplete items can be completed by another Subcontractor, if the responsible Subcontractor fails to do so in a timely manner. Allow two weeks or so for completion of any punchlist work, and if not complete within that time frame, advise the Subcontractor, in writing, that

4.14.3
Punchlist and Warranty Items

Contractor's Letterhead

Addressed to the Subcontractor

Attention: Subcontractor's Project Manager or Owner

Re: Project Name and Number

Subject: Pre-Punchlist Requirement

Dear Mr. (Ms.) (   ),

Prior to your demobilization from the jobsite, we will inspect your work and prepare a punchlist that will be transmitted to you on/about (date). We will expect this work to be completed not later than (date) and prior to your full demobilization.

This punchlist, prepared by our project superintendent in conjunction with your on-site supervisor, is in addition to the architect/engineer's formal document that will be submitted to your office when received from (architect/engineer's name).

With best regards,

Project Manager

other forces can be brought in to complete the unfinished work, creating a backcharge to the underperforming Subcontractor.

- Identify what constitutes punchlist work as opposed to warranty work. For example, repair of a malfunctioning variable air volume (VAV) box is actually a warranty item and not a punchlist. And while the repair or replacement should be diligently pursued, the Architect may not be justified in withholding funds pending prompt replacement of the defective box, once shown that the Subcontractor has ordered the repair or replacement.

4.14.4   Attic Stock and Spare Parts

Requirements for attic stock are very clear, and Subcontractors should be apprised of the need to supply attic stock and spare parts when they are awarded a subcontract agreement. In that way they can order this stock or parts when initially placing orders with their vendors.

Often, in the case of acoustical ceiling tiles, the Subcontractor may use significant amounts of attic stock to replace tiles damaged during construction. They must be advised to replenish this stock prior to project close-out.

A quick review of the specifications will turn up those items that generally require "attic stock":

- Carpet, resilient flooring, vinyl base
- Ceramic tile
- Acoustical ceiling pads and extra grid sections
- Specialty lightbulbs
- Finish hardware items
- HVAC filters, belts, fire damper replacement links
- Sprinkler heads, tools for head replacement
- Electrical wiring devices, cover plates, spare circuit breaker components, fluorescent light fixture ballasts
- Plumbing valves or replacement parts for the valves

4.14.5   Material Safety Data Sheets (MSDSs)

The Occupational Safety and Health Administration (OSHA) requires material safety data sheets for all hazardous materials that will be shipped to the construction site. Both Subcontractors and the company's purchasing personnel need to be reminded of this requirement, since the MSDSs are required to be on-site before the hazardous product arrives. Failure to produce the MSDSs when requested by an OSHA inspector is cause for a citation, so it is not only important to obtain these documents but also to have them readily at hand when required.

If close-out requirements include a complete set of MSDSs issued during construction, one way to collect them is to set aside a three-ring binder so that each MSDS can be put into the binder right after it is received.

4.14.6 Subcontractor Responsibilities

Although the company's subcontract agreements will include provisions dealing with its contract administration, it is often helpful to review some of the more key elements in those agreements to highlight the issues and requirements that will be expected of the Subcontractors and their second- and third-tier contractors during the course of construction.

Selected portions of the contract specifications can be distributed dealing with such topics as:

- Submittal procedures—the number and type of submittal required and the time allowed for review by the Architect/Engineer
- Coordination drawing preparation, the coordination process, and the responsibility of the Subcontractor if it fails to properly coordinate its work with the other trades
- The process for submission of change orders, premium time costs, and winter conditions (if applicable)

The latter, the process for submission of change orders, premium time costs, and winter conditions, may require more clarification even though these topics may be covered in the contract specifications. Contractors may require more clarity than the specifications require, particularly if the contract with the Owner is a Cost-Plus or Guaranteed Maximum Price (GMP) type contract.

The Project Manager may wish to create a more detailed list of procedures that will actually help the Subcontractor and General Contractor to provide the Owner and its Architect with more detailed information to speed up the processing of requests for extras, premium time, and winter condition costs, when submitted.

The "protocol" set forth in Section 4.14.7, if followed, will enhance the approval process.

4.14.7 A Protocol for Submission of Change Orders, Premium Costs, Winter Conditions *(Page 4.99)*

4.14.7
A Protocol for Submission of Change Orders, Premium Costs, Winter Conditions

Change Orders

1. Each proposed change order is to contain a brief explanation of the nature of the change and who has initiated it [Owner, Architect/Engineer (A/E), Contractor, Subcontractor). Attach all supporting documentation, e.g., letters, SK from A/E, and request from General Contractor (GC).

2. If scope of work is increased or decreased, state prior condition and proposed condition and delta between prior and proposed quantities.

3. All costs submitted by Subcontractors, and their second- and third-tier Subcontractors, are to be broken down into Labor (hours i rate) and materials (number if applicable, lineal or square feet if applicable). Overhead and profit (OH&P) to be added to "costs" and percentage of OH&P to conform to contract requirements.

4. Equipment—indicate whether rental is from Subcontractor or independent rental company. List number of hours/days i applicable rate. Provide receipt of delivery, and return. Differentiate between idle and active hours, if applicable.

5. Subcontractor to review its proposal carefully before submission to include scrutiny of labor hourly rates to ensure that those rates and burden are proper and when premium time hours are applied, the correct upcharge with respect to burden items is in order. When Federal Unemployment Tax Authority (FUTA) and State Unemployment Tax Authority (SUTA) limits have been reached, hourly rates should reflect deletion of this item from labor burden computation.

6. If work is Time and Material (T&M), in addition, follow procedures indicated below.

7. If requested by GC, Owner's representative will be present when change order negotiations between GC and Subcontractor(s) take place.

Time and Material Work Authorized by General Contractor

1. Contractor's supervisor to receive Daily Tickets for all T&M work, to include worker's trade category (carpenter, laborer, etc.), number of hours worked, and task performed. Ticket must be signed by Contractor's supervisor validating contents.

2. Receiving tickets for all materials and equipment are to be attached.

3. With submission of invoice, furnish substantiating invoices/bills for material and equipment purchases and rentals.

Premium Costs

1. For Subcontractor's self-performed work, follow procedures outlined above for T&M work, but include reason for premium time work, e.g., weather delays, request by General Contractor to maintain previously agreed-upon schedule, late delivery of critical material due to late approval, lack of response from GC, Owner, or A/E. All tickets are to be approved by GC's superintendent.

2. Subcontractor to accumulate and present all such tickets to the GC's field supervisor on a weekly basis, or on a more frequent basis when requested to do so.

4.14.7
A Protocol for Submission of Change Orders, Premium Costs,
Winter Conditions *(Continued)*

Winter Conditions (When Applicable)

1. Indicate operation taking place requiring winter conditions.
2. Indicate GC's field supervisor who authorized work.
3. Provide log with temperature readings at 7:00 A.M., noon, 2:00 P.M.
4. Provide daily tickets for labor as outlined above.
5. Provide list of materials used, type of fuel consumed, and, with invoice, copies of all bills for same.
6. Provide list of any equipment utilized whether rented from Subcontractor or an equipment rental company.
7. All such tickets to be signed by GC's supervisor on a daily basis.
8. Subcontractor to accumulate and present all such tickets to the GC on a weekly basis, or on a more frequent basis when requested by the GC.

4.15   Securing Lien Waivers

4.15.1   General

Simply stated, a lien is a security interest in the particular real estate that has been improved. It is placed to secure payment for labor and material used in the property's improvements. The lien provides for the right to sell the property to which the lien attaches if the debt is not paid. A lien waiver is a short document executed to waive an individual's or a company's right to assert a lien. The right to assert a lien is not one recognized by common law, and is therefore defined by statute. Lien laws exist in every state and vary greatly in their terms.

This section does not discuss the intricacies of liens and lien laws. For that you must consult with a competent attorney to confirm the specific requirements and details of local law. Besides purely statutory rights and requirements, these intricacies will be further qualified by other material considerations such as:

- Type of property owner, whether state, federal, or municipal agency or private owner
- Presence or lack of payment and performance bonds
- Presence or lack of other available remedies for nonpayment under the terms of the general contract or subcontract

As a contracting professional, it is incumbent upon you to become intimately and completely familiar with all intricacies of liens and lien waivers for every geographic location in which you do business. They will profoundly affect daily operating decisions regarding treatment of Owners and Subcontractors for their appropriate and inappropriate actions regarding all project payments. Have your attorney provide you with the appropriate full and partial lien waiver forms to be used for each project.

This section, then, focuses on *securing* proper lien waivers from Subcontractors and suppliers as a requirement for project payments—after those requirements have been correctly determined.

4.15.2 Full versus Partial Waivers of Lien

Since lien rights are a creature of statute and not of common law, they can be waived. The waivers themselves are necessary, both as a protection of the Company and as an inducement to the Owner to release additional payment.

Many general requisition forms and procedures to the Owner will require the delivery of appropriate lien waivers for all payments in prior periods as a condition of releasing the current payment. If a single sub-subvendor is delinquent in the delivery of a single waiver, it can have the effect of locking up the entire general payment.

Attention must therefore be continually and sharply focused on maintaining *all* required waivers on an absolutely current basis.

Partial Waivers of Lien are documents executed by the payee to waive its right to assert a lien for an amount equal to the respective payment. This is usually a straightforward procedure accepted by most without much objection.

Full Waivers of Lien generally waive *all* rights to every lien on the property that is the subject of the agreement for work completed, *or yet to be completed.*

Although requiring a party to execute a full waiver may appear to be an extreme measure, in reality it may be a convenient mechanism that either can streamline contract procedures without actually compromising rights, or will keep dispute resolution options focused on the actual dispute resolution provisions defined in your subcontracts, for example, without allowing sharp attorneys to pervert the intention of your agreement.

Example: By statute, state-owned property in Connecticut cannot be liened. In place of lien rights, General Contractor payment and performance bonds are provided in order to preserve payment rights of subvendors.

Although there are no lien rights, some contracts may still actually require delivery of lien waivers as a condition of payment. In these cases, requiring a full waiver from all subvendors at the start of the project will relieve the logistical effort necessary to secure each waiver during every payment cycle.

Example: If you as a General Contractor have provided the Owner with a 100% payment bond, and your subcontract agreement provides for arbitration after a certain dispute-resolution procedure is followed, payment rights of the Subcontractor have been substantially protected. Leaving Subcontractor lien rights in place in these conditions only allows them to irritate the Owner and cause other legal problems that are expressly outside the dispute-resolution *intentions* of the subcontract. Requiring a Subcontractor to execute a full waiver puts the problem resolution back into the subcontract procedure and not onto the property.

4.15.3 Securing Subvendor Lien Waivers

1. *Full Lien Waivers.* If the execution of a full lien waiver has been a condition of a subcontract or purchase order, have it executed at the time the subcontract is executed. If for any reason this has not been done, have the waiver executed upon delivery of the *first* progress payment, as a condition of payment.

   Refer to Section 4.7 for related discussion on contract execution and distribution procedures.

2. *Partial Lien Waivers*

    a. The sample Subcontract/Purchase Order Transmittal Form Letter of Section 4.7 is the first written notification of the requirement after the subcontract or purchase order agreement itself.

    b. The sample Letter to Subcontractors Regarding Submittal Requirements of Section 4.10.3 is the second written notification to provide the forms.

    c. In *every* case where it is at all practical (regardless of whether it is "inconvenient" for the payee), have an *authorized* representative of the Company pick up the periodic payments and execute the waiver as the condition of payment. If the waiver is not properly executed for whatever reason, the payment should not be released.

    d. In cases where it is not logistically practical for a payee to pick up a payment (such as a supplier across the country), conditions must be provided that will allow for the next payment to be withheld until the correct waiver for previous payments has been delivered.

    e. As a condition of payment, it is the primary subvendor's responsibility to provide lien waivers for its sub-subvendors for payment for all materials through all previous payments. If these waivers are not provided, the current payment to the primary subvendor must not be released.

    f. Refer to Section 4.15.4 for the sample letter to induce compliance from your subvendors.

4.15.4   Sample Letter to Subcontractors/Suppliers Regarding Failure to Provide Lien Waivers (*page 4.103*)

The sample Letter to Subcontractors/Suppliers Regarding Failure to Provide Lien Waivers that follows is to be used in every case where there is an observed deficiency in any subvendor providing correct lien waivers for itself or for any of its sub-subvendors. It is used to:

1. Summarize the specific waiver forms necessary.

2. Notify that failure to provide the correct forms will result in delay in payment to all affected vendors and subvendors.

3. Notify the party that they will be held responsible for any effects on other parties resulting from the noncompliance.

Refer to Section 4.16 for a form combining these notifications with those for Certified Payroll Reports as discussed there.

4.16   Securing Subvendor Certified Payroll Reports

4.16.1   General

In many states and on federal projects, any project providing for the use of public funds will likely require the payment of prevailing wages under the provisions of the Davis-Bacon Act or other minimum-wage payment requirements.

4.15.4
Sample Letter to Subcontractors/Suppliers Regarding Failure to Provide Lien Waivers

Letterhead

Date: _____

To: _____   Faxed: No: _____   Yes: _____
    _____   Fax #: _____
    _____

ATTN: _____

RE: Project:_____
           Company Project #: _____

SUBJ: Certified Payroll Reports

Mr. (Ms.) _____ :

To date, we have not received properly completed Certified Payroll Reports for the following time periods:

_____   _____
_____   _____
_____   _____
_____   _____

These original documents in correct form and executed by those properly authorized to do so are to be delivered to this office by or before _____, 20 _____. Be sure that the weekly payroll ledgers accompany these submissions in order to support each respective Payroll Report Form.

NO PAYMENT OTHERWISE DUE CAN BE RELEASED UNTIL THESE REQUIREMENTS HAVE BEEN MET.

Failure to comply may also affect the payments of all other subcontractors and suppliers for the period(s) represented, and may also delay the current general payment. In that event, you will be held responsible for all resulting costs and effects.

Thank you for your cooperation.

Very truly yours,

COMPANY

_____
Project Manager   cc: File: Vendor File_____
                                                   Certified Payroll Reports
                                                   CF

On such projects, confirmation of the payment by Subcontractors of such prevailing wages will be done by way of Certified Payroll Reports. These are reporting forms on which the respective employer certifies to the Owner the actual amount of wages, taxes, and fringe benefits paid to the individual employees for hours worked on particular dates.

Section 4.16.2 displays an example of a Certified Payroll Report Form that is commonly used on state-funded projects. Check with your state or federal labor board for the correct forms to be used on the particular project as required.

This section, then, focuses on *securing* Certified Payroll Report Forms from each project Subcontractor on a current basis as a condition of payment.

4.16.2 Example Certified Payroll Report Form
(page 4.105)

An example of a Certified Payroll Report Form that is commonly used on state-funded projects follows. Check with your state or federal labor board for the correct forms to be used on a particular project as required.

4.16.3 Payroll Liabilities on Construction Projects

On projects requiring prevailing wages, such wages are normally required for work performed at the site, but not for fabrication processes performed off-site.

If a subvendor fails to pay such prevailing wages to its employees, and a complaint is filed by any of its employees within the statutory period, the state or federal labor department will normally look for satisfaction of such payment of back wages and benefits in the following order:

1. The subvendor who has been confirmed to be in violation of the wage laws.

2. The General Contractor.

3. The payment bond.

Because the payment bond is usually the last recourse, the project Owner will not normally have any liability. As such, you should not tolerate any efforts by any project Owner to withhold any payment because of an alleged violation of labor laws on bonded projects. Because prevailing wage projects are usually bonded, this should apply to all such projects.

Because of their recognized lack of liability, project Owners vary greatly in their attitudes toward policing wage payments. Some are relentless, while others will go through an entire project without ever asking for them.

The General Contractor, however, should be fully aware of the amount of extreme liability assumed if subvendors are allowed to violate wage laws, or if at least measures are not taken to confirm that the wages certified to be paid have actually been paid.

4.16.2
Example Certified Payroll Report Form

U.S. DEPARTMENT OF LABOR
WAGE AND HOUR DIVISION

PAYROLL
(For Contractor's Optional Use; See Instruction, Form WH–347 Inst.)

Form Approved.
Budget Bureau No. 44–R1093

NAME OF CONTRACTOR ☐ OR SUBCONTRACTOR ☐

PAYROLL NO. _____ FOR WEEK ENDING _____ ADDRESS _____ PROJECT AND LOCATION _____ PROJECT OR CONTRACT NO. _____

(1) NAME, ADDRESS, AND SOCIAL SECURITY NUMBER OF EMPLOYEE	(2) NO. OF WITHHOLDING EXEMPTIONS	(3) WORK CLASSIFICATION	OT. OR ST.	(4) DAY AND DATE — HOURS WORKED EACH DAY	(5) TOTAL HOURS	(6) RATE OF PAY	(7) GROSS AMOUNT EARNED	(8) DEDUCTIONS FICA	WITH-HOLDING TAX	OTHER	TOTAL DEDUCTIONS	(9) NET WAGES PAID FOR WEEK
			O									
			S									
			O									
			S									
			O									
			S									
			O									
			S									
			O									
			S									
			O									
			S									
			O									
			S									

FORM WH–347 (1/68) – FORMERLY SOL 184—PURCHASE THIS FORM DIRECTLY FROM THE SUPT. OF DOCUMENTS

4.105

4.16.2

Example Certified Payroll Report Form (Continued)

Date _____

I, _____
(Name of signatory party)   (Title)

do hereby state:

(1) That I pay or supervise the payment of the persons employed by _____

_____   on the _____
(Contractor or subcontractor)   (Building or work)

; that during the payroll period commencing on the _____

day of _____, 19____, and ending the _____ day of _____, 19____,
all persons employed on said project have been paid the full weekly wages earned, that no rebates
have been or will be made either directly or indirectly to or on behalf of said

_____ from the full
(Contractor or subcontractor)

weekly wages earned by any person and that no deductions have been made either directly or
indirectly from the full wages earned by any person, other than permissible deductions as defined
in Regulations, Part 3 (29 CFR Subtitle A), issued by the Secretary of Labor under the Copeland
Act, as amended (48 Stat. 948, 63 Stat. 108, 72 Stat. 967; 76 Stat. 357; 40 U.S.C. 276c), and described below:

_____

_____

_____

_____

(2) That any payrolls otherwise under this contract required to be submitted for the above
period are correct and complete; that the wage rates for laborers or mechanics contained therein
are not less than the applicable wage rates contained in any wage determination incorporated into
the contract; that the classifications set forth therein for each laborer or mechanic conform with the
work he performed.

(3) That any apprentices employed in the above period are duly registered in a bona fide
apprenticeship program registered with a State apprenticeship agency recognized by the Bureau
of Apprenticeship and Training, United States Department of Labor, or if no such recognized agency
exists in a State, are registered with the Bureau of Apprenticeship and Training, United States Department of Labor.

(4) That:

(a) WHERE FRINGE BENEFITS ARE PAID TO APPROVED PLANS, FUNDS, OR PROGRAMS

☐ —In addition to the basic hourly wage rates paid to each laborer or mechanic
listed in the above referenced payroll, payments of fringe benefits as listed in the

contract have been or will be made to appropriate programs for the benefit of such
employees, except as noted in Section 4(c) below.

(b) WHERE FRINGE BENEFITS ARE PAID IN CASH

☐ —Each laborer or mechanic listed in the above referenced payroll has been paid, as
indicated on the payroll, an amount not less than the sum of the applicable basic
hourly wage rate plus the amount of the required fringe benefits as listed in the
contract, except as noted in Section 4(c) below.

(c) EXCEPTIONS

EXCEPTION (CRAFT)	EXPLANATION

REMARKS

NAME AND TITLE	SIGNATURE

THE WILFUL FALSIFICATION OF ANY OF THE ABOVE STATEMENTS MAY SUBJECT THE CONTRACTOR
OR SUBCONTRACTOR TO CIVIL OR CRIMINAL PROSECUTION. SEE SECTION 1001 OF TITLE 18 AND
SECTION 231 OF TITLE 31 OF THE UNITED STATES CODE.

To help with such periodic confirmation, maintain efforts on all prevailing wage jobsites to continually discuss requirements with the individual workers. Follow up on any hint of a violation. Beyond these efforts, require first in your subcontract agreements, and then in practice, the delivery of the Subcontractor's actual weekly payroll ledger to support the Certified Payroll Report Forms being submitted.

Use the sample form letter of Section 4.16.4 to help in the enforcement of delivery of the proper Payroll Report Forms and payroll ledger, and *release no payments unless the requirements are completely fulfilled on a current basis.*

4.16.4   Sample Letter to Subcontractors Regarding Delivery of Certified Payroll Report Forms *(page 4.108)*

Whenever Certified Payroll Report Forms are required, they will be required to be submitted *weekly*. This is an express condition of payment. No payment can be released until the proper, complete forms have been delivered.

Use the sample Letter to Subcontractors Regarding Delivery of Certified Payroll Report Forms to:

1. Identify the particular weeks of deficiency.

2. Notify the offending party that failure to provide the correct forms is a breach of contract and will cause delay in payment until the condition is corrected.

3. Notify the party that it will be held responsible for any effects on other parties as a result of noncompliance.

Refer to Section 4.15.4 regarding a similar procedure for securing lien waivers, and to Section 4.16.5 for a sample letter combining the forms for Certified Payroll Reports with those for lien waivers.

4.16.5   Sample Form Letter to Subcontractors Regarding Delivery of Lien Waiver Forms and Certified Payroll Report Forms *(page 4.109)*

4.17   Backcharges

4.17.1   General

A backcharge is a charge against a Subcontractor's or supplier's account to cover the cost of having had to perform some portion of their responsibility for them by other means, or to correct some portion of their work. It is the contract mechanism that reimburses the Company for all expenses incurred, including but not limited to supervision (and other management efforts), overhead, and profit.

It is not the objective of this section to necessarily encourage Project Managers, Project Engineers, and Site Superintendents to pursue backcharges aggressively in every possible situation. Rather, it is here to:

4.16.4
Sample Letter to Subcontractors Regarding Delivery of Certified Payroll Report Forms

---

Letterhead

Date: _____

To: _____   Faxed: No: ____   Yes: ____
        _____   Fax #: _____
        _____

ATTN: _____

RE: Project:_____
        Company Project #: _____

SUBJ: Certified Payroll Reports

Mr. (Ms.) _____:

To date, we have not received properly completed Certified Payroll Reports for the following time periods:

_____   _____
_____   _____
_____   _____
_____   _____

These original documents in correct form and executed by those properly authorized to do so are to be delivered to this office by or before _____, 20_____. Be sure that the weekly payroll ledgers accompany these submissions in order to support each respective Payroll Report Form.

NO PAYMENT OTHERWISE DUE CAN BE RELEASED UNTIL THESE REQUIREMENTS HAVE BEEN MET.

Failure to comply may also affect the payments of all other subcontractors and suppliers for the period(s) represented, and may also delay the current general payment. In that event, you will be held responsible for all resulting costs and effects.

Thank you for your cooperation.

Very truly yours,

COMPANY

_____
Project Manager   cc: File: Vendor File_____
                                                             Certified Payroll Reports
                                                             CF

4.16.5
Sample Form Letter to Subcontractors Regarding Delivery of Lien Waiver Forms and Certified Payroll Report Forms

Letterhead

Date: _____

To: _____   Faxed: No: _____   Yes: _____
_____   Fax #: _____
_____

ATTN: _____

RE: Project:_____
           Company Project #: _____

SUBJ: 1) Certified Payroll Reports   2) Subcontractor/Supplier Lien Waivers

Mr. (Ms.) _____:

To date, we have not received:

_____   Properly completed Certified Payroll Reports for the following time periods:
         _____   _____
         _____   _____

_____   Properly executed lien waivers for:
_____   Your Sub-subcontractors: _____
                                         _____
                                         _____
                                         All others as applicable.

_____   Your Material Suppliers: _____
                                         _____
                                         _____
                                         All others as applicable.

These original documents in correct form and executed by those properly authorized to do so are to be delivered to this office by or before _____, 20_____. NO PAYMENT OTHERWISE DUE CAN BE RELEASED UNTIL THESE REQUIREMENTS HAVE BEEN MET.

Failure to comply may also affect the payments of all other subcontractors and suppliers for the period(s) represented, and may also delay the current general payment. In that event, you will be held responsible for all resulting costs and effects.

Thank you for your cooperation.

Very truly yours,
COMPANY

_____
Project Manager   cc: File: Vendor File_____, CF

1. Encourage project management to:
    a. Keep adequate records and substantiation
    b. Provide and maintain adequate notification
    c. Be able to prepare an appropriate backcharge whenever necessary, and to support it properly
2. Establish and maintain project environments in which each Subcontractor and supplier is aware of Company policies and procedures regarding backcharges. In so doing, subvendors who know of the Company's ability to do its work when necessary, support the charges, and their willingness to do so whenever necessary will have less of a tendency to allow matters to deteriorate to that level very often.

4.17.2   Conditions of the Backcharge

1. The Subcontractor or supplier has failed to or refuses to perform the work in accordance with the requirements of its bid package in any respect. This applies to:
    a. Physical work not completed at all
    b. Completed work requiring correction
    c. All administrative requirements of the bid package
2. The Subcontractor or supplier was *notified* that the respective work must be completed or corrected by a particular date.
3. The party cannot or will not comply with the directive to complete or correct the work, or to do so by the time required.

4.17.3   Backcharge Procedure

1. Establish that the work (subject of the proposed backcharge) is required:
    a. To allow the job to progress
    b. To comply with any project requirement
2. Confirm *all* requirements of the contract documents:
    a. The general documents between the construction force and the Owner
    b. The specific subcontract or purchase order of the respective bid package being considered for the charge
3. Notify the party verbally (in a meeting, conversation, or telephone call). Advise:
    a. Of the *specific* work to be done
    b. Of the date by which the work is to be complete
    c. That if the work is not completed as directed, you will arrange to complete it by whatever means necessary, and charge the party for *all* associated costs.
    Be prepared to actually move on this notice. Never threaten to do anything that you're not really prepared to do.

4. Follow up with the sample Backcharge Notice of Section 4.17.4. Such a written notice must be sent immediately.

5. When the work is not completed or the item not complied with by the deadline, *follow through* on your notice. Do not create any environment in which your notifications become viewed as idle gestures.

   Immediately arrange to complete the item by whatever reasonable means necessary. It is often useful to arrange to have the work completed by another Subcontractor, rather than with your own forces. This can go a long way later when the original Subcontractor argues about the size of the actual cost of the corrective work.

6. Take photographs before, during, and after all work. Refer to Section 5.16 for instructions on the proper procedure and identification of such photos.

7. Process the backcharge as a regular change order to the subvendor's contract. Be sure to include all costs associated with:

   a. Soliciting prices, coordination, and other direct efforts necessary to arrange for the work to be done
   b. Field coordination and supervision
   c. Prevailing or other appropriate wage rates
   d. Trucking and transportation charges
   e. Appropriate rates for overhead and profit

8. Use the Backcharge Summary Log of Section 4.17.7 to keep a running record of smaller incidents that can be more conveniently consolidated in the completed change order after the total costs reach an appropriate magnitude.

4.17.4 Use of Backcharge Notice Form Letter

The Backcharge Notice Form Letter in Sections 4.17.5 and 4.17.6 is used to follow through with the notification requirement of the previous discussion. Specifically, it:

1. Establishes the true date and time of notification as of the moment of conversation (*not* as of the receipt of the written notice).

2. Delivers the written notice as soon as possible (by fax and mail), thereby helping to confirm the urgent nature of the notice.

3. Concisely but completely describes the work required, along with the subsequent work that is being immediately affected by the work in question.

4. Gives the firm deadline by which acceptable performance must be achieved.

5. Notifies of your intent to complete the work and charge the subvendor's account.

4.17.5 Sample Backcharge Notice Form Letter—Completed Example

---

Letterhead

Date: _AUGUST 20, 1994_

To: _CRACK PLASTER CO., INC._   Confirmation of Fax:
     519 HEART TPK   Fax #: _984-9920_
     CANTON, CT 06902

ATTN: _PETER CRACK_

RE: Project: _RAINBOW CHILD CARE_
     Proj. #: _9442_

SUBJ: _CONTINUING SLIPPAGE IN PLASTER WORK COMPLETION SCHEDULE_

Mr. ~~Ms~~: _CRACK_   :

Confirming our conversation at _9:10_ (AM) / PM this date, please:
  CORRECT THE PLASTER PATCHING AT ROOMS 203, 204, & 205
  IN ORDER TO ALLOW PAINTING TO CONTINUE.

This work must be completed by _5:00_ AM (PM) on _AUGUST 22_ , 19_94_ .

If the work is not completed by then, we will make arrangements to complete the work for you, and backcharge your company for all costs incurred, including mobilization, supervision, overhead and profit.

Very truly yours,

COMPANY

David Veltko
PROJECT ENGINEER

cc: Accounts Payable
        File: Vendor _PLASTER_
              CF

4.17.6   Sample Backcharge Notice Form Letter—Blank Form

Letterhead

Date: _____

To: _____   Confirmation of Fax:
    _____   Fax #: _____
    _____

ATTN: _____

RE: Project: _____
       Proj. #: _____

SUBJ: _____

Mr. / Ms. _____ :

Confirming our conversation at _____ : _____ AM / PM this date, please:
_____
_____
_____
_____
_____
_____
_____
_____
_____

This work must be completed by _____ : _____ AM / PM on _____ , 20 _____ .

If the work is not completed by then, we will make arrangements to complete the work for you, and backcharge your company for all costs incurred, including mobilization, supervision, overhead, and profit.

Very truly yours,

COMPANY

_____
_____

cc: Accounts Payable
       File: Vendor_____
              CF

4.17.7   Use of Backcharge Summary Log

The Backcharge Summary Log is a form prepared for the individual subvendor upon the initial incident creating any potential for the first backcharge. It will be prepared along with the first Backcharge Notice as described in previous sections, and will remain a part of the permanent Subvendor File.

Large backcharge items will justify the immediate preparation of the appropriate deduct change order to the subvendor's subcontract or purchase order. Small backcharge items can be listed on the Backcharge Summary Log as they occur, and consolidated at any convenient time into a single or a few such change orders.

The log can be kept in the respective subvendor's General File Folder (see Section 2.4) or behind the Subcontractor Summary Form of Section 4.4 in the Subcontractor Summary and Telephone Log book.

4.17.8   Sample Backcharge Summary Log—Completed Example
(page 4.115)

4.17.9   Sample Backcharge Summary Log—Blank Form
(page 4.116)

4.17.10   Site Cleanup—A Special Case

General housekeeping and site cleanup are areas of chronic problems with Subcontractors. If the site is not policed with extreme diligence by the Site Superintendent, it quickly becomes a mess. When that is allowed to happen, each Contractor loses a corresponding amount of initiative to keep its own work clean. The process will continue to degenerate until the site is such a mess that it is everyone's fault—but it will be "no one's fault" because the mess will be a combination of everybody's materials.

Cleanup should be a primary consideration at every Subcontractor's meeting and throughout every workday. No violation is to be tolerated.

Use the Backcharge Notice procedure as described as a *weekly routine* to ensure that by every Friday each Subcontractor has arranged to clean up all materials and debris. If not, you must keep making arrangements to clean up the site for them.

In these cases it is crucial that the associated backcharges be processed *immediately*. This will help give the message to all vendors that you're absolutely serious. In most cases, Subcontractors will then fall in line with their cleanup responsibilities—at least until the next time.

4.17.11   Sample Letter to Subcontractors Regarding
Disregard for Finishes *(page 4.117)*

The sample Letter to Subcontractors Regarding Disregard for Finishes is a special notification to be sent to all Subcontractors at the appropriate stage of project completion when finishes are proceeding.

Although you should have the right to expect that tradespeople would automatically respect finish construction as it is being installed, this unfortunately cannot be relied upon in too many cases.

4.17.8
Sample Backcharge Summary Log—Completed Example

Backcharge Log

Project: NEWTOWN COMMUNITY CNTR   Subcontractor / Supplier:
Proj. #:   9900   BRICKYARD MASONRY CONTR. INC.

Date Notification Sent	Work Req'd To Be Complete By	Actual Date Completed	Completed By (Name)	Total Cost	OH & P	TOTAL
9-9-94	9-11-94	9-12-94	C-HILL	940.00	142.00	1,082.00
10-10-94	10-12-94	10-16-94	MCCI	220.00	33.00	253.00
12-2-94	12-4-94	12-5-94	C-HILL	350.00	53.00	393.00

4.17.9
Sample Backcharge Summary Log—Blank Form

Backcharge Log

Project: _____   Subcontractor / Supplier:

Proj. #: _____   _____

Date Notification Sent	Work Req'd To Be Complete By	Actual Date Completed	Completed By (Name)	Total Cost	OH & P	TOTAL

4.17.11
Sample Letter to Subcontractors Regarding Disregard for Finishes

---

Letterhead

(Date)

To: (List all Project Subcontractors)

RE: (Project)
                (Company Project #)

SUBJ: Care and Regard for Finishes

Gentlemen:

As the project nears completion, finish products are being installed and applied daily. At this time, negligence and disregard for the work of other trades will not be tolerated. Cleaning made necessary and/or damage caused to any finish work will be corrected at the expense of the responsible party(ies). Backcharge costs will include mobilization, preparation, protection, direct costs, supervision, overhead, and profit. Any person guilty of repeated and/or serious violation will be banned from the site.

We trust that everyone will demonstrate due consideration for the work of others, and hope there will be no need for these measures.

Thank you for your cooperation.

Very truly yours,

COMPANY

Project Manager

cc: Owner
        Architect
        Jobsite
        File: Backcharge File
                Vendor File:_____
                CF

The letter first calls attention to the stage of completion of the project and to your expectation of proper consideration. After that, it will serve as your notice of backcharges if such negligence and damage should occur.

4.18 The Punchlist: Expediting Final Completion

4.18.1 General

The "punchlist" should be on everyone's mind throughout the completion of any and every item of work. If anything is observed during the course of construction that you know is not satisfactory as finished work, complete it now—don't wait for it to find its way onto an official distributed list and possibly have some cost assigned to it.

Let the Owner and Architect observe you taking this approach. It will go a long way in building their confidence in you and in the final completed product.

As the project nears substantial completion, prepare your own punchlist, and prosecute its completion with the various Subcontractors. Do this in an effort to keep the official punchlist small.

When the project is sufficiently ready, typically at the point of or immediately after substantial completion, notify the Architect that the project is ready for the punchlist inspection.

Upon receipt of the official punchlist, proceed as directed in the remainder of this section both to confirm to the Owner that you are proceeding with the punchlist completion, and to expedite the completion itself with the various Subcontractors.

4.18.2 Sample Letter to the Architect Regarding Substantial Completion and Punchlist Review *(page 4.119)*

When the project is ready for inspection, use the sample Letter to the Architect Regarding Punchlist Review to notify the architect that:

1. The project is substantially complete.

2. The punchlist should be prepared as soon as possible.

3. Every effort must be made to ensure that the punchlist is complete.

We'd like to see *one* punchlist. It is hoped that the construction force will then be mobilized to complete the punchlist work—we don't want to keep repeating the effort for multiple lists.

The letter will give you at least some basis to call foul if second and third punchlists appear after the work of the first one has been completed.

4.18.3 Punchlist Review and Distribution Procedure

1. Immediately upon receipt of the official punchlist, it should be reviewed to confirm that each item on it is legitimate; that the design professionals and/or the Owner are not attempting to make the construction force responsible for

4.18.2
Sample Letter to the Architect Regarding Substantial Completion and Punchlist Review

---

Letterhead

(Date)

To: (Architect)

RE: (Project)
             (Company Project #)

SUBJ: Substantial Completion and Punchlist Review

Mr. (Ms.)   :

As of (Date), the project is substantially complete.

Please inspect the site as soon as possible to determine your punchlist of any items that you feel are necessary to correct prior to final completion.

It will be greatly appreciated if every effort is made to prepare as complete a punchlist as possible; it will be in everyone's best interest if we can all deal with a single list. This, as you know, will facilitate efficient completion of all items as expeditiously as possible.

Thank you for your consideration.

Very truly yours,

COMPANY

Project Manager

cc: Owner
       Jobsite
       File: Substantial Completion
               Punchlist
               CF

items of work that are not its fault. Items of wall repair and paint touch-up, for example, commonly fall into this category if the punchlist is allowed to be made after all or a portion of the project has been occupied.

2. Immediately advise the Architect of any item that in your opinion is in question. Remember, however, the "Conduit" Principle (Section 3.1.2) and the Pass-Through Clause (Section 3.5.8). Even though you may for the moment disagree with the item(s), they must still be forwarded with the current direction to the respective Subcontractor(s). You may after all discover that they'll get done without argument.

3. Mark each item on the punchlist to identify the specific party responsible to correct or complete it. It is often useful to assign a number to each Subcontractor, with a key written directly on the first page of the punchlist.

4. Review the list to determine that all administrative requirements are included, such as:
   a. Guarantees and warranties
   b. Attic stock
   c. Maintenance manuals
   d. As-built drawings
   e. System balancing reports
   f. System instruction
   g. Any other special requirements that will be necessary to close out the project

   If they have not been listed, add them yourself.

   Note that it will be of great advantage to you if you made arrangements throughout the project to secure these kinds of items well in advance of the punchlist. They typically are difficult and time-consuming to prepare and may be disproportionate in cost if you actually have to wind up doing them yourself.

   Having them completed in advance of the punchlist will also help to avoid any "hostage" situation if you have to arrange to finish items for any noncomplying Subcontractors and backcharge them for costs incurred.

5. Transmit the punchlist to each Subcontractor, requiring completion of all items by a particular date. Indicate the specific problems resulting from noncompletion, and that other measures will be taken to complete the work by other means if necessary. Use the sample Punchlist Notification Form Letters of the following sections to help in this procedure.

6. Refer to Section 4.19 for procedures with regard to securing appropriate guarantees and warranties.

7. After exhausting efforts with any original Subcontractor, consider moving ahead with your notice to complete the work by other means. Proceed carefully and completely, in a manner consistent with the discussion of Section 4.17, Backcharges.

4.18.4   Sample Punchlist Notification Form Letter #1 *(page 4.122)*

Use the sample Punchlist Notification Form Letter #1 to:

1. Transmit the punchlist to each responsible party.
2. Notify each party of the date by which all items must be completed.
3. Notify each party that failure to comply will directly affect all project payments.

Consider using the sample Punchlist Notification Form Letter #2 (Section 4.18.5) as a stronger initial approach. Use Letter #3 (Section 4.18.6) to follow up on Letter #1 with those Subcontractors who have not complied with your direction.

4.18.5   Sample Punchlist Notification Form Letter #2 *(page 4.123)*

The sample Punchlist Notification Form Letter #2 is a stronger version of Letter #1. It goes on to notify that failure to complete the work as directed may result in the Company making immediate arrangements to complete the work by whatever means available, and backcharge the delinquent party for all costs incurred. Consider further changing "may" to "will" to make the letter even stronger.

Note that it advises that the work will be completed by the "most expedient" means available. This is the notification that relieves later criticism of high cost to complete the work. Remember, it is *you* being forced into the position, not the Subcontractor.

Be prepared to actually follow through on the notice, however, if you decide to take this approach. Follow the considerations of Section 4.17, Backcharges, for related procedures.

4.18.6   Sample Punchlist Notification Form Letter #3 *(page 4.124)*

The sample Punchlist Notification Form Letter #3 can be used as a second notification to follow up on Letter #1 for those Contractors failing to complete their work by the specified date.

Letter #1 was a softer initial notification approach. Letter #3 raises the severity of the communication by giving one last and final real deadline, after which costs will start to accumulate for the offender.

4.18.7   Sample Notice of Supervisory Costs for Late Final Completion *(page 4.125)*

The sample Notice of Supervisory Costs for Late Final Completion can be sent to all Subcontractors who for any reason have not completed contract, punchlist, or guarantee work by the dates required. Consider using it as an interim

4.18.4
Sample Punchlist Notification Form Letter #1

Letterhead

Date: _____

To : _____
        _____   CERTIFIED MAIL
        _____   RETURN RECEIPT REQUESTED

ATTN: _____

RE: Project:_____
         Project #:_____

SUBJ: Punchlist Completion

Mr. (Ms.)_____:

Attached is the Architect's Punchlist for the project.

All items pertaining to your work must be complete by or before _____, 20_____. This is necessary in order to avoid delay in final completion and resulting delays in retainage release and final payment.

After you have completed your items, notify this office so that final inspection can be made.

In addition, immediately upon completion of all your items, return a copy of the Punchlist marked to indicate the dates that the respective items were completed.

Thank you for your cooperation.

Very truly yours,

COMPANY

_____
Project Manager

cc: Jobsite
       Vendor File:_____
       Punchlist File
       CF

4.18.5
Sample Punchlist Notification Form Letter #2

Letterhead

Date: _____

To : _____
_____   CERTIFIED MAIL
_____   RETURN RECEIPT REQUESTED

ATTN: _____

RE: Project:_____
       Project #:_____

SUBJ: Punchlist Completion

Mr. (Ms.)_____:

Attached is the Architect's Punchlist for the project.

All items pertaining to your work must be complete by or before _____, 20_____. This is necessary in order to avoid delay in final completion and resulting delays in retainage release and final payment.

If your work is not complete by that date, other arrangements will be made to complete the work by whatever means available that are most expedient. In that event, you will be backcharged for all costs incurred, including mobilization, preparation, protection, direct costs, supervision, overhead, and profit. We accordingly will appreciate your efforts to avoid these possibilities.

After you have completed your items, notify this office so that final inspection can be made.

In addition, immediately upon completion of all your items, return a copy of the Punchlist marked to indicate the dates that the respective items were completed.

Thank you for your cooperation.

Very truly yours,

COMPANY

_____
Project Manager

cc: Jobsite
       Vendor File:_____
       Punchlist File
       CF

4.18.6
Sample Punchlist Notification Form Letter #3

Letterhead

Date: _____

To : _____

_____

_____

ATTN: _____

Confirmation of Fax
Fax#:_____

CERTIFIED MAIL
RETURN RECEIPT REQUESTED

RE: Project:_____
       Project #:_____

SUBJ: Punchlist Completion

Mr. (Ms.)_____:

You have been directed on a number of occasions to complete your punchlist work, but to date items remain to be completed.

As your final notice, be advised that these items must be absolutely complete by or before _____, 20_____. After that date, arrangements will be made to complete your work by whatever means available that are most expedient. In that event, you will be backcharged for all costs incurred, including mobilization, preparation, protection, direct costs, supervision, overhead, and profit.

Please confirm to this office today whether you will complete your work as directed, or not.

Very truly yours,

COMPANY

_____

Project Manager

cc: Jobsite
       Vendor File:_____
       Punchlist File
       CF

4.18.7
Sample Notice of Supervisory Costs for Late Final Completion

Letterhead

(Date)

To :   (List All Project Subcontractors)   (List all respective Fax #'s)

RE: (Project Description)
            (Company Project #)

SUBJ: Supervisory Costs for Late Final Completion

Gentlemen:

As you are aware, the original Substantial Completion Date for this project was (Date). More than suffi-cient time has since been allowed for all trades to complete respective punchlists.

Be advised that after (Date), no (Company) supervisory personnel will be on the jobsite on any regular basis. Any of your contractual work progressing for any reason beyond that date will make it necessary for (Company) management and supervisory personnel to be at the site while your work is being per-formed. In that event, you will be backcharged accordingly.

Very truly yours,

COMPANY

Project Manager

cc: Jobsite
         File: Punchlist
                   Backcharge
                   CF

notification to, in a sense, keep the wheel squeaking, as another prod to keep Subcontractors moving to complete their work, and as another notification that will help the Company recover at least some of the real costs of late completion.

4.19   Securing Subcontractor/Supplier Guarantees and Warranties

4.19.1   General

The process of securing Subcontractor and supplier guarantees and warranties is very similar and closely related to that for the punchlist. In fact, these items will actually be part of the punchlist if they have not been secured by that time.

They are, however, a special case because of the nature of the items and the difficulty in providing acceptable documents to the Owner if they are not, for whatever reason, forthcoming from the original subvendor. If the acceptable documents are not provided, it can have a profound effect on the release of final payment, and even on retainage reduction. It is therefore incumbent upon the General Contractor to take a firm, direct approach to securing these documents as expeditiously as possible, and to take strong measures in the face of non-compliance.

The process described in this section effectively parallels that for the punchlist. Refer also to Section 3.14, Guarantees and Warranties, for important related discussion.

4.19.2   Procedure

1. As the project nears substantial completion, prepare your own list of required documents.
   a. Don't wait for the Architect or Owner representative to formally advise you of the need to provide general and specific guarantees and warranties.
   b. If a list of required guarantees and warranties is given in the Supplementary General Conditions, do not accept it as complete. Research every specification section yourself to confirm that the list is in fact complete. This will go a long way toward oversight of any item that can later affect final payment.
2. Transmit the requirement to each vendor with any responsibility for providing specified guarantees and warranties. Use the sample Request for Guarantees/Warranties Form Letter of Section 4.19.3 to help here.
3. Follow up with "second" and "third" requests if necessary. Write additional notes on the original request letter to stress the urgency; then photocopy and send out.

4. If there still is no acceptable response, contact the president of the Company, or as high an official as you can reach. Advise him or her of the effort spent, time passed, and potential and real effects on the project and related payments. Be sure to send letters confirming every conversation.

5. When all other efforts have failed, consider using the sample Final Notice to Subvendors to Provide Guarantees/Warranties of Section 4.19.5. Be prepared to actually follow through on the notice—and follow through immediately on the prescribed date.

4.19.3 Sample Request for Guarantees/Warranties Form Letter #1 *(page 4.128)*

The sample Request for Guarantees/Warranties Form Letter #1 is another example of a similar form letter provided in Section 3.13.6. It is the first written notification to the Subcontractor or Supplier to provide all proper guarantees and warranties. Specifically, it:

1. Advises that proper form and language are required.

2. Specifies the date when the guarantee or warranty is to take effect.

3. Informs that payments will be affected until the proper submission is received.

Fax and mail the letter.

4.19.4 Sample Final Notice to Subvendors to Provide Guarantees/Warranties *(page 4.129)*

The sample final Notice to Subvendors to Provide Guarantees/Warranties is to be sent to any subvendor who, despite every effort, simply fails to provide any guarantees or warranties, or fails to provide them in proper form or content. Specifically, it notifies the offender that:

1. The Company is now making other arrangements to provide the documents.

2. An appropriate amount of money will be withheld from the subvendor's account to cover the cost of possible guarantee work.

3. The Subcontractor will be backcharged for all the time, effort, and expense involved (since the initial effort).

It is a good idea to consult with the Owner and/or the Architect before proceeding. Advise them of the trouble experienced with the particular subvendor and try to elicit a confirmation that they will actually accept such alternative guarantees or warranties without other legal maneuvering. If they are not agreeable, consult with your attorney to determine *your* rights under the general contract to provide such alternate guarantees and warranties.

4.19.3
Sample Request for Guarantees/Warranties Form Letter #1

Letterhead

(Date)

To: (List All Project Subcontractors)   (List All Respective Fax #'s)

RE: (Project)
           (Company Project #)

SUBJ: Delivery of Guarantees and Warranties

Gentlemen:

It is your responsibility to submit guarantees and warranties in proper form. These must be delivered to this office by or before (Date). Please review your specifications to confirm specific content and language as detailed therein, and comply in every respect.

Release of retainage is contingent upon your compliance with these requirements.

If you have any questions, please contact me immediately.

Very truly yours,

COMPANY

Project Manager

cc: Jobsite
       File: Guarantees/Warranties
                CF

4.19.4
Sample Final Notice to Subvendors to Provide Guarantees/Warranties

---

Letterhead

Date: _____

To : _____
_____
_____

ATTN: _____

RE: Project:_____
Project #:_____

SUBJ: Failure to Provide Proper Guarantees/Warranties

Confirmation of Fax
Fax#:_____

CERTIFIED MAIL
RETURN RECEIPT REQUESTED

Mr. (Ms.)_____:

You were requested on a number of occasions to provide all appropriate guarantes and warranties in accordance with the terms of your contract. To date, you continue to fail to comply.

Accordingly, we are now making arrangements to provide the required guarantes and warranties to the Owner on your behalf in order to allow project close-out.

Until your proper guarantees/warranties are received by this office and accepted by the Owner, an appropriate amount of money as determined in our discretion will be withheld from your current payment as security to cover the cost of possible guarantee/warranty work. We estimate this value at $_____.

In addition, your contract will be backcharged for all costs related to the excessive measures necessary to resolve this issue, including attorney's fees, administrative time, overhead, and profit.

Very truly yours,

COMPANY

Project Manager

cc: Jobsite
Vendor File:_____
Guarantees/Warranties File
CF

4.20   Subvendor Performance Evaluation

4.20.1   Use of Subvendor Performance Evaluation Form

The Subvendor Performance Evaluation Form is to be completed by those with project manager, project engineering, and superintendent responsibilities—those who were closest to the performance itself.

It will be placed in the subvendor's permanent file to assist purchasing efforts in the future, and it can even be consolidated in any narratives or completion reports to be prepared at the end of the project.

Be sure that the form is completed for every subvendor on every project.

4.20.2   Sample Subvendor Performance Evaluation Report— Completed Example *(page 4.131)*

4.20.3   Sample Subvendor Performance Evaluation Report—Blank Form *(page 4.132)*

4.21   Project Close-Out Checklist *(page 4.133)*

The Project Close-Out Checklist is to be considered as the project is nearing its completion. Hopefully, by the time the project is in this stage, many of the items listed will be either complete or in process.

The checklist summarizes the items most typical of conventional construction projects of any size that need to be completed, submitted, turned over, etc., in order to complete all detailed requirements of each bid package. Sections of it can therefore be conceptually repeated for each bid package, with satisfaction of the checklist itself considered to be the consolidation of the respective requirements for each bid package.

It is important, however, to acknowledge that such requirements are being "supplemented" almost daily—to subtle or dramatic degrees. It is therefore crucial that the specifications and any procedure manual for a specific project be studied in detail to uncover any special close-out requirements for the particular project. Add all such to the checklist as appropriate, and reproduce as the specific project close-out checklist that can be used by each member of the project team.

4.22   Productivity

4.22.1   Section Description

The management of construction projects in the lump sum general contracting, construction management, and design-build project delivery formats largely deals with a very high percentage of work being performed by subtrades

4.20.2
Sample Subvendor Performance Evaluation Report—Completed Example

Date: AUGUST 16, 1994
Project: HIGHTOWER APTS
Project #: 9990
Owner: PAUL HIGHTOWER
Architect: GEORGE ARCH
Location: 555 HIGHTOP PL.
Project Manager: MARK LEONARDO
Project Engineer: DAVE JEFFKO
Superintendent(s): PAUL SMALL

X Subcontractor _____ Supplier
Co. Name: STRAIT STEEL CORP
Address: 950 HIGHLAND PKWY
NEWTOWN, CT 06424
Contact: JOHN STRAIT
Phone: (203) 555-1652 Fax: (203) 555-1660

Scope of Work: STRUCTURAL STEEL
Subcontract/P.O. Value: $ 96,000.-
Start: Req'd: 4-20-94 Actual: 4-27-94
Comp.: Req'd: 5-1-94 Actual: 5-1-94

Project Type:
___ State ___ Federal ___ Municipal ___ Private

	PERFORMANCE	Excellent	Good		Poor
1.	Understanding of Subcontract/P.O.		X		
2.	Understanding of Contract Documents		X		
3.	Timely preparation of Approval Submittals	X			
4.	Complete preparation of Approval Submittals		X		
5.	Understanding of/adherence to schedules		X		
6.	Record of material delivery schedules	X			
7.	Quality of communications			X	
8.	Timeliness of communications			X	
9.	Attendance at job meetings				X
10.	Cooperation; willingness to resolve problems		X		
11.	Adequate management		X		
12.	Adequate number and ability of administrative staff		X		
13.	Ability & willingness to comply with project procedures			X	
14.	Timeliness/completeness of Certified Payroll Reports			X	
15.	Timeliness/completeness of lien waivers & releases			X	
16.	Worker payment record		X		
17.	Subcontractor payment record			N/A	
18.	Supplier payment record		X		
19.	Jobsite cleanup record			X	
20.	Ability to minimize punchlist		X		
21.	Overall level of workmanship		X		
22.	Submission of Guarantees & Warranties			X	
23.	Preparation/submission of As-Built documents			X	
24.	Proper submission of all project close-out documents				X
25.	Safety compliances and safety performance		X		
26.	EEO & MBE employment compliances			X	
27.	Avoidance of claims and claims posturing	X			
28.	Avoidance of need for backcharges	X			
29.	TOTAL	4	12	6	5	0

Describe the most significant good and/or poor performances: KNOWS TRADE & WORKMANSHIP WELL. CAN USE IMPROVEMENT IN TIMLINESS OF SUBMISSION OF OVERALL PROJECT DOCUMENTATION - BUT IT EVENTUALLY GETS SUBMITTED CORRECTLY. A PLEASURE TO WORK WITH ON-SITE

Prepared By: Mark Leonardo   Date: 8/16/94

4.20.3
Sample Subvendor Performance Evaluation Report—Blank Form

Date: _____   ___ Subcontractor   ___ Supplier

Project: _____   Co. Name: _____
Project #: _____   Address: _____
Owner: _____   _____
Architect: _____   Contact: _____
Location: _____   Phone: _____ Fax: _____
Project Manager: _____
Project Engineer: _____   Scope of Work: _____
Superintendent(s): _____   Subcontract/P.O. Value: _____
_____   Start: Req'd: _____ Actual: _____
                                           Comp.: Req'd: _____ Actual: _____

Project Type:
___ State   ___ Federal   ___ Municipal   ___ Private

	PERFORMANCE	Excellent		Good		Poor
1.	Understanding of Subcontract/P.O.	___	___	___	___	___
2.	Understanding of Contract Documents	___	___	___	___	___
3.	Timely preparation of Approval Submittals	___	___	___	___	___
4.	Complete preparation of Approval Submittals	___	___	___	___	___
5.	Understanding of/adherence to schedules	___	___	___	___	___
6.	Record of material delivery schedules	___	___	___	___	___
7.	Quality of communications	___	___	___	___	___
8.	Timeliness of communications	___	___	___	___	___
9.	Attendance at job meetings	___	___	___	___	___
10.	Cooperation; willingness to resolve problems	___	___	___	___	___
11.	Adequate management	___	___	___	___	___
12.	Adequate number and ability of administrative staff	___	___	___	___	___
13.	Ability & willingness to comply with project procedures	___	___	___	___	___
14.	Timeliness/completeness of Certified Payroll Reports	___	___	___	___	___
15.	Timeliness/completeness of lien waivers & releases	___	___	___	___	___
16.	Worker payment record	___	___	___	___	___
17.	Subcontractor payment record	___	___	___	___	___
18.	Supplier payment record	___	___	___	___	___
19.	Jobsite cleanup record	___	___	___	___	___
20.	Ability to minimize punchlist	___	___	___	___	___
21.	Overall level of workmanship	___	___	___	___	___
22.	Submission of Guarantees & Warranties	___	___	___	___	___
23.	Preparation/submission of As-Built documents	___	___	___	___	___
24.	Proper submission of all project close-out documents	___	___	___	___	___
25.	Safety compliances and safety performance	___	___	___	___	___
26.	EEO & MBE employment compliances	___	___	___	___	___
27.	Avoidance of claims and claims posturing	___	___	___	___	___
28.	Avoidance of need for backcharges	___	___	___	___	___
29.	TOTAL	___	___	___	___	___

Describe the most significant good and/or poor performances: _____
_____
_____
_____
_____

Prepared By: _____   Date: _____

4.21
Project Close-Out Checklist

1. All systems on operation
   a. Plumbing ____
   b. HVAC ____
   c. Electrical ____
   d. Fire Protection ____
   e. _____ ____
   f. _____ ____

2. Performance tests conducted
   a. Plumbing ____
   b. HVAC ____
   c. Electrical ____
   d. Fire Protection ____
   e. _____ ____
   f. _____ ____

3. O & M Manuals delivered
   a. Plumbing ____
   b. HVAC ____
   c. Electrical ____
   d. Fire Protection ____
   e. _____ ____
   f. _____ ____

4. Operating Instructions to Owner performed
   a. Plumbing ____
   b. HVAC ____
   c. Electrical ____
   d. Fire Protection ____
   e. _____ ____
   f. _____ ____

5. Final completion of physical work
   a. Company Punchlist complete ____
   b. Arch/Owner Punchlist complete ____
   c. All sign-off forms completed ____
   d. Completion Certificate(s) rec'd ____

6. Demobilization complete
   a. Field Offices ____
   b. Equipment & furnishings ____
   c. _____ ____
   d. _____ ____

7. Termination of Temporary Services
   a. Heat, light, power, phone ____
   b. Fire, police, guard service ____
   c. Insurance transfers ____
   d. _____ ____
   e. _____ ____

8. Final cleaning of all areas completed
   a. _____ ____
   b. _____ ____
   c. _____ ____
   d. _____ ____

9. As-Built Drawings
   a. General ____
   b. Site ____
   c. Plumbing ____
   d. HVAC ____
   e. Electrical ____
   f. Fire Protection ____
   g. _____ ____

10. Guarantees & Warranties
   a. General ____
   b. All subcontractor documents ____
   c. Roof & other bonds ____
   d. _____ ____

11. Inspection Certificates
   a. _____ ____
   b. _____ ____

12. Material/Installation Certificates
   a. _____ ____
   b. _____ ____

13. Lien Waivers & General Release
   a. General ____
   b. All subcontractor documents ____
   c. _____ ____

14. Billing & Charges Processed
   a. All acknowledged Change Orders ____
   b. All subcontractor changes ____
   c. All subcontractor backcharges ____
   d. Final billings from all subvendors ____
   e. Final billing submitted to Owner ____

15. Steps taken to finalize outstanding claims
   a. To Owner ____
   b. To subcontractors ____
   c. By subcontractors ____
   d. _____ ____

16. Project Completion Report submitted ____

17. Proj. records transferred to home office ____

18. Forwarding address confirmed
   a. Post Office notified ____
   b. All subs & suppliers notified ____
   c. _____ ____

19. Other general contract, subcontract, purchase order, specification, and Company Procedure items necessary to close out the project:
   a. Attach list ____
   b. _____ ____

and correspondingly reduced amounts of work being performed by the prime contractor or construction manager directly. Although many prime contractors and construction managers certainly continue significant operations with their own direct labor force, such operations are generally confined to specific trades. Some general contractors, for example, preserve the ability to perform at least certain types of concrete work, rough and finish carpentry work, and/or masonry work. Subcontractors, however, are most often precisely the opposite. These companies are the ones who perform the direct labor for the greatest percentage of the entire project. An electrical subcontractor, for example, might provide 10% to 20% of the total labor expended on the project, as may a concrete subcontractor.

In such companies that maintain these types of direct profit centers, their management responsibility divides between two distinct categories. The first is the management of contracts and subcontracts. This discipline addresses all of those things necessary for effective administration of the interactions among all contracting parties. The second responsibility addresses the need for efficient organization and management of one or more direct profit centers. Planning and organization of the materials, personnel, equipment, etc., necessary to construct the particular scope of work within the estimated budget and time frame required by the progress schedule is a distinct—and completely separate type of—management activity. Very often, this separate activity must be performed each day by the same individual(s) who are responsible for effective management of the overall contract administration function.

This section of the Operations Manual addresses certain factors affecting productivity on our construction projects. To many of us in construction, the concept of "productivity" on first consideration appears to be straightforward and simple. When asked to quantify and measure productivity, many project managers, project engineers, superintendents, and cost engineers fall short in their knowledge of such matters.

Productivity is actually a measure of how much can be accomplished with resources available. It is not the intention of this section of the Operations Manual to measure or establish any particular productivity, but to offer concepts and suggestions that will help you improve your labor productivity, whatever your starting point may be. In order to control any item, it is usually necessary to begin with some sort of benchmark against which to measure. In this case, such a "benchmark" against which to measure our productivity improvement efforts might be considered to be the control-estimate labor-hours prepared as a basis of a particular bid.

For the prime contractor, construction manager, or specialty subcontractor, "productivity improvement" can simply be defined as accomplishing more work with little (or no) increase in resources, including labor, money, and time. It means that proverbial "working smarter, not (necessarily) harder." Following the ideas and recommendations of this section will hopefully result in not only the improvement of productivity of specific work activities, but also an improvement in the perspective of the organization as well. It should encourage the preservation of certain methodologies that will continually work to improve the factors affecting productivity in everything we do.

This section will review the particular factors that may affect productivity, will continue with a self-audit that may help identify certain productivity improvement efforts, and follow up with a list of items that will assist in your institution and maintenance of an ongoing productivity improvement program for your company.

4.22.2   Factors Affecting Productivity

There are certainly many things that definitely and potentially can affect "productivity." Labor productivity is a function of several specific variables, many of which can be analyzed directly. The impact of many of these variables can even be estimated before, and followed through on, in application. The first step in the exercise of productivity improvement, then, is to consider the factors affecting productivity in order to have a target to which to devote our attention, and to establish the "baseline productivity" against which to measure our improvement efforts.

Baseline productivity can be thought of as the labor (man-hour) rates that can be expected to be achieved at a particular location under "normal" or "average" conditions. Some factors that affect baseline labor productivity can include the following:

- Project design
- "Normal" or unusual contract arrangements
- Local climate
- Local craft availability
- Local craft composition; union/nonunion
- Degree and manner of supervision
- Amount and manner of equipment utilization
- Use of technology
- Material type and availability
- Project schedule and organization
- Management capacity and competence (prime contractor)
- Management capacity and competence (subvendor)
- Site layout
- Site access

Because every job, every prime contractor, and every subvendor is different, the relative importance of the above factors is sure to change. With such changes also come changes in your ability to control and influence any of these factors.

More specifically, the need to address the adverse effects on labor productivity that result from causes that are beyond the control of the contractor may also become the basis of a compensable change order, if some mechanism can be implemented that will credibly quantify the impact of any of these adverse conditions.

In the checklist in Section 4.22.3 that itemizes the factors affecting labor productivity, the individual items identified cover a variety of conditions. The values included in the column on the right side of the checklist can be considered to be factors to be added to the basic labor costs of work activities that occur under "standard operating conditions" and during normal working hours. These factors are intended to serve as a guide only. The percentages listed are examples of the types of effects that might be experienced by the individual conditions indicated. They are likely to vary from company to company, contract to contract, crew to crew, and jobsite to jobsite. It is therefore important to recognize that individual cases may prove these factors to be either too high or too low.

Use the Factors Affecting Labor Productivity checklist as a reference that should be tested by your own ongoing work experiences. As such, modify the factors as you may deem necessary in your use of them in order to develop your company-specific checklist of the Factors Affecting Labor Productivity that will be backed up by your own company history. To the extent that this can be accomplished, it might later serve as powerful support for future claims of reduced productivity.

4.22.3 FMI Annual Survey on Productivity

FMI Corporation, headquartered in Raleigh, North Carolina, is the largest management consulting firm devoted specifically to the construction and engineering industries. They frequently poll the industry's CEOs, CFOs, COOs, and other upper-level executives in both the general contracting and specialty contracting industries to compile information relating to many topics, productivity being an important one. Their 2012 survey reveals trends and management techniques to improve productivity along with the challenges the industries must overcome to do so. A portion of that survey is presented here:

Exhibit 11	Largest internal challenges to improving productivity
Exhibit 12	Largest external challenges to improving productivity
Exhibit 14	On projects where BIM was used, what was the impact on productivity?
Exhibit 15	In what capacity did you use BIM on these projects?
Exhibit 16	Have you worked on a project where IPD was used?
Exhibit 17	On projects where IPD was used, what was the impact on productivity?
Exhibits 22 & 23	Prefabrication as a strategy for productivity

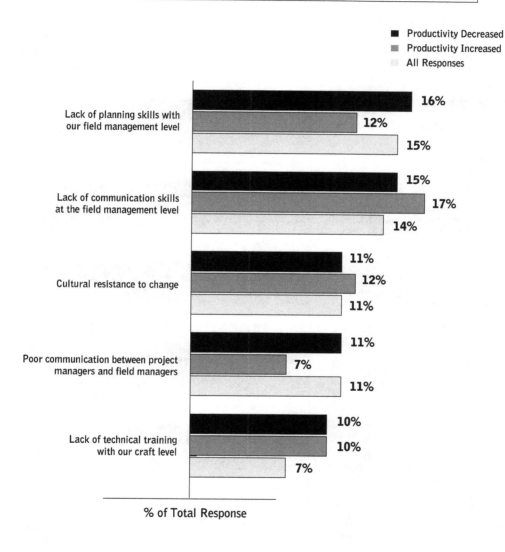

Exhibit 11. Largest Internal Challenges to Improving Productivity

% of Total Response

Exhibit 12. Largest External Challenges to Improving Productivity

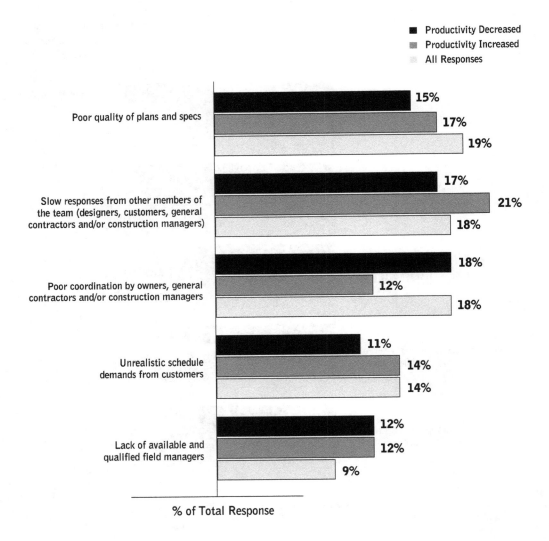

■ Productivity Decreased
■ Productivity Increased
▫ All Responses

Poor quality of plans and specs
15%
17%
19%

Slow responses from other members of the team (designers, customers, general contractors and/or construction managers)
17%
21%
18%

Poor coordination by owners, general contractors and/or construction managers
18%
12%
18%

Unrealistic schedule demands from customers
11%
14%
14%

Lack of available and qualified field managers
12%
12%
9%

% of Total Response

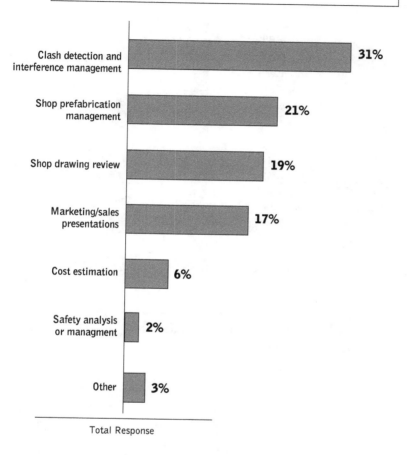

Exhibit 15. In what capacity did you use BIM on these projects?

Total Response

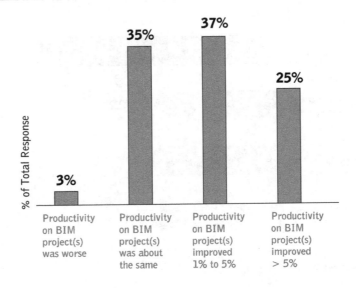

Exhibit 14. On projects where BIM was used, what was the impact on productivity?

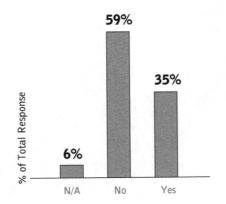

Exhibit 16. Have you worked on a project where IPD was used?

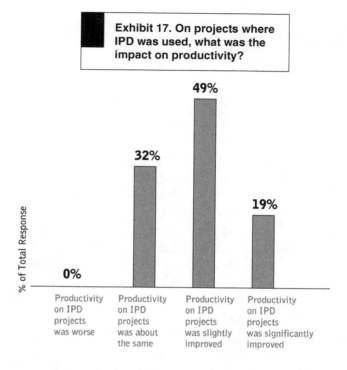

Exhibit 17. On projects where IPD was used, what was the impact on productivity?

% of Total Response

0%
Productivity on IPD projects was worse

32%
Productivity on IPD projects was about the same

49%
Productivity on IPD projects was slightly improved

19%
Productivity on IPD projects was significantly improved

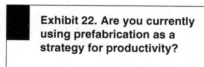

Exhibit 22. Are you currently using prefabrication as a strategy for productivity?

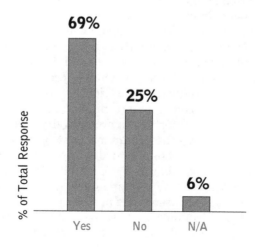

% of Total Response

69%   25%   6%
Yes   No   N/A

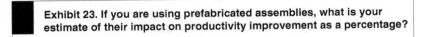

Exhibit 23. If you are using prefabricated assemblies, what is your estimate of their impact on productivity improvement as a percentage?

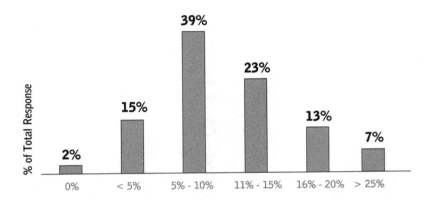

4.22.4   Productivity Improvement Self-Audit

The first step in considering the issue of productivity with the aim of developing information that will help lead to improvement in the productivity is to turn within. We need to study our organizations in order to determine what we are doing and how are doing it. We need to specifically consider the ways in which we organize, direct, and control our own work, and identify those components of our own work effort that need improvement. In general, this self-consideration (self-audit) should not be considered to be a specific, itemized, and definitive checklist of particular do's and don'ts, and is not intended to necessarily criticize your current level of productivity. Instead, such a productivity self-audit is intended to bring to the attention of management those considerations and ideas that will be helpful or necessary in order to improve productivity and to maintain such higher levels of productivity on a sustained basis.

This productivity self-audit, then, is a general guide that is designed to improve management's awareness. As such, it should be referred to periodically in order to keep the concepts and objectives of higher productivity continually in the forefront of every manager's daily consideration.

Each contractor, construction manager, and specialty contractor is sure to have a list of items important to productivity and its improvement that may be different from every other company that operates in similar circles. Certain items, however, can be common to all companies. These common items can therefore boil down to those that follow. Each area identified will contain a list of questions to be considered by management. Carefully considering them will help managers to honestly assess the present position of their companies relative to productivity performance, and realistically identify those areas that can benefit most from concerted efforts toward productivity improvement. The general areas, then, to be considered in the self-audit of productivity improvement are as follows:

1. The level of awareness throughout the company's management and staff of current levels of productivity, productivity ability relative to your industry niche, and level of concern for the need to continually address productivity improvement.

2. The level of knowledge among company management and staff regarding "poor," "acceptable," and "superior" levels of performance, and those factors affecting performance in the critical areas of your own operation.

3. The degree and manner in which efforts throughout the company are directed to those items that have been identified as having the greatest potential for productivity improvement.

4. The manner in which the company's production processes are analyzed and improvement methods are being applied.

5. The degree to which the concept of productivity has or has not been made part of the company's culture; the manner in which responsibility and accountability for productivity and its improvement has or has not been defined throughout the organization.

6. The manner in which the productivity effort has been formalized and established as part of the company's ongoing management responsibilities and efforts, including the degree of employment of specialized tools and the use of other resources.

1. Awareness of current levels of productivity, relative ability, and level of concern for productivity improvement.
   a. Is sufficient time devoted to productivity improvement matters throughout the company?
      i. Are productivity and the concept of productivity improvement regularly discussed in company staff meetings?
      ii. Is sufficient time devoted throughout the company to productivity matters?
      iii. Is specific time regularly devoted to reviewing operational performance, or other productivity issues?
      iv. Is the subject of productivity encouraged to be a regular part of ongoing discussions?
      v. Is "productivity improvement" continually requested from subordinates?
      vi. Do all management, staff members, and labor force members believe that the company is seriously concerned about productivity improvement?
   b. Does each member of the organization accept the idea of the need for productivity improvement?
      i. Is the concept of productivity improvement regularly discussed as an important agenda for the company?
      ii. Does the organization attempt to involve employees in the implementation of specific productivity improvement efforts?

       iii. Does the organization develop specific productivity objectives and follow up on the attainment of those objectives?

       iv. Does the organization distribute information relative to the success of any efforts to improve productivity?

  c. Are any incentives used in the company or on any project in order to encourage participation by everyone in productivity improvement?

       i. Do formal communication vehicles, such as company newsletters, highlight any gains in productivity and give credit to specific individuals involved in productivity successes?

       ii. Are recognition items such as awards, certificates, prizes, or money given for the achievement of certain productivity objectives?

  d. Are long-term and short-term productivity improvement objectives established and formalized?

       i. Are performance improvement objectives developed for each major area of operation in the company?

       ii. Are such objectives quantified in terms that are measurable?

       iii. Is there a mechanism established to follow up on performance and to compare actual performance against the target objectives?

2. The level of knowledge regarding levels of performance, and those factors affecting performance in the critical areas.

  a. Is "efficiency" identified in a way that can be measured and tracked?

       i. Is unit cost information available for all activities to which productivity improvement will be applied?

       ii. Are unit times developed for each significant activity?

       iii. Is trend information available for each of the unit times?

       iv. Is efficiency information reported periodically to management?

       v. Is efficiency information compared to other organizations performing similar work on similar projects?

  b. Does the company establish performance standards for significant items of work?

       i. Have industry standards for particular work outputs been identified?

       ii. What percentage of the total company work activities has been addressed by "industry standard" information?

       iii. Has the company output for previous periods been compared to appropriate industry standards in order to establish the relative performance level of the company?

       iv. If the output has been compared to appropriate industry standards, has the comparison been trended over time?

       v. Are performance reports regularly prepared and distributed to management and staff members who are responsible for the achievement of the productivity improvement objectives?

       vi. Are performance standards used by supervisors in planning, scheduling, and performing daily operations?

       vii. Are comparisons of actual work output to the established performance standards used by the estimating department in the preparation of bids for future work?

    c. Is the effectiveness of the implementation of the productivity improvement program measured in any way?

        i. Are performance indicators established and evaluated in a manner that will reflect the effectiveness of all productivity improvement goals?

        ii. Are measurable goals associated with each performance indicator in order to allow determination of program effectiveness?

    d. Is the quality of work being controlled at an acceptable level consistent with the same effort devoted toward productivity improvement?

        i. Is the subject of "quality" regularly discussed at project staff meetings?

        ii. Do all company management, staff, and supervisory employees clearly understand the definition of appropriate quality and how it will be measured?

        iii. Does the company have any manner of identifying and measuring rework or other problems associated with poor quality?

        iv. Is quality performance reported in any manner on any regular basis?

3. Manner in which efforts are directed to those items having the greatest potential for productivity improvement.

    a. Has an effort to evaluate the potential for productivity improvement been implemented in the company?

        i. Have specific work activities been identified for productivity improvement?

        ii. Are these work activities ranked in any prioritized sequence with respect to their potential for improvement?

        iii. Does the method(s) for selecting potential areas and work activities for improvement address specific items in the measurable way?

    b. Is the identification of items targeted for improvement prioritized based upon their potential for cost savings or other specific benefits?

        i. Is the cost-saving or other benefit potential determined before significant resources and effort are devoted to a particular item?

        ii. Are other considerations, such as the relative importance of an activity or work output to other activities, being given adequate attention?

        iii. Is the list of major problem areas periodically reassessed and reprioritized?

4. Manner in which the company's production processes are analyzed and improvement methods are being applied.

    a. Are the methods employed that relate activities, valuations, and results valid?

        i. Do the original assumptions made in the program determination remain valid?

        ii. Does the company continually consider alternative approaches?

        iii. Is the prioritized list of targets that have been selected for productivity improvement periodically reviewed and evaluated?

        iv. Is the program given attention in the appropriate intervals?

        v. Are the company managers aware of what each other is doing within the organization with respect to productivity improvement?

    vi. Are the company managers aware of what others are doing in companies that are similar to yours?

    vii. Does the company employ productivity specialists or consultants to conduct studies or to offer recommendations?

    viii. Do managers, supervisors, and employees make suggestions with respect to specific actions that can be taken that impact actual work activities?

b. Is the manner in which employees use their time reviewed periodically?

    i. Is work distributed evenly among managers, supervisors, and staff?

    ii. Is "idle time" identified, along with necessary and unnecessary reasons for it?

    iii. Are the unnecessary reasons for idle time targeted as items for elimination?

    iv. Are cleanup operations performed effectively, or do they result in excessive, nonproductive work time?

    v. Are cleanup operations performed insufficiently, resulting in loss of productivity due to "difficult" or even unsafe work areas?

    vi. Are formalized scheduling and/or production planning techniques used properly?

    vii. Do peak periods of work frequently cause imbalances between the work backlog in the available workforce?

c. Are managers, staff, and labor force members motivated toward productivity improvement?

    i. Are all company employees advised with respect to their manner of performance?

    ii. Are any merit increases or awards tied to productivity improvements?

    iii. Are any specific motivational methods being used?

5. Manner in which responsibility and accountability for productivity and its improvement has been defined.

a. Are productivity improvement responsibilities clearly identified and assigned to managers, supervisors, and staff members?

    i. Are supervisors and managers held accountable for productivity improvement?

    ii. Are the responsibilities assigned to company supervisors and managers for productivity improvement commensurate with their levels of authority?

    iii. Are the various efforts implemented throughout the company toward productivity improvement coordinated through a responsibility structure with a single individual ultimately responsible for productivity improvement throughout the organization?

b. Are the different responsibilities for productivity improvement effectiveness specifically identified?

    i. Is the responsibility for program effectiveness assigned to and implemented by senior managers?

    ii. Is the responsibility for operational efficiency and preservation of all quality measures assigned to and implemented by managers and supervisors?

   c. Have adequate procedures been implemented to assure that productivity improvement responsibilities are clearly understood and implemented as instructed?

      i. Is a system in place to ensure that productivity improvements are reported regularly?

      ii. Are actual results in achieving the implementation of productivity improvement efforts reported?

      iii. Is there any method to determine if follow-up efforts improve the attainment of productivity improvement?

6. Manner in which the productivity effort has been formalized.

   a. Have the functions of program evaluation, target activity identification, work measurement, procedures management, and other necessary program components been adequately staffed?

   b. Are the necessary technical disciplines available among company staff, including management, engineering, computer technology, systems research and design, and training?

   c. Are outside resources being used effectively, including:

      i. Productivity improvement consultants?

      ii. Seminars and training?

4.22.5 Productivity Improvement Guide
for Construction Contracts (*page 4.148*)

The Productivity Improvement Guide for Construction Contracts that follows was prepared by the National Construction Employers Council, Washington, D.C., and is reprinted here with permission. The purpose of the guide is not to measure productivity directly, but to improve your current levels, whatever they are, and to do so through the use of the resources that are currently available. It works toward the accomplishment of this by focusing attention on some of the key factors that affect construction productivity.

For the construction contractor, productivity improvement means accomplishing more work with little or no increase in associated necessary resources. The guide that follows is to be used as an aid to help consider a project from this point of view of productivity.

Using the guide every few months can help managers analyze productivity and continually search for ways to improve. If participation by all company project managers, project engineers, supervisors, and foremen is encouraged, the findings of these people may become subjects of discussion for meetings between field and office management. And so this guide can be considered to be a focus on improving management productivity, and not necessarily a method for measuring and motivating the direct labor force directly.

The guide uses the checklist and multiple-choice format in order to help pinpoint actions needed to improve productivity in these areas. Your responses to questions and statements in Parts A, B, and C will create a "productivity profile." Part D, the Tally, will help you analyze this profile in order to determine the best ways to improve your productivity.

4.22.5
Productivity Improvement Guide for Construction Contracts

Part A: Overview

Throughout this guide, you are to evaluate *one* specific job you now have in progress. To give you an overview of that job before you start, respond to the following statements.

1. Job being evaluated: _____
2. Date today: _____
3. Just as a reference point, check one of the boxes that best matches your view of productivity on this job so far.
   A □ very productive
   B □ average productivity
   C □ below average productivity
   D □ very unproductive

	Yes	No
4. Do you have a written schedule?	□	□
5. Does the schedule set clear monthly targets?	□	□
6. Does the schedule set clear weekly targets?	□	□
7. According to the schedule, do you know where you should be as of today?	□	□

8. Whether or not you have a schedule, keeping tabs on job progress is important. Give your best (honest) estimate of your *expected* percent completion of the job as of this week. If you have a schedule, be sure to include any revisions necessary to reflect contract changes.
9. What is your best estimate of the *actual* percent completion of the job as of today?
10. Divide the amount of dollars you've expended so far on the job by the total estimated job cost to get a percentage expenditure to date.
11. Now compare the expected completion percentage in (8), the actual completion in (9) and the actual percent expended in (10). All three percentages provide measures of job progress that may be useful to you. Are the percentages reasonably close? If not, why not? Is this important for this job?
12. Productive construction requires, among other things, a great deal of coordination and cooperation, particularly among contractors, subcontractors and labor organizations, all of whom are closely interdependent. Check here the "supporting trades" (those contractors whose work has the greatest impact on your own) and the labor organizations you most frequently deal with.

Contractors	Labor Organizations
□ General	□ Asbestos workers
□ Boilermakers	□ Bricklayers
□ Ceilings and Interiors	□ Boilermakers
□ Electrical	□ Carpenters
□ Elevator	□ Electrical workers
□ Erectors	□ Elevator constructors
□ Glazing	□ Engineers
□ Insulation	□ Glazers
□ Masonry	□ Ironworkers
□ Mechanical	□ Laborers
□ Painting and Decorating	□ Painters
□ Roofing	□ Plasters
□ Sheet work	□ Plumbers and Pipefitters
□ Other _____	□ Roofers
	□ Sheet Metal workers
	□ Other _____

Part B: Checklist

INSTRUCTIONS: Go down the list of statements in this part, checking the box that best applies to this job. The boxes compare this job to your experience with all other jobs in the past.

4.22.5
Productivity Improvement Guide for Construction Contracts *(Continued)*

	One of My Best	Clearly Above My Average	Within My Average Range	Clearly Below My Average	One of My Worst	Don't Know	Not Applicable
1. Schedule allows for proper work sequence.	□	□	□	□	□	□	□
2. Prejob conference with general and subcontractors to plan job, schedule sequence and coordinate work with others.	□	□	□	□	□	□	□
3. Manning in relation to optimal crew size and work pace.	□	□	□	□	□	□	□
4. Time craftsmen spend waiting for instructions from foremen.	□	□	□	□	□	□	□
5. Appropriate work packages clearly defined with minimum interdependence on other crafts.	□	□	□	□	□	□	□
6. Accuracy of job estimate.	□	□	□	□	□	□	□
7. Worker's knowledge of job progress, problems and plans for upcoming work.	□	□	□	□	□	□	□
8. Adequacy of clean-up locker room and toilet facilities.	□	□	□	□	□	□	□
9. Agreement on responsibility for meeting safety regulations between general and subs.	□	□	□	□	□	□	□
10. Instructions and drawings are coordinated with work of others to minimize conflict.	□	□	□	□	□	□	□
11. Work areas free of clutter and debris.	□	□	□	□	□	□	□
12. Time spent moving up and down and walking to and from work locations.	□	□	□	□	□	□	□
13. Availability of alternative work while waiting on other trades, deliveries or delays.	□	□	□	□	□	□	□
14. Communication between front office and on-site field manager.	□	□	□	□	□	□	□
15. Introduction of new hires to overall job and to their tasks within the job plan.	□	□	□	□	□	□	□

4.22.5
Productivity Improvement Guide for Construction Contracts *(Continued)*

	One of My Best	Clearly Above My Average	Within My Average Range	Clearly Below My Average	One of My Worst	Don't Know	Not Applicable
16. Adequacy of parking for the on-site labor force.	☐	☐	☐	☐	☐	☐	☐
17. Liaison and cooperation with supporting trades and craftsmen.	☐	☐	☐	☐	☐	☐	☐
18. Quality of internal prejob planning and coordination prior to starting.	☐	☐	☐	☐	☐	☐	☐
19. Job safety and accident record.	☐	☐	☐	☐	☐	☐	☐
20. The right tools and equipment available when needed.	☐	☐	☐	☐	☐	☐	☐
21. Minimum handling of materials before final installation.	☐	☐	☐	☐	☐	☐	☐
22. Foremen's knowledge of current schedule and overall job progress.	☐	☐	☐	☐	☐	☐	☐
23. Prejob conference with union(s) to discuss manpower needs and potential jurisdictional problems.	☐	☐	☐	☐	☐	☐	☐
24. Quick recognition of outstanding performance by an individual or a crew.	☐	☐	☐	☐	☐	☐	☐
25. Realistic construction schedule and deadlines.	☐	☐	☐	☐	☐	☐	☐
26. Minimum rework due to outside interference or damage caused by other trades.	☐	☐	☐	☐	☐	☐	☐
27. Tools and equipment operated at peak efficiency and effectiveness.	☐	☐	☐	☐	☐	☐	☐
28. Time spent idle while one (or more) crew member waits for rest of crew.	☐	☐	☐	☐	☐	☐	☐
29. Sufficient on-site field management to observe work, talk with employees, and make timely decisions.	☐	☐	☐	☐	☐	☐	☐
30. Clarity and accuracy of drawings.	☐	☐	☐	☐	☐	☐	☐
31. Cooperation between management and labor organization(s).	☐	☐	☐	☐	☐	☐	☐

4.22.5
Productivity Improvement Guide for Construction Contracts *(Continued)*

	One of My Best	Clearly Above My Average	Within My Average Range	Clearly Below My Average	One of My Worst	Don't Know	Not Applicable
32. Quality of areas for lunch breaks.	☐	☐	☐	☐	☐	☐	☐
33. Working relationship between general and subcontractors.	☐	☐	☐	☐	☐	☐	☐
34. Lost time due to stops and starts because schedule not followed by other contractors.	☐	☐	☐	☐	☐	☐	☐
35. Quality of finished work.	☐	☐	☐	☐	☐	☐	☐
36. Repetitive tasks assigned to specially-organized crews.	☐	☐	☐	☐	☐	☐	☐
37. Materials delivered on time as ordered.	☐	☐	☐	☐	☐	☐	☐
38. Planning includes contingencies for delays and changes normally encountered on jobs of this type.	☐	☐	☐	☐	☐	☐	☐
39. Suggestions from foremen and craftsmen for improving job productivity.	☐	☐	☐	☐	☐	☐	☐
40. Jobsite working conditions (e.g., heat, noise, ventilation, light, etc.)	☐	☐	☐	☐	☐	☐	☐

Part C: Choices

INSTRUCTIONS: For each statement, check the box (or boxes) that best applies to this job. Fill in the blanks as required. If you need to, make notes in the margins to clarify what you mean by your response.

1. Which of the following conditions is the most serious cause of failing to meet your schedule? Rank them, 1 for most serious and 5 for least.
   A ☐ late deliveries of materials and equipment
   B ☐ supporting trades miss their schedule and slow the job
   C ☐ change orders extend the workload but not the schedule
   D ☐ stops due to regulation noncompliance (safety, permits, etc.)
   E ☐ other_____
2. Near the end of the job, when the schedule becomes squeezed and time is running out, most likely you
   A ☐ don't worry too much because you're off the job by then
   B ☐ overman the job to push work faster
   C ☐ work overtime
   D ☐ work out of sequence in order to keep work moving
   E ☐ jump ahead of other trades when possible and pull every trick you can to make sure you are not at fault and to avoid getting stuck
   F ☐ other_____
3. Who has done the most outstanding work for you this past week on the job?_____
   Have you let him/her know you are aware of the high quality of the work?
   A ☐ yes
   B ☐ not yet
   C ☐ do not plan to

4.22.5
Productivity Improvement Guide for Construction Contracts *(Continued)*

4. If you fail to get eight hours work from each individual on the job, it is most likely because the labor force
   A ☐ arrives late
   B ☐ leaves early
   C ☐ takes long lunch
   D ☐ takes long breaks
   E ☐ other_____

5. If you fail to get eight *productive* hours from each individual on the jobsite it is most likely because of
   A ☐ a lack of needed tools and equipment
   B ☐ overmanning relative to the work needs
   C ☐ too much travel on the jobsite
   D ☐ too much waiting on the jobsite
   E ☐ other_____

6. How many hours each week does the on-site field manager spend talking with a union representative about the job?
   A ☐ less than 1
   B ☐ between 1 and 4
   C ☐ more than 4

7. If job site supervision lacks in any one area, it is in
   A ☐ clarity of instructions to work force
   B ☐ coordination with other trades
   C ☐ getting eight hours productive work from crews
   D ☐ planning ahead to avoid problems
   E ☐ other_____

8. If a worker is idle at the job, it is most likely due to
   A ☐ waiting for other trades to complete a task
   B ☐ waiting for materials, tools and/or equipment
   C ☐ waiting for instructions
   D ☐ other_____
   E ☐ don't know

9. How many labor hours were charged to the job this past week? ☐____How many hours did the on-site field manager and the home office supervisor spend walking the job this past week? ☐____Dividing the second number by the first gives a percentage
   A ☐ less than 2%
   B ☐ between 2% and 6%
   C ☐ between 6% and 15%
   D ☐ more than 15%

10. What is the greatest safety hazard on the job site?
   _____

   What have you done to protect your workers against it?
   A ☐ told them about it

   B ☐ placed a warning to isolate it
   C ☐ took steps to eliminate it
   D ☐ nothing yet

11. What has been the workers' biggest gripe on the job so far?
   _____

   What have you done about it?
   A ☐ solved the problem
   B ☐ tried to solve the problem
   C ☐ let everyone know why nothing could be done about it
   D ☐ nothing yet

12. On this job the most counterproductive union work rule is
   _____

   To offset the negative effect of this rule, you have _____

   If you have to think of one thing to trade the union for getting this rule relaxed for *this* job, it would be . . . (don't write it down!)

13. For a good job, how dependent are you on the following items? Rank them, 1 for most dependent and 5 for least
   A ☐ on time delivery of materials and equipment
   B ☐ attitude of labor force at work site
   C ☐ supporting trades completing their work on time
   D ☐ reasonable overall schedule coordinating work of general and subcontractors
   E ☐ other_____

14. Where do you believe your planning and coordinating time is best spent?
   A ☐ communication with suppliers
   B ☐ talking with the labor force
   C ☐ coordinating with foremen and supervisors in supporting trades
   D ☐ meetings between general and subcontractors
   E ☐ other_____

15. (a) List the supporting trades that work directly ahead of you, whose work must be in before you can do the bulk of your work. ____
   _____

   (b) List the supporting trades that work alongside of you, requiring the most coordination.
   _____

   (c) List the supporting trades that follow you and depend on your work being done before they can do the bulk of theirs. _____

   (d) For the trades you follow in (a) above, what one thing can they do to help you

4.22.5
Productivity Improvement Guide for Construction Contracts *(Continued)*

the most? _____

_____

(e) What one thing can they do to hurt you the most? _____
_____

(f) For the trades you work with in (b) above, what most encourages helpful cooperation and coordination among foremen and workers on the jobsite? _____
(g) What most interferes with coordination and a cooperative attitude? _____
_____
(h) For the trades that follow you in (c) above, what one thing can you do to most help them?
_____
_____

(i) What one thing can you do to most hurt them? _____
_____

16. What do you believe sets your company apart from your competitors in the eyes of your customers?

	We are better	We are worse
quality of workmanship	☐	☐
speed of installation	☐	☐
price	☐	☐
customer relations	☐	☐
estimating accuracy	☐	☐
other _____	☐	☐

17. How do you think your fellow contractors see your company? Do you think, given the choice, they would prefer to work on the same job with you?
18. How do you think your labor force sees your company? Given the choice, would they prefer to work for you or someone else?
19. Looking back on the job so far, what one thing do you wish *you* had better planned for? _____
_____

20. Looking back on the job so far, what one thing do you wish *someone else* had better planned for? _____
_____

Part D: The Tally

In this part you will tally your responses to Part B (the checklist) and use the tally plus your responses in Parts A and C to analyze your "productivity profile." Begin by using the tally sheet below to collect your responses in Part B.

Instructions. Tally your responses in Part B (the checklist) by using the boxes on this sheet. For every statement in Part B which you checked in either of the first two columns (Best or Above Average), place a check in the correspondingly numbered box below.

Planning and Coordination	1 ☐	9 ☐	17 ☐	25 ☐	33 ☐
	2 ☐	10 ☐	18 ☐	26 ☐	34 ☐
Management Methods	3 ☐	11 ☐	19 ☐	27 ☐	35 ☐
	4 ☐	12 ☐	20 ☐	28 ☐	36 ☐
	5 ☐	13 ☐	21 ☐	29 ☐	37 ☐
	6 ☐	14 ☐	22 ☐	30 ☐	38 ☐
Labor Relations	7 ☐	15 ☐	23 ☐	31 ☐	39 ☐
	8 ☐	16 ☐	24 ☐	32 ☐	40 ☐

The Tally Sheet. After you have completed the tally, look at the pattern of checks on the sheet. The checks indicate areas where you are doing particularly well. The blanks do not necessarily mean you are doing poorly but do indicate where you ought to turn first for productivity improvement.

Every box on the tally sheet falls into one of the three areas identified in the introduction as important factors in productivity improvement: planning and coordination, management methods, and labor relations. Your pattern of checks and blanks may already show you in which of these three areas your greatest apparent strengths and weaknesses lie.

Don't Know. How many items in Parts B and C did you check "don't know"? To some extent the number of "don't knows" will depend upon your position within the company. At the supervisory level, however, you should have first hand knowledge of most of these areas. More than five checks indicates you are not receiving essential information which could lead to a loss of control.

Not Applicable. How many did you check "not applicable"? If you checked more than five, you may be kidding yourself since nearly every statement is applicable to most jobs. Perhaps you are overlooking areas in which you could make a direct impact on productivity.

Planning and Coordination. One of the most difficult aspects of job management involves planning and coordination with other "supporting trades"—those contractors whose work is most

4.22.5
Productivity Improvement Guide for Construction Contracts *(Continued)*

closely tied into your own. In Part A, question 12, you identified who these supporting trades were; in Part C, question 15, you noted some of the relationships between your work and theirs.

Look back now to your pattern of checks and blanks in the "planning and coordination" field of the tally sheet. These statements all referred to planning and scheduling for the overall job and working relationships among contractors.

On your tally do you show a blank for numbers 17 and 33—liaison and coordination with others? The first two questions in Part C further probe this working relationship. Did you mark choice "B" in Question 1—failure of supporting trades to hit their schedule? What about your own behavior toward other trades as revealed in Question 2? If you are fortunate to be one of the contractors in and out early on the job, you may not have worried before about the accumulation of every small delay as it hurts the contractors at the end of the job. But, over many jobs, don't you believe the ability of the finish trades to stay in business may well determine the health of the entire industry?

Were you involved in planning and scheduling the overall job and are you involved now in following its progress? Numbers 1, 2, 18, 25, and 34 in Part B concern planning and scheduling. Do you show blanks there? How could the schedule be more useful, more accurate? What kind of coordination would help you?

Part of the answer may lie in finding better ways to bring everyone into the job earlier. For example, how could subcontractors be brought into the scheduling process by the general contractor so that the resulting schedule is acceptable from all sides and is realistic? (Recognize that all subcontractors are not on board at bid time, that delays are inevitable, and that the owner will push for the shortest schedule, even if it's impossible to meet.)

Glance back at your responses to questions 4 through 7 in Part A—written scheduling. Do you have four "yes" checks? If not, do you believe better scheduling could help your productivity, particularly if everyone else did it too and all stuck to the same schedule?

Management Methods. While most everything you do (including the planning and scheduling discussed above) can be loosely described as "management methods," the pattern of checks and blanks in the central field on your tally sheet refers to the management of the job once it begins.

Thinking through your part of the job should include preparing contingency plans (numbers 13 and 38 in Part B). Are you prepared with alternative plans when problems arise and can the people in

the field make the decisions necessary to carry out those plans when the need arises?

Getting the right tools and equipment to the job (20) and materials to the right place at the right time (21 and 37) are important factors in keeping work crews productive. Clear instructions are important too. Is work "packaged" (question 5) for clarity and ease of accomplishment?

Nonworking time can be minimized by keeping such facilities as tool rooms and materials supply close to work locations (12), keeping crew sizes small (3 and 28) and keeping foremen up-to-date on what to do next (4 and 22). These same items are covered in questions 5, 7, and 8 in Part C.

Good jobsite supervision, coupled with strong home office support, is essential to productivity improvement. Question 29 probes the adequacy of on-site field management. In Part C, question 9 asks how much time management spends actually walking the job. Do you think it is enough? How important is on-site management to your productivity?

How well are you tying together all of the information you have? Your job estimate (6) can be an important management document. Do the drawings in the field contain the information needed (30)? Are communications between office and site as frequent and complete as they should be (14)?

Lastly, how about the quality of your finished work (35)? Improving productivity does not mean cutting corners and giving your customers less than they should expect from industry professionals.

Labor Relations. Some contractors say that good planning and good management are the keys to good labor relations. They believe that if your prove yourself capable of carrying management's share of the responsibility for producing a good job, labor will contribute their fair share as well.

One of management's responsibilities is to support the labor force. Numbers 8, 16, 32, and 40 speak to supporting physical needs. Part B, numbers 7 and 15 suggest support through shared information as does question 6 in Part C.

Do you involve labor in prejob planning (number 23 in Part B)? Such involvement may not only make for a smoother job but may also produce suggestions for improving productivity on the job (39).

Support for your labor force can also take the form of positive reinforcement for a job well done (Part B, number 24 and Part C, question 3). Support also includes responding to gripes (Part C, question 11) and keeping the workplace safe (Part C, question 10).

Do you limit your concept of labor relations to

4.22.5
Productivity Improvement Guide for Construction Contracts *(Continued)*

what happens at the bargaining table? Or do you believe that labor relations is a part of the whole way you do business and treat people all of the time? Some people think the gap between labor and management will destroy the unionized sector of the construction industry. Others think the gap can be bridged through increased communication and cooperation. What effect do you believe your own attitude has on bridging this gap? (Refer to your response to question 12 in Part C for one clue.)

Summary. By now you can see that this guide will not supply you with ready-made answers to improve your productivity. Ask yourself, however, if it has caused you to discover some things *you* can do to get improvement. How did you answer questions 18 and 19 in Part C? That is one starting point. Is there something else you can do better next time?

From the analysis you may have found that one necessary (but difficult) route to productivity improvement is through better coordination and coop-

eration with other contractors and tradesmen on the job site. Does it seem to you that if everyone supported everyone else, the overall job would be more productive? Improving overall productivity, however, may mean that each individual will need to make some compromises from time to time. No one wants you to sacrifice your own business "for the good of the industry" since the health of your business is an important part of the industry. The real question is what can you do in the short-term to benefit your own business that also benefits the industry in the longer-term?

The search for better ways to work together must also overcome the fear of responsibility for liability (often fostered by lawyers) that now prevents people from cooperating more fully on a job. What can be done about it?

And how can union leadership best define its own share of responsibility in cooperating to produce the best job possible?

No easy answers exist. But can your efforts to improve your own productivity involve others so as to contribute to a solution that benefits us all?

Site Superintendence

5.1   The Site Superintendent Function—Section Description

The Superintendent is the key individual responsible for administration of the actual construction in the field. The physical activity on the site is the source of potentially great successes and extreme liabilities. The Superintendent, accordingly, must not only be a competent constructor, but must develop and maintain a profound respect for the necessary contributions of each member of the project team. Documentation must be treated as a *fundamental component* of the Superintendent's responsibility, and not as an inconvenient secondary effort.

Thus, the Superintendent must be a competent building constructor *and* administrator as well as possessing significant skill as a team builder. He or she must be able to develop confidence among the direct employees and Subcontractors on the site, mold them all into a cohesive workforce, and keep the project always moving forward without actually doing other persons' work for them.

The relationship of the Superintendent function within the organization is outlined in Section 1.5.5 of the Manual. This section details the specific activities that are necessary to actually perform the project Superintendent's responsibilities. Sample full-size forms, checklists, and specific instructions are provided at each step of the way.

This section is very closely coordinated with:

- Section 4, Project Engineering
- Section 6, Safety and Loss Control
- Section 10, Progress Schedules and Funds Analysis

As with Section 4, Project Engineering, it remains most important when reviewing this Section 5, Site Superintendence, to appreciate the genuine lack of any distinction that should be made between systems designed to be used manually and management systems—or components of systems—that are designed to be implemented electronically. Refer to Section 4.1 for discussion of the application of computerized management tools and of the differences between performing certain management and administrative functions manually or electronically.

5.2   Responsibilities of the Site Superintendent

The Site Superintendent is the individual or team that is directly responsible for the expeditious completion of the physical work. As project engineering in a sense builds the project on paper, the Site Superintendent orchestrates the actual construction. The general duties of the Site Superintendent function, as first outlined in Section 1.5, are developed here as follows:

- Reporting to the Project Manager and carrying out his or her directives with respect to field operations
- Working to assure adequate staffing of the workforce, supplies of materials, and complete information as necessary for assembly

- Planning for staffing, materials, and information in advance so as not to interfere with the progress of any one component
- Generating, securing, and otherwise confirming all information needed to create, monitor, and modify the progress schedule on a continuing basis
- Developing the progress schedule with the Project Manager and Project Engineer
- Participating in scope reviews of the various bid packages in order to properly coordinate their respective interfaces and to ensure that nothing is left out and nothing is bought twice
- Working with the Project Manager to develop and administer the site utility plan, site services, security arrangements, safety program, and other facilities and arrangements necessary for appropriate service to the construction effort
- Identifying field construction and work sequence considerations when finalizing bid package purchases
- Monitoring actual versus required performance by all parties; working to bring deviating performances back in line
- Determining whether all Subcontractors are providing sufficient workforce and hours of work to actually achieve committed promises of performance
- Monitoring the performance of the Company's purchasing and project engineering efforts to ensure that all subcontracts, material purchases, submittals, deliveries, clarifications, and changes are processed in time to guarantee their arrival at the jobsite by or before the time needed
- Directing any Company field staff
- Being thoroughly familiar with the requirements of the general contract, and thereby identifying changes, conflicts, etc., that are beyond the scope of Company responsibility
- Preparing daily reports, job diaries, narratives, backcharges, notice documentation, and other special documentation as detailed in this and other sections of the Manual, and as may be determined by the Company and by project needs

Although this section includes those specific activities that fall within the restricted definition of the Site Superintendent function, actual practice will necessarily dictate large amounts of overlap of individual responsibilities across the related disciplines noted in Section 5.1. Particular project staffing assignments will determine the actual individuals responsible to *perform* a given function, such as scheduling, or even completing daily field reports. It is important, however, that the Site Superintendent realize that even if it is not his or her direct responsibility to generate the actual document, the Site Superintendent must know what it is, when and how it must be done, who must be doing it, and exactly what his or her contribution to the complete effort must be.

5.3   Field Organization

5.3.1   Field Staff Considerations

In most instances, projects of moderate size will require a single individual on-site as the Construction Superintendent. Requirements of the function, however, are becoming more demanding every day. Expectations of performance and the actual number of responsibilities are growing well beyond what they were even a few years ago.

Staffing requirements are no longer simply based upon the physical size or relative direct cost of a project. Relatively, the projects themselves may not be becoming any more physically complex, but problems with such factors as

- Quality of design documents
- Subcontractor and supplier performance
- "Creative" contract provisions
- Claims consciousness of *everyone*
- Exponential increases in every type of liability

now demand especially careful consideration of the complete set of responsibilities that must be properly accommodated in the field every day. Nowhere else but in the position of the Site Superintendent is the principle of "false economy" more evident if the position is skimped on in terms of competence and actual expectations of the individuals who must live up to the role.

The result of these considerations may, in the final analysis, actually turn out again to be the assignment of a single individual to fulfill the characteristic position. In other cases it may be more effective to design the site staffing more closely around other significant project considerations. Accordingly, before the site staff is finalized, consider the following ideas:

1. Site logistics
    a. A single building on a small site?
    b. A single building on a large, complex site?
    c. Multiple buildings at a single location?
    d. Multiple buildings at several addresses?

2. Physical size
    a. Multiple stories?
    b. Size and configuration of building footprint?

3. Site complexities
    a. Large open areas (construction ease, material staging, etc.)?
    b. Tight site, adjacent buildings?
    c. Open or restricted site access for workers and operations?
    d. Amount of contact with the public?

4. Project complexity
    a. Simple warehouse?
    b. Tenant fit-up?
    c. Office building?
    d. Medical or scientific facility?
    e. Low- or high-tech factory or assembly plant?
    f. Trade school—complex equipment?
    g. "Smart" buildings—complex computer, electrical, and mechanical systems?

5. Building/site relationship
    a. Building project with moderate site?
    b. Easy location with no problems on adjacent properties?
    c. Large, complicated site?
    d. Unusual site designs and other needs that "force" the building into the site?

6. Personality and talent of the Superintendent candidate(s)
    a. "Pusher" with lower administrative skills?
    b. Good administrative talent with weakness in dealing with Subcontractors?
    c. Project Manager in Superintendent's clothing?

7. Practical and technical administrative requirements of project
    a. Field/home-office logistics?
    b. Subcontract/direct-hire structure?
    c. Level of support expected from home-office?
    d. Daily and other periodic reporting requirements?
    e. Formal and informal meetings and reporting?
    f. Simple or complex inspection procedures?
    g. Straight or bureaucratic project authorities?

8. Contract type
    a. General contract?
    b. Design-build?
    c. Construction management (advisory)?
    d. Construction management with guaranteed maximum price?

9. Apparent quality of construction documents
    a. Do they seem to be complete; have all the right parts?
    b. Have they been skimped on in terms of absolute amount of information?
    c. Do they seem to be properly coordinated?
    d. Are they clear or confusing?
    e. Were sub-bids for the major bid packages consistent, or were prices all over the place?

10. Project risk structure
    a. Are there liquidated damages? To what extent?
    b. Are there heavy or otherwise unique penalties, either described or *implied*?

c. Is there an available bonus for early completion?

d. Is there "heavy" or "sneaky" contract language giving clues to an unfavorable Owner disposition?

11. Value engineering arrangement

a. Are there contract incentives for cost or time reductions? What kind of effort will be necessary?

12. Project purchasing procedure

a. Is purchasing 100% home-office centralized?

b. Will key purchases be made from the field?

c. Will field purchases remain only for incidental items?

5.3.2 Example Field Staff Arrangements

The following examples are listed as possible field staff arrangements that might accommodate the considerations of the previous section. Although they are most common, they are certainly not conclusive or exhaustive. The project itself and the individuals finally selected will determine the best fit. The examples are arranged in order of increasing complexity:

1. Single Superintendent on-site

a. Off-site Project Manager, Project Engineer, Scheduler, and Project Accountant

b. Supplemented with area and/or Assistant Superintendents as necessary

c. Moderate administrative skills and disposition

d. Strong field tendencies; good with direct work and Subcontractor communications

2. On-site Project Engineer–Superintendent team

a. Off-site Project Manager and Project Accountant

b. On- or off-site scheduling; preferably on-site

c. Allowing the Superintendent to focus totally on field activities with moderate reporting, while major administration is conducted by Project Engineer

3. On-site Project Engineer, Superintendent, and Secretary

a. Advantages of Example 2, with more complete administrative capability added

4. On-site Project Manager/Engineer, General Superintendent, and Secretary

a. Single individual as Project Manager/Project Engineer to allow project management attention for single project

b. Moves more direct authority onto the site, along with complete administrative responsibility

c. Scheduling done on-site

5. On-site Project Manager, Project Engineer, General Superintendent, and Secretary

a. Intensified arrangement of Example 4

      b. Office Engineers, Area/Assistant Superintendents, and Time Keepers added as needed

      c. Off-site project accounting authority

6. On-site Project Manager, Project Engineer, General Superintendent, Project Accountant, and Secretary

      a. Final conversion of Example 5 into a complete operational center with profit/loss responsibilities

      b. Purchasing on- or off-site

5.4   Site Utilization Program

5.4.1   General

The site utilization program identifies the comprehensive treatment of every use of the site throughout the construction period. Its purpose is to develop, coordinate, and police compliance with all site usages related to the construction, administration, safety, neighborhood relations, and any other physical administrative consideration necessary for the proper logistics of meeting all the practical, legal, and simply considerate measures necessary to work the project.

The complexity, public relations, neighborhood, site, political, and other characteristics of the project may be significant determinants of the approach that will be necessary to confirm the site utilization program that will finally be implemented. In most cases, development of the complete site utilization program will appropriately be left in the hands of the Company project team, with certain courtesies extended to the Owner or major Subcontractors for at least their input. In other cases, major portions of the program may be dictated by the permit, specific Owner concerns, or by the contract documents themselves.

5.4.2   Program Components

Components of the complete site utilization program include everything necessary to administer the management, inspection, safety, loss control, security, material staging and handling, site access, and personnel, and all construction support facilities and operations.

Ideally, each facility itself, its location and operation/utilization, will be determined in such a way as to enhance its coordination with the project, and to minimize or eliminate its interference with any part of the project. We do not, for example, want to place the field office in a location that will eventually have to be dug up for a utility line.

Consider each of the items listed here for their immediate need and for your long-term plans. Although each item in the list may appear to be separate and distinct from the others, each is directly affected by every other; so each idea must be considered in complete consideration of all others together. Add to the list as appropriate for a specific project.

1. Configuration of field offices
    a. Quantity and sizes: separate or combined offices for Company, inspectors, design professionals, major Subcontractors
    b. Configurations and composition: office/conference/storage areas, locations of doors and windows, views of site, storage, parking areas
    c. Access to storage rooms
    d. Life safety provision: walkways, entry platforms, handrails, perimeter lighting
    e. Sanitary provision: hookup of trailer facility, use of rental space, separate portable facility

2. Location of field offices
    a. Proximity to construction: close enough, but not too close
    b. Availability of utilities: electric, gas, oil, phone, sanitary; close or need to be brought in
    c. Proximity to adjacent properties: next to the property line, operating businesses, other construction
    d. Proximity to existing parking or temporary construction parking
    e. Proximity to each other: consolidate location of office complex for efficiency in communication, utilities management, etc., or spread out because of narrow site constraints

3. Location of storage trailers; location and configuration of material staging areas
    a. Proximity to construction: close enough, but not too close
    b. Fit within planned security arrangements
    c. Need to be altered at any point during construction
    d. Volume developments and changing needs as construction rates change: starts, increases, and winds down
    e. Ability to consolidate (security?), or need/advantage in arranging in different areas (construction ease?)

4. Major stockpile and staging areas
    a. Loam stockpile (left on-site or moved off?)
    b. Structural steel or concrete staging areas
    c. Special materials or equipment
    d. Proximity to hoisting arrangements

5. Site security
    a. Gates and perimeter fencing
    b. Existing, temporary, or new site lighting (which will be incorporated into the project)
    c. Any existing security arrangements

6. Site access
    a. Existing permanent roads and parking
    b. Temporary construction access (into, around, through site): construction materials and equipment, construction administration, construction personnel

    c. Future permanent access (to be incorporated into project): roads, parking areas, walks

    d. Time limitations: available hours of day for certain areas, heavy traffic during rush hour, availability on weekends and holidays

  7. Safety

    a. Protection of public: covered walkways (with lighting?), barricades, fencing, signs and notices, special lighting, provision for visually or physically handicapped

    b. Protection of Owner or adjacent properties: tree fencing, surface protections

    c. Protection of workers: temporary walkways, barricades, personnel distribution

  8. Environmental

    a. Protection of wetland areas: silt fence, buffer zones

    b. Management of hazardous waste

  9. Construction waste management

    a. Dumpster locations

    b. Trash chutes

    c. Site-work debris

  10. Material hoisting and movement

    a. Truck, tower, or combination cranes: stationary positioning or movement

    b. Material hoists (combined with personnel hoists or separate?)

    c. Changing needs and uses at different stages of construction

  11. Restricted areas

    a. Off limits to any construction personnel at any time

    b. Partially restricted: hours of day

  12. Interferences with construction

    a. Will any items need to be moved or modified at any point to allow construction to proceed: construction of site underground/overhead utilities (power, storm, sanitary, communications, etc.), parking or roadway areas, walks, playing fields, access to otherwise closed-off areas

5.4.3 Sample Site Utilization Plan *(page 5.13)*

On projects of small to moderate size it may be sufficient to represent components of the site utilization program on the site plan, site utilities plan, or even landscaping plan of the contract set.

Preparing a separate site utilization plan, however, has distinct advantages. Its development will:

1. Aid in the development and coordination of the program components themselves.

2. Provide the vehicle to communicate the program to the Owner, design professionals, all Subcontractors, and all authorities having jurisdiction over the work.

5.4.3
Sample Site Utilization Plan

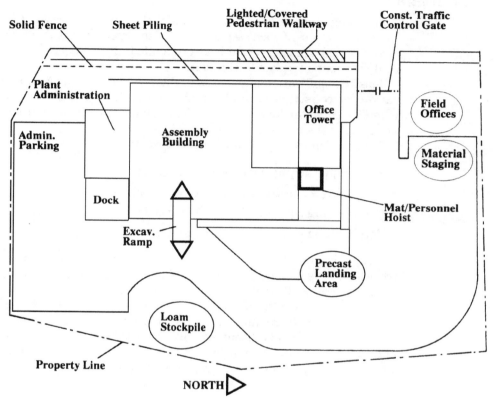

The sample Site Utilization Plan that follows is an example of a way to illustrate the properly coordinated program components of the site utilization program. Two points to note in its preparation include:

1. For clarity, it is most often useful to prepare a drawing or sketch of the site and major building components separate from any of the contract drawings that include much extraneous information. This drawing might almost be prepared as a schematic.

2. There is no need to have the program on a full-size drawing. Prepare the document on an 11″ by 17″ sheet of paper. Once it is completed, copies and distribution can be made with a regular photocopy machine; no need to have prints made.

   a. The plan can even be reduced to 8½″ by 11″ if such a reduction does not interfere with reading the information.

   b. Use of these reduced sizes makes posting and distribution of the program easy and inexpensive.

5.5   Field Office Mobilization

5.5.1   General

There is much work to be done to establish Company presence in the field. The relative amounts of each step are dependent upon the actual project and staffing considerations, as described in Section 5.3, but the specific procedure is essentially the same for each mobilization.

Jobsite mobilization is essentially divided between the actual physical preparation for construction and its administration—what we typically think of when we consider "mobilization." But it includes also preparation of the management of the construction *contract*.

This section, then, details the specifics of:

- Setting up the physical space of the field office
- Arranging for utilities and other site services
- Establishing contact with Owner and design representatives
- Establishing contact with enforcement authorities

Setting up the administration of the office in preparation for the correct administration of the general contract itself, and of each subcontract package, is dealt with in Section 5.6, Jobsite Administrative Mobilization. This section is directly related to and should be closely coordinated with Section 5.4, Site Utilization Program. There the relationship of the field office with the entire site is developed.

Finally, this section discusses only the establishment of Company presence, and gives no consideration to any facilities that may be required for Owner, agency, or design professional representatives.

5.5.2   Establishing the Field Office and Facilities

Having coordinated the configuration and location of the field office in accordance with Section 5.4, Site Utilization Program, proceed with establishing the Company site presence with the following considerations:

1. Field office space
    a. Is space available in an existing building (renovation project)? If so, consider problems with construction interference by the office presence.
    b. Is retail space available close by?
        (1) Size?
        (2) Cost?
        (3) Proximity to site?
        (4) Level of convenience or inconvenience?
        (5) Expansion or storage possibilities?
    c. For items *a* or *b*, will a temporary or conditional Certificate of Occupancy be necessary?
        (1) For construction field staff use?
        (2) If administrative, secretarial, or staff personnel are added to the site?

    d. Will office trailer(s) be used?
- (1) Owned or rented, real or assigned cost?
- (2) Size and condition?
- (3) Configuration: office, conference, equipment, or supply storage areas?
- (4) Adequate lighting?
- (5) Safe, OSHA-approved stairs, handrails, etc.?
- (6) Adequate heat, power, and air-conditioning?
- (7) Any minimum requirements listed in the specifications?

2. Office furniture
   - *a.* Desks, chairs
   - *b.* Conference table
   - *c.* Folding chairs (for meetings)
   - *d.* Plan rack
   - *e.* Plan table
   - *f.* File cabinets (locking)
   - *g.* Bookshelves
   - *h.* Tackboards (schedule and notice postings)
   - *i.* Plan storage cabinets or files
   - *j.* Supply cabinets

3. Office equipment and supplies
   - *a.* Copier
   - *b.* Fax
   - *c.* Typewriter
   - *d.* Communication equipment
   - *e.* Computer and printer (modem?)
   - *f.* Blueprint machine or printing arrangements
   - *g.* Refrigerator, coffee machine, supplies
   - *h.* Bottled water
   - *i.* Beepers or pagers
   - *j.* Copy, computer, and fax paper
   - *k.* Maintenance items for all equipment

4. Safety and security equipment
   - *a.* Fire and intrusion alarm system
   - *b.* Fire extinguishers
   - *c.* Hard hats—company personnel and visitors
   - *d.* First-aid kit and supplies
   - *e.* Emergency phone numbers
   - *f.* Stretcher

5. Signs and notices
   - *a.* Company "Field Office" signs
   - *b.* Visitor "Sign-In" notices
   - *c.* Prevailing wage notifications
   - *d.* "Hard Hat" signs
   - *e.* EEO notifications
   - *f.* "Keep Out" and "Restricted Area" notices

6. Office utilities (independent or related to overall site utilities?)
   a. Heat—electric, natural gas, propane, oil?
   b. Light and power
   c. Phone
   d. Sanitary
      (1) Use of existing
      (2) Trailer hookup
      (3) Portable units

Once the office space and project administrative areas have been established, start the habit *now* of keeping the entire area *clean and presentable* at all times. This is not just a "good idea"; it is absolutely necessary to:

- Keep personnel work areas efficient
- Preserve office equipment (copiers, faxes, computers, etc.)
- Maintain a safe, healthy environment
- Offer courtesy to others using the space
- Enhance the Company image

5.5.3 Establishing Company Presence

During the development of the site utilization plan of Section 5.4, it is a good idea to contact the building officials to at least request the "benefit of their experience" with respect to office location. This is particularly appropriate with the local fire marshal in consideration of life-safety issues.

Once you are at the point of actually establishing the field presence, establish contact with all appropriate individuals with whom you will be living over the project's duration. Let these people know who you are, what we're doing, that we care about doing it right, and that we value their concerns, opinions, and overall relationship. These individuals include:

1. Enforcement authorities
   a. General building inspector
   b. Plumbing, HVAC, and fire protection inspectors
   c. Electrical inspector
   d. Fire marshal
   e. Special inspectors
2. Design professionals
   a. Architect, office and field representatives
   b. Plumbing, HVAC, and fire protection engineers
   c. Electrical engineer
   d. Structural and civil engineers

     e. Landscape architect

     f. Special engineers and consultants

  3. Owner representatives

     a. Field inspectors

     b. District supervisors

     c. Administrative offices

5.5.4 Visitor Control

Establish visitor control procedures immediately.

1. Have the Visitor Sign-In Forms (sample provided in Section 5.5.5) placed in a clipboard hung on the wall adjacent to the field office door.

2. Post the Visitor Sign-In Notice outside the field office, at every site entrance, and at several appropriate areas throughout the jobsite.

3. *Police the procedure.* It is there to be followed. Follow it.

4. Turn in copies of the sign-in sheet *weekly* to the home-office to be placed in the General Project File. It will become an important record not only of who has been on the jobsite and when, but also of Company efforts to enforce appropriate management and safety procedures.

It is a good idea to require that the sign-in forms be returned habitually with any field payroll reporting. In this way, the routine is easy to establish and maintain.

5.5.5 Sample Jobsite Visitor Sign-In and General Release
(page 5.18)

The sample Jobsite Visitor Sign-In and General Release may be considered as a substitute for a Visitor Sign-In Form. This form serves the objectives of a simple visitor sign-in form, but goes an important step further.

Integral with the form is a General Release. The release is intended to include language that will release and discharge the Company from claims, actions, or hazards that may occur to the visitor or the visitor's property while on the premises of the project. The language included on the sample form is an example that must be reviewed and edited by a competent attorney before any language is finalized or the form itself is put to any use.

The heading at the top of the form (including all project information) should be completed initially, so that all a visitor needs to do is to add his or her name to the list, along with the visitor's signature and the visit to date.

If competent legal counsel determines that the form or some version of it is appropriate, it should be implemented on all Company jobsites. If any visitor to the jobsite refuses to sign the finalized form, that visitor should then be advised that he or she is not allowed beyond the Company field office.

5.5.5
Sample Jobsite Visitor Sign-In and General Release

Company: _____

Project: Name: _____

                Location: _____

THE UNDERSIGNED hereby acknowledges and agrees that permission to enter upon the premises of the Project indicated above for any purposes whatsoever, including observation, inspection, delivery of materials and/or equipment, posting notices, delivering information of any type, performing any work, or for any other reason whatsoever is granted by the company named above for the sole and exclusive benefit of the undersigned. I hereby attest that in consideration of such permission, I agree for myself and any heirs to any claims or actions that I now have or may have any time hereafter related in any way to death, bodily injury, or damage to property which may occur to me and/or my property while on the premises of the Project, whether or not such death, injury, or property damage is or may be due wholly or in part to any act or negligence on the part of the above named company. Accordingly, I fully understand, agree with, and accept full responsibility for all risks associated with the above.

Name *(Print):*	Signature:	Date:	Time In:	Time Out:

5.5.6   Sample Visitor Sign-In Notice

Reproduce this Visitor Sign-In Notice to include the Company's complete name. Post the notice outside the Company field office, at each site entrance, and at every appropriate area around the jobsite.

ALL   VISITORS

Please Report to
The (COMPANY) Field Office

BEFORE Proceeding
Anywhere in the Building
Or on the Site

Thank You For Your Cooperation

5.6   Jobsite Administrative Mobilization

5.6.1   General

Section 5.5 describes the procedure for what we commonly consider to be "jobsite mobilization," but it is only the establishment of the field presence. Here mobilization is completed by setting up the field office administratively to prepare the field staff for the management of the construction.

The order of the items listed here is not important, but all the components are.

These procedures are consolidated in Section 5.7, Jobsite Mobilization Checklist, with those listed in Section 5.5, Field Office Mobilization, as a more convenient mechanism to help push through the procedure as methodically as possible.

5.6.2   Construction Administration

This section lists the supply of job forms to be assembled at the job start-up, which will be necessary to routinely complete all records and reporting as required by the respective sections of this Manual, and provide for the establishment of the project filing system. Each item listed here and its use are described in detail in the appropriate sections of the Manual:

1. *Project manuals and log books*
   a. Company Operations Manual
   b. Any specific project operations manual
   c. Subcontractor Summary and Telephone Log
   d. Submittal Log (if submittals processed on-site)
   e. Change Order Summary Log (if change order administration is done on site)
   f. RFI Log

2. *Supply of administrative forms.* Most, if not all, of the forms listed here are provided in the Company Operations Manual. This list is produced as a convenience to ensure that an appropriate supply of each form is provided at job start-up.
   a. Daily Field Reports
   b. Visitor Sign-In and General Release Forms
   c. Change Order Forms
   d. Quotation and Phone Quotation Forms
   e. Field and Administrative Payroll Forms
   f. Time and Material Tickets
   g. Job Meeting Forms
   h. Record of Meeting/Conversation Forms
   i. Memos
   j. Photograph Record Forms
   k. Excavation Notification Checklists
   l. Equipment Use Release Forms
   m. Full and Partial Lien Waiver Forms
   n. Transmittal Forms
   o. Fax Memo/Transmittal Forms
   p. Field Purchase Order Forms
   q. Request for Information (RFI) Forms
   r. Certified Payroll Report Forms
   s. Schedule Status Report Forms
   t. Backcharge Forms
   u. Subtrade Performance Evaluation Forms

3. *Start-up project files.* Set up the files as described in Section 2.4, Files and File Management. Perform the filing procedure strictly in accordance with the requirements of that section. The Project Manager or his or her designee at the home-office is to be copied on *all* correspondence generated from the field. File designations for the respective item at the home-office will then be determined by the person there.

   Although maintenance of the field files is a fundamental responsibility of the field staff, file management must proceed with the idea that the home-office files *must* be maintained in a current and absolutely complete manner at all times.

The jobsite files are to be kept in *locking* file cabinets. After the specific folders have been set up, start-up files will include:

a. Contract and correspondence files

b. Subcontracts and purchase orders for all bid packages

c. Certificates of insurance for *everyone* allowed to work on the site

d. Approved submittals for any construction about to proceed

e. Executed Equipment Use Release Forms as appropriate

f. Any relevant correspondence in existence at start-up

4. *Construction and contract documents.* Arrange for copies of all documents relevant to every aspect of the technical construction, including:

a. The plans, specifications, and all addenda—Construction Set

b. The plans, specifications, and all addenda—As-Built Set

c. Project Manual or working procedure instruction

d. Current copies of relevant building codes

e. Copies of referenced standard specifications

f. All other special project documentation as required

5. *Contract management documents.* Arrange for copies and be aware of the specific details of all contractual responsibilities of the Company—in addition or as a supplement to the technical construction responsibilities:

a. The Owner or Company agreement

b. Bid documents

6. *Overall contract considerations.* Be aware of the general complexion of the Owner or Company agreement. Know the contract type and overall parameters of expected performance, including:

a. Contract type (General Contractor, Design-Build, CM, CM w/GMP, etc.)

b. Contract execution date

c. Construction start date

d. Completion date(s)

e. Number of calendar days

f. Number of working days

g. Presence of liquidated or other "damage" provisions; if so, what?

h. Presence of bonus provisions

i. Unusual restrictions (physical or contractual)

7. *Contract execution.* Be completely familiar with all project start-up requirements and procedures, and those to be routinely managed throughout the project, including:

a. General building permit—who secures, who pays?

b. Plumbing, HVAC, fire protection, electrical, and other permits—who secures, who pays?

c. Billing procedure:

(1) Date Subcontractor requisitions due

(2) Date general requisition due to Owner

(3) Status of the General Schedule of Values

        (4) Status of subvendor Schedules of Values

        (5) Requisition review/approval procedure

    d. Change order procedure

        (1) Change clause present

        (2) Procedure and forms required

    e. EEO requirements

        (1) Mandatory

        (2) Good faith

    f. Independent testing laboratories

        (1) Areas and work types required

        (2) Payment responsibilities

    g. Baselines and benchmark

        (1) Responsibility to provide

        (2) Responsibility for engineering expense

    h. Job meeting arrangement and schedule

        (1) Who provides the minutes

    i. Dispute resolution provisions

        (1) Dispute resolution clause or prescribed procedure present

        (2) Arbitration provision

        (3) Notice requirements

    j. Notice requirements for various circumstances

8. *Job cost and production control.* Be aware of all project cost considerations and your complete responsibilities relative to the achievement of all cost/progress targets. Relevant documents include:

    a. General project budget

    b. Resource estimates, labor and materials

    c. Job cost report

    d. Change order estimates

    e. Baseline construction schedule

    f. Baseline cash-flow projection

9. *Documentation of the site utilization program.* The development of the site utilization program and of the plan itself is treated in Section 5.4. Here the program is to be posted and distributed, and arrangements are to be made to secure the relevant provisions of the plan. These are site considerations, as distinguished from the field office considerations of Section 5.5.

    a. Temporary fences and other protective and security measures

    b. Guard service, security lighting, security systems, or other appropriate security determinations

    c. Temporary electric

    d. Temporary water

    e. Dumpster, waste-disposal arrangements

    f. Arrangements for management of hazardous materials

    g. Progress photograph arrangements

    h. Testing laboratory service arrangements
       (1) Soils
       (2) Concrete
       (3) Steel and welding
       (4) Other/special

5.7   Jobsite Mobilization Checklist (*page 5.24*)

The checklist that follows summarizes the discussions and outlines of Section 5.4, Site Utilization Program; Section 5.5, Field Office Mobilization; and Section 5.6, Jobsite Administrative Mobilization. It is provided to facilitate following through on the arrangements for the many decisions, services, suppliers, and facilities described throughout the complete mobilization process.

Some of the forms listed are described in later sections of the Manual. For these and all others, refer to the individual sections for more complete discussions of the respective items and their relevance to all others. Make a copy of the checklist and use it throughout the start-up of every project.

5.8   Subcontractor Summary and Jobsite Phone Log Book

5.8.1 General

Section 2.4.8 describes the basic preparation of the Jobsite Subcontractor Performance Summary and Telephone Log and its coordination with the other project documentation mechanisms. This section details its use.

The binder will be the location in which the summaries of each Subcontractor will be noted and performance compliances with respect to their jobsite responsibilities (as opposed to their contract and engineering responsibilities of Section 4.4) will be consolidated. It will provide at a glance the status of a particular vendor's performance and, more importantly, its nonperformance.

If the scheduling documentation is being correctly maintained, and the Daily Field Reports are being kept as current and as comprehensive as they should be, it may not become necessary to use the phone log feature of this book fully, as it would become a redundant effort. If, however, these kinds of notes are not being added to the Daily Field Reports routinely and consistently, the notes feature of the log book must be used. The procedure is as follows:

1. Prepare an appropriately labeled 1-inch three-ring binder in a color to match the project set. Insert a set of alphabetically tabbed dividers.

2. Prepare a Jobsite Subcontractor Summary Form, as described hereafter, for each vendor and place it in the appropriate alphabetical location.

3. As completed Subcontractor/Supplier Reference Forms (see Section 4.5) are received for each vendor, insert them behind the respective Summary Forms.

5.7
Jobsite Mobilization Checklist

SITE & SITE SERVICES
1. Site Utilization Plan prepared _____
   Approved by: _____ _____
2. Temp. fences, protection (see safety) _____
3. Guard service _____
4. Temporary electric _____
5. Temporary water _____
6. Dumpster, disposal arrangements _____
   Who pays: _____ _____
7. Progress Photograph service _____
   Who pays: _____ _____
8. Testing Laboratories _____
   a. Soils _____
   b. Concrete _____
   c. Steel/welding _____
   d. Other:_____ _____
   Who pays: _____ _____
9. Weather information phone numbers _____
10. "Call-Before-You-Dig" or other One-Call
    System in effect: _____ _____

FIELD OFFICES & OFFICE EQUIPMENT
1. Facility Type
   a. Trailers _____
   b. Retail space _____
   c. Space within the construction area _____
   d. Other: _____ _____
2. Number of Offices
   a. Company _____
      1) Project Manager _____
      2) Superintendent _____
      3) Other: _____ _____
   b. Owner representative(s) _____
   c. Design professionals _____
   d. Subcontractors _____
      1) _____ _____
      2) _____ _____
      3) _____ _____
      4) _____ _____
      5) _____ _____
      6) _____ _____
3. Temporary Facilities
   a. Heat (Type:_____) _____
   b. Lighting & power _____
   c. Telephones/portable phones _____
   d. Site communications equipment _____
   e. Lavatories
      1) Use of existing _____
      2) Hookup of trailers _____
      3) Portable (Quantity: _____) _____
   f. Water
      1) Use of existing _____
      2) Hookup of trailers _____
      3) Bottled water _____
   g. Other:_____ _____

4. Office Furniture
   a. Desks, chairs, stools _____
   b. Conference table _____
   c. Folding chairs _____
   d. Plan rack _____
   e. Plan table _____
   f. File cabinets
      1) Regular _____
      2) Locking _____
      3) Fireproof _____
   g. Bookshelves _____
   h. Plan storage cabinets _____
   i. Supply cabinets _____
   j. Plan edge reinforcing machine _____
5. Office Equipment & Supplies
   a. Copier _____
   b. Fax _____
   c. Typewriter _____
   d. Communication equipment _____
   e. Computer, printer & modem _____
   f. Software
      1) Word processing _____
      2) Scheduling _____
      3) Spreadsheet _____
      4) Database _____
      5) Other: _____ _____
   g. Printing arrangements
      1) Blueprint machine _____
      2) Printing arrangements _____
   h. Refrig., coffee mach., supplies _____
   i. Beepers, pagers _____
   j. Copy, computer, fax paper _____
   k. Maintenance items for all equipment _____
6. Office Safety & Security Equipment
   a. Fire & intrusion alarm _____
   b. Fire extinguishers _____
   c. Hard hats
      1) Company personnel _____
      2) Visitors _____
   d. First-aid kit & supplies _____
   e. Emergency phone numbers _____
   f. Stretcher _____
   g. Names of any employees with
      medical training
      1) _____
      2) _____
      3) _____
7. Signs & Notices
   a. Company "Field Office" signs _____
   b. Visitor "Sign-In" notices _____
   c. Prevailing wage or other
      Labor Department notifications _____
   d. "Hard Hat" signs _____
   e. EEO notices _____
   f. "Keep Out," "Danger," and
      "Restricted Area" notices _____

5.7
Jobsite Mobilization Checklist *(Continued)*

ADMINISTRATION

1. Project Manuals & Log Books
   a. Company Operations Manual _____
   b. Project Operations Manual _____
   c. Sub. Summary & Phone Log _____
   d. Submittal Log _____
   e. Change Order Summary Log _____
   f. RFI Log _____
2. Supply of Job Forms
   a. Daily Field Reports _____
   b. Visitor Sign-In Sheets & clipboard _____
   c. Change Order Forms _____
   d. Quotation & Phone Quote Forms _____
   e. Field Payroll Forms _____
   f. Administrative Payroll Forms _____
   g. Time & Material Tickets _____
   h. Job Meeting Forms _____
   i. Record of Mtg/Conversation Forms _____
   j. Memos _____
   k. Photograph Record Forms _____
   l. Excavation Notification Checklists _____
   m. Equipment Use Release Forms _____
   n. Full & Partial Lien Waiver Forms _____
   o. Certified Payroll Report Forms _____
   p. Transmittal Forms _____
   q. Fax Memo/Transmittal Forms _____
   r. Field Purchase Order Forms _____
   s. Request for Information (RFI) Forms _____
   t. Backcharge Notices _____
   u. Backcharge Forms _____
   v. Schedule Status Report Forms _____
   w. Sub Performance Evaluation Forms _____
3. Start-Up Project Files
   a. Contract & Correspondence Files _____
   b. Submittal Files _____
   c. Special Files _____
4. Start-Up Subcontractor Submissions
   a. Subcontracts & Purchase Orders _____
   b. Certificates of Insurance _____
   c. Sub Payment & Perform. Bonds _____
   d. Executed Equip. Use Release Forms _____
   e. Approved Submittals _____
   f. Other: _____ _____
5. Project Directory _____

CONSTRUCTION & CONTRACT DOCUMENTS
1. Plans, Specs, Addenda—Const. Set _____
2. Plans, Specs, Addenda—As-Built Set _____
3. Project Manual/Working Procedure _____
4. Building Codes _____
5. Referenced Standard Specifications _____
6. Other:
   1) _____ _____
   2) _____ _____
   3) _____ _____

CONTRACT MANAGEMENT DOCUMENTS & GENERAL INFORMATION
1. Owner/Company Agreement _____
2. Bid Documents _____
3. Contract Type (GC, CM, CMw/GMP, DB) _____
4. Contract Execution Date: _____
5. Construction Start Date: _____
6. Substantial Compl. Date: _____
7. Final Completion Date: _____
8. Number of Calendar Days: _____
9. Number of Working Days: _____
10. Liquidated Damages
    a. Value/Day: $ _____
11. Bonus
    a. Value/Day: $ _____
12. Special Considerations:
    _____
    _____
    _____

CONTRACT EXECUTION
1. Permits   Who Pays   Rec'd
   a. General Building _____ _____
   b. Plumbing _____ _____
   c. HVAC _____ _____
   d. Fire Protection _____ _____
   e. Electrical _____ _____
   f. _____ _____ _____
   g. _____ _____ _____
2. Billing Procedure
   a. Date Subcontractor requisitions due _____
   b. Date General requisitions due _____
   c. Requisition review/appr. procedure _____
3. Change Order Procedure
   a. Change Clause: Sect. _____
   b. Procedure/forms required _____
4. EEO Requirements
   a. Mandatory _____
   b. Good faith _____
5. Independent Testing Laboratories
   a. Areas & work types required:
      1) Soils _____
      2) Concrete _____
      3) Steel _____
      4) Other: _____ _____
   b. Payment responsibility:
      _____
6. Baselines & Benchmark
   a. Responsibility to provide:
      _____
   b. Payment responsibility:
      _____
7. Job Meetings
   a. Preconstruction Meeting Date: _____
   b. Regular Meeting Schedule _____
   c. Who provides official minutes: _____

5.7
Jobsite Mobilization Checklist *(Continued)*

8. Dispute Resolution
    a. Dispute resolution clause:
        Gen. Cond. Section _____
        Suppl. Cond. Section _____
    b. Arbitration Provision:
        Gen. Cond. Section _____
        Suppl. Cond. Section _____
    c. Notice Period: _____
    d. Special considerations:
        _____
        _____
        _____

JOB COST & PRODUCTION CONTROL
1. General project budget   _____
2. Resource estimates: material & labor   _____
3. Job Cost Report   _____
4. Change Order estimates   _____
5. Baseline Construction Schedule   _____
6. Baseline Cash-Flow Projection   _____

PROJECT CONTACTS
1. Enforcement Authorities
    a. General Building Inspector:
        _____
    b. Plbg./HVAC/Fire Protect. Inspector:
        _____
    c. Electrical Inspector:
        _____
    d. Fire Marshal:
        _____
    e. Special Inspectors:
        _____
        _____
        _____
        _____

2. Design Professionals
    a. Architect: Office Representative:
        _____
    b. Architect: Field Representative:
        _____
    c. Plumbing & HVAC Engineer(s):
        _____
        _____
    d. Fire Protection Engineer:
        _____
    e. Electrical Engineer:
        _____
    f. Structural Engineer:
        _____
    g. Civil Engineer:
        _____
    h. Landscape Architect:
        _____
    i. Special Engineers & Consultants:
        _____
        _____
        _____
        _____

3. Owner Representatives
    a. Field Inspector:
        _____
    b. District Supervisor:
        _____
    c. Other:
        _____
        _____
        _____

4. Security / Life Safety
    a. Police: _____
    b. Fire: _____
    c. Hospital: _____
    d. Emergency: _____
    e. Alarm Service: _____
5. Jobsite personnel home phone numbers   _____

Note that the absence of a form by the time any vendor is on the site is your cue that either the home-office has for some reason neglected to forward you your copy, or the vendor has not yet completed it and sent it in. In either case, whenever you notice a form to be missing, this is the time to get it.

4. Place a small supply of Telephone Log Forms as described hereafter behind the Summary and Reference Forms for each vendor.

5. Follow through with the use of each form as described throughout the remainder of this section.

5.8.2   Purpose

Each project will have dozens of vendors, and each vendor will have dozens of jobsite requirements. A precious few Subcontractors and suppliers will take the appropriate initiative and shoulder their responsibilities in a well-coordinated, responsible manner. Most, however, will fit somewhere into a wide range of capabilities, tendencies, and intentions.

These vendors are the ones who will need constant pushing, pulling, and chronic follow-up to ensure not only that they do everything they're supposed to, but also that things are done when and in the manner they're needed by *you*.

The first purpose, then, of the Summary Forms is to consolidate the routine items that must be provided or complied with by each subvendor. Summarized in this way in a checklist fashion, it becomes very convenient to review performance and compliance, and to identify those cracks through which something might have fallen.

The second purpose is to increase the ease with which the chronology of communication with the various subvendors can be reviewed and used. Daily Field Reports (see Section 5.9) are the traditional means to document not only on-site performance, but also nonperformances, your expediting efforts, notices given, conversations, and so on. If the daily reports are maintained as instructed in that section, documentation will at least be adequate.

The difficulty with providing records of conversations, notices, and the like on the daily reports is that:

1. The information is consolidated with all other project information for that particular day. It may be buried deep and be difficult to locate.

2. All relevant information for a specific vendor is spread out over possibly hundreds of pages over the life of the project. It will be there to support your contentions in a claim or quasi-claim situation, but it will never be researched effectively (or at least very conveniently) for practical purposes so that it can be used during the regular daily communication cycles.

Providing this kind of information in the Log Book solves these problems. The Daily Field Reports will speak for the actual field performance. The Log Book will catalog the communication history and consolidate it in a way that will provide at a glance:

1. Everything that needs to be complied with at the moment or in the near future.
2. Previous requirements that have not yet been properly met.
3. The short- and long-term histories of easy or tedious and strained communications.

5.8.3   Procedure and Use

1. Use the standard form. Add any specific requirements unique to the project and photocopy a supply of *project* forms.
2. Add any requirements specific to a particular Subcontractor or supplier to that vendor's forms. Try to do this as the form is initially being prepared, and develop the habit of considering this idea every time you look at any of the forms.
3. *The most important idea of all is to develop the habit of always having the binder open and in front of you.* This aspect of the Log Book's use is crucial to the realization of its full potential as a management tool and as a documentation vehicle.
4. Use the binder as your regular telephone and address directory. In this way you'll automatically have it open to that vendor's summary *before you pick up the phone.*
5. Log all relevant information *as it occurs.* Since the binder is already open, it will take less than a moment. Waiting for the end of the week, or even later in the day, is your guarantee that the information will never be recorded.
6. Enter *all* information, especially when nothing has happened. If, for example, all you get for your efforts is "he's not in; he'll call you back," note those remarks along with who gave them, date, and time. A list of six of these in a row, for example, goes a long way in demonstrating the performance failures of the other party.
7. Use every opportunity to "subtly" make everyone aware of these records.
   a. *Having* the records is the first step toward getting the upper hand in every situation every time.
   b. If the other parties are *aware* of the existence of such detailed records, they're more likely to learn fast that it is in their best interest to avoid games. When they see your ability to simplify greatly the usually extreme complexity of project relationships, they'll take another look. They'll hopefully decide that it's best not to fool with *this* relationship.

Some ways to do this include:

a. Whenever a subvendor phones to complain, to question, or for any other reason:

    (1) Review the vendor's performance summary *before* you pick up the phone.

    (2) Before the party gets a chance to say what his or her problem is, fire off *every* piece of paper that has not been delivered properly, and every item that you're still waiting for. Then move into any specific jobsite performance issue that *you* have. Be sure to advise him or her of all those dates (and times) that you've been trying to reach him or her about *your* issues. After you have gone through all *your* issues and documented the latest round of performance commitments relative to them, *then* you can ask what *their* problem was.

b. In any formal or informal meeting:

    (1) Open the binder whenever a relevant issue is raised.

    (2) Let everyone *see* you making your notes in their file *as the discussion is occurring.* This action gives profound legitimacy at a later date to *your* information as being the *right* information. It will make everyone think before they speak, and force them to consider their actions (and inactions) very seriously.

8. If more elaborate confirmation or an impromptu meeting has taken place, consider in addition to your notes using the Meeting/Conversation Record Form of Section 4.13.12.

5.8.4 Sample Jobsite Subvendor Summary Form—Completed Example (*page 5.30*)

The sample Jobsite Subvendor Summary Form included here is the form to be used to follow through on the procedure described in this section. The completed form is provided as an example of its proper use. Section 5.8.5 provides the blank form to be photocopied for actual use.

5.8.5 Sample Jobsite Subvendor Summary Form—Blank Form (*Page 5.31*)

5.8.6 Sample Telephone Log Form—Completed Example (*page 5.32*)

The sample Telephone Log Form included here is the same form that was used in Section 4.4 by the Project Engineer. It is to be used to follow through on the procedure described in this section. A completed form is provided as an example of its proper use. Section 5.8.7 provides the blank form to be photocopied for actual use.

5.8.4
Sample Jobsite Subvendor Summary Form—Completed Example

PROJECT: _TRADESMAN SCHOOL_   Sub: _FOUNDATIONS, INC._   Sect.: _O3200_
219 AMITE BLVD.   _O3300_
Company #: _9890_   _COLWELL, CT 06902_
Owner #: _BI·Q·177(B)_   Phone: _555-1625_
Subvendor #: _—_   Name: _JERRY GRANT_   Fax: _555-1670_

1.	Subcontractor Approval Received	Date:	_6·4·94_	
2.	Subcontract Executed / On-Site	Date:	_6·10·94_	
3.	Insurance Certificate On-Site	Date:	_6·10·94_	
4.	Information Request Form On-Site	Date:	_6·10·94_	
5.	Schedule of Values Approved / On-Site	Date:	_6·24·94_	
6.	Plans & Specs Delivered	Date:	_5·15·94_	
7.	Submittal Requirements Letter Sent	Date:	_5·15·94_	
8.	Shop Drawings	Received:		Approved:
9.	Construction Schedule	Delivered: _5·15·94_		Confirmed: _6·10·94_

10. Required Material Delivery Date   Date: _6·14·94_
11. Material Delivery Confirmation Date   Date: _6·14·94_
     Correspondent: _J. GRANT_
     Qualifications: _SITE START CONFIRMED, FOOTING, FNDTN WALL & S·O·G._
     START DATES CONFIRMED; ALL ALLOWED DURATION ACCEPTABLE.

12. Material Received:
     Ahead of Schedule   Date:
     _x_ On Schedule   Date: _6·14·94_
     Behind Schedule   Date:
     Remarks:
     CONC. DELIVERIES PROGRESSED AS NEEDED

13. As-Built Drawings Prepared / Updated   Date: _6·24·94_
                                            Date: _7·12·94_
                                            Date: _7·19·94_
                                            Date:

14. Equipment Use Release Forms Executed   Date: _6·13·94_

15. Backcharges:
     Date: Amount: Description / Remarks:
     _8·10·94_   _$1,422.-_   _REPAIR CRACKED WALK; DAMAGED BY CONC. TRUCK._

16. Notes & Remarks (Use Additional Pages as Required):
     QUALITY OF WORK & SCHEDULE PERFORMANCE ACCEPTABLE. COMPETENT
     SUPERVISION & WILLINGNESS TO COOPERATE.
     NEEDED HELP CONFIRMING LAYOUT OF THEIR WORK.
     GENERALLY ON·TIME W/ PAPERWORK

5.8.5
Sample Jobsite Subvendor Summary Form—Blank Form

PROJECT: _____ Sub: _____ Sect.: _____

Company #: _____ _____ _____
Owner #: _____ _____ Phone: _____
Subvendor #: _____ Name: _____ Fax: _____

1. Subcontractor Approval Received   Date: _____
2. Subcontract Executed / On-Site   Date: _____
3. Insurance Certificate On-Site   Date: _____
4. Information Request Form On-Site   Date: _____
5. Schedule of Values Approved / On-Site   Date: _____
6. Plans & Specs Delivered   Date: _____
7. Submittal Requirements Letter Sent   Date: _____
8. Shop Drawings   Received: _____ Approved: _____
9. Construction Schedule   Delivered: _____ Confirmed: _____

10. Required Material Delivery Date   Date: _____
11. Material Delivery Confirmation Date   Date: _____
        Correspondent: _____
        Qualifications: _____
        _____
        _____

12. Material Received:
    _____ Ahead of Schedule   Date: _____
    _____ On Schedule   Date: _____
    _____ Behind Schedule   Date: _____
          Remarks: _____
          _____
          _____

13. As-Built Drawings Prepared / Updated   Date: _____
                                              Date: _____
                                              Date: _____
                                              Date: _____

14. Equipment Use Release Forms Executed   Date: _____

15. Backcharges:
    Date: Amount: Description / Remarks:
    _____ _____ _____
    _____ _____ _____
    _____ _____ _____
    _____ _____ _____
    _____ _____ _____

16. Notes & Remarks (Use Additional Pages as Required):
    _____
    _____
    _____
    _____
    _____
    _____

5.8.6
Sample Telephone Log Form—Completed Example

Project: WESTSTREET SCHOOL RENOVATION   Proj. #: 96200

Date	Time	Discussion
5.19.94	8:30 Am	CALLED FOR RAY DEERING TO ASK WHEN WE WILL RECEIVE SUBMITTALS - NOT IN; WILL CALL BACK
	2:20 Pm	" - " " " " "
5.20.94	9:10 Am	" - RAY IS EXPECTED BY NOON
	1:05 Pm	" - RAY CONFIRMED THAT ALL SUBMITTALS WILL BE IN THIS OFFICE BY 5.26.94
5.25.94	11:25 Am	CALLED RAY DEERING TO CONFIRM 5/26 DEL. OF SUBMITTALS - RAY SAID HE DOESN'T EXPECT FROM HIS SUPPLIER "FOR ANOTHER WEEK" · I ADVISED RAY THAT EVEN THE 5.26.94 DATE WAS A WEEK AFTER HIS SUBCONTRACT COMMITMENT, & THIS ADD'L SLIP IS TOTALLY UNACCEPTABLE - ITS PUSHING ALL CONCRETE FOOTING WORK BACK BY AS MANY DAYS. · RAY SAID "HE'LL SEE WHAT HE CAN DO, & GET BACK" (REFER TO COMPANY MEMO TO RAY THIS DATE.)
5.26.94	9:15 am	CALLED FOR RAY - NOT IN; EXPECTED BY 2:00 Pm
	2:30 Pm	" " " " " "
	4:00 Pm	" " " " " "
5.27.94	8:30 Am	" " " " " · ASKED FOR JOHN SUTER (RAY'S REC.) -EXPLAINED THE ENTIRE SITUATION & DELAY TO JOHN -JOHN SAID "HE'LL GET TO THE BOTTOM OF IT & GET BACK TO ME ASAP - HOPEFULLY TODAY"
	11:05 am	· J. SUTER CALLED BACK; ADVISED THAT THE SUPPLIER WILL EXPRESS MAIL DIRECT - WE SHOULD HAVE IN OUR OFFICE BY 10:30 AM TOMORROW - I ASKED FOR SUPPLIER'S PHONE # TO VERIFY (802) 555-1625 FRED DAVIS.
	11:15 am	· CALLED FRED, WHO CONFIRMED JOHN'S REMARKS. (REFER TO CO. MEMO THIS DAY TO BOTH FRED DAVIS & JOHN SUTER)

5.8.7   Sample Telephone Log Form—Blank Form

Project:_____   **Proj. #:** _____

Date	Time	Discussion

5.9   Daily Field Report

5.9.1   Description and Responsibility

Maintaining complete, comprehensive, detailed records of every facet of performance at the jobsite is much more than just a good idea. It is fundamentally essential to the efficient control of the work, to the achievement of all Company and project objectives, and to the control of the potentially extreme liabilities lurking in every corner of the site—including the Owner, design professionals, and all Subcontractors.

The Daily Field Report is the most important project documentation mechanism at the jobsite used to accomplish these objectives. It is to be filled out every day. On most projects the report will be completed by the Project Superintendent. On very large projects, the report may be completed by the Project Manager, with field information supplemented by the Superintendent. On very small projects, the report should be maintained by whoever has absolute project responsibility.

5.9.2   Purpose

The Daily Field Reports will be the detailed record of precisely what did—or did not—happen on a given day or throughout a particular period. Such detailed history is prepared in such a manner that it may be consulted to confirm the particulars of the facts whenever any portion of the work is questioned, or the performance of any party becomes an issue. It will become the key basis for support in the prosecution or defense of claims in both directions involving the Owner, design professionals, and Subcontractors.

The purposes of the Daily Field Report are therefore to:

1. Provide a chronological, day-to-day account of the workforce, the respective activities planned, and the actual activities performed.

2. Record all visitors to the site, with any significant comments.

3. Provide easily retrievable weather information, pertinent data that may support activity or inactivity.

4. Isolate by description specific work to be charged to and otherwise identified with any change order.

5. Record materials and equipment being received on or sent from the site by the Company or by any Subcontractor.

6. Call attention to particular problems or situations.

7. Provide a written request for materials, information, or other assistance from the Company home-office or senior management.

5.9.3   Need for Proper, Consistent Attention

It is important to recognize from the onset that the greatest problem with Daily Field Reports is simply that the information in them is not needed until there is a problem—but then it is needed *badly*. Characteristics that contribute to the problem include the following:

1. Because the reports are not consulted every day (or even periodically if the project has not been plagued with problems), the process can too easily get looked upon by those who prepare the report as one about which the home-office or senior management is really not all that concerned.

2. When busy activities around the site begin to steal precious time from the Superintendent, it will become tempting to begin to simplify the information on the reports. If allowed, the process will continue to degenerate:

   a. First, they won't really be prepared daily. They'll be "caught up with" near the end of the week, or at some other later period.

   b. The accuracy of the information will become compromised. Because of time lag, the edge will be off (was the concrete placed along column line K or J? Was it placed on Monday or Tuesday?).

   c. The notes themselves will become short and marginally useful. When a large amount of information needs to be caught up on, it will all be shortened and compromised.

   To combat these common traps, the Superintendent and all other individuals even marginally associated with the effort must begin with the profound realization that this reporting responsibility is a *fundamental* requirement of the position. If it is not being done correctly, it is to be considered a *major* performance deficiency.

5.9.4 General Procedure

1. The Project Superintendent must fill out the report every day. Keep the day's report on a clipboard, and carry it with you wherever you go. In this way you will have it handy to jot down all relevant information *as it is observed*. As this habit becomes well developed, the daily completion of the report will become almost automatic; by the end of the day it will already have been done.

2. Each report is to be signed by the person initiating the information. If the Project Manager is an on-site position, he or she must also sign the report.

3. Copies of the reports *must* be distributed on a current basis to the home-office for periodic review by senior management, and to be filed in the General Project File. Two optional ways to effectively police this procedure are to:

   a. Attach copies of the reports to the jobsite Payroll Report Forms being returned to the office each week. Instruct those responsible for processing the payroll to contact the site immediately whenever the Daily Field Reports are not attached, and make every effort to enforce the policy.

b. Require the Superintendent to mail the copy of the Daily Field Report to the home-office every day. This really does not add much expense, but has the significant benefit of ensuring that the reports are in fact being prepared daily. This is the preferred procedure.

In either case, it is equally the responsibility of the Project Manager or other appropriate senior management to develop their own habit of routinely reviewing these reports. It not only is a great way to become currently aware of the project details, but it will help you to assist the field staff in improving their information recording methods, and in calling attention to missed or otherwise incompleted reports.

5.9.5   Report Preparation Guidelines

The following general ideas are to be used throughout the preparation of each Daily Field Report.

1. Be aware of the Rules of Effective Project Correspondence as detailed in Section 2.3.2.

2. Report all facts. Keep the report with you at all times, and record *all* relevant information. If there is not enough space in any category, use as many continuation sheets as necessary. Never abbreviate notes because of lack of space—get more space.

3. Report only facts. Keep frustrations, innuendo, and any remarks that do not belong in a professional communication out of it altogether.

4. Draw only appropriate conclusions. Do not draw speculative conjecture. Include only those conclusions that are the result of a direct cause-and-effect relationship, and that lead to or require some action (correction of work, backcharge, etc.).

5. Avoid buzzwords and trade jargon. Use ordinary language to describe everything.

6. Brief is acceptable, but be complete. Outline statements are fine if they still manage to include all facts and complete descriptions.

7. *Be precise.* Note specific locations, limits of work, and whatever information is necessary to identify the work and processes described positively without question. "Poured concrete," for example, does no one any good. You may as well not bother to make a report at all. Instead, a note such as "Place 30 cy conc. footing along north wall between col. lines H & I, and 22 cy conc. foundation wall along north wall between col. lines F & H," will more positively identify the actual work performed.

Treatment of your remarks in this manner is probably the most important consideration of all in the correct preparation of the Daily Field Report.

8. Identify all sources of information positively. Any information other than your own observations must have its source(s) identified. Refer to companies and *name names*.

9. Include a Field Report for every workday. If, for any reason, no work is performed on a specific day, fill out the day's report, indicating "no work." Be sure to include a statement of the reason for the condition, such as "still waiting for Walkway Change Order (033) approval before any work can proceed," or "HVAC Subcontractor no-show."

10. Include a report for every non-workday. Provide a report to account for each weekend and holiday. If no work was done, days (such as Saturday and Sunday) may be combined in a single report.

5.9.6   Report Information Guidelines

1. *Title Box*. Indicate project name, Company job number, project location, and the name of the individual completing the report.

2. *Date/Page*. There will always be a minimum of two pages available for a day's report. The first sheet is the identification and administrative portion, with at least one Field Work Report Form included to describe the physical work. As many Field Work Report sheets are to be used as necessary for a comprehensive report. Never use the back of any report form.

3. *Weather*
   a. Include a short remark on the typical condition of the day (cloudy, rain, heavy snow, clear, etc.).
   b. Record the temperature in degrees Fahrenheit at the beginning, middle, and end of the day.

4. *Staff (S/V)*. List all Company staff on the site that day. Begin with all permanent staff, indicating "S" (site) after their names. Follow with Company personnel visiting the site that day, noting "V" after their names.

5. *Equipment (C/S)*
   a. List all equipment on the jobsite that day. Indicate:
      (1) Quantity and type
      (2) Whether it is a Company piece (C) or the responsibility of a particular Subcontractor (S); name the Subcontractor.
      (3) Include every piece every day.
   b. Describe the work performed with a brief note that will be presumably elaborated upon in the Field Work Report portion of the report. Include the estimated number of hours worked, or check whether the equipment was idle that day.
   c. If it arrived or departed that day, indicate the time in the area provided. Otherwise leave the lines blank.

6. *Visitors / Conversations / Meetings / Photos*

   Use the jobsite Telephone Log of Section 5.8 as a supplement.

   a. List every visitor to the jobsite that day, and include a brief remark as to the reason for their visit.

   b. Note every relevant conversation with anyone at the site or on the phone.

   c. If the comments, conclusions, direction, etc., are short enough to be included completely here, do so. If not, identify the fact and refer to the appropriate detailed record (job meeting minutes, Company record of Phone Conversation/Meeting Form, a confirming memo or fax, etc.).

   d. Use this area to summarize new or continuing problems, or to indicate any miscellaneous remarks that are relevant to the work.

   e. Indicate if regular or special photos have been taken and the corresponding reason.

7. *Required Materials / Information*

   a. Summarize your requests both to Subcontractors and suppliers and to Company personnel for any information or materials, and the date promised. Again, name names.

   b. This is *not* to be considered your complete documentation effort. It is only a summary note, and should only be there to catalog your more elaborate notifications, memos, faxes, etc., that have been properly supported, documented, and distributed in strict accordance with the requirements of those respective portions of the Manual.

8. *Field Work Report (Page 2+)*

   a. Record all on-site activity in accordance with the recommendations of Section 5.9.5, Report Preparation Guidelines.

   b. Start each entry with a Subcontractor's company name or the Company name, underlined.

   c. Indicate the number of each classification of worker, separating foremen. Indicate the correct EEO code for each individual.

   d. Include complete and comprehensive descriptions of specific work performed for each company and for each group of workers. Include enough information so that anyone who has not been on the site that day can understand and identify precisely what went on.

   e. On the right side of the form indicate if the work described is the subject of a change order, proposed change, possible claim, done under protest, etc. Include the Company file number and Owner change order number whenever there is one available. If there is no Company number available, contact the Project Engineer and secure a new number assignment immediately.

9. *Distribution.* Attach any Subcontractor or Owner reports if they are available, and either forward to the office with your payroll records or mail in daily, as determined by the Project Manager.

Follow the completed example of the sample Daily Field Report Form in Section 5.9.7 as further guidance in the preparation of these critical project documents. A blank form is given in Section 5.9.8.

5.9.7 Sample Daily Field Report Form—Completed Example
(page 5.40)

5.9.8 Sample Daily Field Report Form—Blank Form
(page 5.42)

5.10 Equipment Use Release Form

5.10.1 General

The Equipment Use Release Form is to be used in all cases where any Subcontractor is allowed to use any Company equipment or facility. Its purposes are to:

1. Describe the equipment authorized to be used.
2. Confirm the specific use allowed, including scope, time frame(s), and location(s).
3. Confirm the acknowledgment or responsibility for all costs, effects, damages, etc., resulting from the use of the equipment by non-Company personnel.
4. Secure the legal release of Company liability, and the user's agreement to hold the Company harmless for all loss and damage, including property damage, accident, dismemberment, death, or any result whatsoever.

Note: Check with your own attorney to confirm the specific language necessary and/or allowed in your area. The form provided in Section 5.10.3 is an example only, and is not to be considered as legal advice or the specific legal document to be used.

5.10.2 Use and Procedure

1. The form's most common application is in the case of scaffolding provided by the Company for use by several trades. It is, however, to be used in any case in which any non-Company person or company is allowed to use any Company equipment or facility.
2. The form must be filled out completely, and in duplicate. Each copy must be executed by a person authorized by his or her company to do so.

5.9.7
Sample Daily Field Report Form—Completed Example

Project: FIREHOUSE ADDITION   No: 9924   Date: MAY 15, 1994   Page ___ of 2
Location: NEW CITY, CT   Weather: CLEAR / BREEZY
Superintendent: D. JEFFKO   Temp: 8 AM 71° 1 PM 79° 4 PM 77°

STAFF

Name	Classification	Name	Classification
D. JEFFKO	SUPER		
M. LEONARDO	P.E.		

EQUIPMENT

Quant	Co/Sub	Type/Size	Work/Idle	Work Performed	Arrive	Depart
1	S	CJ7 DOZER	/ ✓			
1	S	920 LOADER	/ ✓			
1	S	770 HIGH-LIFT	✓ /	EXTER. BRICK VENEER		

VISITORS / CONVERSATIONS / MEETINGS / SAFETY REVIEWS

D.J. 9:20 AM TELECON W/RAY SMITH (ARCHITECT); RAY EXPECTS TO
HAVE A-LINE CLARIFICATION (ANSWER TO RFI #12) BT 5-16-94

J. CARRIGAN (OWNER REP) CONDUCTED GENERAL SITE REVIEW
BETWEEN 11:10 AM & 12:00 NOON

B. KROY (KROY ROOFING) STOPPED BY AT 2:40 PM TO CONFIRM
THAT SITE IS READY FOR KROY'S MAT. DELIVERIES ON 5-18-94

REQUIRED MATERIAL / INFORMATION

Item	Requested From	Company	Promised By
RFI #12 ANSWER	R. SMITH	A & S ARCHITECTS	5-16-94
TEMP. PROTECT. MAT'L FOR H.M.	P. DAVIES	D & U MASONRY	5-16-94

Project Manager (Signature)   Superintendent (Signature)

5.9.7
Sample Daily Field Report Form—Completed Example *(Continued)*

FIELD WORK REPORT

Project: FIREHOUSE ADDITION   # 9424   Date: MAY 15, 1994

# Workers	Classification	EEO Code	Complete Description and Location of Work	CO#
			STATESIDE FRAMING	
3	CARP.	Wm	CONTINUE FRAMING EXTER 16 GA. STL. STUD &	
1	CARP. FOREMAN	Bm	GYP SHEATHING @ SOUTH ELEVATION	
			HEAT & AC, INC.	
1	FOREMAN	Wm	· COMPLETE CHANGES TO F-C UNITS	#1
1	PLUMBER	Wm	· CONTINUE DUCT BRANCHES IN ROOMS	
1	"	WF	219, 220, 224, & 227	
			SHORE ELECTRIC	
1	FOREMAN	Hm	· COMPLETE BRANCH WIRING @ 2ND FLOOR	
1	ELECTRICIAN	Bm	· CONTINUE LIGHT FIXTURE INSTALLATION @ 1ST FLOOR	
1	"	Wm	ROOMS 112, 112A, & 113	
			D & W MASONRY	
1	FOREMAN	Wm	· CONTINUE BRICK VENEER AT NORTH & EAST	
2	MASONS	Bm	EXTERIOR ELEVATION	
2	MASONS	Wm	· COMPLETE 2ND FLOOR (FINAL) ELEV. SHAFT	
2	LABORERS	Wm		
17	TOTAL FORCE			

5.9.8
Sample Daily Field Report Form—Blank Form

Project:_____ No: _____ Date:_____ Page____ of____
Location:_____ Weather:_____
Superintendent:_____ Temp: 8 AM _____ 1 PM_____ 4 PM_____

STAFF

Name	Classification	Name	Classification

EQUIPMENT

Quant	Co/Sub	Type/Size	Work/Idle	Work Performed	Arrive	Depart
			/			
			/			
			/			
			/			
			/			
			/			
			/			
			/			
			/			

VISITORS / CONVERSATIONS / MEETINGS / SAFETY REVIEWS

REQUIRED MATERIAL / INFORMATION

Item	Requested From	Company	Promised By

Project Manager (Signature)   Superintendent (Signature)

5.9.8
Sample Daily Field Report Form—Blank Form *(Continued)*

FIELD WORK REPORT

Page _____ of _____

Project:_____ #_____   Date:_____

# Workers	Classification	EEO Code	Complete Description and Location of Work	CO#

3. Return one executed copy to the home-office to be filed in the respective subvendor's bid package file; retain the other at the jobsite.

4. Have a new form prepared and executed for any use and/or time frame not specifically described on any previous form.

5.10.3 Sample Equipment Use Release Form (*page 5.45*)

5.11 Preconstruction Survey

5.11.1 General

The preconstruction survey is the documentation effort that establishes to the best degree possible the precise existing conditions prior to the start of *any* work at the site. It should begin prior to actual mobilization, and continue as appropriate throughout the mobilization period. When the initial preconstruction survey effort is completed, it may become necessary at various points of construction (particularly in renovation projects) to supplement the survey with new information for areas that are becoming available.

The survey effort will include:

1. Preconstruction photographs.

2. Preconstruction video.

3. General verification of existing site information, both for properties directly adjacent to the site, and for the site itself within the contract limit lines.

These processes are closely related to and should be coordinated with Section 5.16, Construction Photographs, and Section 5.12, Field Engineering, Layout, and Survey Control.

5.11.2 Preconstruction Photographs

Ideally the preconstruction photo set should be taken even before the field office has been mobilized. Every area of the site, every adjacent property, and all surrounding private and public properties must be photographed extensively in every detail. Procedures that will efficiently accomplish this include the following:

1. *Use a 35-mm camera of good quality.*
   - This is necessary in order to preserve the ability to select specific photos later, and to maintain acceptable quality in enlargements.
   - Autofocus is available on many cameras with little or no additional cost. It is not required, but highly desirable.

5.10.3
Sample Equipment Use Release Form

Date: _____

Project: _____   # _____

Location: _____

Company: _____

_____

_____

Authorized Representative: _____

Equipment to be used by personnel of the company indicated above
(Completely describe all equipment, including specific locations at all areas of the jobsite and summary descriptions of purposes):

_____

_____

_____

_____

_____

_____

_____

Date(s) and time(s) to be used (Include every possible instance):

_____

_____

_____

_____

_____

_____

_____

As an authorized representative of _____ (Company indicated above), we agree to use only the equipment specifically described above in the locations indicated, on the dates listed, and within the hours indicated.

We agree to strictly abide by all _____ Company, project, safety, and government rules and regulations, and to comply with all requirements of OSHA and all other entities having jurisdiction over the work.

We agree to be completely responsible for any and all direct, indirect, and consequential damages caused by any act, omission or negligence on the part of _____ (Company indicated above), and to hold _____ Company, including its officers, agents, servants, and employees harmless from any damages of any nature or kind whatsoever.

AGREED TO BY:

Company:

Authorized Individual (Signature): _____

Date: _____

Executed in the Presence of (Signature): _____

- Many cameras have optional camera backs available that will stamp the specific photo optically with the date. These are reasonably priced and should be used if possible.
- Autoflash systems are available for little or no additional cost on some cameras, or at least at little additional cost compared to the price of a conventional flash. Although not required, it is a great feature, which saves time while guaranteeing correct exposure.

2. *Organize your approach.*   The photo sets will be divided between:
   - The site itself (within the contract limit lines)
   - All properties immediately adjacent to the site
   - Approach routes to the site

3. *Keep each area as a distinct set.*
   - Start in a logical place and proceed methodically through the entire area. Do the entire site first; then photograph every property adjacent to the site. Pay particular attention to areas along the property lines, and to all physical construction (buildings, parking areas, fences, etc.).
   - Make particular notes of all existing damage—cracks, settlement, damaged surface finishes, etc.
   - Refer to the Site Utilization Program of Section 5.4 to determine the general routes of all construction and worker vehicle traffic into and out of the site. Include photo sets of the approaches of these routes into the site.

4. *Secure permission for complete photo sets.*   Whenever there is any existing building or structure that may be affected by any construction operation such as blasting, pile driving, or dewatering, contact the owner of such property in an attempt to secure permission for a detailed preconstruction photo survey of the complete property. If such permission is obtained, photosurvey the entire premises. If permission is not obtained, photograph as many views and features as are possible from within project property lines. In either event, pay particular attention to all existing apparent damage, such as foundation cracks, badly maintained landscaping and grounds, and settlement of older structures.

5. *When in doubt, shoot first and ask questions later.*   Film is the least expensive part of project documentation that will have the most dramatic effect on settling disputes and saving big dollars. *Never* skimp on film, even with the noble intention of saving a few dollars in the process. Keep your priority on ensuring the most complete, thorough, and comprehensive photo survey possible.

6. *Identify and date the survey.*   Note the survey on appropriate Daily Field Reports and include a copy of the reports in the survey record file. Include any narrative or other appropriate description to make the file complete.

7. *Have all photographs developed immediately.*   Do not leave undeveloped film as the photo record. Not only is it cumbersome to deal with and difficult to file properly, but you will not be able to research the file. There is

also no guarantee that the photos all came out (or that they are actually not of some stranger's vacation . . .).

8. *File correctly.* Forward all developed photos with the field reports and other relevant documentation to be filed permanently at the home-office.

5.11.3   Preconstruction Video

It is becoming increasingly more desirable to supplement the preconstruction photo survey with a preconstruction video. Even if the Company does not own its own video camera, they are so readily available from individuals or by rental that there is really no excuse not to use one.

The preconstruction video is *not* a substitute for the preconstruction photos. It is a supplement. Difficulties with the video lie principally in the limited ability to select, reproduce, and enlarge specific photographs for review and/or demonstration.

The major advantage of the video is that the record picks up so much more on the tape than the photographer is actually observing at the time. It ties the survey together to clearly demonstrate relationships of each area.

Follow the guidelines for preconstruction photographs in Section 5.11.2 to organize your approach and take the video. Additional recommendations to help an acceptable production of the video include:

1. Walk slowly and hold the camera *steady*.

2. Proceed along a planned route. Pan the camera very slowly.

3. Narrate.
   - Start the tape with the complete project identification, date, photographer's name, and any other relevant information.
   - Describe each view. Try to do so in a manner that would allow a person unfamiliar with the site to locate the area, identify the point of view, and understand what he or she is observing.

4. Return the completed video to the home-office to supplement the regular preconstruction photograph record file.

5.11.4   General Verification of Existing Site Information

The final component of the preconstruction survey is a general review of the various conditions of the property as compared to those actually represented in the contract documents and/or observable during a prebid site review. This review is not complicated, but can disclose a surprising amount of issues and questions before the first cubic yard of material is removed. It will protect you, the Company, and even Subcontractors from exposures ranging from the repair of simple damages that existed prior to construction start to problems as significant as removing hundreds or even thousands of cubic yards of earth off the site.

The considerations are again divided between properties adjacent to the construction area and the construction site itself within the contract limit lines.

5.11.5   Adjacent Properties

The characteristics of the properties immediately adjacent to the construction site that are or have the potential to become significant may not be easily apparent. For many of them to become visible, it may boil down to being in the right place at the right time—to be in a specific area just as the problem is occurring.

The kinds of problems that will fall in this category may have been present during the bidding process, but because of their nature have been effectively hidden during the prebid site investigation (and even from the design process itself). The Owner may have been aware of the condition, but not aware of any possible effect on construction, or there may be other reasons why their inclusion in the bid documents may have been considered unnecessary or inadvisable at the time. On the other hand, even the Owner and the design professionals may be completely justified in not being aware of the adjacent properties' characteristics due to their cryptic nature.

Examples of the kinds of conditions that have the potential of impacting construction operations seriously include:

1. A seasonal watercourse that drains several acres of property directly into your footing excavation.
2. Heavy traffic that restricts perimeter mobility during key hours of the day.
3. Construction on an adjacent site that no one was properly aware of, which had begun before this one, presenting unanticipated problems for your own operation (shoring, dewatering, access, staging, etc.).
4. Undisclosed unusual ordinances or other regulations limiting noise during regular working hours.

It can take a great deal of imagination to actually identify these kinds of conditions early on. They are the most difficult of all conditions to spot before they become painfully apparent. The purpose of their identification here, then, is first to make the Superintendent aware of the survey category. From this awareness the Superintendent must make an effort to:

1. Do his or her best to actually identify these conditions as early as best possible.
2. Be constantly aware of the potential for these kinds of conditions; and be quick to identify them for what they are if any should appear at any point during construction.

Whenever such an actual or possible condition is observed, the Project Manager must immediately be notified. Fast action will become critical to:

1. Confirm the actual circumstance.
2. Prepare and deliver all appropriate notifications to the Owner, design professionals, and all potentially affected Subcontractors.

3. Develop at least one (or preferably optional) corrective action plan(s) to be presented to the construction team for consideration as to the most desirable or sensible way to proceed, considering all general cost and schedule implications.

4. When the best plan is decided upon, develop a complete analysis of all costs and schedule impacts to be used for ultimate presentation to the Owner in order to secure appropriate related contract modifications of cost and time.

Consider the Pass-Through Clause of Section 3.5.8. Know in advance whether it is the Company's immediate responsibility to deal with the specific problem (heavy traffic, for example) or whether the problem really belongs to a particular Subcontractor (water draining into an excavation).

If it is determined to be the responsibility of a Subcontractor, advise it of the condition for its own determination of how the Subcontractor should proceed. Always be aware of the Company's contractual responsibilities, both to the Owner and to respective Subcontractors, and have a clear understanding of the relationships before proceeding.

If it is determined to be a Company issue, consider the sample Letter to Owner Regarding Unanticipated Effects of Adjacent Properties of Section 5.11.6.

5.11.6   Sample Letter to Owner Regarding Unanticipated Effects of Adjacent Properties *(page 5.50)*

The sample Letter to Owner Regarding Unanticipated Effects of Adjacent Properties that follows is an example of the kind of notification to be sent to the Owner upon the discovery of some condition on an adjacent property that:

1. Did not exist or was not apparent at the time of bid.

2. Affects construction directly in an adverse way by delaying the project and/or increasing cost.

It is intended, first, to guarantee compliance with any contractual notification requirement by confirming such notice immediately upon the discovery of even a potential condition and, second, to push the resolution procedure into motion.

Before using the letter or anything similar, consider the Pass-Through Clause of Section 3.5.8 to make sure that dealing with the condition is really not a Subcontractor's obligation (instead of the Company's).

5.11.7   Verification of Grades, Elevations, and Contours

It is ideal to begin this component of the preconstruction survey prior to mobilization, but as a practical matter, the procedure will continue into the actual mobilization period and even into the actual site start-up period. In any event, it must be conducted prior to the start of any work in the affected areas.

Before any existing site conditions are disturbed, the actual state of affairs must be verified physically and compared to the conditions represented in the

5.11.6
Sample Letter to Owner Regarding Unanticipated Effects of Adjacent Properties

---

Letterhead

(Date)   Confirmation of Fax
                                                Fax#:_____

To :   (Owner)

RE: (Project Description)
              (Company Project #)

SUBJ: Unanticipated Effects of Adjacent Property:
              Watercourse at North Property Line

Mr. (Ms.)   :

On (date); one day after heavy rain, it was observed that the several acres directly north of the project property line (contract limit line) drain into a natural swale. This swale creates a watercourse that directs the flow of the surface drainage of that property directly into the footing excavation along the north property line.

We have begun dewatering, but significant water continues to drain into the excavation even two days after the rain. Our growing concern is that we are moving into April; characteristically the rainy season. We therefore now unfortunately anticipate that this will become a chronic and growing problem.

Because the project was bid in August of last year, the natural watercourse could not be apparent to anyone during any prebid site investigation, and no such water problem is indicated anywhere in the contract documents. We are accordingly advising you that the project is being impacted in a manner that will affect both cost and time.

As time is of the essence, we request that you immediately have your design professionals review the condition, and advise us on the specific way in which you wish us to proceed. When that information is received, we will submit the complete proposal for the work and its effects.

Very truly yours,

COMPANY

Project Manager

cc: Jobsite
    Architect
    Engineer
    File: Site, Change File (   ), CF

contract documents. Two major categories of site information to be verified include:

1. Grades, elevations, and contours (discussed here).
2. Existing site constructions (discussed in Section 5.11.9).

Grades, Elevations, and Contours. The contours of the new site design are normally prepared by either a Registered Land Surveyor or a Professional Engineer. At times, existing data are verified to some extent before the new design proceeds, but at times they are not. In these instances, grades and contours may, for example, have been lifted from existing land surveys or prior as-built drawings.

During the bid preparation, the contractor has no choice but to rely on the accuracy of this design information. Accordingly it affects the estimates for nearly everything on the site, including, for example:

- Earthwork cuts and fills
- Demolitions and alterations
- Trench excavation lengths, widths, and depths
- Sequence of site activities
- Planned locations of material stockpiles, staging areas, and field offices
- Provisions for temporary power, phone, and water
- Surface water control
- Time and extent of road excavations and tie-ins
- Pavement cutting and patching
- Excavation shoring
- Treatment of mass and trench rock

Action

1. Before any work is done, the photographs and possible video will have been made in accordance with the recommendations of Sections 5.11.2 and 5.11.3.
2. Spot-check existing grades at several easily identifiable locations. If these checks prove to be accurate, note them on the as-built drawings, and continue to the point where you are reasonably confident in the remaining information. If no significant discrepancies or errors are discovered at this step, the work should be allowed to proceed.
3. If *any* discrepancies are discovered, this should be your signal to step up the level of detail of your verification efforts. Identify a key area of the site, for example, and conduct a detailed check of the contour information.
4. If this detailed check confirms multiple or significant discrepancies, notify the Project Manager immediately. Reconsider the validity of your own information,

and determine if those checks themselves should be confirmed before letting the world know of the problem. Time is truly of the essence.

5. If the potential extent of the contour discrepancy is large, consider immediately securing the services of a Registered Land Surveyor or Professional Engineer to substantiate and detail the precise actual condition. This decision must be made with the Project Manager.

6. Whether or not a Land Surveyor or a Professional Engineer becomes involved at this point, the Owner and the design professionals must be notified at the earliest moment when the discrepancies are positively confirmed. Consider using the sample Letter to Owner Regarding Discrepancies in Existing Grades and Elevations of Section 5.11.8 as an example of the kind of required notification.

7. Depending upon the extent of the discrepancy and the potential value of corrective work and/or other project effects, stopping work in the affected area might be considered. This, however, is a very serious action, and should only be considered if for some reason preservation of the area is necessary for prompt and/or equitable resolution of the problem. No stop-work action may be taken without confirmation by Company senior management as to the appropriate procedure.

8. Rather than the Company assuming such profound responsibility for decision to stop, proceed, or take any other action, it will likely be more advisable, as part of the Owner notification, to *request specific direction* as to whether to stop or to proceed. Request such information by a specific date, after which other action may be necessary to mitigate the complete effects on the project.

5.11.8   Sample Letter to Owner Regarding Discrepancies in Existing Grades and Elevations *(page 5.53)*

The sample Letter to Owner Regarding Discrepancies in Existing Grades and Elevations that follows is an example of the notification discussed in step 6 of the previous section and should be sent by the Project Manager. Specifically, it notifies the Owner that:

1. It has been confirmed that discrepancies exist in a specific area of the site.

2. Research is continuing, and engineering data will be provided.

3. Complete total costs, effects on activity sequences and durations, and overall job impact will be calculated. When all this information is known, it will be packaged and submitted for payment.

4. It *requests direction* from the Owner to either stop or continue work. The direction is requested by a specific date.

5. Rights are reserved to claim additional costs for effects that are unforeseen at this time.

5.11.8
Sample Letter to Owner Regarding Discrepancies in Existing Grades and Elevations

Letterhead

(Date)

To: (Owner)

RE: (Project)
             (Company Project #)

SUBJ: Discrepancies in Existing Grades and Elevations

Mr. (Ms)   :

It has been discovered that the existing site grades and elevations differ materially from those shown on Drawing L-1. Per my conversation with your (Name) on (Date), we have retained a registered land surveyor to prepare a complete analysis. We expect the information to be complete by (Date) and will immediately forward it to you.

Upon completion of the survey, an analysis of the changed work, including any effects on activity durations and sequences, will be completed, and you will be advised of any changes in cost and/or construction program.

Your direction is therefore requested to either stop work, or to proceed with any related additional work. Your response is required by (Date) in order to minimize interferences caused by these conditions.

The resolution of this problem will profoundly affect our ability to properly proceed with the work. The utmost urgency is therefore stressed. We accordingly must reserve all rights to claim all additional costs that are unforeseen at this time.

Very truly yours,

COMPANY

Project Manager

cc: Architect
        Jobsite
        File: Site, Change File (   ), CF

Strongly consider the inclusion of the last sentence of the letter at this point. Be aware that this letter will be among the very first of such communications to the Owner, and as such will play a major role in setting the tone for the project. It may instead be more advisable to leave this last phrase off and save it for the second letter if the Owner does not give you an appropriate response by the required date. Let someone else be the first to shoot himself or herself in the foot.

5.11.9   Verifications of Existing Site Constructions

The considerations for the verification of existing site constructions are identical to those for grades, elevations, and contours of Section 5.11.7. For the same reasons as listed before, engineering data relative to the actual configuration of existing site constructions (as opposed to those configurations included in the contract documents) may not have been properly verified before the new designs were prepared. Problems could, for example, exist relating to situations such as the following:

- Manhole locations and storm and sanitary line invert elevations originally provided by the City Engineer's office or other authority may be based upon aging as-built information that is not accurate.

- Telephone and power line locations were verified with the respective utility companies; or maybe they haven't been.

- Transformer pad locations and configurations have or have not been coordinated with the utility company.

Action

1. Again, photographs and a possible video should have already been taken per Sections 5.11.2 and 5.11.3.

2. Check the locations of manholes, markers, or any available indicator to help verify the locations of existing telephone, water, sewer, and gas lines and fuel tanks.

3. Open a few manholes. Spot-check actual invert elevations and compare them with those shown on the plans.

4. Check the locations of telephone poles, street signs, pole guys, and any other constructions. Confirm that they do not interfere with new roads, walks, pavement, excavations, or other site improvements.

5. Check the actual horizontal distances among telephone poles, light poles, manholes, drainage structures, and so on. Compare them to those indicated on the plans.

6. Significant and/or numerous discrepancies may be a strong indicator of the overall lack of quality of the plans themselves, and may thereby strongly suggest additional detailed investigations. In this instance, verify your own information and notify the Project Manager immediately.

7. As with the condition for discrepancies in site contours and elevations, the Owner must be notified immediately upon your confirmation of the accuracy of your own information.

It is important to be aware of the potential effects on the project, which may be limited and immediately apparent:

Example: CB #2 needs to be raised 1 ft to make the drainage work...,

Or the effects may in the final analysis turn out to be profound:

Example: When CB #2 is raised, CB #1 no longer works; when that's raised, the parking lot must be picked up, affecting adjacent grades, that may affect permit considerations...

The potential for these kinds of extended effects may or may not even be apparent at the early stages of discovery and review.

It is therefore crucial that all actions, notifications, and considerations be taken with utmost care to avoid assuming unnecessarily the responsibility for the ripple effects of subsequent necessary design changes. It will *only* be the Company's job to notify the Owner of the problem, leaving complete determination of the solution with the Owner (and with the Owner's engineer's professional liability insurance).

Consider using the sample Letters to Owner Regarding Changed Site Conditions of Sections 5.11.10 and 5.11.11 as examples of such a notification:

- Letter #1 can be used in the instance where a changed condition is very easily identifiable as one that can be corrected without significant effect on surrounding or contiguous construction.

- Letter #2 can be used in the instance where a changed condition has the possibility for significant effect on surrounding or contiguous construction.

5.11.10   Sample Letter #1 to Owner Regarding Changed Site Conditions—Simple Condition *(page 5.56)*

The sample Letter #1 to Owner Regarding Changed Site Conditions that follows is an example of the first of such notifications discussed in step 7 of Section 5.11.9 and should be sent by the Project Manager. Its use should be considered in those cases where it is easily apparent that correction of the respective situation has little to no effect on surrounding construction. (If this is not the case, consider Letter #2.) Specifically, it notifies the Owner that:

1. It has been confirmed that discrepancies exist between the actual conditions of certain items and those conditions indicated in the contract.

2. All substantiating calculations are complete, and the costs are included in the attached change order proposal.

3. Owner approval is required before the changed work can proceed.

5.11.10
Sample Letter #1 to Owner Regarding Changed Site Conditions—Simple Condition

---

Letterhead

(Date)

To: (Owner)

RE: (Project)
           (Company Project #)

SUBJ: Changed Site Condition due to
           Discrepancies in Existing Elevations

Mr. (Ms)   :

Drawing SU-1 indicates the pipe invert at Existing MH #2 to be at 35.5'. The actual invert elevation is 25.5'. The additional 10' in depth of the invert places the new pipe into rock.

The corresponding increase in the contract price to cover the cost for the additional work is ($), computed in accordance with the procedures for Earth and Rock Excavation General Conditions Article 4. The details of this calculation are attached.

It seems that simply lowering the pipe to meet the bottom of MH #2 will not affect surrounding construction. We cannot, however, assume design responsibility for the design change. We therefore offer this solution as suggestion only, to be considered by your engineer; the final design must be provided by your design professionals. Any change between the final design and those conditions represented here will cause the cost proposal to change accordingly.

Your confirmation of the design and approval of the change is required by (Date) in order to allow the work to proceed without additional interruption. All rights are reserved to claim all additional costs that are unforeseen at this time.

Very truly yours,

COMPANY

Project Manager

cc: Architect
       Jobsite
       File: Site, Change File (   ), CF

4. No responsibility is assumed for the new design; approval is therefore construed to confirm that the project design engineer has incorporated the new design into the project and as such is covered by his or her own design liability.

5. Rights are reserved to claim costs and damages that are unforeseen at this time.

As with the other letters of this section, consider this last phrase carefully. Be aware that this letter will be among the very first communications between the Company and the Owner, and as such will play a major role in setting the tone for the entire project. It may instead be more advisable to leave the last phrase off, saving it for the next letter if the Owner's actions become inappropriate.

5.11.11   Sample Letter #2 to Owner Regarding Changed Site Conditions—Complex Condition (*page 5.58*)

The sample Letter #2 to Owner Regarding Changed Site Conditions that follows is an example of the second of such notifications discussed in step 7 of Section 5.11.9 and should be sent by the Project Manager. Its use should be considered in those cases where it is a possibility that correction of the respective immediate situation definitely has or may have *any degree* of effect on surrounding construction. (If this is not the case, consider Letter #1.) Specifically, it notifies the Owner that:

1. It has been confirmed that discrepancies exist between the actual conditions of certain items and those conditions indicated in the contract.

2. The conditions affect surrounding work, and will likely cause some degree of redesign in contiguous construction.

3. Work in the affected areas cannot proceed until the completed redesign is confirmed and the associated change order is prepared, reviewed, and approved.

4. The project is dramatically impacted. Because the new design does not yet exist, and the change order has not been finalized, there is no way to anticipate the net effect on the project. Accordingly, when all these resolutions have been finalized, the complete effect on the project will be determined, along with the resulting effects on cost and time.

5. Rights are reserved to claim costs and damages that are unforeseen at this time.

As with the other letters of this section, consider this last phrase carefully. Be aware that this letter will be among the very first communications between the Company and the Owner, and as such will play a major role in setting the tone for the entire project. It may instead be more advisable to leave the last phrase off, saving it for the next letter if the Owner's actions become inappropriate.

5.11.11
Sample Letter #2 to Owner Regarding Changed Site Conditions—Complex Condition

---

Letterhead

(Date)

To: (Owner)

RE: (Project)
              (Company Project #)

SUBJ: Changed Site Condition due to
              Discrepancies in Existing Elevations

Mr. (Ms)   :

Drawing SU-1 indicates the pipe invert at Existing MH #2 to be at 25.5'. The actual invert elevation is 35.5'. The discrepancy may be the result of an unfortunate typographical or transpositional error.

Simply raising the invert is not possible, as it will cause all surrounding drainage based upon the incorrect invert to fail. The drainage throughout the entire area must therefore be redesigned, hopefully in such a way that will minimize resulting effects on the current design for the new parking area.

This is a very serious problem that has stopped all work throughout the entire area of the site as of (Date). Utmost urgency is stressed to resolve the redesign and to approve the subsequent change order as soon as possible in order to mitigate the potentially extreme consequences to the project. Time is truly of the essence.

We are available to meet at your convenience to assist in the expedient resolution of this problem in any way that you may require. When the redesign is finalized, you will be advised of the anticipated effects on the project's cost and time.

Very truly yours,

COMPANY

Project Manager

cc: Architect
       Jobsite
       File: Site, Change File (   ), CF

5.12   Field Engineering, Layout, and Survey Control

5.12.1   Responsibility, Organization, and Description of Work

The Superintendent must first be aware of the direct responsibility to lay out the work physically. This begins with determining the responsibility to identify and to physically establish the starting baselines and benchmark, and follows through the work of each respective bid package.

The categories of field engineering are therefore divided between the physical layout and the coordination of:

1. The overall building and site.
2. The work directly performed by the Company.
3. The work performed by Subcontractors or trade contractors.

As an industry, there is a pronounced trend among General Contractors in the movement of their company structures toward larger percentages of subcontracted work, until the project is effectively 100% subcontracted. This form of management has distinct advantages in cost performance and management efficiency. There are, however, obvious and subtle approaches that must be taken in many areas—field engineering and coordination of work among them—if these advantages are to be realized, and not simply an inordinate amount of risk assumed on the part of the Company.

There are clear, distinct responsibilities for field engineering, layout, and coordination of the various parts of the work that are very well represented in the respective general contract and subcontract agreements. The successful Superintendent will not, however, assume that every party to the contract:

- Is aware of its responsibilities
- Intends to follow through on them
- Is capable and qualified to complete them in every respect
- Is competent in its efforts to coordinate with the layout work of other trades

The Superintendent must:

- Know in every case what the specific responsibilities of each party to the contract are
- Be able to immediately recognize the presence or lack of intention, competence, and qualifications
- Constantly devote an *appropriate* amount of attention to policing the effort and *verifying the information*
- Take immediate, appropriate action when small or serious deficiencies are observed

5.12.2 Baselines and Benchmark

It is most common for the Owner to be responsible for providing a minimum of two baselines and a benchmark from which the Contractors will lay out the remainder of the work.

Know what the language of the general contract specifically requires. It is not usually enough for the Owner's representative to point out the locations of the baselines and benchmark on the plans and leave you with the responsibility to establish them in the field. It is more likely to be the Owner's responsibility to *physically establish* these fundamental construction starting points.

5.12.3 Site and Building Layout and Procedure

Although the Site and Concrete Subcontractors are technically responsible (as are the other Subcontractors and trade contractors) to lay out their own work (refer to the Pass-Through Clause of Section 3.5.8), the effects on the project resulting from even a small error in layout in these areas can be too catastrophic to leave this function completely up to someone else. These, then, are the two bid packages in which exceptions to the Pass-Through Clause must be made to a significant degree.

In the past it has been most common for the General Contractor to lay out the work with its own forces. Too many of these Contractors, however, have compiled their stories of foundations, parking areas, and other structures being in the wrong location. Be sure that *you are* the Superintendent who through consistent, proper attention doesn't have one of these stories of your own.

The procedure, then, to provide the building and site layout is as follows:

1. Ensure that the Owner has physically established a minimum of two baselines and one benchmark at the site.

2. Arrange to have a Professional Engineer or a Registered Land Surveyor perform all point establishments as described in the remainder of this section.
   a. These must be certified professionals with appropriate professional liability insurance.
   b. There are no exceptions. General site and building layouts must not be performed by anyone without appropriate certification and liability insurance.
   c. Prior to beginning such a layout, have the survey company and/or individuals submit copies of their professional liability insurance policies to the Project Manager for confirmation.

3. Arrange a coordination meeting with the survey crew, the Concrete Subcontractor, the Site Subcontractor, and any other Subcontractor or trade contractor immediately affected by the location of survey information (stakes, points, etc.).
   a. Review excavation limits, construction access patterns, the site utilization program (see Section 5.4), and all other relevant temporary and permanent site constructions and movements to identify methods of establishing

survey points that won't be destroyed by or during construction operations. These considerations will determine things such as:

- Setback distances for excavation stakes
- Sequence of surveys (if, for example, certain areas of the site must be established to allow room to proceed with other areas)
- Quantities of survey points needed for the respective Subcontractors and trade contractors to properly lay out the remainder of the work

4. Prior to beginning any actual construction layout, the survey firm should begin with the verification of basic site information. This verification can be considered as really a portion of the preconstruction survey of Section 5.11, and should include as a minimum the verification of:

    a. Property boundary lines
    b. The Owner's correct establishment of the baselines and benchmark

5. The survey crew will then proceed to establish all control lines as discussed in step 3. As part of this initial survey effort, arrange for the survey crew to transfer the benchmark elevations of as many areas around the site as possible onto stable existing land features or constructions. Remember, once the building begins to go up, the original points will be obstructed. Be sure to have baseline elevations at every convenient location around the site.

6. The building footings and major site components will be laid out by the respective trades responsible for those portions of the work.

    a. The actual layout work will be done by the responsible trades.
    b. The Company will initially and periodically confirm that the procedure being used by the subtrades is appropriate and is being performed correctly.
    c. The actual layouts themselves will be checked as often as feasible by Company site personnel.

7. At the start of construction, after the initial site effort, verify alignment with the established building lines and benchmark elevations.

8. After footing placement and before foundation-wall construction, arrange for the return of the survey crew to pin the foundation and thereby physically establish the actual major foundation-wall locations. The remainder of the foundation walls can be laid out from those points.

9. With the exception of only the simplest and smallest layouts, arrange for the survey crew to verify, at appropriate points of construction:

- Anchor bolt locations
- Bearing plate elevations
- Beam pocket settings
- Additional concrete structures
- Slab elevations (spot-check several points to confirm specified flatness)

10. Throughout construction, Company personnel should periodically (often) verify for themselves:
    - Foundation centerlines and locations
    - Square points
    - Anchor bolts, bearing plates, and beam pockets
    - Plumbness and squareness of structural steel and all concrete structures
    - Alignment of beams, walls, columns, etc.—visually and with survey equipment
    - Slab elevations and flatness
    - Apparent relationship of actual contiguous construction to that anticipated from the documents

11. At the completion of the foundation, arrange for a certified survey of the building location. This information is now first needed to confirm error-free building location. From there it will be used to:
    - Provide as-built location confirmation (see Section 5.20) necessary for the Certificate of Occupancy and for the as-built requirement itself.
    - Establish the first portion of what will become the completed certified survey of the building when the remaining structure is completed, if one is deemed necessary. Review this requirement with the Project Manager.

12. At the completion of the building structure itself, including all major components (overhangs, towers, etc.), it may become necessary to provide a certified survey of the entire building. In some cases the final height of the building must be included in this certification. Review the requirement with the Project Manager and ensure that arrangements are made to provide all required information.

13. As a general note, the project characteristics themselves will greatly determine the amount of professional survey effort advisable throughout the construction period. Considerations for project size, complexity, proximity to property lines and other structures, and so on, may indicate the necessity to have professional survey crews periodically (monthly or weekly) return to the site to perform and confirm the Superintendent's checks described in item 10. This level of attention must be determined on an overall basis by Company senior project management at the project's onset, and must be effectively supervised and managed by the Superintendent throughout construction.

14. All original survey file information is to be forwarded to the home-office for record, with a copy retained at the jobsite file.

5.13   Excavations—Special Precautions

5.13.1   General

Having performed the preconstruction survey in all the detail of Section 5.11, thereby at least generally confirming the overall accuracy of the existing visible information, the probability remains that:

- The actual configurations of underground constructions (pipes, conduits, wires, ductbanks, etc.) are different than those shown.

- Even if "substantially" correct, the *precise* locations of these items will vary to greater or lesser degrees from information on the plan.
- There will be items below the ground (hopefully not a long-distance phone cable) that are not shown at all.

Before any excavation proceeds in any previously undisturbed area of the site, the superintendent is to:

1. Contact "Call-Before-You-Dig" or any other one-call system available for securing services necessary to positively locate all underground utility and communication structures.
2. Use the Notification Checklist Prior to Proposed Excavation in Section 5.13.3, first to ensure that all appropriate individuals and companies are notified of the planned excavation, and second to document the effort.

5.13.2 "Call-Before-You-Dig"

In every metropolitan area and in most other locations throughout the country, one-call systems have been put into place and maintained by associations of utility companies in the respective geographic areas. Generally, it will be an 800 number listed under the name "Call-Before-You-Dig."

This phone number should be available from every utility in the area and from most insurance companies. Find out what the number is, and write it on the Notification Checklist Prior to Proposed Excavation prior to reproducing the job supply of the form.

5.13.3 Sample Notification Checklist Prior to Proposed Excavation
(page 5.64)

The sample Notification Checklist Prior to Proposed Excavation that follows must be completed by the Superintendent prior to any excavation in any previously undisturbed area of the project:

1. All underground systems must be physically located in the field.
2. Every individual or entity possibly affected by the construction must be notified prior to the work proceeding.
3. The effort to perform this research and notification must be properly documented and filed in the correct locations.

The purpose of the checklist, then, is to:

1. Provide a vehicle to aid in the coordination of this important effort, and to help eliminate oversight.
2. Permanently record the names and phone numbers of all authorized parties contacted, along with the dates they were contacted and their inaction or actions taken.

5.13.3
Sample Notification Checklist Prior to Proposed Excavation

Project: _____ # _____
Location: _____

OPERATOR	PHONE	PERSON'S NAME	TIME NOTIFIED	DATE NOTIFIED	SITE ARRIVAL	ACTION TAKEN
Gas—Local						
Gas—Transmission						
Electric						
Telephone—Local						
Telephone—Lg Dist.						
State Maintenance						
County Maintenance						
Local Water Dept.						
Local Sewer Dept.						
Local Highway Dept						
Local Police Dept.						
Local Fire Dept.						

"Call-Before-You-Dig" or other one-call system in effect: _____

INFORMATION TO BE TRANSMITTED TO THE ABOVE OPERATORS:

Company Name: _____   Company Field Rep. _____
Company Address: _____   Co. Field Address _____

Company Telephone: _____   Co. Field Telephone _____

Project Owner: _____   Location of Work: _____
Plans Available At: _____

Dates of Excavation: _____
Purpose of Excavation: _____

_____
_____
_____
_____
_____

Record of Conversations, Meetings, Notes, & Remarks:

_____
_____
_____
_____
_____
_____
_____
_____
_____
_____
_____
_____
_____

Checklist/Information prepared by: _____

3. Transmit all necessary information regarding the project, the work to be done, and the dates when the work will proceed to each operator of every underground system.

Complete the form in its entirety prior to beginning excavation in any new area. Forward the original to be filed at the home-office, and retain a copy in the jobsite Site Notification File.

5.14 Cutting Structural Elements

5.14.1 General

If it should ever become necessary or desirable to cut even a small portion of any structural element, *no* action is to be taken by Company personnel and *no* authorization is to be given to any Subcontractor without first fulfilling the requirements of the sample structural Modification Authorization Form of Section 5.14.2.

Types of structural elements falling under this consideration can include:

- Steel beams, columns, joists, and deck systems
- Steel struts, braces, and supports of any kind
- Concrete grade beams, bond beams, beams, wall panels, columns, supports, and bracing
- Structural studs, composite wall constructions, self-supporting walls, cantilever designs, any kind of spandrel construction
- Any sizable penetration or modification to masonry
- Anything that is unusual in any way, or that you are unfamiliar with

5.14.2 Sample Structural Modification Authorization Form
(*Page 5.66*)

Before any such modification work can be allowed to proceed, *written* authorization must be secured from all authorized entities, specifically allowing the particular modification. This is necessary for *all* modifications—from those as complex as the relocation of structural components to the ones as simple as a 1.5-inch pipe penetration through the 60-inch web of a wide-flange steel beam.

In complex situations it is likely that the issue is so serious that it has or will warrant a large amount of attention with appropriate elaborate documentation of the design change and likely change order. In these cases the documentation and authority should be well taken care of.

If, however, this should for any reason not be the case, or if you find yourself with a "simple field condition" that would otherwise be handled quickly, verbally, and without the full contract procedure, the sample Structural Modification Authorization Form must be used. It will:

1. Describe the proposed modification in complete detail.
2. Secure the approving signatures of all appropriate authorities.
3. Provide complete documentation of the modification and authority.

5.14.2   Sample Structural Modification Authorization Form

Project:_____ #_____ Date:_____

This form summarizes modifications necessary to accommodate job conditions in order to allow construction to proceed in affected areas.

1. Description of need and proposed changes:

_____
_____
_____
_____
_____
_____
_____
_____
_____
_____

2. Indicate in drawing below or on attached drawings to be referenced below all structural components and proposed modifications. Include all dimensions and descriptions of materials:

Authorization and Approval:
The work described above is not to proceed until this Structural Modification Form is executed below by all authorized individuals as indicated:

Owner:_____   Architect:_____   Engineer:_____
By:_____   By:_____   By:_____
Signature:_____   Signature:_____   Signature:_____
Date:_____   Date:_____   Date:_____

When the form is used, be sure to:

1. Describe the modification in a way that cannot be misunderstood. Include a sketch with all dimensions either on the modification form itself if possible, or as an attached supplement.

2. Tape a photocopy of the Modification Authorization Form on the jobsite and as-built sets of plans as the permanent record of the modification. Do this immediately after the form has been completed and signed.

3. Have the form signed in duplicate. Forward an original to the home-office file of the respective bid package, and retain one at the corresponding jobsite file.

5.15   Control of Materials Embedded in Concrete

5.15.1   General

On projects where the Company is placing its own concrete, the considerations of this section are crucial to the avoidance of costly errors and rework due to the oversight of items required to be embedded or cast into concrete.

In those instances where the concrete work on the project is being performed by a Subcontractor or trade contractor, the responsibility for the respective coordination is with those Subcontractors and trade contractors. It does remain, however, in the Company's best interest to take whatever steps are possible to collectively help minimize the possibility of error. This is not just a service for those who are actually performing the work, but a realization that any success in this area translates directly into project success in terms of avoidance of arguments and, most importantly, avoidance of problems that delay work on the critical path.

The remainder of this section assumes concrete placement by the Company and coordination efforts to be directly assumed by your own operation.

5.15.2   Sample Concrete Placement Checklist and Sign-Off Form (*page 5.68*)

Depending upon the actual jobsite staff, the Superintendent, Concrete Foreman, or Field Engineer will be charged with the responsibility to complete the Concrete Placement Checklist and Sign-Off Form prior to each placement of concrete:

1. Completing the form will aid in the coordination of your own work.

2. Requiring the signatures of the foremen or other authorized Subcontractors and trade contractors will force everyone to take close looks and to double-check their own work. Subsequent problems will accordingly be minimized.

3. Securing the signatures of these authorized representatives will directly release the Company from later criticisms of "failure to coordinate" in the event related problems do arise.

5.15.2
Sample Concrete Placement Checklist and Sign-Off Form

Project:_____ # _____ Date: _____

1.   Concrete Contractor/Subcontractor_____

2.   Concrete Supplier:_____

3.   Design Mix:_____ Required PSI:_____

4   Quantity to be placed: _____ CY

5.   Description & location of placement:
_____
_____
_____
_____
_____

6.   Special material requirements (reinforcing, color, topping, finish, etc.):
_____
_____
_____
_____

7.   Significant observations, remarks, and notes:
_____
_____
_____
_____
_____

8.   Coordination of Work:
The signature below of the authorized representative of the respective company performing the work certifies that the work complies with the current requirements of the contract documents in every respect, and has been thoroughly coordinated with the work of all other potentially affected trades; considering dimensional locations, clearances and tolerances, operation, anchoring methods, materials used, that all materials used in their configurations are as approved in accordance with the Contract, and all other coordination issues have been properly accommodated:

TRADE	COMPANY	INDIVIDUAL	SIGNATURE	DATE
Concrete				
Steel Reinforcing				
Masonry				
Structural Steel				
Misc. Steel				
Sitework				
Plumbing				
HVAC				
Fire Protection				
Electrical				
Communications				
Other:_____				
Other:_____				
Other:_____				
Project Superintendent				

Insist on the procedure's being instituted at the very start of the project in even the simplest footing placements. It:

- Makes everyone familiar with the procedure

- Clarifies how important we consider each subtrade's responsibility for the coordination of its own work

- Establishes a routine so that everyone will know what is expected of them, and will be prepared to deal with it in each case in stride.

5.16   Construction Photographs

5.16.1   Description and Requirements

The job photo record begins with the preconstruction photo survey as described in Section 5.11.2. Ongoing construction photographs are the subject of this section. They are divided into regular progress photos and those required in special situations. There should always be a camera on-site and ready. Whenever there is any situation involving a significant question or the potential for changes, problems, etc., taking a photo of the area before any further discussion—and certainly before the area becomes further disturbed—must become the habit of the field staff. Film will always be the least expensive but most powerful agent of every negotiation, resolution, and settlement effort.

Resist the temptation to use wide-angle lenses in all photographs. While they make it easier to get more into the picture, they distort shapes, relative sizes, and perspectives. Remember that the objective is to provide an accurate record. Use lenses with "normal" focal lengths, and your photos will remain truer representations of reality.

As the use of digital photography has become more common, so too has the proliferation of various electronic recordkeeping systems that can best make use of the digital technology. In such systems, the regular project photographs and special photographs can be stored and organized in easily retrievable electronic systems. The photo record for the entire project can in many cases be burned onto a single CD. The photos can then be keyed either directly or through some database program linking the photographs to other key project information. Such links can include correlations between photographs and electronically filed Daily Field Reports, Meeting Minutes, and almost any other component of the official project record.

5.16.2   Regular Progress Photographs

Regular progress photographs may or may not be required by the Owner as part of the General Conditions. If they are, the number of views, size of prints, and other criteria may be specified. If progress photos are not an Owner requirement, arrange for regular progress photos anyway as a Company record. In this effort:

1. Set aside a regular day each month to conduct the photo record. Use the first Monday or first Wednesday, for example. This will help establish a routine that will be easy to monitor.

2. Use a 35-mm camera of good quality. Autofocus and autoflash systems are highly desirable, but not necessary. If at all possible, using a camera back that automatically dates the film is one of the greatest available features.

3. Generally, follow the guidelines for preconstruction photographs in Section 5.11.2. There should be no need for the level of detail of the preconstruction set. These regular photos are intended to show the state of progress, and not necessarily aimed at catching every problem. The latter is discussed in Section 5.16.3.

4. Consider supplementing the photo set with a regular video. As with the preconstruction effort, the video is not a substitute, but will be a thorough record of the entire project.

5. Identify and date each regular survey. Include a copy of the Daily Field Report that records the photo effort.

6. Have each photo set developed immediately and sent to the home-office Progress Photo File.

5.16.3   Special Situations

"Before-During-After" photos are necessary throughout each situation involving actual or *potential*:

- Change orders
- Claims
- Backcharges
- Insurance claims
- The "surprise du jour"
- Any other special situation

In these kinds of events, fast action is necessary to gain the maximum advantage possible. A "precondition" photo or series of photos will be the best record to confirm the actual state of affairs prior to a changed work sequence.

If the duration of the anticipated sequence or activity is relatively long, progress photos of the specific sequence should be taken.

An "instant" camera may prove to be a valuable addition to the project photo effort. Again, it is to be used as a supplement and can only become a substitute for better-quality photos in the simplest of situations (a cleanup backcharge?).

The principal advantages of instant pictures is that they immediately display the specific information that has been recorded in the photo. It is therefore immediately evident if your photos indicate everything intended, or if additional photos, perhaps from other angles, will be necessary. This may be particularly important if conditions are likely to change quickly, leaving little time for conventional photos to be reviewed. Even so, one-hour developing is becoming readily available.

5.16.4   Use of Photograph Layout Form

Regular progress photographs for the records will not require any identification other than that described in this section and in Section 5.11.2 for the preconstruction photo survey.

Special photographs are, however, taken for a specific purpose and must be so identified. In the worst conditions of arbitration or litigation, each photo must be properly and completely identified according to criteria. If it is not, even the best photo will not be allowed consideration. Even if the photo's uses will not become so extreme, references to special photos most often come at points in time long after the original incident. Memories fade, orientations get confused, and so on.

It is therefore important that each significant photo be properly identified if it is to be used to support a charge, contention, or any special circumstance.

Using the sample Photograph Layout Form that follows will guarantee that each photo that will immediately or eventually be used in any presentation or as support for any kind of charge or contention will be correctly identified with:

1. Names of project and Owner, Company project number.
2. Photo location, area identification.
3. Photo orientation, direction of view.
4. References to any appropriate correspondence, field reports, or anything else that ties it directly to the detailed project record.

5.16.5   Sample Photograph Layout Form—Completed Example (*page 5.72*)

5.16.6   Sample Photograph Layout Form—Blank Form (*page 5.73*)

5.17   Managing Time and Material

5.17.1   General

Time and material (T&M) considerations apply in the performance of:

1. Company work to be charged directly to the Owner.
2. Subcontractor work to be charged to the Company that will be passed through as an Owner change order.
3. Subcontractor work to be charged to the Company that will not be passed through as an Owner change order.

5.16.5
Sample Photograph Layout Form—Completed Example

Project: ___FIREHOUSE ADDITION___   # __9424__   Date: __JUNE 16, 1994__
Taken By: ___MARK LEONARDO___   Time: __9:20__   (AM) PM

Location: __NORTH END OF CORRIDOR 224 - CRACK AT DUCT PENETRATION__
Orientation (Indicate on Key Plan): __VIEWING NORTH__
Remarks: __PHOTO #1 - FROM SOUTH END OF CORRIDOR; SHOWS SURROUNDING CONST.__
__PHOTO #2 - FROM CORRIDOR MID-POINT; CLOSE-UP OF WALL SECTION__

Place photos /Assign numbers to multiple photos / Attach this form to large photos.

(Photo)   ①

(Photo)   ②

Key Plan of Photo Area(s):

DOOR 201   224   DOOR 216
①→   ②→

5.16.6
Sample Photograph Layout Form—Blank Form

Project: _____ # _____   Date: _____
Taken By: _____   Time: _____ AM PM

Location: _____
Orientation (Indicate on Key Plan): _____
Remarks: _____
_____
_____

Place photos /Assign numbers to multiple photos / Attach this form to large photos.

Key Plan of Photo Area(s):

The procedure will be followed in all instances where:

1. The work in question is an agreed extra in any of the preceding categories.
2. The work in question either is in dispute as an extra or is still subject to further review and/or reconsideration.

In cases of a Company charge to the Owner, T&M is a final option normally included in the Change Clause of the general contract. (Refer to Section 3.5.7 for a complete description.) It should be pursued immediately as an approved pricing form in any instance where lump sum or unit-price methods fail to secure agreement on costs.

For the company performing the T&M work, the advantages include:

1. Little or no risk of cost overruns. Everything is billed at cost-plus. Coordination and efficiency problems are "absorbed" by the work.
2. The work can proceed with less delay than that associated with change order approval.

For the Owner or Company who must pay for the charges, the disadvantages include:

1. Assumption of nearly all risk of coordination effectiveness and performance efficiency, under conditions where:
   a. Incentives to do these things well are effectively eliminated. With little risk of failure, there is reduced incentive for adequate attention.
   b. There is a financial conflict of interest. The more efficient the working force is, the less will be the final change order value.
2. Complete open-ended nature of both final cost and time.

These are *significant* disadvantages that are not readily overcome.

5.17.2 Field Staff and Company Responsibility

Responsibility of the field staff and of the Company itself with respect to T&M is therefore divided between:

- Controlling T&M work by Subcontractors on behalf of the Company
- Effectively managing T&M work to be charged to the Owner in a responsible, equitable manner

Subcontractor T&M Work. On balance, it is a *very* rare situation that truly justifies *any* T&M work by any Subcontractor to the Company. As a Company policy, Subcontractor T&M should only be considered in those instances where for very real reasons complete pricing and time estimates are truly not possible or advisable.

If you find yourself considering the approval of any T&M work, take another hard look. Be aware that you will be expected to explain to senior management why the pricing could not be secured properly before any changed and/or additional work on the part of a Subcontractor was authorized to proceed. Realize at the start that:

1. If you restructure the proposed T&M work to a lump sum arrangement, you:
   a. Return the risk of success, along with all coordination responsibilities and performance incentives, back to the Subcontractor
   b. Free yourself from the extreme responsibilities and daily efforts necessary for management and verification of the work
2. If you allow proposed T&M work to proceed, you and the Company directly assume these substantial performance issues.

In any event, the Superintendent is not authorized to allow any T&M work without express approval by the Project Manager.

T&M to Be Passed Through to the. In situations where there is any delay in the resolution of a change order's final cost, or if there is any real reason why complete pricing is not feasible, T&M may turn out to be the option of choice for the Company. The principal reason would be the ability to proceed with the work, thereby mitigating the net effect on the project.

In every approved T&M situation, whether the work is being done by Subcontractors and passed through to the Owner, or the work is being done by Company forces and charged directly, there is a primary obligation on the part of the entire field staff and of the Company to shoulder our own responsibilities completely. In this regard, any T&M work will be approached as if it were lump sum:

1. With the same considerations and attention given to coordination, management efficiency, and every cost-control effort.
2. Performed with the overall idea of performing a service for the Owner.

5.17.3 T&M Procedure

In all cases where T&M work has been authorized to proceed, the following procedures and precautions can be taken to improve control:

1. *Have effective standard subcontract language.* Be sure that your subcontracts contain a standard clause disallowing any T&M work unless:
   a. It has been approved in writing in advance
   b. T&M tickets are signed *daily*; otherwise the T&M work expressly will not be recognized
2. *Daily means daily.* A 2-inch stack of T&M tickets presented near the end of the altered work is impossible to decipher with any degree of effectiveness.
   a. If they had been prepared as a group, you can guarantee that each ticket will likely be "generous" in its assignment of man-hours to extra (as opposed to contract) work.

b. Even if you've got your own detailed records with which to check, the process will be long, tedious, and error-prone.

In contrast, strictly enforce the requirement for *daily* review. Verify all information on a current basis. Always take the presenter to task in explaining each day why any portion of the work represented to be extra is not actually part of its contract.

Emphatically emphasize the fact that T&M tickets that have not been signed off *on the day that the work was performed* will not be recognized as extra work. Consider using the sample Letter to Subcontractors Regarding T&M Submission Requirements in Section 5.17.4 for help in placing emphasis on the requirement.

3. *Verify actual hours.*   If multiple items are being billed for on a T&M basis, or if the same individuals are being used concurrently for contract work, check the total number of hours billed on a given day. Do they add up to twelve hours for an eight-hour day? If so, such abuses must be detected early. You will have every right to be offended and to let everyone know that advantage is being taken. Get control over the situation at the start.

4. *Verify labor classifications.*   Are you getting billed for a 50% apprentice at a journeyman's rate? Is a high wage rate from another local being added to the stack of forms? Are prevailing wages truly being paid to the workers?

5. *Verify overhead and profit application.*   Are overhead and profit rates being billed at the subcontract (general contract) change order rates, or are they being creative? If you have agreed to pick up the cost of premium time charges, is overhead and profit being applied to the premium-time portion of the bill without your authorization?

6. *Take photographs as appropriate.*   Film is cheap. In every situation involving any question, immediately take photographs. Refer to Section 5.11, Preconstruction Survey, and Section 5.16, Construction Photographs, for more in this regard.

7. *Evaluate production rates.*   As soon as possible, and at points throughout the T&M work, review rates of progress and production efficiency and compare to those you can reasonably expect.

   a. Even the smallest liberties being taken at the start must be put back into perspective immediately.

   b. Significant and/or chronic abuses must be dealt with decisively.

   Tolerate *no* abuse. Get help from the home-office if necessary to put appropriate pressure on the offending party. Consider:

   - Arranging for Company senior management to resolve the problem with the offender's senior management
   - Rescinding the authorization to proceed
   - Notifying the party that certain portions of the work will not be recognized for payment

In so doing, be prepared with specific criticisms backed up with hard documentation. Be able to definitively support every contention.

When presenting Company T&M tickets to the Owner, it is important to understand that the Owner has the same concerns toward the Company as you have toward Subcontractors. Owners, however, can experience even greater feelings of loss of control. From their perspective:

- The work is difficult or impossible to check
- The work needs to be watched with a microscope
- The totals *always* seem to add up to more than expected (or hoped for)
- You are operating with little or no risk, and therefore without any motivation for cost control or production efficiency

5.17.4 Sample Letter to Subcontractors Regarding T&M Submission Requirements (*page 5.78*)

The sample Letter to Subcontractors Regarding T&M Submission Requirements that follows is designed to follow through with the recommendations of item 2 of the procedure described in order to support the effort to police each T&M Subcontractor's requirement to submit the tickets in proper, complete form and on time. Specifically, it notifies the respective Subcontractor that:

1. *Daily* authorization and acknowledgment of work performed is absolutely the requirement.
2. T&M tickets not approved on the day that the work was performed *will not be recognized* as additional cost.
3. Labor and material information requires detailed breakdowns and complete substantiation of all costs.
4. Signatures of field personnel confirm only the fact that certain work was actually performed with certain forces on that day. This does not acknowledge or agree in any way that the work itself is extra or that the rates charged are accepted. These require agreement at another level.

The real purpose of the letter is to set the tone for the type of attention that will be given the T&M work. If the proper routine is established from the start, the entire process will move ahead much more efficiently and equitably.

5.17.5 Sample T&M Form (Daily Report of Extra Work) (*page 5.79*)

The sample T&M Form (Daily Report of Extra Work) that follows is to be used to fulfill the requirements of this section. A supply should be given to each Subcontractor about to work under any T&M arrangement in order to create uniformity in T&M reporting and to help make their reviews more routine.

5.17.4
Sample Letter to Subcontractors Regarding T&M Submission Requirements

---

Letterhead

(Date)

To: (Subcontractor)

RE: (Project)
           (Company Project #)

SUBJ: (Company Change File No.)
           (Change Description)

Mr. (Ms)   :

On (Date), you were directed to proceed with the subject work on a Time & Material basis in accordance with the provisions of (insert the appropriate general or subcontract reference). Conditions of the arrangement are confirmed as follows:

1.   All T&M tickets are to be signed daily by authorized personnel. Tickets not so approved on the day that the work was actually performed will not be recognized as an additional expense.

2.   Precise labor classifications are to be described on each ticket.

3.   Material invoices are to be included.

4.   The signatures of field personnel confirm the fact that certain work was performed by certain forces on that day. They do not acknowledge or agree that the work itself is in addition to your contract, or that the rates charged are accepted without further review and confirmation.

Thank you for your cooperation.

Very truly yours,

COMPANY

Project Manager

cc: Jobsite
       File: Vendor File: _____
                Change File (   )
                CF

5.17.5

Sample T&M Form (Daily Report of Extra Work)

DAILY REPORT OF EXTRA WORK

Date: _____
Project: _____ #: _____

Charged To: _____

Contract Work: _____
Change Order: _____
Company CO File #: _____

Work Done Under Protest: ___ YES ___ NO
Work Part of a Claim: ___ YES ___ NO

Bid Package: _____
Owner CO #: _____

Description of Work: _____

(A) LABOR

Class	Quan	Hrs	Rate*	Amount

Total Labor (A) $_____

* Includes applicable health, welfare, pension, insurance, and taxes.

(B) MATERIAL

Description	Quan	Unit Price	Amount

Subtotal $_____
Sales Tax $_____
Total Material (B) $_____

(C) EQUIPMENT

Size/Class	Quan	Hrs	Rate	Amount

Total Equipment (C) $_____

GRAND TOTAL (A+B+C): $_____

Prepared By: _____
Approved By: _____

No matter which forms are finally used for any work, be certain that each one is prepared with *all* relevant information and secured *daily*.

5.17.6   Using T&M Records to Support Changes and Claims

An important use of the T&M procedure occurs in those instances where the responsibility for certain work as part of the Owner/Contractor agreement is not called for, is not clearly extra, or is clearly extra but is, at least for the moment, in dispute.

If circumstances lead project management to decide to proceed with the disputed or questioned work, either pending a final decision or in anticipation of a later claim, use the T&M procedure and forms to keep accurate, indisputable records. In order to do this:

1. Prepare the T&M forms as you would if the Owner were clearly paying for the work.

2. Arrange with the Owner's site representative to have the daily forms signed only to acknowledge the fact that certain work was performed by certain workers on that day.

   a. Clarify that it is the documentation of fact only, and that you are aware that the Owner has not agreed to any payments for the work at this time.

   b. If necessary, use the sample Letter to Owner Regarding Acknowledgment of Actual Work Performed of Section 5.17.7 to help accomplish this.

3. If for any reason the Owner representative refuses to even provide this basic acknowledgment, present the forms to him or her *on a daily basis* anyway. Upon each refusal, write the precise circumstances on the respective T&M form, including the fact that it was presented for acknowledgment, or that the work itself was reviewed by the representative who refused to sign.

This effort will become extremely valuable in later Company efforts to settle the issue. The problem will be narrowed to the issue itself, with little time and energy necessary to substantiate the costs.

5.17.7   Sample Letter to Owner Regarding Acknowledgment of Actual Work Performed (*page 5.81*)

The sample Letter to Owner Regarding Acknowledgment of Actual Work Performed that follows is an example of the confirmation that follows through on the recommendations of the preceding section. It is used to advise the Owner that you will expect daily verification of actual work performed for any item of work that is in dispute for any reason. It is to be used in every instance where:

- The decision for the responsibility of the work is pending.

- You have been directed to proceed with work that is definitely or may be extra to the contract.

5.17.7
Sample Letter to Owner Regarding Acknowledgment of Actual Work Performed

---

Letterhead

(Date)

To: (Owner)

RE: (Project)
                (Company Project #)

SUBJ: (Company Change File No.)
                (Change Description)
                Acknowledgment of Actual Work Performed

Mr. (Ms)   :

On (Date), you directed the subject work to proceed without agreeing to a change order.

Please be advised that we are proceeding with the subject work under protest, strictly in the interest of minimizing impact on the project.

We will be preparing daily Time & Material tickets to document the actual work performed, along with all resources used. They will be presented each day to your on-site representative for signature. We recognize that this signature will acknowledge only the facts of the information contained in the respective T&M tickets, and does not at this time indicate acceptance by you of the responsibility for the work.

Very truly yours,

COMPANY

Project Manager

cc: Jobsite
        Owner Field Representative
        Architect
        File: Change File (   )
                CF

- Project management has decided that it is in the best interest of the project to proceed with the work in question, rather than to stop until the issue is completely resolved.

The letter advises the Owner that:

1. You are proceeding under protest per the Owner's direction, and do not agree to be responsible for the extra costs.
2. You will be preparing T&M forms on a daily basis for the purpose of confirming actual work performed.
3. You expect the Owner's representative at the site to sign the daily tickets; specifically to acknowledge the accuracy of the reported information regarding labor used and resources consumed.
4. You acknowledge that the Owner's signature in no way agrees to the idea of an extra to the contract at this time.

5.18   Field Purchases Procedure

5.8.1   General

The field purchases procedure is to be followed in all instances where minor purchases are made directly by jobsite personnel from locations near the field. Only Company employees are to make any purchases on behalf of the Company.

The purposes of the procedure are to:

1. Confirm that the item(s) considered for purchase are not already available from the home-office or another jobsite.
2. Control the amount of materials purchased and buy only those actually needed for the work item in question.
3. Limit "supplementary" or otherwise inappropriate items (such as small tools, blades, bits, and anything else that is the responsibility of someone else).
4. Ensure that the materials purchased are properly identified and clearly assigned to those ultimately responsible for the cost (change orders, backcharges).

5.18.2   Procedure

The Project Engineer, the Superintendent, and the individual actually making the purchase all have responsibilities to fulfill in the procedure:

1. Project Engineer
    a. The Field Purchase Order Log is established for the individual project and placed in the Telecon Log Book (see Section 5.18.3).

b. The Project Engineer or designate will be responsible to assign field purchase order numbers at the time the appropriate Company on-site individual calls for authorization.

c. Prior arrangements must be made with the Site Superintendent before any field purchase order number is given. At that time, the following information must be entered into the Field Purchase Order Log:

(1) Description of purchase

(2) Where the purchase will be made

(3) Reason for purchase, including adequate description of cost responsibility and reference to appropriate change order or backcharge file.

d. The field purchase order number is assigned to the purchase.

e. The Field Purchase Order Form is then completed (see Section 5.18.4). Information provided on the form includes:

(1) Reason for purchase

(2) Complete change order or backcharge description, file reference, and all other appropriate information and references

f. A copy of the Field Purchase Order Form is distributed to:

(1) Accounts Payable

(2) Any referenced change order file

(3) Any referenced backcharge file

(4) Any affected Subcontractor or supplier file

2. Superintendent and field personnel

a. Before leaving the site to purchase anything, the purchase must be set up with the Project Engineer or designate. Include information as required in 1c and get a field purchase order number.

b. While at the purchase site, a call from the vendor is required to confirm the field purchase order number. At that time the complete description of the items purchased and the total cost of the purchase must be available.

3. Accounts Payable

a. After the Field Purchase Order Form has been received from the Project Engineer, it is to be placed into the Payable File for that project.

b. When the respective invoices are received from the vendors, the Field Purchase Order Form is attached to them. The invoice now has the complete detailed information included with it.

5.18.3 Sample Field Purchase Order Log (*page 5.84*)

5.18.4 Sample Field Purchase Order Form (*page 5.85*)

5.18.3
Sample Field Purchase Order Log

PURCHASE ORDER LOG

Project: _____ No: _____

P.O #	AMOUNT	APPR BY	DATE	FIRM	DESCRIPTION	BACK CHARGE	CO FILE #

5.18.4

Sample Field Purchase Order Form

FIELD PURCHASE ORDER

TO: _____
_____
_____
_____

PROJECT: _____
PROJECT NO.: _____

ORDER NO.: _____

ORDER DATE: _____

Confirmation of verbal order: _____

Ship Via: _____
F.O.B. _____

Point: _____
Tax Exempt: _____ YES _____ NO   Tax Exempt No: _____

DESCRIPTION OF NEED: _____

Bid Package Responsibility: _____

Change Order #: _____   Company File #: _____

Item #	Description	Quantity	Unit Price	TOTAL

Sub Total: $_____

Sales Tax: $_____

TOTAL: $_____

Authorized By: _____

5.19   Winter Precautions

5.19.1   General

Winter precautions for projects in freezing climates are too often gone about on an uncoordinated, as-problems-become-apparent basis. The purpose of this section is to identify the issues and provide a straightforward method to organize a coordinated approach to this potentially expensive jobsite condition.

The considerations for winter precautions generally boil down to:

1. The physical constructions and services needed to properly protect the work, and to maintain such precautions.
2. Determination of the complete responsibility to pay for the work, fuel, and materials involved.

5.19.2   Subcontractor, General Contractor, and Owner Responsibilities

It is rare that the complete responsibility for all winter precautions lies only with the General Contractor. It is important to be aware from the start that the responsibility to provide any winter precautions at all is specifically related to the situation that can be anticipated from the bid documents, particularly with respect to the originally anticipated project duration at the time the contract was executed. This is the only circumstance under which the immediately following remarks apply.

Where it has clearly been anticipated as part of the original project cost and time, the General Contractor will generally provide those precautions and maintenance items necessary, as associated with the overall protection of the facility itself, including items such as:

- Closing in of open floors
- Providing space heaters
- Maintaining fuel supply to space heaters

Each respective Subcontractor will generally be responsible for *all* specific precautions and protections associated with any particular item of work or piece of equipment. Refer to the Pass-Through Clause of Section 3.5.8 for related discussion.

In any instance of delay, interference with, and/or disruption of planned sequences, the entire responsibility picture is likely to change:

- A Subcontractor's delay in delivering permanent heating equipment or in securing system operation may force the need for unanticipated temporary heating equipment.

- Any Subcontractor's delay in the work of the critical path may place portions of the work into freezing conditions that would otherwise have been done without the need for such protection by the General Contractor or another Subcontractor.

   Example: The Concrete Subcontractor delays the foundation for exterior architectural masonry, causing the need to provide protection and heat for the masonry.

- The Owner (or design professionals) delay the work in any way (change orders, lack of timely action, etc.), placing work into freezing conditions.

   These types of circumstances alter dramatically the responsibility to pay for establishing, maintaining, and removing winter precautions. The complete responsibility, then, for each anticipated item must be confirmed with the Project Manager prior to proceeding with any such work as part of any direct Company expense.

5.19.3   Winter Precautions Checklist (*page 5.88*)

The Winter Precautions Checklist that follows is to be used as an aid in:

- Reviewing job conditions
- Assessing responsibilities for specific precautions
- Determining adequate precautions necessary for specific areas of work
- Confirming that appropriate arrangements have been made to provide for each required precaution

   Use the checklist:

1. First as a meeting agenda between the Superintendent and the Project Manager to confirm all conditions.
2. Second as a Subcontractor meeting agenda to confirm all that will be done by the respective trades, including timetables.

File the completed checklist in:

- Home and field office files for winter precautions
- All related change order files
- Any related backcharge files
- Every affected Subcontractor or trade contractor file
- The Correspondence File (CF)

5.19.3
Winter Precautions Checklist

YES   NO

A. GENERAL PROJECT STATUS
As of _____
1. Building portions satisfactorily closed to weather:
   a. Roofs & Flashings   ___ ___
   b. Doors & Windows   ___ ___
   c. Building Skin   ___ ___
   d. _____   ___ ___
   e. _____   ___ ___
2. Permanent heating system usable for temporary heat:
   a. Electrical   ___ ___
   b. HVAC   ___ ___
3. Interior pipes/systems subject to freezing:
   a. Remarks:_____
   _____
4. Permanent source of power available: ___ ___
5. Temporary power necessary: ___ ___
   a. Remarks:_____
   _____
6. Permanent source of fuel available: ___ ___
7. Temporary fuel necessary: ___ ___
   a. Remarks:_____
   _____

B. CONTRACT ASSESSMENT
1. Temporary heat required between (dates):
   _____and_____
2. Responsibility to provide temporary heat:
   a. Owner   ___ ___
   b. Prime Contractor or Const. Manager ___ ___
   c. Sub or Trade Contractor(s)
      1)_____   ___ ___
      2)_____   ___ ___
      3)_____   ___ ___
3. Responsibility to provide temp. protect.
   a. Owner   ___ ___
   b. Prime Contractor or Const. Manager ___ ___
   c. Sub or Trade Contractor(s)
      1)_____   ___ ___
      2)_____   ___ ___
      3)_____   ___ ___
4. Temporary heat/protection now required because of delay: ___ ___
5. If (4) yes, who is responsible:
   a. Owner   ___ ___
   b. Prime Contractor or Const. Manager ___ ___
   c. Sub or Trade Contractor(s)
      1)_____   ___ ___
      2)_____   ___ ___
      3)_____   ___ ___
   d. Reasons/Remarks:_____
   _____
   _____
   _____
   _____

YES   NO

6. If (5) is Owner:
   a. Change Order File established   ___ ___
   b. C.O. acknowledged by the Owner ___ ___
   c. Is a claim necessary (denied C.O.) ___ ___
   d. If (6.c.) yes:
      1) Written notification made: ___ ___
         Date: _____
         To: _____
      2) Documentation provided: ___ ___
         _____
         _____
         _____
7. If (5) is Subcontractor or trade contractor:
   a. Has backcharge procedure begun ___ ___
   b. Written Backcharge Notice sent ___ ___
   c. Responsibility accepted   ___ ___
8. Estimated cost of temporary services (Attach detailed estimate forms)
   a. Protection $_____
   b. Heating equip. $_____
   c. Heating fuel $_____
   d. Light & power $_____
   e. Total $_____

C. OVERALL JOB PRECAUTIONS
1. Arrangements made to secure:
   a. Temporary protection materials   ___ ___
   b. Temporary enclosure materials   ___ ___
   c. Continuous fuel supply   ___ ___
2. Temporary heating equipment is:
   a. Of adequate size & type   ___ ___
   b. Is maintained / fully operational ___ ___
   c. Of type(s) allowed by codes   ___ ___
   d. Situated in safe manner relative to pedestrians, traffic, building materials, and ventilation   ___ ___
   e. On a service/maintenance schedule ___ ___
3. Temporary fuel is:
   a. On hand and in adequate supply   ___ ___
   b. Properly and safely stored   ___ ___
   c. On a set refueling schedule   ___ ___
4. All water pockets have been eliminated:
   a. Roof areas   ___ ___
   b. Pavement and graded areas   ___ ___
   c. Sleeves, inserts, chases & openings ___ ___
   d. Other:_____   ___ ___
5. Arrangements have been made for:
   a. Snow plowing/removal   ___ ___
   b. Equipment cold weather protection ___ ___
   c. Vehicle maintenance   ___ ___
6. Precautions taken to protect exposed work:
   a. Exposed piping protected, drained, or heat traced   ___ ___
   b. Recently placed work (concrete, formwork, reinf. steel, masonry, etc.) ___ ___

5.19.3
Winter Precautions Checklist *(Continued)*

                                                    YES   NO

7. All project areas have been adequately
    marked to avoid damage during snow
    removal:
    a. Parking areas   ____   ____
    b. Entrances, exits, gates, passageways ____   ____
    c. Pedestrian traffic areas   ____   ____
    d. Material and fuel storage areas   ____   ____
8. Any necessary photographs of all pre-
    winter jobsite conditions taken for record ____   ____

D. SPECIFIC WINTER PRECAUTIONS

1. Item of Work: _____

    Location: _____

    Party responsible: _____
    Specific precautions taken: _____
    _____
    _____

    Precaution start date: _____
    Anticipated end date: _____
    Remarks: _____
    _____
    _____

2. Item of Work: _____

    Location: _____

    Party responsible: _____
    Specific precautions taken: _____
    _____
    _____

    Precaution start date: _____
    Anticipated end date: _____
    Remarks: _____
    _____

3. Item of Work: _____

    Location: _____

    Party responsible: _____
    Specific precautions taken: _____
    _____
    _____

    Precaution start date: _____
    Anticipated end date: _____
    Remarks: _____
    _____

4. Item of Work: _____

    Location: _____

    Party responsible: _____
    Specific precautions taken: _____
    _____
    _____

    Precaution start date: _____
    Anticipated end date: _____
    Remarks: _____
    _____

5. Item of Work: _____

    Location: _____

    Party responsible: _____
    Specific precautions taken: _____
    _____
    _____

    Precaution start date: _____
    Anticipated end date: _____
    Remarks: _____
    _____

6. Item of Work: _____

    Location: _____

    Party responsible: _____
    Specific precautions taken: _____
    _____
    _____

    Precaution start date: _____
    Anticipated end date: _____
    Remarks: _____
    _____

5.20   As-Built Drawings

5.20.1   General

As-Built Drawings are required by nearly every specification for projects of any size. In those rare conditions where the As-Built Drawings are not specifically called for, they will be provided anyway as a Company requirement.

Their purpose is to serve as a permanent record for the Owner regarding all actual conditions relative to those originally designed, to note dimensional deviations not documented anywhere else, and to consolidate the *identification* of the modifications that have occurred throughout the construction period. The information is used to aid in future design, construction, and maintenance. As-Built Drawings are *not* there to repeat the detailed information of any change that is properly documented in the respective files.

5.20.2   Procedure

1. Immediately at the start of the project, one complete set of plans, specifications, and addenda is to be sent to the jobsite clearly marked as "As-Builts." There is no need to "post" the addenda; just include it as part of the set. These documents are *not* to be used for construction. They are to be properly filed and kept in good condition.

2. Review the contract documents to determine any specific conditions required by the Owner for preparation, maintenance, and delivery of the As-Built Drawings. Comply in every respect.

3. It is the responsibility of the Project Engineer, Site Superintendent, and any other field staff to verify that any deviation between actual construction and that as originally designed is in fact properly authorized and documented as such prior to allowing such deviating work to proceed. These will include items such as:
   - Approved change orders
   - "Clarifications" not involving cost or time
   - Accommodations of field conditions that are slightly different than those originally anticipated
   - Actual locations and configurations of existing underground lines and construction as they are uncovered during the course of the work
   - Actual locations of new underground work if at all different from the plan locations

4. The Superintendent is to record in sufficient detail all dimensional deviations and all references to the detailed change records on *both* the jobsite document set and the As-Built Drawings as they occur. All such additions, deletions, or changes are to:
   a. Be indicated in red pencil or pen
   b. Be dated
   c. Include *clear* reference to appropriate authority for the modification, such as:
      - Change order file number

- Job meeting item number
- Conversation and confirming memo with name, conditions, etc.
- Structural Modification Authorization Forms (see Section 5.14.2).

5. In the case of change orders and detailed clarifications, it is not necessary to redraft the detail of the change. Cloud the area affected by the respective change or clarification, and reference the appropriate change order number or other complete reference.

6. Whenever possible, tape a photocopy of any "SK" or other available sketches on the contract set. It is most likely that the only room will be on the back of the previous page. In this case, simply note "Taped Opposite" on the modified plan. Include copies of any Structural Modification Authorization Forms (see Section 5.14.2).

7. It is the Project Engineer's and the Superintendent's responsibility to police each major Subcontractor or trade contractor to include their own as-built information on the Company field set and As-Built Drawings on a weekly basis. This information should be confirmed monthly by the Project Manager as an express condition of payment. These contractors include at a minimum:
   - Concrete
   - Structural steel
   - Plumbing
   - HVAC
   - Fire protection
   - Electrical
   - Controls
   - Communications

8. Confirm final as-built configurations required by the contract prior to delivery to the Owner. It may, for example, be required to transfer the information to a set of mylars provided by the Owner. In such cases it is not necessary to transfer any supplemental documents taped to the plans as discussed in item 6. The references will be adequate.

9. Include all engineered layouts, confirmations, and certifications provided (see Section 5.12.3) and all certified As-Built Drawings by all trades required to provide them (fire protection, for example).

10. Hand deliver completed as-built documents to the Owner, and have them signed for by an individual authorized to receive them.

5.21   The Site Superintendent as Project Leader

5.21.1   Section Description

In many ways, the Site Superintendent is the most front-line ambassador of the company. He or she is looked upon by the outside world as the personification of the company insofar as activity on the site is concerned. Officially, the Site

Superintendent is the "representative" of the company and is therefore empowered to commit the organization.

In a very real sense, it has been said that the Project Engineering function builds the job "on paper," while the Site Superintendent's function builds the physical construction in the field. Together, these two coordinated counterparts work synergistically to make the "work" happen while the "administration" is being taking care of. Everyone—from the president of the company, to the Project Manager, to the Project Engineer, to the direct-higher laborers, to each of these individuals in every Subcontractor's company—looks to the Site Superintendent to provide the strong shoulders necessary to support the entire effort in the field. To this point, this section has dealt with those administrative components that are to be managed by the Site Superintendent as part of his or her responsibilities. The success that the Site Superintendent may have in completing all of these objectives is rooted not only in the knowledge of their composition and competence with their implementation, but in the way that the Site Superintendent builds coalitions, develops cooperation, and—in a word—provides leadership.

The items that follow are a collection of short summaries of general suggestions that may assist the Site Superintendent in fulfilling his or her leadership role while conducting daily business. They do not necessarily add up to a neat checklist of "dos and don'ts," but will hopefully provide some help with working your way through these issues and relationships each day.

5.21.2 Simple Expediting Techniques

The need to expedite surprise jobs is a daily part of the business. It often helps to recognize that as a person moves up the ladder, increasing percentages of that individual's work are generated by that person's judgment—with decreasing percentages left to be filled by all those comfortable routines that change little from day-to-day. After all, if an individual's judgment to deal with "special" situations isn't necessary, that individual will not likely justify the highest salary.

Use some of the following suggestions as an aid to help you expedite resolution of your daily surprises, while minimizing interruption to your smoothly planned, well-organized daily workflow:

1. Do not start running in all directions at once. Sit down, relax, and gather your thoughts. Remaining calm will be your greatest method of determining the best way of tackling the new job.

2. Using your experience on similar projects that you have accomplished in the past, analyze the new item. Don't jump to be too concerned too quickly about "how" the new job will be accomplished just yet. Begin with deciding simply what genuinely has to be completed, and the date or times by which it has to be done. Begin at the end, and work yourself backwards in sequence through the activities needed to accomplish it until you arrive at the first task.

3. Determine which items or information will take the longest to collect or produce. Even though certain items may not be needed until later in sequence, this is construction; you can be sure that something needs to be done now in order to accommodate the occurrence of certain items later and improper sequence.

4. Think backwards. Visualize the new work item in reverse fashion. Determine the end product that needs to be accomplished by the last date. Work your way back in organized and logical sequence, identifying each step that needs to be completed in order to allow the particular component to be accomplished. Work your way all the way back through the determination of "step #1."

5. Once you determine all the steps that need to be done, along with their relevant time periods, it is time to look at the "how" of completing each step. A good supervisor knows which of his or her employees can most effectively handle each part of the project. The most sensitive or higher-level components of the task should be assigned to the most capable and reliable subordinates. Know which parts of the process can tolerate various "levels" of "quality," and which components might be subjected to any type of "contingency."

6. Whenever possible, use teams. Use group leaders to undertake some of the responsibility. Be sure to remind them—clearly—that the regular work still has to get out on time. Have these individuals regularly report back to you on the progress (or lack thereof). Keep their feet to the fire; as has been said in certain management circles, there are really two fires: the regular work and the special, expediting assignment.

7. Try to visualize new ways of accomplishing the objective. Do not be afraid to use novel or unusual methods. A "special" problem may require a "special" approach.

8. Keep up the pressure, and constantly follow up on each subordinate who has been given any component of any assignment. Timing of such constant follow-up, however, can mean the difference between "professional urgency" and frantic panic.

9. Maintain all project or assignment milestones.

10. Prepare yourself for new and unexpected problems. Try to expect the unexpected, and don't panic when it occurs. In such an event, pull from within to be innovative and to demonstrate initiative. Inspire your subordinates to act the same.

11. Schedule catch-up time to recover from little unexpected blitzes in your organized program.

5.21.3 Improving Supervisory Communication

The Site Superintendent has to communicate each day with his or her superiors, subordinates, various representatives of each Subcontractor and supplier, the

Owner, Architect, engineers, inspectors, and even the general public. Difficulties with any of these lines of communication create unnecessary strains, delay resolution of important action items, and make the Superintendent's job more difficult.

The complexities of most operations on construction sites today require that the Superintendent simply must be a good communicator in order to get the work out, to build cooperation among those who are responsible for performance of the work, and ultimately to compel performance.

The introduction of the smartphone and tablets has been of immeasurable help in allowing the Site Superintendent to communicate instantly with members of the construction team, be they other Company or office field personnel, vendors, Subcontractors, sub-subcontractors, design consultants, or Owner representatives. The advantage of having such easily portable devices readily at hand to communicate via text messaging, voice communication, or digital photographs greatly enhances the field Superintendent's or Project Manager's ability to remain in close contact with all members of the team.

The Superintendent and Project Manager can create an electronic record of all of outgoing and incoming communications using this equipment (as well as laptop or desktop computers). This gives them the ability to make hard copies or transfer critical notes and observations onto a hard drive, flash drive, or compact disc (CD).

Face-to-Face Communication.   The most desirable method of communication is face-to-face, a personal oral exchange. Not only does face-to-face communication include an element that a common interchange of information is shared, but there is a chance to see, with proper timing, whether the choice of a topic, tone, and wording has achieved the desired objective. Facial expressions and body language provide numerous insights into the opinions and even intentions of both parties. It provides the opportunity for give-and-take analysis of the effects of the information as it is being communicated.

Telephone Communications.   A telephone call provides much of the give-and-take situation of a face-to-face interchange. A telephone call does not, however, permit immediate observation of any physical reaction of the other party; it only allows audible perception of the other person's reaction. Facial expressions, posture, nervous movement, and other items are totally lost when communication is made over a telephone. Personality is much easier to disguise in such an abstract communication than it may be in face-to-face communications.

The Written Word.   In the construction industry, the written word is by far the form of communication most relied upon. This Operations Manual is devoted to the various written forms of communication necessary to plan, execute, monitor, enforce, compel, retrieve, present, document, and support. It is such a fundamental part of the process that it can to a very large extent define each of our jobs.

Interoffice Memos.   As another example of a written communication, an interoffice memo can be a test of anyone's real ability to communicate. Memos can be quick, clear, to the point, and unmistakable in their purpose, or they can be long, confusing, and with a cloudy agenda. Keep your interoffice memos sharply targeted to their objectives by being sure that each memo contains these three basic components:

1. Opening: begin with a statement that identifies the single purpose of the memo.

2. Facts: include a brief outline of the situation in a logical sequence of events.

3. Conclusion: finish with a brief summary of the facts and an expectation of what is expected of the reader. Include a timetable.

Reports.   While reports are generally more formal than most other forms of written communication, that does not at all mean that they are any more difficult to write. Reports are difficult to write only if they are begun without a clear concept of their purpose, content, and conclusion. Take the time to think through the entire report. Organize the report concepts, and develop those concepts into an outline. The more effort spent on creating a detailed and comprehensive outline, the easier the report preparation will ultimately be. It is normally much faster and easier to deal with a series of short, well-organized topics than to try to create some cohesive sense out of a long, rambling dissertation.

Letters.   Generally, it is a good idea to create a draft of the letter first. Let some time pass, if possible, between initial draft and your subsequent review. Review the draft carefully, looking for ways to remove unnecessary, extraneous language. Consider each phrase and remove any ambiguity. Look at it one more time with an effort to edit out unnecessary wording. Be sure that the letter leads logically to a conclusion. Be sure to clearly state the action that you require in response to the letter, as well as the deadline for any such action.

Are Electronic Records Admissible as Evidence? When a disagreement or dispute is in the offing and the electronic records you have stored on your smartphone, tablet, or laptop are necessary to back up your claim, a legal question arises: "Are these electronic records admissible in court as evidence?" The first step in answering this question should be a discussion with the Company's attorney, in which you review all of the facts surrounding the dispute or claim and the electronic documentation you have accumulated.

The Business Records Act, 28 U.S.C. § 1732 (2000), states that a reproduction made "in the regular course of business" of a record made "in the regular course of business" by "any department or agency of government" is "as admissible in evidence as the original itself in any judicial or administrative proceeding whether the original is in existence or not." Although this statute refers specifically to a government agency, it may apply to the private sector as well; case history will be built as more and more claims utilizing electronic records as documentation are thrashed out in the courts.

5.21.4 Reducing Clerical and Administrative Errors at the Jobsite

A Superintendent can take several steps to improve the quality of information provided to others outside the company. This is true for information generated by himself or herself, as well as information generated by all subordinates. Consider the following items to help reduce errors and improve the quality of your information:

1. Train your subordinates to check, recheck, and correct improper work products.

2. Review all written instruction and provide steps that can be taken for error correction and detection.

3. Provide cross-checking procedures, whereby all work is to be rechecked by a separate individual.

4. Assign responsibility for accuracy. Consider implementing a procedure whereby specific individuals will be required to initially check documents and items.

5. Coach subordinates in techniques that will help them recognize errors. Identify different "tests of reasonableness" that will help subordinates conceptualize expected information—against which to compare documented information.

6. Be sure subordinates are familiar with the tools given to them to do their jobs. Calculators, computers, printers, and all manner of equipment are useful only to someone who is able to operate them.

7. Consider having the human resources department include relevant tests for different types of subordinates. Math tests, for example, might be given to those who are aspiring to estimating and bidding responsibilities, writing tests might be given, etc.

5.22 Hot Work Permit

(Note that the sample policy herein may be considered very differently in different companies, and in different geographic locations. Consider the self-explanatory statements in this section as well as the specific language that is included on the sample form itself, research the issue, and make your own decision with respect to the manner in which you intend your own company to deal with these conditions. Consider describing the use of the Hot Work Permit directly in your subcontracts in order to remove any doubt as to the responsibility of each Subcontractor to comply with this critical requirement.)

The Hot Work Permit Form given in Section 5.22.1 should be used whenever any type of "friendly fire" is intended to be used on any company jobsite, whether that work is being performed by company employees or by the employees of any Subcontractor. Such friendly fire includes any manner of welding, cutting, burning, and other operations that may directly or indirectly result in an open flame, sparks, hot slag, the production of combustible gases, or similar conditions.

A separate permit form should be required for each hot work operation each day. Make it an official company policy that after some short, uniform period of time, say two hours, after completion of the hot work, the area is to be inspected in order to assure that the activity is completed, and there is no further fire, or risk of fire.

5.22.1   Sample Hot Work Permit Form (*page 5.98*)

5.23   Jobsite Equipment Use, Operation, and Maintenance

5.23.1   Section Use

This Section of the Operations Manual is intended as an example of company instruction to employees with respect to the manner in which company-owned equipment is to be used and maintained while on the jobsite. It is intended to provide a summary of general rules that apply to the use of all company equipment, and includes checklists of preventative maintenance considerations that may be followed by equipment operators in order to assist those operators in ensuring the proper ongoing usability of all equipment.

The language and forms of this section are to be considered in the context of your own company and its manner of operation. The information presented is of a general nature and arranged in a form as company-to-employee communication. Carefully consider these objectives and use the information as a guide to prepare specific information relative to the particular equipment being used by your company.

5.23.2   Equipment Prestart, Operation, and Maintenance

Equipment safety, reliability, and operating efficiency are no accident. These things do not happen by themselves or without a prescribed program. Following simple rules about providing such organized and reasonable care will reduce equipment downtime, increase operating efficiency, and improve operator well-being.

Every equipment operator should be thoroughly familiar with the checklists of this section, and follow them each day for each piece of equipment that is being operated.

The rules and checklists that follow are generic; they apply generally to all equipment. Accordingly, each equipment operator must also be thoroughly familiar with the specific safety and maintenance checklists provided for each specific piece of equipment being used, and apply each of them daily—in addition to the rules and checklists of this section.

5.22.1
Sample Hot Work Permit Form

(COMPANY) Hot Work Permit

Date of Permit: ____/____/____   Time: ____:____ AM/PM

Permit Duration: **FROM:**
____/____/____   Time: ____:____ AM/PM

TO:
____/____/____   Time: ____:____ AM/PM

Project:

Number: _____

Name: _____

Location: _____

ALL COMBUSTIBLE MATERIALS ARE TO BE REMOVED
FROM THE WORK AREA AND ITS PERIMETER

1.   Description of Work Performed:

☐ **WELDING**   ☐ **TORCH CUTTING**   ☐ **SOLDERING**   ☐ **OTHER**

Description: _____

_____

_____

_____

2.   Work Area(s) Affected: _____

_____

_____

_____

3.   FIRE WATCH EQUIPPED WITH: _____

_____

_____

PERMIT EXPIRES AT END OF WORK DAY ON DATE INDICATED ABOVE

Prepared by: _____   Title: _____

cc: ☐ Safety & Health Manager   ☐ Project Manager   ☐ _____
       ☐ Human Resources Manager   ☐ _____   ☐ _____

5.23.3   Sample General Equipment Safety Checklist

1. No persons other than Company personnel are allowed to operate
    any company equipment under any circumstances.   _____

2. Only qualified operators are allowed to operate the equipment.   _____

3. Learn the location and purpose of all controls, instruments, indicator light, and labels.   _____

4. Be sure that a first aid kit is fastened to all major equipment.   _____

5. Keep a fully charged fire extinguisher mounted conveniently. Learn to use it correctly.   _____

6. Wear fairly tight, properly fitting clothing and safety equipment.   _____

7. Avoid high-pressure fluids that can penetrate skin and cause injury.   _____

8. Relieve pressure before disconnecting hydraulic or other lines.   _____

9. Tighten all connections before applying pressure.   _____

10. Keep hands and body away from pinholes and nozzles
    that eject fluids under high pressure.   _____

11. Use a piece of cardboard to search for leaks. Do not use your hand.   _____

12. *If ANY fluid is ejected into the skin, it must be surgically removed*
    within a few hours by a doctor with this type of injury, or gangrene may result.   _____

13. Wear suitable hearing protective device such as earmuffs or earplugs,
    in order to protect against loud noise.   _____

5.23.4   Sample Equipment Safe Operation Checklist

1. Use handrails and steps to enter and leave the operator's station.
   Do not use the steering wheel. _____

2. Keep handrails, steps, floor, and controls free of water, grease, and dirt. _____

3. Do not operate any equipment in an unsafe condition.
   Put a tag on the steering wheel or other appropriate high-visibility location. _____

4. Before starting or operating any equipment:
   a. Check the condition of the equipment (See Prestart Inspection Checklist below). _____
   b. Be sure there is enough ventilation. _____
   c. Know the correct starting and stopping procedure. _____
   d. Sit in the operator's seat. _____
   e. Clear the work area of people and obstacles. _____
   f. Check the service brakes and parking brake. _____

5. Be sure engine is running and foot brakes are operating before releasing the park brake. _____

6. Do not allow riders on the equipment. _____

7. Drive slowly in congested area, over rough ground, and on slopes and curves. _____

8. Do not drive near the edges of ditches and excavations. _____

9. Keep loading areas smooth. _____

10. Check locations of utilities, cables, gas lines, water mains, etc., before digging. _____

11. Keep away from power lines at all costs.
    Do not touch power lines with any part of the equipment. _____

12. Carry buckets and loads as low as possible for better stability and visibility. _____

13. Keep equipment in gear when going down steep grades. _____

14. Use accessory lights and devices to warn operators of other vehicles. _____

15. Position backhoe booms on uphill side when driving across hillsides. _____

16. Set stabilizers before operating any backhoe equipment. _____

17. Use care when raising stabilizers; they may be the only restraint preventing movement. _____

18. Do not dig under stabilizers. _____

19. Avoid swinging any backhoe bucket in the downhill direction. _____

20. Before dismounting the equipment:
    a. Engage parking brake. _____
    b. Lower all equipment to the ground. _____
    c. Stop engine. _____
    d. Release hydraulic pressure; turn steering wheel back and forth,
        move hydraulic control levers until the equipment does not move. _____

5.23.5   Sample Prestart Inspection Checklist

Follow the checklist below to inspect all equipment before you start it each day that the equipment is used.

1. Vandalism. Check to see that:
    a. Smokestacks and exhaust pipes are clear of debris and obstructions. _____
    b. Fuel, water, gas, and oil filters have not been tampered with. _____
    c. Lights and glass are not broken or loosened. _____
    d. Gauges are not damaged. _____
    e. Wires have not been cut. _____
    f. The equipment looks in good overall condition. _____

2. Tires and Wheels:
    a. Inspect for loose or missing bolts. _____
    b. Check tire pressure. _____

3. Operator's Station:
    a. Be sure it's clean. _____
    b. Check pedals for freedom of movement. _____

4. Hydraulic System:
    a. Check oil levels _____
    b. Check for leaks, kinked lines, and lines
       or hoses that rub against each other or other parts. _____

5. Engine Compartment:
    a. Check engine oil level. _____
    b. Check transmission oil level. _____
    c. Check fuel filter for sediments. _____
    d. Check air cleaner. _____
    e. Check radiator coolant level and clean radiator. _____

6. Lubrication:
    a. Check lubrication points shown in the respective equipment service manual. _____

7. Electrical System:
    a. Check for worn or frayed wires and loose connections. _____

8. Protective Devices:
    a. Check guards, canopy, shields, seat belt. _____

9. Booms, Buckets, Structural Components:
    a. Check for bent, broken, or missing parts. _____

5.23.6 Sample Equipment Service Safety Checklist

1. Put a support under all raised equipment.   _____

2. Before beginning service: _____
    a. Review requirements for hard hat, safety shoes, safety glasses or goggles,
       gloves, reflective vest, ear protectors, and respirator.   _____
    b. Be sure that service is approved.   _____
    c. Understand all procedures.   _____
    d. Stop all equipment.   _____
    e. Stop the engine (unless necessary for service).   _____

3. Disconnect the battery ground wires
    before welding or working on the engine or electrical systems.   _____

4. Before working on the hydraulic system:
    a. Release all pressure.   _____
    b. Loosen fittings slowly.   _____

5. Do not smoke:
    a. When you fill the fuel tank.   _____
    b. When you work on the fuel system.   _____
    c. When you handle fuels or lubricants.   _____

6. Do not fill the fuel tank when the engine is running.   _____

7. Guard against eye injury when hammering.   _____

8. Do not lubricate or work on equipment when it is in motion.   _____

9. Avoid high-pressure fluids:
    a. Relieve pressure before disconnecting hydraulic or other lines.   _____
    b. Tighten all connections before applying pressure.   _____
    c. Keep hands and body away from pinholes and nozzles
       that eject fluids under high pressure.   _____
    d. Use a piece of cardboard to search for leaks; do not use your hand.   _____

10. Engine coolant:
    a. Only add coolant to the radiator when the engine is stopped or running at slow idle.   _____
    b. Do not remove cap unless engine is cool.   _____
    c. Release all pressure before removing cap and loosen slowly.   _____

11. Tires:
    a. Do not attempt to mount tires without the proper equipment and experience
       to perform the job safely.   Failure to follow proper procedure can produce an
       explosion that may result in serious bodily injury or death.   _____
    b. Be sure all tire rims are correctly assembled and interlocking before inflating tires.

    c. Use an inflation cage, safety cables, or some such protective device during inflation.   _____

5.23.7 Sample Equipment Fire Prevention Maintenance Checklist

1. Daily Prestart Maintenance:

    a. Check fire extinguisher for correct charge.   _____

    b. Open all access hoods and shields.
        Remove all trash from all areas inside these compartments from:

        i.   Exhaust manifold, turbocharger, and muffler.   _____

        ii.   Bottom guards and under engine.   _____

        iii. Sides of engine.   _____

        iv.   Radiator and oil cooler.   _____

        v.   Batteries.   _____

        vi.   Hydraulic lines.   _____

        vii. Fuel tank.   _____

2. Check for leaking fuel lines, hydraulic lines, or fittings.

    a. Tighten loose fittings.   _____

    b. Replace bent or kinked lines.   _____

3. Clean trash from grilles.   _____

4. Clean trash from cab areas.   _____

5. Be sure all doors and grilles are in place.   _____

6. Shut-Down:

    a. Be on guard for fires; especially when refueling.
        Temperature in engine compartment may go up immediately after stopping engine.   _____

    b. Wait until the engine has cooled before filling the fuel tank.   _____

    c. Do not smoke when refueling.   _____

5.24 Building Commissioning and Turnover to the Owner

As the construction project nears completion, a series of requirements are generally included in the Specifications Manual to create a smooth transition as the building or facility is turned over from the construction team to the Owner's team. Submission of As-Built Drawings, warranties and guarantees, spare parts, and Operation & Maintenance (O&M) Manuals are submitted as part of the building turnover or commissioning process.

The word "commissioning" originally came from the shipbuilding industry, referring to the fact that a ship was required to pass through various stages before being turned over to the Owner. The approach adapted by shipbuilders is similar to the commissioning process employed in the commissioning of a building: the product passes through various stages before being accepted by the Owner.

- The building's systems—electrical, plumbing, HVAC, electronics, security, etc.—are certified by the design team as being installed per the contract requirements.
- These systems are tested, any problems that arise are identified and solved, and the designers certify that their operation meets the contract specifications.
- The Owner's maintenance crew is familiarized with the operation and maintenance of the equipment, aided by the manufacturer's instructions and the O&M Manuals.

5.24.1 National Institute of Building Sciences Introduction to Building Commissioning

It is often said that commissioning is all about good project documentation. The purpose of commissioning documentation is to record the standards of performance for building systems, and to verify that what is designed and constructed meets those standards. Commissioning is a team effort to document the continuity of the project as it moves from one project phase to the next. In the Planning and Development . . . phase of a project, planning and programming documents begin to define an owner's requirements for building performance. When the entire project delivery process is documented in a consistent manner, a historical perspective is created that explains the iterative process of determining the agreed-to project requirements at each step of the development process. Commissioning documentation becomes the road map for the success criteria to be met by facilities that are put in service.

At post-occupancy, commissioning documentation becomes the benchmark to ensure that the building can be maintained, retuned, or renovated to meet future needs. It documents the Owner's Project Requirements (OPR) in the beginning of the project and records compliance, acceptance, and operations throughout the facility's life.

This WBDG page provides information on common commissioning documentation practices and resources related to commissioning specific systems and assemblies.

By permission: National Institute of Building Sciences and the Whole Building Design Guide.

5.24.2 Recommendations

Document all Levels of Project Development and Acceptance. Requiring documentation of results and findings of the commissioning process at all project delivery stages and phases provides a record of the benefits received from commissioning. It also provides documentation to be used in the future to troubleshoot problems and optimize operating strategies. Decision making is an "iterative" process made over the course of a project through analysis of options, selection of alternatives, refinement of application, and integration of the design components. As each decision is made, commissioning documentation provides the basis for evaluation and acceptance to proceed to the next development level.

Emphasize Inspection, Testing, and Training on Commissioned Systems. An essential element of the commissioning process is field verification inspection and testing of commissioned systems, assemblies, and features. The Commissioning Authority coordinates and witnesses commissioned systems verification tests to verify that the systems operate in accordance with the design intent. The Commissioning Authority may be tasked with conducting special testing of commissioned systems beyond what is required in specification requirements. Deficiencies discovered during verification testing are documented and logged by the Commissioning Authority in corrective-action reports. Retesting specific systems and/or system components takes place once the respective deficiencies discovered during the first test are resolved.

A draft set of system readiness checklists (SRCs) and verification test procedures (VTPs) is included in the commissioning specification to communicate to the bidding contractor the level of rigor that can be expected during the testing phase of the commissioning process. The SRCs are detailed checklists for documenting that each system is prepared for testing. The VTPs are a detailed set of instructions and acceptable results for thoroughly testing each system.

During functional performance testing and operator training, the commissioning team moves to the forefront. The team verifies the performance of building systems based on detailed test procedures developed by the commissioning team and determines the most efficient equipment settings. Testing must be performed not only in normal operating modes but also under all possible circumstances and sequences of operation, with real-life conditions simulated as much as possible. Further, integrated systems testing should examine systems as a whole in order to evaluate overall design and compatibility.

The team also supervises operations staff training on commissioned systems and equipment, and organizes warranty information. Ultimately, the team prepares extensive documentation on systems, including benchmarks for energy use and equipment efficiencies, seasonal operational issues, start-up and shutdown procedures, diagnostic tools, and guidelines for energy accounting.

By permission: National Institute of Building Sciences and the Whole Building Design Guide.

5.24.3   Key Commissioning Documentation: Predesign, Design, Construction

Compile Key Commissioning Documentation.   Commissioning documentation is generated throughout the project delivery process, and key documentation such as OPR, BOD, Cx Plans, schedules, and inspections and test results are included in a Commissioning Report. Commissioning documentation that will be included in the Commissioning Report is normally shown in a table format with responsibilities of individual team members who will prepare, review, and accept the results and documentation. A partial list and descriptions of key commissioning documentation includes:

- *Owner's Project Requirements (OPR)*—For commissioning to be successful programming documentation must summarize the OPR that is both general and specific to critical requirements. The OPR is a summary of critical planning and programming requirements and owner expectations that is updated by the commissioning team as the project evolves. If program or mission elements change during the span of project delivery, the OPR should be updated to reflect changes in building performance requirements.

- *Basis of Design (BOD)*—The BOD is a narrative and analytical documentation prepared by the design A-E along with design submissions to explain how the Owner's Project Requirements are met by the proposed design. It describes the technical approach used for systems selections, integration, and sequence of operations, focusing on design features critical to overall building performance. An OPR is developed for an owner/user audience while the BOD is typically developed in more technical terms.

- *Design Review Comments*—Comprehensive reviews targeted to critical systems at all design phase submissions are an important aspect of commissioning documentation. Reviews for code compliance and constructibility will pertain to all systems of all projects, while commissioning reviews are focused to commissioned systems, equipment, and building assemblies and building components they are interfaced with.

- *Certification Documentation*—Owners sometimes require their facilities to achieve certifications such as Energy Star, LEED (http://www.usgbc.org/Displaypage.aspx?CategoryID=19), Green Globes (http://www.thegbi.org/commercial/about-green-globes/questionnaire.asp), or governmental agency testing and inspection. When such performance certifications are required as part of a design or construction contract, they become critical to an owner's project expectations and may be included as commissionable elements.

- *Submittal Review Comments*—Concurrent with the design team and owner review, a designated commissioning team member reviews products and systems submittals for compliance with the Owner's Project Requirements. Special attention should be given to substitutions and proposed deviations from the contract documents and Basis of Design documentation. Submittal review comments on commissioned systems will often generate issues for coordination between integrated systems, equipment, and technologies.

- *Inspection Reports*—Commissioning Inspection Reports should be prepared regularly to document progress of the work on commissioned building systems. These reports will normally produce functional issues, integration issues or operational issues that are then captured in Issues Logs for discussion and clarification of performance expectations, integration issues, or operational issues. The construction delivery team (and owner's representative (CM), if applicable) will also prepare inspection reports pertaining to all building systems and components.

- *Test Data Reports*—Test Data reports contain results of the Testing and Inspection Plans and include Pre-Functional Test (PFT) reports, Functional Test Reports (FTP), and other test results specified for the commissioned systems.

- *Issue Logs and Reports*—Issues Logs and Reports are a formal and ongoing record of problems or concerns—and their resolution—that have been raised by members of the Commissioning Team during the course of the Commissioning Process. Issues Logs should be included in Commissioning Reports because, along with minutes, design review comments, and Inspection Reports, they explain the thought process and rationale for key decisions in the commissioning process.

- *Commissioning Reports*—The commissioning requirements, process, documentation, and findings are incorporated in a Commissioning Report that accompanies the construction contractor's turn-over documentation. *ASHRAE Guideline-0* recommends that the Commissioning Report be included with O&M manuals in a Systems Manual. Commissioning Report contents should be clearly defined in Commissioning Plans and include a narrative of the commissioning process, the design intent document, design review comments—and resolution, meeting minutes from all commissioning-related meetings, corrective action reports, blank verification test reports for future use, completed training forms, completed system readiness checklists, and tests and inspection reports for commissioned systems, equipment, assemblies, and building features.

- *Systems Manuals*—The Commissioning Authority CA reviews the project operations and maintenance (O&M) manuals to verify that commissioned systems and equipment information and documentation are included. The Commissioning Authority also reviews the as-built drawings, in particular the sequences of operations documentation for automated systems that are commissioned, to verify that the documents turned over to the owner are accurate and reflect what was installed and tested. *ASHRAE Guideline-0* recommends that O&M manuals, submittals, as-built drawings, specifications, certifications, training documents and commissioning documentation be organized by building systems in a "Systems Manuals" for ease of access and use by building management staff (see the ASHRAE table on page 5-108). Some owners find it is efficient to have the Commissioning Authority compile Systems Manuals for all systems—both commissioned and non-commissioned.

- *O&M Training Documentation*—During the Design Phase, training requirements for operations and maintenance personnel and occupants must be identified relative to commissioned systems, building features, and equipment. It is critical that the operations and maintenance personnel have the knowledge and skills required to operate a facility in accordance with the owner's functional plan and the designed intent.

- *Post Seasonal Testing*—Due to weather conditions, not all systems can be tested at or near full load during the construction phase. For example, testing a boiler system might be difficult in the summer and testing a chiller and cooling tower might be difficult in the winter. The performance and testing of active solar systems is also dependent on seasonal conditions. Commissioning plans should therefore provide for off-season testing to allow testing, balancing, and optimization of integrated systems under the best conditions.

TABLE D-1 Documentation Matrix Phase	Document	Input By	Provided By	Reviewed / Approved By	Used By	Notes
Pre-Design	Owner's Project Requirements	O&M, Users, Capital Projects, Design Team	CA or Designer	Owner	CA, Design Team	Design Team may not be hired yet.
	Commissioning Plan	Owner, Design Team, CA	CA	Owner	CA, Owner, Design Team	Design Team may not be hired yet.
	Systems Manual Outline	O&M, CA	Owner or CA	Owner	Design Team	May be included in OPR
	Training Requirements Outline	O&M, Users, CA, Design Team	Owner or CA	Owner	Design Team	May be included in OPR
	Issues Log	CA	CA	N/A	CA, Design Team	May be only format at this phase
	Issues Report	CA	CA	Owner	Design Team, Owner	
	Pre-Design Phase Commissioning Process Report	CA	CA	Owner	Owner	Close of Phase report
Design	Owner's Project Requirements Update	O&M, Users, Capital Projects, Design Team	CA or Designer	Owner	CA, Design Team	
	Basis Of Design	Design Team	Design Team	Owner, CA	Design Team, CA	
	Construction Specifications for Commissioning	Design Team, CA, Owner	Design Team or CA	Owner	Contractors, CA, Design Team	May also be provided by Project Manager / Owner's Rep.
	Systems Manual Outline-Expanded	Design Team, CA, O&M, Contractor	Design Team or CA	Owner, CA	Design Team, Contractor	Contractor may not be hired yet.
	Training Requirements In Specifications	O&M, Users, CA, Design Team	Owner or CA	Owner	Design Team	Contractor may not be hired yet.
	Design Review Comments	CA	CA	Owner	Design Team	
	Issues Log	CA	CA	N/A	CA, Design Team	
	Issues Report	CA	CA	Owner	Design Team, Owner	
	Design Phase Commissioning Process Report	CA	CA	Owner	Owner	Close of Phase report
Construction	Owner's Project Requirements Update	O&M, Users, Capital Projects, Design Team	CA or Designer	Owner	CA, Design Team, Contractors	
	Basis of Design Update	Design Team	Design Team	CA, Owner	Design Team, CA	
	Commissioning Plan Update	Design Team, CA, Owner, Contractor	CA	CA, Owner, Design Team, Contractor	CA, Owner, Design Team, Contractors	
	Submittal Review Comments	CA	Design Team	Design Team	Contractor	

ASHRAE GL-0 Table D-1 Documentation Matrix

By permission: National Institute of Building Sciences and the Whole Building Design Guide.

Safety and Loss Control

6.1 Section Description and Company Policy

6.1.1 Section Description

Safety and Loss Control. "Safety" has become another one of those buzzwords getting much use in the construction industry at increasing rates. It has certainly been used correctly in many circles; unfortunately, it can become a cliche if it is allowed to. "Safety" and "loss control" are, however, words with real meaning—and equity—that vary from company to company and from individual to individual. This may be true perhaps more for these than for any other words in the business; these are words that deserve direct and serious attention on a daily basis.

As an industry, providers of construction services have been forced in recent years to wrestle with a large number of serious dilemmas in ways that are strikingly different from those dealing with similar issues in other industries. Until relatively recently, history has confirmed the construction field to be one of the most consistently dangerous for its employees. While there may be a great distance to go with respect to continuing improvements in our business's safety record, it is at least equally true that as an industry, we've achieved more safety gains than any other in recent years. These gains are not merely in statistics and real safety achievement, but are in attitude, perspective, and genuine intent as well.

Like so many other categories of responsibility, the issue of overall safety and loss control in construction is now prescribed, directed, recommended—and mandated—in ways that genuinely affect each member of every company that provides construction services. We've been told that "worker safety and health protection is a decisive factor in reducing work-related injuries and illnesses," and "Every employer should develop and maintain a safety program . . . that should provide policies, procedures, and practices that protect all employees from safety and health hazards." Like it or not, we as constructors are all part of the program, have certain rights within the system, and carry significant responsibilities with us from the office or shop to the jobsite and back each day. For those of us who are doing what we can to take these realities seriously, they mean different things at different times.

Administrators have the responsibility to determine and adequately provide for the requirements of a proper, company-wide, ongoing safety program; managers are responsible to implement and enforce compliance by every member of the organization; and employees bear the burden of complying with the clear set of company directives as communicated.

Employees have a right to a safe working environment—at least to the minimum standards prescribed by law, and employers have certain rights with respect to expectations that qualified employees will obey laws and act with the responsibility that their position requires.

Beyond all of these things, there is a finite set of reasons why any of us in the construction industry are ultimately motivated to provide for safe working conditions in our jobsite and office environments. The relative importance within the total mix that an individual will place on any one component will vary to wide degrees, but intelligent, responsible people should ultimately conclude that giving safety regular, significant attention is not only the right thing to do, but is good business as well.

All motivations for giving safety proper attention boil down to three categories: moral, economic, and legal. These are those "components" within the total mix that are given different levels of importance by different individuals. All preaching aside, it almost doesn't matter where any of us are along the total safety-perspective continuum, as long as we manage to reach the correct conclusion for ourselves—that genuine attention to safety is necessary, correct, and even profitable.

The Moral Component.   Morally, we all know the reasons for safety. Not only do we know that providing safe working environments for all employees is the right thing to do, but it is also intuitively clear that none of us have any right under any circumstances to knowingly and willingly subject another individual—whether an employee, peer, or even a family member—to conditions that may or will adversely affect the health of that individual. If any of us need examples of case law or specific government guidelines to convince us of that general principle, we will likely have other difficulties with the "moral" component as well.

"Doing the right thing" with respect to jobsite safety can often turn out to be an easy principle in cases that are black and white (should there be a handrail or any kind of fall protection at the edge of that fifth-floor slab?). Other cases, however, can become confusing in those gray areas that permeate the complete safety and loss control issue (am I satisfied with those minimum published government guidelines, or is there something about the situation that still leaves me uncomfortable?).

As it turns out, once the basic legal requirements have been determined and provided for in a specific safety situation, there is a simple test that can routinely be applied to every questionable condition that will help to leave a concerned person with clear direction—and a clear conscience. The test can be considered at two levels. First, consider the condition from the point of view of a safety-and-health-conscious individual, and determine if you would be satisfied with placing yourself under the total set of working conditions in which you are about to place another individual. Even if for some reason you are overly casual about the level of peril that you're willing to tolerate personally, consider the second level; simply ask yourself if you would be satisfied to allow your own child or spouse (of adequate age, training, and experience) to work in the same condition. While not specifically satisfying (or violating) any legal requirement that I am aware of, this simple test has always helped me personally to instantly transform those shades of gray clearly to black or white. In other words, if the conditions are not good enough for my kids, they're not good enough for my employees or peers.

The Economics of Safety and Loss Control.   On the economic level of consideration, safety is a set of numbers related to an individual's compensation, and to a company's profit and loss statement. Studies and statistics can and have cataloged the direct and indirect economic effects on individuals and organizations. The direct effect on an employee begins with loss of wages. Even if compensated by insurance, the reduced physical capacity related to injury or illness affects every aspect of the individual's life. The indirect effects can multiply exponentially.

For the company, economic losses related to a jobsite accident divide still further among added cost related to interruption, rework, and increased costs. A serious accident on a jobsite stops the project—either partially or entirely—until the condition is dealt with on its immediate (emergency), intermediate (secure area), and long-term (corrective measure) phases. The work is interrupted, and wages continue for every employee involved with each phase of the accident. Jobsite supervision is removed from production considerations and instead is forced to focus on the emergency. Office administrators—and even company officers—are pulled from their ongoing work and rushed to deal with the emergency condition.

After consideration of the jobsite and company interruption is the realization that certain preventable safety emergencies can result in damage to the work, which must then be corrected. An example might be a trench collapse that damages shoring, formwork, or equipment in the vicinity. Simply stated, all this means is that after the emergency has been dealt with, the construction area must be reworked at the added expense of removal and replacement of any damaged areas.

Legal Considerations.   Legal considerations are intertwined with all others every step of the way. Minimum requirements for company safety programs and specific criteria for jobsite conditions can be clearly spelled out sufficiently to allow for scheduled or surprise inspections that can result in large fines for perceived violations. Once cited for a safety violation, be prepared to add (of course) those legal fees—even if the company is ultimately found to have been in compliance all along.

Insurance Impact.   Finally (but not least) is the added insurance costs to a company that are guaranteed to be assessed due to the worsening of the company's safety record as the result of a serious accident. These costs are often not routinely dealt with by anyone other than senior company management and certain administrators, and can have their significance diminished in the eyes of managers and supervisors unless the cost increases are brought to their attention with the full impact that they deserve. Such increases can easily run into the tens of thousands of dollars and can continue for years until an improved company safety record is re-established.

When all the labor, equipment, administrative, and overhead costs to a company are added up for the entire scenario to deal with even a single incident of a serious jobsite safety emergency, it is not difficult to see that the costs that would have been saved through prevention of the incident would likely surpass the cost of implementing an entire comprehensive, company-wide safety program. If that implemented program then goes on to prevent a second serious emergency, it becomes clear that the idea of attention to safety moves decisively into the category of a direct major contributor to the company's bottom line. It is not smoke-and-mirror accounting, but profit/loss reality. It is genuinely profitable to be safe.

It is all these considerations that are alluded to in the phrase "safety is good business." Safety does turn out to be "good business" for all the right reasons.

6.1.2 Most Frequently Cited OSHA Standards Violations as of September 2012 (*page 6.7*)

6.1.3 Ten Most Frequently Cited OSHA Violations for the Fiscal Year 2011

The Deputy Director for OSHA's Directorate of Environmental Programs, Patrick Kapust, provided the new list at the National Safety Council's annual Congress. The list includes a mix of general industry and construction standards:

1. **Fall protection in construction** (1926.501): 7,139 violations; the Bureau of Labor Statistics (BLS) says 260 workers died in fiscal year 2010 due to violations of this standard.

2. **Scaffolding in construction** (1926.451): 7,069 violations; BLS says in FY 2010, 37 workers died because of incidents involving faulty scaffolding.

3. **Hazard communication** (1910.1200): 6,538 violations; Kapust says in one fatal incident involving hazard communication, an employee lit a lighter to see the level of material inside a barrel, and the substance ignited. Proper labeling should have indicated how full the barrel was and that the material inside was combustible.

4. **Respiratory protection** (1910.134): 3,944 violations; five million workers in the U.S. are covered under this standard and are required to wear respirators at their jobs.

5. **Lockout/tagout** (1910.147): 3,639 violations; the average days away from work for employees injured in incidents connected to this standard is 24.

6. **Electrical wiring methods** (1910.305): 3,584 violations; employees affected by this standard range from engineers, electricians, and other professionals who work with electricity directly, to office workers and administrative staff who use any type of electrical equipment.

7. **Powered industrial trucks** (1910.178): 3,432 violations; BLS says in FY 2010, there were 8,410 injuries connected to use of powered industrial trucks, such as forklifts.

8. **Ladders in construction** (1926.1053): 3,244 violations; falls are consistently one of the top three causes of worker fatalities.

9. **Electrical general requirements** (1910.303): 2,863 violations; this standard seeks to prevent injuries and deaths from electric shock, fires and explosions.

10. **Machine guarding** (1910.212): 2,748 violations; this standard also covers anchoring of equipment.

This list covers the period from October 1, 2010 to September 30, 2011.

Source: Adapted from U.S. Department of Labor, Occupational Safety and Health Administration; www.osha.gov/dcsp/compliance_assistance/frequent_standards .html

6.1.2
Most Frequently Cited OSHA Standards Violations
as of September 2012

The following were the top 10 most frequently cited standards in fiscal year 2012 (October 1, 2011 through September 30, 2012):

1. Fall protection, construction (29 CFR 1926.501)
2. Hazard communication standard, general industry (29 CFR 1910.1200)
3. Scaffolding, general requirements, construction (29 CFR 1926.451)
4. Respiratory protection, general industry (29 CFR 1910.134)
5. Control of hazardous energy (lockout/tagout), general industry (29 CFR 1910.147)
6. Powered industrial trucks, general industry (29 CFR 1910.178)
7. Electrical, wiring methods, components and equipment, general industry (29 CFR 1910.305)
8. Ladders, construction (29 CFR 1926.1053)
9. Machines, general requirements, general industry (29 CFR 1910.212)
10. Electrical systems design, general requirements, general industry (29 CFR 1910.303)

The following are the standards for which OSHA assessed the highest penalties in fiscal year 2012 (October 1, 2011, through September 30, 2012):

1. Fall protection, construction (29 CFR 1926.501)
2. Scaffolding, general requirements, construction (29 CFR 1926.451)
3. Control of hazardous energy (lockout/tagout), general industry (29 CFR 1910.147)
4. Machines, general requirements, general industry (29 CFR 1910.212)
5. Powered industrial trucks, general industry (29 CFR 1910.178)
6. Ladders, construction (29 CFR 1926.1053)
7. Electrical, wiring methods, components and equipment, general industry (29 CFR 1910.305)
8. Process safety management of highly hazardous chemicals (29 CFR 1910.119)
9. Hazard communication standard, general industry (29 CFR 1910.1200)
10. Electrical systems design, general requirements, general industry (29 CFR 1910.303)

For more detailed information, visit Frequently Cited OSHA Standards. At that site, you can generate a report on the most frequently cited federal or state OSHA standards by your SIC code and the number of employees in your establishment.

Source: Adapted from U.S. Department of Labor, Occupational Safety and Health.

6.1.4 Company Safety Policy

No single aspect of the work must take on greater importance than safety and loss control. Consider the field condition. Place yourself on your own jobsites on a regular basis and be truthful in evaluation of the safety quality of the area. Now imagine your children or other members of your own immediate family working daily in the same environment. Are you still comfortable with the situation? What are the things that you would do in order to improve the work environment so that you would be satisfied with it? Finally, imagine your children or family members as pedestrians or sidewalk superintendents. Are you satisfied that the existing provisions surrounding the jobsite will eliminate all risk to them?

By now, it will hopefully be clear that it should be a policy of your company and each individual employed by it to always:

- Provide safe work environments.
- Conduct each operation in the manner that reduces risk to themselves and other workers.
- Maintain all conditions in a way that eliminates all risk to visitors and to the public.
- Eliminate risk of damage to property adjacent to the site.

To attain these objectives and clearly establish communication to each company employee in a way that cannot be misunderstood, it is recommended that a formal company Safety and Loss Control Policy be written and distributed to each employee. Its terms should be completely unambiguous. To this end, a sample Company Safety and Loss Control policy follows in Section 6.1.5. Consider the suggested language carefully and specifically. Have it reviewed by your appropriate safety, legal, and insurance professionals. Finalize the precise language, adopt it as your own, and distribute it throughout the organization.

6.1.5 Sample Company Safety and Loss Control Policy

As (Company Name) continues through our (number) year of providing construction services, we've taken great pride in our consistent record of safety and in our concern for the safety, health, and welfare of all of our employees. There is no other aspect of our work that takes on greater priority.

It is the policy of Company to:

- Provide safe working environments and conduct all company activities in a manner that reduces risk to all employees and to all other workers.
- Maintain all office, equipment, and jobsite conditions in ways that eliminate risk to visitors and the public, and eliminate risk of damage to property and equipment.
- Provide safe, well-maintained equipment, proper training for all equipment operators, and instruction on safe methods and procedures.

- Comply with all federal, state, and local laws and regulations as they apply to all work performed.
- Never accept any unsafe working condition for any reason, and take immediate corrective action when any safety violation is observed.

Correspondingly, we expect that:

- Each Company employee will be responsible for safety and health.
- All safety equipment issued must always be used as intended.
- Company equipment must not be damaged, removed, altered, or abused, or operated when observed to be in such condition.
- Employees must be familiar with and observe all safety rules, procedures, and policies, as they may change from time to time.
- Supervisors will see that all rules and procedures are observed by their crews and immediately enforce appropriate corrective measures whenever violations are observed.

The collective results of our safety efforts will affect our overall company success. Conduct with respect to safety will affect the manner in which the performance of all employees will be measured.

Although we enjoy a safety record to be proud of, our goal is 100% accident-free work, while ensuring our tradition of quality construction and client satisfaction. The good intentions, cooperation, and good judgment by all employees in the use of safe and responsible work practices is the path toward continued personal and company improvement.

6.1.6   Sample Jobsite Safety Policy Notice (*page 6.10*)

A notice similar to the following sample should be placed in a prominent location at the field office and in any bulletin area provided at the site for such notices. It is a statement of safety policy, and of the conduct expected to be displayed by every person on the site.

Each jobsite has its own unique arrangements and circumstances. The language of the notice can be considered a guide that can be modified as appropriate.

Like all such notices, it gets no power from the paper on which it is printed, but directly from the conduct of the Company personnel on a daily basis. Your conduct with respect to your own work and with respect to routinely policing compliance by others will be returned with equal amounts of attention.

6.1.7   Sample Disciplinary Action Policy and Procedure for Safety Violations

Sample Disciplinary Action Policy.   As with the balance of this section, consider the issues discussed below and any sample language provided carefully and specifically. Have the information reviewed by your appropriate safety, legal, and

6.1.6
Sample Jobsite Safety Policy Notice

Letterhead

It is the policy of (   ) Company to at all times:

- Provide safe work environments,
- Conduct all operations in a manner that eliminates risk to any tradesperson,
- Maintain all conditions in a way that eliminates all risk to visitors and to the public, and
- Eliminate risk of damage to property on and adjacent to the site.

This is the fundamental responsibility of every individual on the site. All supervisors must routinely accept complete responsibility for prevention of accidents and for the safety of all work under their direction.

By contract and by law, every company on the site is obligated at a minimum to conform to the Federal Occupational Safety and Health Act, and to the laws of every entity having jurisdiction over the work.

Any company or individual refusing to correct observed safety violations will be banned from the site at least until such violations are corrected, and will be held completely responsible for all resulting effects.

There is no magic to safety. With proper attention, awareness, and cooperation by everyone, we will achieve an accident-free job, and the pride that goes with it.

(   ) Company

President

insurance professionals. Finalize the precise language, adopt it as your own, and include it as part of your ongoing company Safety and Loss Control Program.

Compliance with all company safety rules and procedures should be made a condition of employment. All employees should be given the responsibility to familiarize themselves with company safety rules and procedures, and to comply with them in every respect. Supervisory, administrative, and management personnel at all levels should be given the direct responsibility for taking immediate corrective action when a violation of the Safety and Loss Control Policy is observed. Foremen should be made responsible for compliance by each member of their crews.

Any employee causing or knowingly allowing an unsafe condition to remain should be subjected to a warning. Severe and/or repeated instances could lead to the possibility of dismissal. Employees guilty of intentional serious and/or repeated violations should be considered to be made subject to dismissal without pay.

Disciplinary Action.   If a safety violation is observed, or comes to the attention of any supervisor, administrator, or management personnel, action must be taken immediately to correct the violation. Immediately thereafter, the Safety and Health Manager, or other individual with this specific authority, is to be notified. The person fulfilling the functions of a company Safety and Health Manager should then follow the procedures patterned after the items below for necessary disciplinary action. More than ever, human resource, legal, and safety professionals should be consulted before any of the following actions are implemented to any degree.

FIRST WARNING.   The first warning to a person responsible for an intentional safety violation may be verbal and written, with a copy of the sample Company Safety Violation Warning Notice of the following section given to the employee and distributed to the employee's personnel record.

SECOND WARNING.   A written warning similar to the first warning might be given to the employee, be retained by the company individual fulfilling the responsibilities of a Safety and Health Manager, and be distributed to the employee's personnel record. A meeting should be held with the employee, foreman, and Safety and Health Manager in order to determine why the employee is not willing to comply with the rules and regulations of the company. Any further action taken at this time should be directly addressed by management and be based upon the severity of the violation.

THIRD WARNING.   A written warning should be prepared with a copy given to the employee, be retained by the individual filling the responsibilities of the company Safety and Health Manager, and be distributed to the employee's personnel record. Three warnings for safety violations within a twelve-month period should be cause enough to result in suspension or termination of employment. Three safety violations within a six-month period might even result in termination of employment without pay.

The actions listed above are intended to be taken when a violation is observed. The company should not tolerate actions or negligence that may result in an injury.

6.1.8   Sample Safety Violation Warning Notice (*page 6.13*)

The sample Safety Violation Warning Notice on the following page is intended to be used as prescribed in the preceding section whenever a safety violation is observed:

6.2   Safety and Loss Control Responsibilities and Employee Participation

The form and content of this Section are intended as further examples of how the company might provide instruction to its employees with respect to the material presented. It is intended to provide for each employee:

- The function of a company Safety Committee
- The participation of each management level and of every employee, and other efforts that are rewarded in each other
- The responsibilities of each employee and the conduct expected by the company

The material that follows is an example of a company Safety and Loss Control Program approach to the objectives indicated above. The sample language and forms are to be considered in the context of your own company and its manner of operation. Consider the function of a Safety Committee specifically as the company may actually intend to operate one, and edit the final material based upon those genuine intentions.

Carefully consider these objectives during the review and editing process of the section material in order to customize the section content to the specific needs of your organization. Prepare final language that is consistent with decisions made by your own management in the determination of your own company program.

It is important to understand that safety rules and regulations change from time to time. Be sure to review the final draft of the Section and the entire program manual with appropriate safety, insurance, and legal professionals prior to issuance and use. Periodically review your program with company management and all appropriate professionals to be certain that the company program remains current with all requirements.

6.2.1   Responsibility Assignment

Whether at the jobsite, warehouse, shop, or office, every employee has a responsibility to work safely. That responsibility includes the elimination of hazards wherever possible, and immediately upon their observation to report actual or potential hazards to management or to any individual given the responsibility of a Safety and Health Manager.

6.1.8
Sample Safety Violation Warning Notice

(COMPANY) Safety Violation Warning Notice

Notice Date: _____/_____/_____   Time: _____:_____ AM/PM

Employee Name: _____

Employee Number: _____

Date of Violation: _____   Time of Violation: _____ AM/PM

Location: _____
_____

Description of Violation:

_____
_____
_____
_____
_____
_____
_____

Course of Action:

_____
_____
_____
_____
_____
_____
_____
_____

Employee Warned of: ☐ Suspension   ☐ Dismissal   ☐ Dismissal Without Pay

Signed:
Safety & Health Manager: _____

Employee: _____

cc: ☐ Supervisor ☐ Employee ☐ Personnel File

Supervisors should be held accountable for compliance with all safety objectives in the performance of every activity. This is especially so in the areas of training and hazard control. Any performance evaluation(s) should reflect their involvement and active participation in accident prevention.

The fact that the company may have assigned a person to a formal position of Safety and Health Manager does not release supervisors from their safety and health responsibilities in any way. Any individual operating in the capacity of a Safety and Health Manager should be available to assist supervisors in implementing the company Safety and Health Program, but that person should not be expected to perform the respective supervisors' duties.

6.2.2   Company Safety Committee

A company Safety Committee can be established that is made up of both employer and employee representatives who are charged with the responsibility of general oversight of the company Safety and Loss Control Program. If so established, the committee may convene quarterly and perform the important functions described in Section 6.2.3.

6.2.3   Safety Committee Function and Objectives

If the company determines to establish a formal Safety Committee, consider assigning the following responsibilities to be carried out by it:

- Meet regularly to review the overall Safety and Loss Control Program and specific operating issues that may arise during the period.
- Serve in an advisory capacity to the Safety and Health Manager and to management.
- Familiarize themselves with applicable construction safety standards.
- Review established company procedures and evaluate the effectiveness of their implementation.
- Recommend corrections by company management to all company employees. Ensure that all updates and changes to company procedure as approved by company management are adopted, properly coordinated with the entire company Safety and Loss Control Program, and distributed.
- Participate in communication procedures by which the company shall train committee members and company employees.
- Prepare and distribute minutes of committee meetings. Make records of committee activities and communications available to company employees.

6.2.4   Safety Committee Structure and Duties

Again, if the company has decided to establish a formal Safety Committee, and has charged the committee with the responsibility for the functions in objectives in a manner similar to those outlined in Section 6.2.3, the Safety Committee Structure Duties might then be considered as follows:

Safety Committee Duties

- Develop company-wide action plans for compliance with safety regulations.
- Develop an action program for project accountability.
- Provide suggestions and direction for efficient implementation of prescribed corrective measures.
- Direct the distribution of safety regulations, programs, and other informational materials.

Safety Committee Chairperson Duties

- Schedule and enforce participation in all Safety Committee meetings.
- Prepare and distribute Safety Committee meeting agendas.
- Report to senior company management regarding the status of all recommendations.
- Ensure preparation and distribution of minutes of meetings.
- In the absence of the Chairperson, the Cochairperson is authorized to assume all responsibilities and authority of the Chairperson.

Safety Committee Member Duties.   (Superintendents, foremen, managers, and employees)

- Participate in all Safety Meetings.
- Assist with safety inspections.
- Apply and enforce OSHA and company regulations.
- Report unsafe conditions.
- Report all accidents or near accidents.
- Contribute ideas and suggestions for improvement of safety and better communication of safety programs and information.

6.2.5   The Safety and Health Manager

The company should assign an individual responsible to carry out the functions of a Safety and Health Manager. That person will be responsible for:

- The administration and routine dissemination of all company safety and loss control information,
- The administration of all company safety and loss control procedures, and
- Monitoring the compliance of company employees with all stated policies and procedures.

The individual fulfilling the functions of the Safety and Health Manager should be recognized as a deputy of each Senior Project Manager and should report directly to each of them. In matters of imminent safety concern, the Safety and Health Manager should simultaneously report directly to company senior management. Any recommendations issued by the Safety and Health Manager with respect to safety and loss control should be considered as if they had been issued by a Senior Project Manager. In situations where imminent danger or serious hazards exist, the Safety and Health Manager should be given the authority to order a work suspension.

The Safety and Health Manager should be placed in the position that will allow him or her to assist the Project Managers and Site Superintendents in all matters pertaining to safety and loss control.

Duties of the Safety and Health Manager should then include but may not be limited to the following:

- Assist each company employee with compliance with all company and OSHA safety and health policies and regulations.
- Implement the company Safety, Health, and Loss Control Programs, and monitor compliance.
- Coordinate all company safety activities in ways that facilitate their implementation.
- Advise on the purchase of safety and health materials to ensure compliance with all safety standards.
- Advise management regarding proposed and/or necessary changes in safety standards and regulations.
- Conduct field inspections in efforts to identify unsafe conditions and/or actions of jobsite personnel. Make verbal and written recommendations for both immediate and future correction. Follow up to assure correction. Issue warnings of persistent, uncorrected unsafe conditions.
- Organize and conduct training of supervisory and hourly employees in safe work procedures.
- Be familiar with applicable local, state, and OSHA rules, regulations, and standards. Be able to assist management in interpretations, determination of policy, and implementation.
- Accompany any local, state, insurance, or OSHA inspectors in every jobsite situation to represent company management during inspections.
- Coordinate the deployment of emergency care systems, such as first aid, medical, fire protection, evacuation, and fire alarms.
- Be knowledgeable of health-related and hygiene-related activities, and pursue continuing education with respect to same.
- Coordinate any/all site security procedures and personnel, when required for the specific project.

- Assist all Project Managers in the implementation of the company Safety Programs as they relate to the specific projects. Maintain communication with company management.

- Respond to company managers and supervisors with answers to their requests for information.

- Coordinate jobsite meetings with company Project Managers. Assist or lead weekly jobsite Tool Box Talks for company trades to the extent available to do so. Assist all project managers with organization and presentation of weekly Tool Box Talks.

- Assist the Project Managers in the establishment and maintenance of all job-site notices, signs, etc., and maintenance of each "# Days Safely Worked" sign.

- Coordinate the organization and implementation of all emergency care warning and evacuation systems.

- Maintain the accident record system, including preparation of reports, investigation of accidents, and prompt notification of company management whenever necessary.

- Assist in the preparation and presentation of the monthly Supervisory Safety Meeting and quarterly Employee Safety Meetings.

- Coordinate and participate in disciplinary actions and procedures.

- Review all accident reports for thoroughness.

- Periodically evaluate the effectiveness of all Safety, Health, and Loss Control Programs, and make appropriate recommendations to management.

6.2.6 Responsibilities of the Senior Project Manager

The Senior Project Manager should be responsible for supervising and monitoring the Safety and Loss Control Programs for all projects under his or her direct supervision, and through each Project Manager for those projects being reported to him or her. Safety inspections should be conducted by the Senior Project Manager during all routine jobsite tours.

6.2.7 Responsibilities of the Project Manager

The Project Manager should be ultimately responsible for the complete safety and loss control effort on the projects under his or her direct authority. All reporting should be made to the Project Manager, who will be responsible to:

- Plan production so that all work will be performed in accordance with established safety regulations.

- Administer the Safety and Loss Control Program on all projects under his or her control.

- Disseminate all safety information to all appropriate personnel and Subcontractors.

- Receive and process all reports, surveys, accident reports, and other information relating to safety and loss control that are to be submitted to the Project Manager. The Project Manager is responsible for determining the need for corrective action and implementing such action.

The Project Manager should coordinate the specific responsibilities as listed in Section 6.2.7 with the Site Superintendent and determine the most effective implementation of all safety and loss control responsibilities between the Project Manager and Site Superintendent.

6.2.8   Responsibilities of the Site Superintendent

The Site Superintendent, as coordinated with and under the direction of the Project Manager, should be responsible for the active administration and control of all aspects of the jobsite safety program. The responsibilities of the Site Superintendent should be coordinated with and/or shared by the Project Manager as the Project Manager should determine, and in any event should include but may not be limited to:

- Plan production so that all work will be performed in accordance with established safety regulations.
- Assist in the preparation of the Jobsite Safety Program.
- Establish and maintain all jobsite notices, signs, etc.
- Regularly conduct inspections to identify unsafe conditions and/or acts of jobsite personnel. Make verbal and written recommendations for both immediate and future correction. Follow through to assure correction.
- Be familiar with applicable local, state, and OSHA rules, regulations, and standards; be able to assist management in its interpretations and implementation. Be prepared and available to accompany any local, state, insurance, or OSHA inspectors to represent company management whenever requested to do so.
- Coordinate emergency care systems, such as first aid, medical, fire protection, evacuation, and fire alarms-with the Safety Officer.
- Be knowledgeable in control of health-related and hygiene-related activities.
- When required for the specific project, coordinate site security procedures and personnel.
- Lead weekly jobsite Tool Box Talks for company trades. Assist the Safety Officer in the presentation of Tool Box Talks when he or she is available as coordinated.
- Coordinate emergency care warning and evacuation systems.
- Ensure that proper safety material and protective devices are available, in correct working order, and used whenever required.
- Instruct foremen in safety requirements and make certain that each foreman passes the instruction on to their crews.

- Take advantage of offered safety training and be aware of all safety rules.
- Review all accidents, oversee correction of all unsafe practices, and file accident reports.
- Conduct Jobsite Safety Meetings and provide all employees with proper instruction on safety requirements.
- Require conformance to safety standards by all Subcontractors.
- Notify the company home-office of all actual or alleged safety violations by any/all parties.
- Provide for the protection of the public from company operations.
- Attempt to ensure safe performance by others present on the site, including Owner and Architect/Engineer representatives, the general public, visitors, and the employees of other contractors.

6.2.9 Responsibilities of the Job Foreman

The Job Foreman, as coordinated with and under the direction of the Site Superintendent, should be responsible for the active administration and control of all aspects of the jobsite safety program as they relate to the crews under their direction. The responsibilities should include but may not be limited to the following:

- Carry out the Safety Program at the work level.
- Be aware of all safety requirements and safe working practices.
- Plan all work activities with adherence to safe working practices.
- Instruct new employees and existing employees performing new tasks in safe working practices, provide their crews with continuing instruction on safety requirements, and conduct regular Tool Box Talks.
- Install and maintain devices to protect the public from company operations.
- Make certain that protection equipment is available, maintained in operating condition, and used.
- Act without delay to correct all hazards, including unsafe acts and conditions that are within the scope of your crews' work.
- Properly secure prompt medical attention for any injured persons.
- Immediately report all actual, perceived, or alleged safety violations to your immediate supervisor.

6.2.10 Responsibilities of All Company Employees

Each company employee should be responsible for learning and abiding by those rules and regulations that are applicable to the assigned tasks, and for reporting observed violations, suspected violations, and anticipated hazards to his or her immediate supervisor. If such reported conditions are not corrected, the employee should be required to report the failed correction to the individual serving as the company Safety and Health Manager.

All members of each jobsite labor force should be responsible to act as follows:

- Work safely and in such a manner as to ensure their own safety, as well as the safety of coworkers and all others.
- Request help when unsure of how to perform any task safely.
- Correct unsafe acts or conditions within the scope of immediate work.
- Report any uncorrected, unsafe acts or conditions to supervision.
- Report for work in good physical and mental condition to safely carry out assigned duties.
- Avail themselves of company- and industry-sponsored safety programs.
- Use and maintain all safety devices provided.
- Maintain and properly use all tools under their control.
- Follow all safety rules.
- Provide fellow employees with help in implementing safety procedures and complying with safety requirements.

All company personnel should be responsible to conduct themselves as follows:

- Strive to make all work environments and company operations safe.
- Maintain mental and physical health conducive to working safely.
- Keep all work areas clean and free of debris and obstacles.
- Assess results of their actions on the entire workplace.
- Replace or repair safety precautions removed or altered before leaving work area.
- Abide by all safety rules and regulations of the owner and every legal authority on each jobsite.
- Work in strict conformance with OSHA regulations.

No company employee shall be required or knowingly be permitted to work in an unsafe environment, except for the purpose of making safety corrections, and then only after proper precautions have been taken for their protection.

6.2.11   Responsibilities of Subcontractors and Trade Contractors

Each company Subcontractor or trade contractor should be responsible for the safety and security of employees under their control and should be required to abide by the company Safety Program as incorporated into their original subcontract or purchase agreement, with all regulations and requirements of OSHA, the Owner, and every other agency having jurisdiction over the work.

Each company Subcontractor or trade contractor should be required to submit to the company Project Manager copies of their respective safety programs prior to the start of their work. Each Subcontractor or trade contractor should then be responsible to:

- Abide by all safety rules.
- Participate in site safety meetings and weekly site safety inspections.
- Appoint one qualified representative and one alternate to attend safety meetings and to take actions on safety violations.
- Provide a list of employees qualified in first aid who will be available to administer first aid to injured workers.
- Provide an adequate first aid kit for use by its employees.
- Notify all other contractors when their activities could possibly affect the health or safety of employees of other companies.
- Report all injuries or any unsafe conditions that may become apparent.

6.2.12 Responsibilities of Architects, Engineers, Owners, and Visitors

Prior to being allowed to proceed anywhere on a jobsite, all personnel and representatives of Architects, Engineers, Owners, and Visitors should be required to:

- Abide by all safety rules.
- Check in with the Project Superintendent so that protective equipment may be confirmed or provided, such as hard hats, eye, or respirator protection.
- Refrain from entering any construction area without the knowledge of company jobsite supervision and confirmation that all required and appropriate precautions and protections have been made.

6.3 Jobsite Safety Program

6.3.1 Section Description

The form and content of this Section of the Operations Manual is intended to illustrate an example of a company's instruction to its employees with respect to the material presented. It is intended to provide information to each employee regarding:

- Jobsite safety planning
- Jobsite safety meetings
- Jobsite safety inspections
- Provisions for basic jobsite first aid materials and training

This Section is an example of a company Safety and Loss Control Program approach to the objectives indicated above. The sample language and forms are to be considered in the context of your own company and its manner of operation.

Carefully consider these objectives during the review and editing process of the Section material in order to customize the Section content to the specific needs of your organization. Prepare final language that is consistent with decisions made by your own management in the determination of your own company program.

It is important to understand that safety rules and regulations change from time to time. Be sure to review the final draft of the Section and the entire program manual with appropriate safety, insurance, and legal professionals prior to issue and use. Periodically review your program with company management and all appropriate professionals to be certain that the company program remains current with all requirements.

6.3.2   Sample Jobsite Safety Inspection Report Form

Each jobsite must be regularly analyzed on a continuing basis in order to identify existing actual or potentially hazardous conditions. It is the responsibility of the Project Manager to administer the complete Jobsite Safety Program through the Site Superintendent and in a manner coordinated with the individual within the company who is fulfilling the responsibilities of the Safety and Health Manager.

The consideration of the Jobsite Safety Program begins the site utilization planning stage and continues throughout the life of the project. Consider the Jobsite Safety Planning Checklist of Section 6.3.6 as a starting point to help organize the particular project's Jobsite Safety Program.

6.3.3   Jobsite Inspection Program

The company should consider establishing the formalized jobsite inspection program that would be designed to identify safety and health violations and hazards that may exist in the workplace. All supervisors should begin in the specific responsibility to observe their jobsites, work areas, tools, and equipment daily, with the objectives to eliminate or control any hazards that are identified. If eliminating or controlling the hazard goes beyond the authority of the supervisor, the individual fulfilling the responsibilities of the Safety and Health Manager should be asked for assistance. Both the company employees and noncompany personnel should be kept away from the hazard.

The frequency of inspections should be based upon potential hazards and the type and complexity of the project. The frequency determination should be made by the Project Manager, Superintendent, or the Safety and Health Manager.

It is important to note that more accidents result from unsafe acts or actions than from unsafe conditions. In addition to inspecting jobsites and work areas for unsafe conditions, supervisors should observe operations, work procedures, and employee actions. Unsafe activities must be eliminated and replaced with safe procedures.

All company employees committing unsafe acts or violating safe working procedures should be reprimanded according to the Disciplinary Action Policy of Section 6.1.7. All hazards identified during daily observations or periodic inspections should be corrected or controlled immediately. The report of such inspections may be included in the project's Daily Field Report that includes all activities on the jobsite, or may be completed using the Jobsite Safety Inspection Report Form of Section 6.3.5.

All field reports and inspection reports noting any safety deficiency or safety concern should then be forwarded to the person fulfilling the responsibilities of the Safety and Health Manager for review and follow-up action when required.

The Safety and Health Manager should make both scheduled and surprise inspections. The results of his or her inspections should then be discussed with the project management and supervisory staff immediately following such inspections. Supervisors should be required to take immediate action to eliminate, correct, or control the hazard or ensure that project management has taken such action.

6.3.4 Field Safety and Loss Control Inspection Procedures

The manner in which a field safety and loss control inspection should be conducted depends upon its purpose and scope. The following is a list of guidelines and suggestions that will help ensure that each field safety and loss control inspection is comprehensive and efficient:

1. Be aware in advance of what you plan to inspect.

2. Review applicable regulations. Familiarize yourself with the hazards that are associated with the operation or equipment that you intend to inspect.

3. Schedule the inspection at a time that will allow a maximum opportunity to view operations and work practices. Midmorning or early afternoon are often good times.

4. Be alert to all hazards, and do not merely run through the checklist. A checklist is only a reminder. Hazards unique to a specific situation should not be overlooked.

5. Choose a systematic inspection route. Cover the entire area footprint and leave nothing out. When reinspecting a work area, approach from a different direction or use a different route to gain a different perspective of jobsite conditions.

6. Take notes, and be sure to note the exact description and/or location of every hazard when observed. Include ideas for corrective action. Do not wait until after the inspection to record hazardous conditions or unsafe actions; details might get forgotten.

7. Look for the source cause(s) of adverse conditions and practices. Think in terms of correction action only. Do not focus on fixing blame.

8. If an unsafe piece of equipment or condition is observed, the supervisor shall warn the employees of the hazard(s) involved. If a life-threatening hazard exists, the operation must be immediately suspended. The work should then be allowed to resume only after the supervisor is satisfied that the hazard has been thoroughly corrected.

Any unsafe conditions on the part of a Subcontractor must be brought to that company's immediate attention. All such unsafe conditions should be immediately reported in writing and noted in the Daily Field Report or in any field Safety Inspection Report. All reports should be forwarded to the Subcontractor, with a copy sent to the company Safety and Health Manager. Always remember that you and the company could be held accountable for the actions and negligence of every Subcontractor associated with the work that is being carried out under your direction.

6.3.5   Sample Jobsite Safety Inspection Report Form (*page 6.25*)

The report of all jobsite inspections may be included in the project's Daily Field Report, or may be completed using the sample Jobsite Safety Inspection Report Form of this section. Check with your own insurance provider to see if they have or can recommend a specific form to be used for these purposes.

6.3.6   Sample Jobsite Safety Planning Checklist (*page 6.26*)

Prior to the start-up of every construction project, consideration should be given to the basic logistics of safety, both in and around the jobsite. As the project develops in the field, the initial determinations and arrangements for each of the items may require periodic or special modification in order to ensure that all safety considerations remain adequate as conditions on the jobsite change. At a minimum, the items listed below should be addressed. Where necessary, action should be taken during all appropriate stages of job development—prior to, during, and after construction.

Page 6.26 shows a checklist of specific items to be considered during the development of the specific jobsite safety program. It is a helpful aid but is not necessarily conclusive.

6.3.7   Vendor Insurance

Coverages.   All contractors, subcontractors, trade contractors, and independent individuals performing any work on the site should be covered by all forms of insurance to the specified limits as enumerated in the Contract Documents, as described in their specific agreements for services, or as otherwise required by the company.

It is at least the responsibility of the Project Manager and the Site Superintendent to confirm the presence of all such forms of insurance and to secure the correct and complete certificates of such forms of insurance from each entity on the site before allowing any work to proceed. There should be no exceptions.

Additionally Insured.   It should be a standard requirement of the company that all subvendor agreements provide that the company is specifically named as additionally insured on the respective policy. It should then be the responsibility

6.3.5
Sample Jobsite Safety Inspection Report Form

(COMPANY) Jobsite Safety Inspection Report

Date: _____   S M T W T F S

Project: _____

Project Number: _____

Location: _____

Project Manager: _____

Superintendent: _____

Observed Hazard:

_____

_____

_____

_____

_____

Responsibility of: ☐ (Company)   ☐ Trade/Subcontractor: _____

Hazard Location:

_____

_____

_____

_____

_____

Recommended Course of Action:

_____

_____

_____

_____

_____

Work Recommended to be Stopped in the Hazard Location: ☐ YES   ☐ NO

Work Stopped in the Hazard Location: ☐ YES   ☐ NO

Corrective Action Required: Date: _____ Time: _____

Inspection Performed By: _____

cc: ☐ Safety & Health Manager   ☐ Project Manager   ☐ Superintendent   File: ☐ PM_____

☐ CF_____

6.3.6
Sample Jobsite Safety Planning Checklist

1. Administration: Establish and maintain on file adequate supplies of:
   a. Site Safety Program _____
   b. Safety Review Checklists _____
   c. Posted listings of emergency services _____
   d. Accident Investigation Report Forms _____
   e. Accident Eyewitness Statement Outline _____
   f. Tailgate Safety Meeting Agendas _____
   g. OSHA forms 101 & 200 _____

2. Forward "Job Start-up Notice" to company insurance carrier _____

3. Identify all formal and available informal safety personnel:
   a. Company Safety Officer _____
   b. Company on-site safety representative _____
   c. Owner safety representative _____
   d. All company employees with first aid, CPR, or other safety training _____
   e. All Owner employees with first aid, CPR, or other safety training _____
   f. All subvendor employees with first aid, CPR, or other safety training _____

4. Identify Safety & Emergency Services:
   a. Notify of the presence of the jobsite, anticipated workforce, and duration of the project:
      i. Fire Department _____
      ii. Police Department _____
      iii. Medical facilities _____
      iv. Identify locations and emergency travel routes to:
         (a) Emergency rooms _____
         (b) Nonemergency medical facilities _____
         (c) Fire Department _____
         (d) Police Department _____

5. Determine Listings of Local Services:
   a. Physicians _____
   b. Eye experts _____
   c. Orthopedics _____
   d. Paramedics _____

6. Unique Owner requirements (In addition to "usual" considerations)
   a. Construction parking _____
   b. Occupied area parking _____
   c. Strict security _____
   d. Hot work or other special work permits _____
   e. Full- or part-time safety person _____
   f. Insurance wrap-up policy _____

6.3.6
Sample Jobsite Safety Planning Checklist (Continued)

7. Traffic Control—a plan to address:
    a. Entering/leaving of the construction workforce   _____
    b. Material/Equipment deliveries   _____
    c. Equipment access & storage   _____
    d. Street & site cleanup & maintenance   _____
    e. Adequate warning signs & other postings   _____

8. Utility Location Protection—Locate & Protect/Relocate:
    a. Water   _____
    b. Gas   _____
    c. Electric   _____
    d. Telephone   _____
    e. Telex/Cable   _____
    f. Communication   _____
    g. Storm/Sanitary sewer   _____

9. Preexisting Condition Survey

    Prior to the start of any construction, record all preexisting damage by whatever
    means appropriate, including engineered surveys, photographs, or videotape recordings.

    a. Initial survey should be made by the Site Superintendent.   _____
    b. Determine the need for additional detailed inspection(s).   _____
    c. If sufficient evidence of damage is present, consider retaining
        a professional photographer to record it.   _____
    d. If extreme or unique circumstances, determine if a professional engineer
        is or may be needed to produce surveys or other documentation.   _____
    e. Prior to proceeding with (c) or (d), secure approval from
        the company Senior Project Manager.   _____
    f. In every case, photograph and videotape the entire site,
        as well as all approach roads and routes prior to site mobilization
        and start of any construction activity, Include:
        i.   Surrounding buildings
        ii. Roads   _____
        iii. Utilities   _____
                                                                      _____

10. Plan the locations and configurations of site services and their relative proximities:
    a. Field offices   _____
    b. Material staging and storage areas   _____
    c. Fuel storage and fuel distribution arrangements   _____
    d. Traffic control   _____
    e. Administrative and worker parking   _____
    f. Pedestrian access   _____
    g. Location & configuration of temporary utilities   _____
    h. Ongoing temporary power arrangement   _____
    i. Temporary lighting—Site   _____
    j. Temporary lighting—Other construction areas   _____
    k. Temporary heat   _____
    l. Temporary power   _____
    m. Welding & cutting torches (friendly fire)   _____

6.3.6
Sample Jobsite Safety Planning Checklist (*Continued*)

11. Identify specific Owner requirements that may exceed customary considerations, such as:
    a. Special access requirements or needs   _____
    b. Unique security requirements   _____
    c. Involvement of Owner or other designated safety or security personnel   _____
    d. Special forms of insurance or legal considerations   _____
    e. Special Notices or other required communications   _____

12. Determine necessary provisions for protection of the public:
    a. Site fencing   _____
    b. Lighting   _____
    c. Signs & notices   _____
    d. Traffic control   _____
    e. Guardrails   _____
    f. Walkways (covered/uncovered)   _____

13. Determine all fire protection needs - Project-specific planning should include:
    a. Appropriate ABC-type fire extinguishers:
        i. Sizes   _____
        ii. Quantities   _____
        iii. Locations   _____
    b. Fire-protected storage cabinets and/or areas for vital, sensitive,
       or special files or materials   _____
    c. Establishment and maintenance of all appropriate fire protection
       measures in accordance with OSHA requirements for specific
       work operations   _____

14. Consider necessary protection of the site and building (if there is one)
    a. Signs and notices posted   _____
    b. Special walkways and traffic provisions   _____
    c. Barricades and other safety barriers   _____
    d. Fences, guardrails, and canopies   _____
    e. Parking, traffic, & walkway lighting   _____
    f. All provisions for handicap access   _____
    g. Security & safety personnel   _____

of the Project Manager—through the Project Engineer if one is assigned to the project—that the company is actually named as additionally insured, and that this stated requirement is not eventually "overlooked" in actual practice. Check with your own insurance professional, but note that the specific language "additionally insured" is likely to be the only acceptable language. Terms such as "certificate holder" or other language not specifically noting "additionally insured" are not likely to be acceptable. Check with your own insurance carrier for specific language and documentation required to identify and confirm minimum coverage requirements acceptable to your company. Describe those specifics in this section of your program.

6.3.8   First Aid Kit

The company should provide a complete medical first aid kit on each jobsite for use by company personnel. All contractors, subcontractors, trade contractors, and individuals (noncompany employees) on the site should be required to provide and maintain their own first aid kits in a manner adequate for the type of work being performed. It should then be the responsibility of the company Project Manager, Site Superintendent, and each trade foreman to enforce this requirement on the part of each subvendor on every jobsite.

6.3.9   Safety and First Aid Training

Basic First Aid and CPR.   All company personnel should be encouraged to take basic first aid training and to become certified in first aid and CPR (cardiopulmonary resuscitation). The Superintendent, Project Manager, all trade foremen, and everyone in any supervisory capacity should feel a particular responsibility in this area.

The company should consider arranging for in-house company training in basic first aid and CPR from time to time. It is not necessary for any company employee to wait for this convenience, however. Contact the local Red Cross, area hospitals, and local fire departments to identify medical first aid training programs available. Participation by every company employee throughout every jobsite and office should be strongly encouraged.

Reimbursement for Training Expenses.   In most cases, it is probably a good idea to be sure that the cost of certain first aid training programs is reimbursed to the company employee, if the training program of interest had been first coordinated with the company Safety and Health Manager and approved in advance by company senior management.

If any company employee wishes to arrange for such basic or advanced first aid training on his or her own, consider a policy of the company that would reimburse the employee for the tuition and direct expenses associated with the training program. With such a program, the employee would then need to present the program certificate or other written documentation that he or she has successfully completed the program.

6.4   Jobsite Safety Meetings

6.4.1   Section Description

This Section of the Operations Manual is intended to provide an example of a company policy with respect to the manner in which regular jobsite safety meetings might be conducted by the Site Superintendent, trade foreman, and/or the Safety and Health Manager. The objectives of such jobsite safety meetings would be to:

- Orient all the company and subvendor tradespeople to project requirements and objectives.
- Identify areas on the site that present special problems or concerns and determine the best corrective action for each situation.
- Swap or trade safety and health topics of both a general and specific nature. Develop conversation, discussion, and documentation that will further the safety education and awareness of all project participants.

The sample company Jobsite Safety Meeting program series is arranged in three levels:

1. The Initial Project Safety Orientation Meeting.
2. Ongoing Company Safety Meetings.
3. Periodic (weekly) Jobsite "Tailgate" Meetings.

6.4.2   Initial Project Safety Orientation Meeting

The conduct and documentation of a certain minimum number of safety meetings may be prescribed in special programs offered by insurance carriers. Check with your own insurance provider for specific help or instruction in the final arrangement and configurations of all company safety meetings.

The sample Project Safety Orientation Meeting described in this section is intended to be conducted by the Site Superintendent at the very onset of the project. It can be its own stand-alone meeting, or it can be combined with an early Superintendent's Meeting with all the company and subvendor trade foremen.

The meeting should be given the complete attention by all project participants that it deserves. All subvendors should be required to participate. Attendance should be documented.

At a minimum, items to be reviewed at the meeting should include:

1. All safety requirements of the company contract, and the stated requirements of each vendor's contracts and subcontracts.
2. Company, project, and special rules as they relate to the specific project.

3. The jobsite utilization program, specifically with regard to arrangements for stored materials, proper materials handling, traffic, access, security, communication, etc.

4. Fire protection requirements and procedures; general consideration and specific "hot-work" areas.

5. Evacuation procedures; alarms, routes, communication, personnel, etc.

6. Posting of all emergency phone numbers.

7. Designation of each company's Safety Officer and Alternate; phone numbers, contact procedure, etc.

8. Identification of all jobsite personnel with any medical training. Issue of hard hat red crosses.

9. Review of required first aid and medical supplies; inventory current jobsite supplies and necessary supplements.

10. Specific jobsite precautions with respect to protection of workers and protection of the public; temporary protection, fall protection, hard hat, safety shoes, gloves, eye protection, etc.

11. Jobsite security issues and arrangements; fences, site and field office area lighting, alarms, etc.

12. Notification to all the company and subvendor participants that willful and/or repeated violations may be grounds for layoffs or ejection from the jobsite.

13. Notification of the expectation of regular participation in periodic safety meetings, and of the need for all trade foremen to periodically conduct their own Tailgate Safety Meetings—and to immediately forward minutes of those meetings to the company.

6.4.3   Regular Jobsite Safety Meetings

Regular, ongoing Jobsite Safety Meetings can be conducted as their own standalone arrangements. It may, however, be more effective and more easily accomplished if these types of meetings are combined with the regular Superintendent's meetings with all trade foremen as the first routine item on each meeting agenda.

Meeting agenda items regarding safety and health should at a minimum include:

a. Review of any observed actual or possible safety violations. Determination and implementation of corrective measures, and responsibility for implementing them.

b. Correction.

c. Any planned or other changes in the jobsite utilization program as previously established.

d. Jobsite housekeeping and cleanup status; protection, safety, and security.

e. Review of stored materials and equipment.

f. Review of past or upcoming hot work, and the handling of burning and welding equipment.

g. Review of hot work permits requirements.

h. Review of temporary power and maintenance, and any necessary corrections or upcoming changes.

i. Review of temporary heat and temporary fuel requirements and handling.

j. Review all safety performances throughout the project.

Consider combining this meeting with a Tailgate Safety Meeting as described in Section 6.4.4. You already have everyone's attention; use it.

6.4.4 Weekly Tailgate Safety Meetings

Note that the content and format of the sample Weekly Tailgate Safety Meetings as described in this Section may be prescribed or suggested by government agencies, your insurance carrier, or other safety authority. Take advantage of available assistance to edit the requirements and suggestions of this sample Section.

Each Site Superintendent—or Crew Foreman if that is the senior site position—should be responsible to conduct periodic—weekly—safety training sessions with all field personnel under that individual's authority. Such meetings may be coordinated with any individual fulfilling the requirements of a company Safety and Health Manager for assistance in such presentations, but the ultimate responsibility should remain with the senior company site representative.

Topics should be relevant and timely with respect to the perceived current needs and all general concerns. Attendance should be recorded by signature and be kept on file by the Safety and Health Manager. Persistent absences by any company employee should be recorded and the information distributed to that employee's personnel record.

The Tailgate Safety Meeting is most often a 10–15 minute discussion, conducted for the purposes of:

- Disseminating information included in the regular Safety Meetings.
- Reviewing jobsite safety and cleanup conditions and establishment of corrective measures.
- Discussing specific safety items for the training and education of all tradespeople.

When possible, the items should be relevant to some observed job condition and ongoing work practices, but they can be of a random nature as well.

6.4.5 Sample Tailgate Safety Meeting Topics and Outlines

The following list can be used as an aid in the organization and development of the jobsite Tailgate Safety Meeting Program described in the preceding sections. One topic should be covered per meeting. Suggested outlines can include:

1. Emergency Care and Procedures to Obtain Aid. Determination of qualifications for administering first aid.
2. Controlling and restoring breathing. Aiding a choking victim.
3. Aiding a burning victim. Aiding a shocked victim.
4. Aiding a victim with a broken bone.
5. Fire Regulations at the Work Site
    a. Smoking and nonsmoking areas
    b. Location and use of firefighting equipment
    c. Periodic check of fire extinguisher "charges"
    d. Storage and use of combustible materials
6. Chains and Slings
    a. Care and proper use
7. On-Site Accidents
    a. Falls—causes and prevention
    b. Jewelry—rings, chains, etc.
    c. How to manually lift loads safely
8. Trench Safety
    a. Shoring requirements
    b. Cave-ins
    c. Rescue procedure
9. Compressed Gas Cylinders
    a. Dangers of compressed air
    b. Handling and storage of cylinders
10. Use of Friendly Fire
    a. Temporary heat and fuel
    b. Temporary power
    c. Proper use of cutting and welding equipment
    d. Jobsite hot work permit requirements
11. Personal Safety Rules and Equipment
    a. Hard hat
    b. Safety shoes
    c. Gloves
    d. Proper clothing
    e. Proper use of safety belts and nets
    f. Goggles and eye injuries
12. Safe Handling of Power Tools
    a. Rip saws
    b. Bench grinders

    c. Drills

    d. Power-actuated tools

13. Heavy Equipment

    a. General safe use

    b. General equipment safe operating procedures

14. Cranes and Rigging

    a. General safe use

15. Earth Moving Equipment

    a. General safe use

    b. Operation near embankments

    c. Operation in confined areas

16. Hazardous Materials—Industrial Hygiene

    a. Asbestos—detection and reaction

    b. Treatment and disposal of hazardous or controlled materials

    c. PCP

    d. Carbon monoxide

    e. Lead

17. Confined-area activity

    a. Definitions of confined area

    b. Egress

    c. Ventilation

    d. Evacuation

18. Scaffolding and Staging

    a. Erecting procedures

    b. Working on it

    c. Use arrangements

19. Ladders

    a. Types and uses

    b. Areas of use

20. Protection of the Public

    a. Walkways

    b. Lighting

    c. Barricades

    d. Signs and notices

21. General Emergency Safety Procedures

    a. General procedures

    b. Accidents involving serious injury or death

    c. Property damage accidents

22. Hazard Communication

    a. The company HAZ-COM manual

    b. Hazardous chemicals on the jobsite

    c. MSDSs

    d. Labels and warning systems

23. Basic First Aid Topics
    a. Identification of on-site personnel with safety training
    b. Availability of safety training programs

6.4.6 Sample Jobsite Safety Review Checklist (*page 6.36*)

The sample Jobsite Safety Review Checklist that follows is prepared as a convenience to aid the Site Superintendent, the Safety and Health Manager, and the Project Manager in their periodic and routine reviews of each jobsite in order to help identify possible conditions that may cause or contribute to an accident.

It is by no means any guarantee that all possible hazards will be identified or accommodated, nor is it a list that specifically complies with any regulatory requirements, but it is a list of those items that are frequently observed and easily identified.

Check OSHA regulations, those of all appropriate authorities having jurisdiction over the jobsite, and, through the company home-office, with the insurance carrier.

The checklist may include many items that may not necessarily apply to a particular project, or that otherwise might apply to projects of a nature other than the specific jobsite being considered each time. Even so, reviewing the entire list periodically may help to identify issues previously overlooked, and that may now apply.

Finally, the use of the checklist and any other site safety reviews should not be confined strictly to the company operations:

- For practical purposes, and to aid in the most comprehensive perspective, consider the entire site as if it is in the custody of the company, whether or not it is legally the case.

- Observe the conditions being provided for by any general contractor, construction manager, other subcontractors, and any separate contractors that may also be operating on the site.

- Report observed violations or possible violations immediately to your company supervisor, in order that the entire site may be made more safe for the company, all other workers, and the public.

6.4.7 Jobsite Safety Inspections by Outside Officials

Consider implementing some type of procedure patterned after the following, to be used in the event an outside official enters the company jobsite with the intent of inspecting or investigating jobsite safety. Such outside officials can include:

- OSHA inspectors
- State inspectors
- Fire inspectors
- Worker's Compensation investigators
- Insurance investigators

6.4.6
Sample Jobsite Safety Review Checklist

1. Signs, Notices, & Notifications
   a. Safety signs in place _____
   b. Emergency phone numbers posted _____
   c. Evacuation plan appropriate/posted _____
   d. Warnings & instructions to public posted _____
   e. Restricted access areas _____
   f. Exits _____
   g. No Smoking _____
   h. Electrical dangers _____
   i. Personal protective equipment needed _____
   j. Operating instructions _____
   k. Flammable materials _____
   l. Hazardous materials _____
   m. Danger areas _____
   n. Trenches _____
   o. All personnel & occupants notified to
      expect loud noises _____
   p. _____ _____
   q. _____ _____

2. Overhead Protection
   a. Entrances _____
   b. Warnings _____
   c. Construction _____
   d. _____ _____
   e. _____ _____

3. Hoisting Equipment
   a. Guys _____
   b. Cables & sheaves _____
   c. Turnbuckles _____
   d. Signals _____
   e. Carcover & enclosure _____
   f. Ladder _____
   g. Car arresting device _____
   h. Base barricade _____
   i. Platforms _____
   j. Clear staging areas _____
   k. _____ _____
   l. _____ _____

4. Walkways & Ramps
   a. Adequate construction _____
   b. Width _____
   c. Railings _____
   d. Curbs _____
   e. Slope & rise limit _____
   f. Nonslip treads & tactile areas _____
   g. _____ _____
   h. _____ _____

5. Ladders
   a. Construction _____
   b. Secure placement _____
   c. Cleats _____
   d. Landings _____
   e. Hand-holds _____
   f. Cages _____
   g. _____ _____
   h. _____ _____

6. Excavations & Trenches
   a. Shoring _____
   b. Slope repose _____
   c. Ladders _____
   d. Stockpile of excavated material _____
   e. Removal of excavated material _____
   f. Barricades & railings _____
   g. Tunnels _____
   h. Blasting arrangements _____
   i. Approved shoring designs _____
   j. Excavations properly dewatered _____
   k. Proper ventilation; free of toxic fumes _____
   l. _____ _____
   m. _____ _____

7. Fire Protection
   a. Storage of flammable materials _____
   b. Container markings _____
   c. Temporary heaters _____
   d. Compressed gas cylinders _____
   e. Tar kettles _____
   f. Welding equipment _____
   g. Welding operations _____
   h. Fire extinguishers (correct quant/type) _____
   i. Fire safety equipment _____
   j. _____ _____
   k. _____ _____

8. Openings—Walls, Floors, Roofs
   a. Perimeter railings _____
   b. Tight covers _____
   c. Flaggings _____
   d. _____ _____
   e. _____ _____

9. Scaffolds
   a. Construction _____
   b. Secure placement _____
   c. Railings _____
   d. Toe boards _____
   e. Rigging _____
   f. Safety lines, belts, rope guards _____
   g. _____ _____
   h. _____ _____

10 Stairs & Landings
   a. Adequate construction _____
   b. Temporary treads _____
   c. Clear of debris _____
   d. Proper rise/run _____
   e. Railings _____
   f. _____ _____
   g. _____ _____

11. Material Handling
   a. Size/bulk _____
   b. No sharp edges _____
   c. Weight limits _____
   d. Team lifting _____
   e. _____ _____
   f. _____ _____

6.4.6
Sample Jobsite Safety Review Checklist (*Continued*)

12. Housekeeping
    a. Nails, debris ____
    b. Tool storage & staging ____
    c. Containers ____
    d. Clear aisles & walkways ____
    e. Clean site ____
    f. Dumpster(s) location/condition ____
    g. Proximity of waste storage to hazardous conditions ____
    h. _____ ____
    i. _____ ____
13. Lighting & Temporary Wiring
    a. Lighting ____
    b. Wire height ____
    c. Proper grounding ____
    d. Wire connection ____
    e. Overcurrent protection ____
    f. Extension cords in good repair ____
    g. All extension cords & temp. power receptacles using GFIs ____
    h. Temp. power closed to weather ____
    i. _____ ____
    j. _____ ____
14. Grounding & Electrical Equipment
    a. Correct grounding ____
    b. Ground-fault interrupters ____
    c. _____ ____
    d. _____ ____
15. Portable & Power Saws
    a. In good condition ____
    b. Guards ____
    c. Kickback protection ____
    d. Ventilation ____
    e. Safe fuel procedures ____
    f. _____ ____
    g. _____ ____
16. Hand Tools
    a. In good condition ____
    b. Insulated and/or grounded ____
    c. Projectile tools ____
    d. Powder-actuated tools? ____
    e. Operators trained in proper use ____
    f. _____ ____
    g. _____ ____
17. First Aid
    a. Proper kit size & contents ____
    b. Kit supply maintained ____
    c. Trained employees ____
    d. Emergency numbers posted ____
    e. Hospital routes known ____
    f. _____ ____
    g. _____ ____
18. Traffic Control
    a. Parking ____
    b. Speed control ____
    c. Barricades ____
    d. Separation of haul roads ____
    e. _____ ____
    f. _____ ____

19. Personal Protective Equipment
    a. Hard hats ____
    b. Goggles / safety glasses ____
    c. Gloves ____
    d. Respirators ____
    e. Hearing protection ____
    f. Safety shoes ____
    g. No loose clothing ____
    h. All work areas sanitary ____
    i. _____ ____
    j. _____ ____
20. Heavy Equipment
    a. Guards ____
    b. Warning bells ____
    c. Fueling ____
    d. Ground slope ____
    e. Rough terrain ____
    f. Cab protection ____
    g. Operator qualifications ____
    h. _____ ____
    i. _____ ____
21. Security
    a. Fencing ____
    b. Lighting ____
    c. Alarm systems ____
    d. Monitoring arrangements ____
    e. Guard service ____
    f. Target equipment ____
    g. Secure equipment practices ____
    h. Police notification procedure ____
    i. _____ ____
    j. _____ ____
22. Liability
    a. Release forms executed/delivered for all trades using:
       Hoists
       Elevators ____
       Scaffolding ____
       Equipment ____
    b. Arrange for jobsite inspection by insurance carrier ____
    c. _____ ____
    d. _____ ____
23. Other
    a. _____ ____
    b. _____ ____
    c. _____ ____
    d. _____ ____
    e. _____ ____
    f. _____ ____
    g. _____ ____
    h. _____ ____
    i. _____ ____
    j. _____ ____
    k. _____ ____
    l. _____ ____
    m. _____ ____
    n. _____ ____
    o. _____ ____

Note that the intent of this or any similar procedure is not in any way to obstruct or otherwise make any inspection more difficult for authorized individuals. The actions outlined in this sample section are only intended to illustrate one manner in which the company might exercise its legal right to proper representation at various government inspections, and to assure that all facts and circumstances are properly considered in any such investigation. Again, check with your own appropriate safety, insurance, and legal professionals for specific instruction in this regard.

Considering the above remark, a company procedure upon such a visit may be as follows:

1. Direct the inspector to the highest-ranking company employee present on the jobsite at that time.

2. The highest-ranking company employee will then:
   a. Ask the inspector of his or her complete intentions, and for proper identification.
   b. Write down the inspector's name, title, organization, ID or badge number, date, and the time of day.
   c. Request the inspector to delay further activity until the company Safety and Health Manager or other appropriate company official can be dispatched to the jobsite.
   d. Contact the company Safety and Health Manager and the company Senior Project Manager, and provide the information obtained to that point. If either individual is unavailable for any reason, so notify the other. If both individuals are unavailable for any reason, immediately contact company senior management.
   e. Offer a place for the inspector to sit and wait, if possible. Should the inspector insist on proceeding without the Safety and Health Manager or other company official present, note the time and accompany him or her through the entire inspection. Take comprehensive notes of the entire activity.

3. Above all, every company employee is to be courteous, helpful, and fully cooperative with every authorized inspector at all times.

4. Contact the company Safety and Health Manager if there are any questions regarding these situations.

6.5   Accident Investigation, Reporting, and Records

6.5.1   Reporting Requirements

All accidents must be reported. In the case of a serious accident, the Company central office must be notified as soon as practical, immediately *after* appropriate emergency measures have been taken.

For any injury to a Company employee while working, a Worker's Compensation Report Form must be completed and filed immediately with the appropriate insurance carrier.

All reporting is to be done by the Superintendent or other designated Safety Officer. It is this individual's responsibility to be aware of and strictly comply with all reporting requirements. If this function is not performed correctly, the Company will be forced to assume extreme and disproportionate amounts of liability with respect to the injured party.

6.5.2 Investigation Requirements

All serious accidents must be investigated. These include serious accidents to all individuals employed by the Company and any Subcontractor, and every accident involving any amount of property damage. Immediately after all appropriate emergency measures, first aid, and damage containment measures have been taken, every effort must be made to immediately:

1. Preserve physical evidence.
2. Take photographs and secure other evidence as appropriate.
3. Take statements from the accident victim, any eyewitnesses, and anyone who may have any knowledge of definite or possible cause(s) of the accident.

If the accident is serious enough that the insurance carrier will be investigating, assist their investigation in any way they need.

6.5.3 Investigation Procedure

The investigation procedure begins as soon as all immediate danger to people and property has been brought under control, and all Company and appropriate authorities have been notified of the incident. The investigation is to be conducted by the Superintendent or designated Safety Officer.

The purpose of the investigation procedure is to secure and confirm as many specific facts as possible, not to place blame. Once the procedure is completed, the causes and conditions can be thoroughly analyzed and determined at a later date.

Realize at the onset that the information secured at the scene may or may not be complete or accurate for any number of reasons. These can range from sincere problems in recollection or even correct perception of the event to deliberate attempts to change, conceal, or omit information. For these reasons, it is essential to:

- Keep asking who, what, where, when, how, and why.
- Work in repeated questions of the same item to the same individual for key considerations. Phrase the question in different ways and spread the repeated questions among other questions and conversation.
- Ask the same questions to as many different individuals as possible.
- Interview each individual in a separate location, keep independent answers truly independent, and eliminate the possibility of two or more individuals consciously or unconsciously "coordinating" their versions of the incident.

The procedure then will be as follows:

1. Identify all individuals who were definitely or possibly in the vicinity of the accident immediately prior to or during the accident.
2. Catalog those individuals in the order of importance:
   a. Those involved in the accident
   b. Those who may have caused or contributed to the accident
   c. Eyewitnesses
   d. Those who were in or around the proximity of the accident
3. Photograph the complete accident area.
4. Immediately summarize your own understanding of the accident, if any. Include a diagram with as much relevant information as possible, including appropriate distances and dimensions.
5. Use the sample Accident Investigation Report Form of Section 6.5.4 as the baseline for your investigation, and complete it after all individuals' statements have been secured.
6. Use the sample Accident Eyewitness Statement Outline of Section 6.5.5 for each of the individuals identified in items 1 or 2.
7. Upon completion, turn all accident investigation reports, photos, and information over to the Project Manager for delivery to appropriate Company personnel.

6.5.4 Sample Accident Investigation Report Form (*page 6.41*)

The sample Accident Investigation Report Form that follows is an example of the report that should be completed whenever there is any significant incident. Contact your own insurance carrier to obtain any actual Accident Investigation Report Form required by them, and use it generally in accordance with the procedures developed in this Section and with any other guidelines that they may give. Review these procedures with your insurance carrier to identify any additions or modifications that he or she would consider appropriate.

6.5.5 Sample Accident Eyewitness Statement Outline (*page 6.42*)

The sample Accident Eyewitness Statement Outline is to be used as an aid to help each eyewitness keep his or her statements complete, organized, and focused. It is in no way to be used to guide or force any respective statement, or as any means to add, delete, or modify any information. To the contrary, all eyewitnesses must be repeatedly encouraged to put any descriptions strictly into their *own* words.

The statement outline includes categories for all minimum information requirements and should therefore *not* be considered to be all-inclusive. Each eyewitness must be encouraged to add *all* relevant information beyond the minimums outlined.

6.5.4
Sample Accident Investigation Report Form

Project: _____ Proj.#:_____

Company: Name: _____
            Address: _____
            Phone: _____

Injured: Name: _____
            Home Address: _____
            Home Phone: _____
            Trade: _____ Position:_____
            Length of Employment: _____

Date of Accident:_____ Time: _____ AM/PM
Date of Report: _____ Reported By:_____

Type of Accident (Check One):   ( ) Vehicular   ( ) Personal   ( ) Other
Did the injured lose any time?: _____ If so, how much (Days/Hours): _____

Was safety equipment in use at the time of the accident (hard hat, safety glasses, gloves, respirator, etc.)? _____
_____
(If not, it is the EMPLOYEE's sole responsibility to process his/her claim through his/her Health & Welfare Fund.)

Description of the Accident (Attach additional pages if necessary):_____
_____
_____
_____

What caused the Accident?:_____
_____
_____
_____

What has been done or will be done to correct the situation and prevent recurrence?:_____
_____
_____

Who is responsible for correction?: _____
Has corrective action been taken?:   ( ) YES   ( ) NO
If not, why?:_____
_____

Indicate streets, street names, vehicle descriptions, and north arrow (use separate sheet, if necessary):

	Injured (#1)	Driver (#2)	Driver (#3)
Insurance Carrier:			
Driver Name:			
Address:			
Operator License #:			
Vehicle License #:			
Vehicle Owner Name:			

6.5.5
Sample Accident Eyewitness Statement Outline

Name: _____

Residence: _____

Home Phone: _____   Age: _____   Soc. Sec. #: _____

Employed By: _____   Position: _____   # Years: _____

Employer's Address: _____

My relationship and acquaintance with any of the parties involved in the accident are as follows:

_____

_____

The accident happened on (date) _____ at (time) _____ AM/PM

The weather was (clear, foggy, rainy, etc.) _____

The road was (sandy, wet, dry, potholed, etc.) _____

Immediately before the accident:

    I saw _____

    I heard _____

    I did/was doing _____

During the accident:

    I saw _____

    I heard _____

    I did/was doing _____

Immediately after the accident:

    I saw _____

    I heard _____

    I did/was doing _____

I have examined the diagram attached to this statement, and have shown my position(s), the position(s) of all parties involved, and all relevant details to the best of my knowledge.

Other persons who might have witnessed or may have knowledge relating to the accident are:

_____

I (did) (did not) make a statement to the police, opposing counsel, insurance, or private investigator, etc.:

_____   (If yes, obtain copy)

Supplementary information:

_____

_____

_____

_____

_____

_____

_____

_____

_____

_____

_____

_____

_____

_____

_____

_____

_____

_____

I have read these _____ pages that make up my complete statement. I have made all remarks voluntarily, to the best of my knowledge, and believe them to be true.

Signed: _____   Date: _____

6.6   OSHA's Hazard Communication Standard and Safety Data Sheets

There are about 650,000 existing hazardous chemicals in use in industry today, many of them in the construction industry. The Hazard Communication Standard (HCS) was established by OSHA in 2012 to ensure that the hazards of all chemicals imported into, or produced for, or used in the U.S. workplace are evaluated and any information relating to these hazardous chemicals is passed on to affected employers and employees. This includes construction workers who may be exposed to hazardous chemicals in both new construction projects and existing buildings. These hazards include some common materials such as muriatic acid (hydrochloric acid), some metal degreasers, and toxic gases caused by exposure to fumes from welding and soldering operations.

6.6.1   The Hazard Communication Standard: Fact Sheet No. OSHA 93-26 (*page 6.44*)

6.6.2   A Hazard Communications Checklist (*page 6.46*)

6.6.3   OSHA Hazard Communication Safety Data Sheets (*page 6.47*)

Before hazardous materials are shipped to the jobsite, they will be preceded by Material Safety Data Sheets (MSDS); this designation will change to Safety Data Sheets (SDS) when a new system is implemented as of June 1, 2015. Each Safety Data Sheet will contain 16 sections listing information such as the hazard, firefighting methods, accidental release measures, and means for proper handling and storage. The SDS are to contain a symbol on a white background within a red border, which represents a distinct hazard. This marking is to appear on hazardous materials as of June 1, 2015.

The following OSHA "QuickCard" is a summary reference list of the parts of the uniform SDS.

6.6.4   HCS Pictograms Required on Products When Implemented (*page 6.48*)

The following are the required SDS symbols representing distinct hazards. One of these markings is to appear on hazardous materials as of June 1, 2015.

6.6.1
The Hazard Communication Standard: Fact Sheet No. OSHA 93-26

U.S. Department of Labor
Program Highlights

Fact Sheet No. OSHA 93-26

HAZARD COMMUNICATION STANDARD

SUMMARY

Protection under OSHA's Hazard Communication Standard (HCS) includes all workers exposed to hazardous chemicals in all industrial sectors. This standard is based on a simple concept - that employees have both a need and a right to know the hazards and the identities of the chemicals they are exposed to when working. They also need to know what protective measures are available to prevent adverse effects from occurring.

SCOPE OF COVERAGE

More than 30 million workers are potentially exposed to one or more chemical hazards. There are an estimated 650,000 existing hazardous chemical products, and hundreds of new ones are being introduced annually. This poses a serious problem for exposed workers and their employers.

BENEFITS

The HCS covers both physical hazards (such as flammability or the potential for explosions), and health hazards (including both acute and chronic effects). By making information available to employers and employees about these hazards, and recommended precautions for safe use, proper implementation of the HCS will result in a reduction of illnesses and injuries caused by chemicals. Employers will have the information they need to design an appropriate protective program. Employees will be better able to participate in these programs effectively when they understand the hazards involved, and to take steps to protect themselves. Together, these employer and employee actions will prevent the occurrence of adverse effects caused by the use of chemicals in the workplace.

REQUIREMENTS

The HCS established uniform requirements to make sure that the hazards of all chemicals imported into, produced, or used in U.S. workplaces are evaluated and that this hazard information is transmitted to affected employers and exposed employees.

Chemical manufacturers and importers must convey the hazard information they learn from their evaluations to downstream employers by means of labels on containers and material safety data sheets (MSDS's). In addition, all covered employers must have a hazard communication program to get this information to their employees through labels on containers, MSDS's, and training.

This program ensures that all employers receive the information they need to inform and train their employees properly and to design and put in place employee protection programs. It also provides necessary hazard information to employees so they can participate in, and support, the protective measures in place at their workplaces.

All employers in addition to those in manufacturing and importing are responsible for informing and training workers about the hazards in their workplaces, retaining warning labels, and making available MSDS's with hazardous chemicals.

Source: U.S. Department of Labor, Occupational Safety and Health Administration; https://www.osha.gov/pls/oshaweb/owadisp.show_document? p_table=FACT_SHEETS&p_id=151

6.6.1
The Hazard Communication Standard: Fact Sheet No. OSHA 93-26 (*Continued*)

Some employees deal with chemicals in sealed containers under normal conditions of use (such as in the retail trades, warehousing and truck and marine cargo handling). Employers of these employees must assure that labels affixed to incoming containers of hazardous chemicals are kept in place. They must maintain and provide access to MSDS's received, or obtain MSDS's if requested by an employee. And they must train workers on what to do in the event of a spill or leak. However, written hazard communication programs will not be required for this type of operation.

All workplaces where employees are exposed to hazardous chemicals must have a written plan which describes how the standard will be implemented in that facility. The only work operations which do not have to comply with the written plan requirements are laboratories and work operations where employees only handle chemicals in sealed containers.

The written program must reflect what employees are doing in a particular workplace. For example, the written plan must list the chemicals present at the site, indicate who is responsible for the various aspects of the program in that facility and where written materials will be made available to employees.

The written program must describe how the requirements for labels and other forms of warning, material safety data sheets, and employee information and training are going to be met in the facility.

EFFECT ON STATE RIGHT-TO-KNOW LAWS

The HCS preempts all state (in states without OSHA-approved job safety and health programs) or local laws which relate to an issue covered by HCS without regard to whether the state law would conflict with, complement, or supplement the federal standard, and without regard to whether the state law appears to be "at least as effective as" the federal standard.

The only state worker right-to-know laws authorized would be those established in states and jurisdictions that have OSHA-approved state programs.

These states and jurisdictions include: Alaska, Arizona, California, Connecticut (state and municipal employees only), Hawaii, Indiana, Iowa, Kentucky, Maryland, Michigan, Minnesota, Nevada, New Mexico, New York (state and municipal employees only), North Carolina, Oregon, Puerto Rico, South Carolina, Tennessee, Utah, Vermont, Virgin Islands, Virginia, Washington, and Wyoming.

FEDERAL WORKERS

Under the hazard communication standard federal workers are covered by executive order.

This is one of a series of fact sheets highlighting U.S. Department of Labor programs. It is intended as a general description only and does not carry the force of legal opinion. This information will be made available to sensory impaired individuals upon request. Voice phone: (202) 219-8151. TDD message referral phone: 1-800-326-2577.

6.6.2
A Hazard Communications Checklist

HAZARD COMMUNICATION CHECKLIST

	Yes	No
Has a program for hazard communication training been established?		
Has a program for hazard communication procedures been established and is the program reviewed on an annual basis?		
Are chemical injuries tracked for program improvement?		
Have chemical hazard control procedures [been] developed for each job?		
Has a chemical inventory of the facility been conducted?		
Are the procedures reviewed on an annual basis?		
Do the hazard communication procedures include the following:		
• A statement of the intended use?		
• Steps for labeling of containers?		
• Steps for safe issuance, use, transfer and disposal of chemicals?		
Are control procedures inspected at least annually?		
Are periodic inspections conducted by a competent employee?		
Is the inspection designed to correct deviations or inadequacies?		
Is the inspection documented?		
Have MSDSs been produced in accordance with 29 CFR 1910.1200?		
Have employees been informed of:		
• The requirements of 29 CFR 1910.1200?		
• Any operations in their work area where hazardous chemicals are present?		
• The location and availability of the written HAZCOM program?		
• The location and availability of the lists of hazardous chemicals?		
Does employee training include at least:		
• Methods & means necessary to detect the presence or release of a chemical?		
• The physical and health hazards of the chemicals in the work area?		
• The steps employees can take to protect themselves from the chemicals?		
• The details of the written program?		
Have criteria for recurrent training been developed?		
Is the training documented?		
Is the training conducted by a competent person?		
Is retraining required whenever there is a change in job assignments?		

Source: a;/hazcom.

6.6.3
OSHA Hazard Communication Safety Data Sheets

Hazard Communication Safety Data Sheets

The Hazard Communication Standard (HCS) requires chemical manufacturers, distributors, or importers to provide Safety Data Sheets (SDSs) (formerly known as Material Safety Data Sheets or MSDSs) to communicate the hazards of hazardous chemical products. As of June 1, 2015, the HCS will require new SDSs to be in a uniform format, and include the section numbers, the headings, and associated information under the headings below:

Section 1, Identification includes product identifier; manufacturer or distributor name, address, phone number; emergency phone number; recommended use; restrictions on use.

Section 2, Hazard(s) identification includes all hazards regarding the chemical; required label elements.

Section 3, Composition/information on ingredients includes information on chemical ingredients; trade secret claims.

Section 4, First aid measures includes important symptoms/ effects, acute, delayed; required treatment.

Section 5, Fire fighting measures lists suitable extinguishing techniques, equipment; chemical hazards from fire.

Section 6, Accidental release measures lists emergency procedures; protective equipment; proper methods of containment and cleanup.

Section 7, Handling and storage lists precautions for safe handling and storage, including incompatibilities.

Section 8, Exposure controls/personal protection lists OSHA's Permissible Exposure Limits (PELs); Threshold Limit Values (TLVs); appropriate engineering controls; personal protective equipment (PPE).

Section 9, Physical and chemical properties lists the chemical's characteristics.

Section 10, Stability and reactivity lists chemical stability and possibility of hazardous reactions.

Section 11, Toxicological information includes routes of exposure; related symptoms, acute and chronic effects; numerical measures of toxicity.

Section 12, Ecological information*

Section 13, Disposal considerations*

Section 14, Transport information*

Section 15, Regulatory information*

Section 16, Other information, includes the date of preparation or last revision.

*Note: Since other Agencies regulate this information, OSHA will not be enforcing Sections 12 through 15(29 CFR 1910.1200(g)(2)).

Employers must ensure that SDSs are readily accessible to employees.
See Appendix D of 1910.1200 for a detailed description of SDS contents.

For more information: www.osha.gov

(800) 321-OSHA (6742)

Source: U.S. Department of Labor, Occupational Safety and Health Administration; https://www.osha.gov/ Publications/HazComm_QuickCard_ SafetyData.html

6.6.4
HCS Pictograms Required on Products When Implemented

Hazard Communication Standard Pictogram

As of June 1, 2015, the Hazard Communication Standard (HCS) will require pictograms on labels to alert users of the chemical hazards to which they may be exposed. Each pictogram consists of a symbol on a white background framed within a red border and represents a distinct hazard(s). The pictogram on the label is determined by the chemical hazard classification.

HCS Pictograms and Hazards

Health Hazard	Flame	Exclamation Mark
• Carcinogen • Mutagenicity • Reproductive Toxicity • Respiratory Sensitizer • Target Organ Toxicity • Aspiration Toxicity	• Flammables • Pyrophorics • Self-Heating • Emits Flammable Gas • Self-Reactives • Organic Peroxides	• Irritant (skin and eye) • Skin Sensitizer • Acute Toxicity • Narcotic Effects • Respiratory Tract Irritant • Hazardous to Ozone Layer (Non-Mandatory)
Gas Cylinder	**Corrosion**	**Exploding Bomb**
• Gases Under Pressure	• Skin Corrosion/Burns • Eye Damage • Corrosive to Metals	• Explosives • Self-Reactives • Organic Peroxides
Flame Over Circle	**Environment** (Non-Mandatory)	**Skull and Crossbones**
• Oxidizers	• Aquatic Toxicity	• Acute Toxicity (fatal or toxic)

For more information:

Occupational
Safety and Health
Administration
U.S. Department of Labor
www.osha.gov (800) 321-OSHA (6742)

OSHA 3491-02 2012

Source: U.S. Department of Labor, Occupational Safety and Health Administration; https://www.osha.gov/Publications/HazComm_QuickCard_ Pictogram.html

Design-Build Project Administration

7.1 Section Description

Design-build had its beginnings in the Master Builder concept that can be traced back to the building of the pyramids in Egypt. It has been recognized for eons that this process of vesting the production of architectural and engineering design with construction affords an owner a single-point responsibility for both activities.

Several studies over the past 10 years have shown that design-build dramatically shortens the entire project delivery cycle, produces fewer change orders, and, taking all factors into consideration, returns savings to the owner while allowing the design-build team potential for increased profits. No wonder both the private and public sectors have embraced design-build in recent years.

In a design-build project, both Project Managers and Project Superintendents may be called upon early in the design stage of the project to contribute their knowledge and experience of constructability issues and "how things go together," thus becoming important members of the team. With the application of design-build in both the private and public sectors growing, Project Managers and their Superintendents need to become more familiar with the process.

A study sponsored by the Design-Build Institute of America (DBIA) in the spring of 2011, produced by RSMeans Reed Construction Data Market Intelligence, found that design-build had achieved a 40% share of the construction market as of May 2011. Design-build has been utilized in nearly 80% of military construction, according to the study.

That RSMeans report also reveals geographic preferences for the design-build project delivery system. The New England area (Connecticut, Maine, Massachusetts, New Hampshire, Rhode Island, and Vermont) and the West North Central Division (Iowa, Kansas, Minnesota, Missouri, Nebraska, North and South Dakota) are less likely to use design-build than other sections of the country, such as the Pacific Census Regional Division (Alaska, California, Hawaii, Oregon, Washington) and the South Atlantic area (Delaware, Florida, Georgia, Maryland, North and South Carolina, Virginia and West Virginia).

Project Managers administering a design-build project will find that there are significant differences between this system and the more conventional design-bid-build approach.

7.1.1 Administrative Differences

These differences can be categorized as follows:

1. Institutional.
2. Licensing and legal.
3. Liability, insurance, and bonding.
4. Contractual.
5. Administration.

7.1.2 Institutional Changes

The role of the Architect as the Owner's agent or watchdog is no longer present. Some Owners may have professional staff on hand to monitor design and construction, but others without trained personnel may feel the need for some form of checks and balances to ensure that their program is being met. In such cases an Owner may engage an outside Owner's representative to look out for his or her interests. Both the design team and the contractor must now view their roles as members of this design-build process in a collaborative manner, not the adversarial manner that was often created when the Architect and builder were defending their own turf. This ability to work together as a team is one of the basic building blocks of the design-build process.

7.1.3 Licensing and Legal Issues

Some states have legislated restrictions on the formation of a design-build team. Until the recent passage of the Architect's Licensure Law, the Commonwealth of Pennsylvania would only allow architects to practice architecture and therefore a contractor-led design-build team would be illegal.

In Arizona, a design-build firm does not have to be licensed to do construction work as long as the builder is properly licensed. In Georgia, a General Contractor can be engaged in design-build as long as a properly licensed Architect creates the design. So it is important to check the licensing laws dealing with design-build in the state in which the entity is planning to work.

The legal aspects of design-build set it apart from other project delivery systems, highlighting the ways in which the role of each party to the process changes, in some cases significantly, from the conventional arm's-length design and construct method.

Some of the legal issues facing the design-build team, based upon several court decisions, are:

- *Liability for design errors, statute of limitation limits*. In the case of *Kishwaukee Community Health Services v Hospital Building and Equipment*, 638 F. Supp. 1492 (N.D. Ill. 1986), the court ruled that liability for design errors, in a design-build project, commences at the *completion* of the project, *not* at the completion of design. A New Jersey court ruling appeared to confirm this North Dakota decision.

- *Ownership of documents*. A Michigan court ruled that progress drawings used by a design-build team cannot be used by the Owner after they have dismissed the design-builder from the project. In another case, this one ruled on by the U.S. Court of Appeals, title of the drawings passed to the Owner because the contract with that Owner stipulated that the design-builder would provide schematic drawings and did not state that the Architect would be retained as architect of record for the completed project.

- *Implied promises*. The Owner can expect certain results from the design-build team; this is known as an *implied promise* and includes such items as:

  Good faith and fair dealing.

  An Owner must pass judgment on acceptable work within a reasonable period of time.

  An Owner must provide sufficient information to a design-build team to allow it to design a reasonably complete program.

  A design-build team, once the design development scheme is approved, must not substantially deviate from that scheme in the final project design.

  A design-build team has a certain fiduciary responsibility with respect to protecting an Owner's interests.

- *Other ownership and use considerations*. Event though an Owner pays for the design, what control does a design-build team have over future use of their drawings? Some design-build contracts include a clause to prohibit the use of their design documents for future use. The design-build team may not want its design used on any Owner promotional literature, either because of security reasons or exclusivity. These types of issues can be handled by contract language.

- *Design error liability*. Design consultants will share some of the responsibility for design errors and so must the builder if it fails to act in a professional manner, so say the courts. In another case, major defects in design were found that required significant changes and increased the cost of the project. The design team requested additional money, stating that these changes produced value and its additional costs would have been incorporated into the Guaranteed Maximum Price (GMP). The Owner countersued, and the court stated that the Owner had a tight budget and if the Owner had been advised of these added costs, the Owner would have abandoned the project. The design-build team could not collect those extra costs.

- *Compliance with code*. The design-build team cannot claim ignorance of any building code. The court in *Tips v. Hartland Developers Inc., 961 S.W.2d 618 (Tex. Civ. App. 1998)*, ruled there is an *implied covenant* that the design-build team will comply with all building codes and provide a building that can be occupied for its intended purpose.

- *The Uniform Commercial Code (UCC)*. The UCC was enacted for the purpose of applying to the sale of goods and products, and the courts are split on whether a design-builder is a *merchant of goods* and therefore liable under the premise that it sells a product. This is an emerging concept, and design-builders need to keep track of developing court decisions.

- *Americans with Disabilities Act (ADA)*. There have been a few court cases challenging whether architects and contractors can be held liable for violation of what is essentially a civil rights law. Some court rulings seem to apply to design-builders who may be sued for violations of ADA because they fail to comply with that act when designing new or remodeled projects.

7.1.4 Liability, Bonding, and Insurance Issues

When responsibility for design and construction are combined in one entity, the question of bonding, insurance, and liability becomes somewhat cloudy. Contractors working for an Owner via an arm's-length, build-only contract are responsible for completing the work in accordance with the contract documents—the plans and specifications prepared by the Owner's design team. However, the new design-build entity bridges the gap between design responsibility and construction responsibility, which is one of the reasons why Owners look favorably upon this single-point project delivery system.

This combination of design and construction responsibility raises a number of legal issues:

- Which party, architect or builder, bears responsibility for latent defects, i.e., structural failure after the normal one-year builder's warranty expires?

- Who is responsible for failure to meet the performance specifications set forth by the Owner as part of its program?

- Does the Uniform Commercial Code (UCC), a law that basically applies to product warranty, apply to the design-build process?

- What is the implied warranty behind a design-build contract? Can the design-build team be held liable if the finished product fails to meet the specific purpose for which it was hired?

- Who owns the contract documents? And once the current project has been completed, is the Owner free to use those drawings for a second, similar building?

Surety Bonds.   When requesting a Payment and Performance bond for a design-bid-build project, the bonding company is being requested to provide guarantees for project performance and completion, but in a design-build project, the bonding company is being requested to include design, and sureties tend to handle this in one of two methods:

1. Insert explicit language and disclaimers limiting their exposure to design.

2. Require the design-builder to include language in the contract with the Owner stipulating that the Payment and Performance (P&P) bond does not cover the design portion of that contract. A typical clause would read thusly:

> The bond does not cover any responsibility for negligence, errors, and/or omissions in design. Coverage under this bond is limited solely to the construction phase and postconstruction phase of this contract. The bond premium is based upon the value of construction and postconstruction portions of the contract and does not include the design aspect of the contract.

Insurance.   The matter of liability also changes when design-build is in effect. The standard architect and engineer Errors and Omissions (E&O) insurance,

Commercial General Liability (CGL), and Contractors Professional Liability (CPL) insurance apply to the more conventional third-party construction contract.

Traditional insurance products don't address the new risks that may be encountered in design-build.

- Traditional CGL policies exclude professional E&O coverage.
- Traditional CPL policies exclude design-build, most environmental issues, and equity interests.

So the design-build team must redirect its insurance policies to fit the risks it will be exposed to when operating in a design-build mode and seek policies that:

- Delete traditional exclusions
- Include the design-builder's errors and omissions exposure
- Provide a wrap-up insurance program

7.1.5   Contract Provisions Unique to Design-Build

The "Everything Is Included" Provision.   The design-builder may include an extensive Exclusion or Inclusion List as a contract exhibit, limiting its scope of responsibility to some degree, but some Owners may object to these kinds of limitations and require a more definitive clause to define what is or what is not included in the project.

> The intent of the contract documents is to include all of the work required to complete the project except as specifically excluded. It is acknowledged that as of the date of the contract, the plans and specifications are not complete but define the scope and nature of the work and are sufficient to establish the contract sum. No adjustment shall be made in the contract sum if, as a prudent design-builder, design-builder should have been aware of or anticipated such additional work as may be required to produce a first-class (office building or whatever the nature of the project is).

Use of Design Documents.   Who owns the design documents upon completion of the project and receipt of final payment? Can the Owner use these documents to contract with another builder to construct a duplicate building? A standard design-build contract provision addresses such issues as follows:

> The owner may use and reproduce design-build plans and specifications for subsequent renovation and remodeling work but shall not have authority to use these design-build documents for other projects without the express, written authority of the design-builder, who shall not unreasonably withhold consent.

The Standard of Care Provision.   Although architects and engineers, in the eyes of the court, are not expected to produce a perfect set of plans and specifications, they must complete their design in a manner that meets accepted "standards of care" and skills associated with their profession. This Standard of Care provision is included in Associated General Contractors (AGC) and Design-Build Institute of America (DBIA) design-build contracts. The AGC provision stipulates:

> The A/E is required to use the same care and skill ordinarily used by members of the architecture and engineering community practicing under similar circumstances at the same time and locality.

The DBIA provision stipulates:

> The designer is expected to use the care and skill ordinarily exercised by members of the architecture and engineering community at the same time and locality unless the owner and the design-builder have set forth certain specific performance standards in an exhibit in the design-build contract.

7.1.6   Administration of the Design-Build Contract

Administrative and management responsibilities during the design and construction process will vary considerably from the more conventional design-bid-build process. The designers and contractor will work together to develop a design concept based upon the Owner's program needs and budget limitations. Both the Architect/Engineer and builder must agree on project administration responsibilities during design and construction. Who will be responsible to track design and budget during design development? Who will inspect during construction, who will direct forces, who will accept/reject work, how will shop drawings be processed, how will design issues be resolved in the field?

The process of developing a conceptual design and preliminary budget will require an in-depth review of the Owner's program and involve quite a bit of the assigned Project Manager's time. This interface between designers and project managers on constructability issues, budget restraints, scheduling, and Owner's program compliance will require lots of meetings.

This concentrated up-front activity is one reason why design-builders must include sufficient general conditions, slightly higher overhead, and increased profit due to a higher risk factor and the added workload they will incur.

7.2   Creating the Design-Build Team

The combining of an architectural and an engineering firm with construction professionals into one operational team presents any number of challenges.

There must be a team leader, and one of the first considerations is whether the design-build entity should be contractor- or architect/engineer-led. The decision can be based simply upon which member of the team is better equipped to secure a contract with the Owner. A favored architect may have a long-term relationship with an Owner who is contemplating a new project and would be more receptive to entertaining a design-build project as long as the architect assumes the lead role in the venture; the same would be true of a builder who has a similar good relationship with a long-term client.

7.2.1 Formation of the Team

The formation of a design-build team not only requires legal assistance, but also requires each party to consider a number of issues relating to the division of responsibility, such as:

- Who is best equipped to assume the lead in extracting and developing the Owner's program?

- Who will be responsible for design errors and omissions, and how will this responsibility be covered in any forthcoming design-builder agreement?

- If Value Engineering is required to meet the Owner's budget, what authority and responsibility does the design team have in reviewing, approving, or rejecting any value engineering proposals?

- How is compensation to be divided, and if any up-front money is required to provide a conceptual design and estimate, how will these costs be apportioned?

- During construction, what role does the contractor play in supervising the project and what supervisory role does the design team play?

- How will any nonconforming or substandard work be identified, and what authority will the architect/engineer team have in rejecting and remediating this work?

- How will profits, and losses, be apportioned when the project has been completed?

- Postconstruction, how will the correction of defective work, design errors, warranty issues, and statute of limitations matters be apportioned and resolved?

7.2.2 Financial Strength

Substantial lines of credit necessary for construction projects may dictate that the contractor should be the lead member of the team, having the financial resources to pay subcontractors and vendors if payments from the Owner are occasionally late.

A joint venture business agreement is one vehicle for forming a design-build team, an agreement whereby each participant's duties, obligations, and responsibilities are spelled out along with the share of profits (and losses). A contractor can subcontract design to an architect, and, conversely, an architect can subcontract construction.

If the project requires a Payment and Performance bond, in the case where public works projects are involved, the contractor, with its long experience dealing with bonding companies, may be the preferred team leader.

In a process engineering project, such as design and construction of a chemical plant or waste or water treatment facility, the engineer will, more than likely, assume the role of project leader.

7.3   Consideration of a New Business Entity

Possibly the first consideration facing this proposed design-build venture is the type of legal entity to be created to actually form the new business venture. Will it be a corporation, partnership, limited liability corporation (LLC), or joint venture? Each business form has its own advantages and disadvantages, and both professional accounting and legal advice is required to assist in making that determination. A brief overview of each type may be helpful.

7.3.1   The Partnership

A partnership can consist of two or more members. The partnership prepares a documented list of rights, responsibilities, and obligations for each of the participating partners. Unlike the corporation, there is no legal protection when claims are brought against any of the partners or against the partnership. This protection must be obtained via insurance—both partnership and individual, both types being rather expensive. Profits and losses from the partnership pass directly to each partner in an amount equal to the percentage specified in the partnership agreement.

7.3.2   The Corporation

This is a business form familiar to all. The corporation is an entity into itself and is recognized by the law as such. Officers are appointed, along with a Board of Directors, which is to hold periodic meetings and prepare and issue minutes of those meetings to the stockholders. The corporation can issue either voting or nonvoting stock to its stockholders. Each state has slightly different requirements and restrictions on the formation and operation of companies incorporated within its jurisdiction.

7.3.3   The S Corporation

This "S" status is conferred by the Internal Revenue Service after ruling on the applicant's request. The S Corporation allows the business to be taxed similarly to a partnership or sole proprietor as opposed to a corporation.

The profits and/or losses of the S Corporation pass through to the Owners who report any gains or losses on their individual tax returns. This avoids the double taxation that a corporation incurs—a corporate tax and the tax paid by the individual receiving dividends from the corporation. There are, however, several disadvantages of an S Corporation:

- Officials of S Corporations can be held personally liable for some of their corporate actions.
- Only one class of stock can be issued.
- The S Corporation is less attractive to outside investors, who may not like the pass-through tax benefits afforded this type of business entity.

- This corporation can have no more than 75 stockholders (which is probably not a problem for a design-build venture).
- The S Corporation is, in fact, a corporation and must conduct regular meetings and maintain corporate minutes of those meetings.
- All shareholders must be U.S. citizens.

7.3.4 The Limited Liability Corporation (LLC)

The LLC offers the limited liability protection of a corporation, existing as a separate and distinct entity generally created for a specific purpose. When used for design-build purposes, each project would necessitate creating a different LLC. Applicants for LLCs are usually required to submit their application to the state where the corporation will be operating. Articles of Organization are submitted with this application, and an application fee is also levied. Some states require an operating agreement, a document similar to a corporation's bylaws or partnership agreement. As with other business forms, the LLC has certain advantages and disadvantages:

The advantages are:

- No formal meetings are required, and no meeting minutes are needed.
- No corporate resolutions are required, unlike a conventional corporation.
- Distribution of profits can be tailored as required.
- All profits, losses, and expenses flow through the corporation to the individual members of the corporation, thereby avoiding the double taxation of paying corporate and individual taxes on money earned.

The disadvantages are:

- The LLC is dissolved when a member dies or undergoes bankruptcy, whereas a conventional corporation can theoretically live forever.
- Because of the "on-off" nature of the LLC, lending institutions are often reluctant to provide funds without personal guarantees from its officers.
- Owners of projects may be reluctant to do business with an LLC because they recognize its single-subject nature.

7.3.5 The Joint Venture

A joint venture (JV) between an architect and a contractor is another avenue to explore when considering a design-build project. The joint venture agreement will spell out the management responsibilities, the scope of work assigned to each member, capital contributions, banking and accounting functions, division of profits or losses, and so forth. A first consideration is a question of licensing for both the designer and contractor. Will a particular state allow the joint venture to practice architecture or to perform construction? If this does not present a problem, the joint venture must be specific in dealing with some rather basic but important considerations:

- Which JV member will take the lead in developing the Owner's program?
- How will design issues and budget restraints be resolved?
- Will it be the designer's obligation to design or redesign to a budget without increasing the cost of its services to the JV?
- Who is responsible for design errors, and how will this be covered by insurance?
- If design errors occur, in what amount(s) and to whom are the proceeds paid?
- If the contractor provides Value Engineering in order to meet the budget, what responsibility does the design team have in reviewing, approving, or rejecting the VE proposals?
- How will compensation be divided? If any up-front money is required for design or project management, who will pay, and in what amounts?
- During construction, does the builder provide all field supervision?
- What inspection authority does the architect and engineer have to reject non-conforming or substandard work?
- Postconstruction, who is responsible to correct defective work or design errors and how are funds dispersed for that purpose and also for punchlist work?

This list of division of work can apply to other documents created by the design-build team, including the Teaming Agreement discussed in detail in Section 7.4.

7.4   Creating the Design-Build Team

If the design-build team has been formed to respond to an RFP that is to be a competitive bid, this decision may be delayed until such time as the team has won the competition. Even if the project is being negotiated with an Owner, the expense and time to create the new business entity might wait until such time as a design-build contract is imminent. But there is one document that ought to be prepared at the time that the architect/engineer firm and the contractor have made the decision to work together as a design-build team. That document is known as a Teaming Agreement.

7.4.1   The Teaming Agreement

This agreement establishes the division of responsibility between the design team and the builder in order to provide the design and construction services necessary to prepare and submit a design-build proposal to an Owner.

There are actually two parts to a Teaming Agreement: one to define responsibilities during the proposal stage, and one that takes effect after the proposal has been accepted by an Owner and a contract is to follow. We'll call the first one Part A and the second one Part B.

7.4.2   AGC's Document No. 499

The Associated General Contractors of America publish a clear and concise teaming agreement, AGC Document No. 499—Standard Form of Teaming Agreement for Design-Build Project. As shown in Section 7.4.3, it is basically a two-part document. Articles 1 through 4 relate to the proposal stage, and Article 5 dictates the measures to be taken if the Owner accepts the design-build proposal.

The basic elements to consider as designer and contractor prepare to execute this Teaming Agreement are similar to those raised when a joint venture agreement is being entertained.

7.4.3   AGC Document No. 499—Standard Form of Teaming Agreement for Design-Build Project *(page 7.14)*

7.4.4   Teaming Agreement Part 1 or Part A

Part 1 or Part A will deal with the Owner's Request for Proposal stage as architect and builder consider creating an entity to respond to that owner's RFP. So the first part of the Teaming Agreement will consider:

1. Is the architect (and/or builder) sufficiently experienced to proceed with the design (or construction) of this type of project?

2. Do the architect/engineer and contractor have like goals and compatible personnel that can work effectively and harmoniously together, and have they worked with each other on previous projects?

3. When this team is assembled, will it have a good chance of winning the competition and ultimately obtain a contract for the work?

4. Does each member of the team have the financial wherewithal to provide the services required to submit the proposal and, if they don't win, be able to absorb all associated costs?

5. Does either party have a positive past relationship with the Owner that will stand them in good stead in this competition?

6. Is the other party in agreement as to the type of business entity that is planned if an award is made?

7. How will costs be apportioned during the bidding process, and how will these costs be repaid if the proposal is successful or not successful?

8. If an award is made, are all parties committed to maintaining a positive relationship?

If the Part 1 or A Teaming Agreement and the Team's subsequent response to the RFP results in a contract award by an Owner, the second part of this document must now be addressed.

7.4.3
AGC Document No. 499—Standard Form of Teaming Agreement
for Design-Build Project
(Reproduced with the expressed written permission of the Associated General Contractors of America under License No. 131.)

THE ASSOCIATED GENERAL CONTRACTORS OF AMERICA

AGC DOCUMENT NO. 499
STANDARD FORM OF TEAMING AGREEMENT
FOR DESIGN-BUILD PROJECT

This Agreement is made this _____ day of _____ in the year _____ , ◆

by and between

TEAM LEADER _____ ◆
                                        (Name and Address)

and **TEAM MEMBER** _____ ◆
                                        (Name and Address)

and **TEAM MEMBER** (if applicable) _____ ◆
                                        (Name and Address)

and **TEAM MEMBER** (if applicable) _____ ◆
                                        (Name and Address)

the parties collectively referred to as the **TEAM** for services in connection with the following **PROJECT**

_____ ◆
                                (Name, Location and Brief Description)

for **OWNER** _____ ◆
                                        (Name and Address)

---

* To order AGC contract documents, phone 1-800-282-1423. These documents are also available in electronic form through AGC DocuBuilder® software by going to www.agcdocubuilder.org.

7.4.3
AGC Document No. 499—Standard Form of Teaming Agreement for Design-Build Project (Continued)

ARTICLE 1

TEAM RELATIONSHIP AND RESPONSIBILITIES

1.1   This Agreement shall define the respective responsibilities of the Team Members for the preparation of responses to the Owner's request for qualifications and request for proposals for the Project. Each Team Member agrees to proceed with this Agreement on the basis of mutual trust, good faith and fair dealing and to use its best efforts in the preparation of the statement of qualifications and proposal for the Project, as required by the Owner, and any contract arising from the proposal.

1.2   The Team Leader, _____◆

shall provide overall direction and leadership for the Team and be the conduit for all communication with the Owner. In addition the Team Leader shall provide expertise in the areas of (a) construction management and construction; (b) the procurement of equipment, materials and supplies; (c) the coordination and tracking of equipment and materials shipping and receiving; (d) construction scheduling, budgeting and materials tracking; and (e) administrative support. The Team Leader's representative shall be: _____

1.3   The principal design professional is Team Member, _____
who shall perform the following design and engineering services required for the Project: _____

In addition this Team Member shall coordinate the design activities of the remaining design professionals, if any. This Team Member's representative shall be: _____◆

1.4   Team Member, _____◆
shall provide expertise in the following areas: _____◆

This Team Member's representative shall be: _____◆

1.5   Team Member, _____◆
shall provide expertise in the following areas: _____◆

This Team Member's representative shall be: _____◆

1.6   Each Team Member shall be responsible for its own costs and expenses incurred in the preparation of materials for the statement of qualifications and the proposal and in the negotiation of any contracts arising from the proposal, except as specifically described herein: _____◆

Any stipends provided by the Owner to the Team shall be shared on the following basis: _____◆

1.7   EXCLUSIVITY No Team Member shall participate in Owner's selection process except as a member of the Team, or participate in the submission of a competing statement of qualifications or proposal, except as otherwise mutually agreed by all Team Members.

7.4.3

AGC Document No. 499—Standard Form of Teaming Agreement
for Design-Build Project (Continued)

ARTICLE 2

STATEMENT OF QUALIFICATIONS
AND PROPOSAL

2.1   The Team Members shall use their best efforts to prepare a statement of qualifications in response to the request of the Owner. Each Team Member shall submit to the Team Leader appropriate data and information concerning its area or areas of professional expertise. Each Team Member shall make available appropriate and qualified personnel to work on its portion of the statement of qualifications in the time frame prescribed, and shall provide reasonable assistance to the Team Leader in preparation of the statement of qualifications.

2.2   The Team Leader shall integrate the information provided by the Team Members, prepare the statement of qualifications and submit it to the Owner. The Team Leader has responsibility for the form and content of the statement of qualifications and agrees to consult with each Team Member, before submission to the Owner, on all matters concerning such Team Member's area of professional expertise. The Team Leader shall represent accurately the qualifications and professional expertise of each Team Member as stated in the submitted materials.

2.3   If requested by the Owner, the Team Members shall prepare and submit a proposal for the Project to the Owner. Each Team Member shall support the Team Leader with a level of effort and personnel, licensed as required by law, sufficient to complete and submit the proposal in the time frame allowed by the Owner. A clear and concise statement of the division of responsibilities between the Team Members will be prepared by the Team Leader. The Team Leader shall make all final determinations as to the form and content of the proposal. The Team Leader shall use its best efforts, after the Team has qualified for the Project, to obtain the contract award, and each Team Member shall assist in such efforts as the Team Leader may reasonably request.

ARTICLE 3

CONFIDENTIAL INFORMATION

3.1   The Team Members may receive from one another Confidential Information, including proprietary information, as is necessary to prepare the statement of qualifications and the proposal. Confidential Information shall be designated as such in writing by the Team Member supplying such information. If required by the Team Member supplying the Confidential Information, a Team Member receiving such information shall execute an appropriate confidentiality agreement. A Team Member receiving Confidential Information shall not use such information or disclose it to third

parties except as is consistent with the terms of any executed confidentiality agreement and for the purposes of preparing the statement of qualifications, the proposal and in performing any contract awarded to the Team as a result of the proposal, or as required by law. Unless otherwise provided by the terms of an executed confidentiality agreement, if a contract is not awarded to the Team or upon the termination or completion of an contract awarded to the Team, each Team Member will return any Confidential Information supplied to it.

ARTICLE 4

OWNERSHIP OF DOCUMENTS

4.1   Each Team Member shall retain ownership of property rights, including copyrights, to all documents, drawings, specifications, electronic data and information prepared, provided or procured by it in furtherance of this Agreement or any contract awarded as a result of a successful proposal. In the event the Owner chooses to award a contract to the Team Leader on the condition that a Team Member not be involved in the Project, that Team Member shall transfer in writing to the Team Leader, upon the payment of an amount to be negotiated by the parties in good faith, ownership of the property rights, except copyright, of all documents, drawings, specifications, electronic data and information prepared, provided or procured by the Team Member pursuant to this Agreement and shall grant to the Team Leader a license for this Project alone, in accordance with Paragraph 4.2.

4.2   The Team Leader may use, reproduce and make derivative works from such documents in the performance of any contract. The Team Leader's use of such documents shall be at the Team Leader's sole risk, except that the Team Member shall be obligated to indemnify the Team Leader for any claims of royalty, patent or copyright infringement arising out of the selection of any patented or copyrighted materials, methods or systems by the Team Member.

ARTICLE 5

POST AWARD CONSIDERATIONS

5.1   Following notice from the Owner that the Team has been awarded a contract, the Team Leader shall prepare and submit to the Team Members a proposal for a Project-specific agreement of association among them. (Such agreement may take the form of a design-builder/subcontractor agreement, a joint venture agreement, a limited partnership agreement or an operating agreement for a limited liability company.) The Team Members shall negotiate in good faith such Project-specific agreement of association so

7.4.3
AGC Document No. 499—Standard Form of Teaming Agreement
for Design-Build Project *(Continued)*

that a written agreement may be executed by the Team Members on a schedule as determined by the Team Leader or by the Owner, if required by the request for proposal. The Team Leader shall use its best efforts, with the cooperation of all Team Members, to negotiate and achieve a written contract with the Owner for the Project.

ARTICLE 6

OTHER PROVISIONS   ◆

This Agreement is entered into as of the date set forth above.

WITNESS: TEAM LEADER: _____ ◆

_____ ◆   BY: _____ ◆

                                                    PRINT NAME: _____ ◆

                                                    PRINT TITLE: _____ ◆

WITNESS: TEAM MEMBER: _____ ◆

_____   BY: _____ ◆

                                                    PRINT NAME: _____ ◆

                                                    PRINT TITLE: _____ ◆

WITNESS: TEAM MEMBER: _____ ◆

_____   BY: _____ ◆

                                                    PRINT NAME: _____ ◆

                                                    PRINT TITLE: _____ ◆

WITNESS: TEAM MEMBER: _____ ◆

_____ ◆   BY: _____ ◆

                                                    PRINT NAME: _____ ◆

                                                    PRINT TITLE: _____ ◆

7.4.3
AGC Document No. 499—Standard Form of Teaming Agreement
for Design-Build Project *(Continued)*

AGC 499

The date of the Agreement and identification of all the team members, the Project and the owner are essential information to be accurately inserted.

Article 1 TEAM RELATIONSHIP AND RESPONSIBILITIES

This Paragraph describes the responsibilities of the Team Members for preparation of responses to the Owner's request for qualifications and for proposals for the Project.

1.2   The Team Leader provides overall direction and leadership and expertise in the areas described. The Team Leader may be any person or entity so long as they assume all of the responsibilities described.

1.3-1.5 Here the remaining Team Members, including the principal design professional, provide their name and the name of their representative.

1.6   Each Team Member shall be responsible for its own costs and expenses incurred in the preparation of materials for the statement of qualifications and the proposal. Exceptions are to be provided here. Also, how the parties share stipends is indicated here.

1.7   No Team Member shall participate in the Owner's selection process except as a member of the Team. Also, no Team Member shall participate in a competing statement of qualifications or proposal.

Article 2   STATEMENT OF QUALIFICATIONS AND PROPOSAL

The Team Members shall prepare a statement of qualifications in response to the request of the Owner. The Team Leader shall integrate the information provided by the Team Members, prepare the statement of qualifications and submit it to the Owner. If requested by the Owner, the Team Members shall prepare and submit a proposal for the Project to the Owner.

Article 3 CONFIDENTIAL INFORMATION

The Team Members may receive from one another Confidential Information, including proprietary information, as is necessary to prepare the statement of qualifications and the proposal. Confidential Information shall be designated as such in writing by the Team Member supplying such information. Unless otherwise provided by an executed confidentiality agreement, if a contract is not awarded to the Team or upon the termination or completion of a contract, each Team Member will return any Confidential Information supplied to it.

Article 4 OWNERSHIP OF DOCUMENTS

4.1   Except as noted, each Team Member shall retain ownership of proprietary rights, including copyrights, to all documents, drawings, specifications, electronic data and information.

4.2   The Team Leader may use, reproduce and make derivative works from such documents in the performance of any contract.

Article 5 POST AWARD CONSIDERATIONS

Following notice of award, the Team Leader shall prepare and submit to the Team Members a proposal for a Project-specific agreement or association among them. Such agreement may take the form of a design-builder/subcontractor agreement, a joint venture agreement, a limited partnership agreement or an operating agreement for a limited liability company.

Article 6 OTHER PROVISIONS

Here other provisions are incorporated, if needed, by the parties.

7.4.5 Teaming Agreement Part 2 or Part B

Part B of the Teaming Agreement must now be completed concurrent with the legal entity to be formed to perform the work. Let's look at the second part, Part B of the Teaming Agreement, at this time.

This portion of the Teaming Agreement will define the design team and the builder's obligations, responsibilities, and rights during their entire working relationship on a particular design-build project:

- Team relationships and responsibilities
- A noncompete clause that prohibits a team member from acting in an independent capacity with the Owner
- A requirement for each member of the team to prepare a Statement of Qualifications when requested by the Owner
- A Confidentiality Agreement between team members to prevent any confidential information being passed on or disclosed to a third party
- The right of ownership of design-build documents and how and when passage of title to those documents will occur
- The contractual formation of the design-build team upon a contract award by the Owner

7.4.6 The Contract Format between Owner and Design-Builder

The actual form of contract between the design-build team and the Owner can be a stipulated or lump sum agreement, a cost-plus GMP contract, or a Construction Management (CM) contract. Standard forms are available from the Associated General Contractors of America (AGC), the Construction Management Association of America (CMAA), the American Institute of Architects (AIA), and the Engineers Joint Contract Documents (EJCD).

7.5 The Bridging Approach to Design

In the bridging process, the Owner contracts with a design professional to provide a set of partially complete design documents that will be used to solicit competitive bids in a rather oblique way. By presenting a design concept to the bidders, the Owner may be able to obtain more definitive pricing information since the Owner has already established the basic design. The bid documents may invite design-build firms to offer variations on the basic design and value engineering suggestions at a stage that will not incur significant redesign costs, if accepted. The bridging consultant will work closely with the Owner to prepare these conceptual drawings and may also provide budget numbers and a milestone schedule.

If the Owner proceeds with the project after receiving the responses to the RFP, the Owner can either hire the bridging design consultant to complete the initial design phase or develop an entirely new one based upon information received during the bid process or turn the bridging design over to the selected design-build team to modify or discard. An architect preparing these conceptual design documents in his or her role as a *bridging design consultant* must consider certain liability issues when accepting this type of commission:

- Who owns the design if it is picked up by the selected design-build team?
- Who is the architect of record in such a case?
- Does the selected architect, if other than the bridging designer, get the credit for the evolving design, or is it shared with the bridger?

The American Institute of Architects Code of Ethics and Professional Conduct Rule 4.201 refers to credit for design, and states that credit for work performed by a member is to be recognized as such and other participants in a project are to be given their share of credit. A design-build team presented with a bridging concept will need to investigate matters a little further before accepting this design as part of its project responsibility. When no bridging proposal has been included in the Owner's solicitation for a design-build proposal, upon selection, the team must be diligent in extracting that Owner's program.

7.6   Extracting the Owner's Program

The process of extracting the Owner's program, while tracking design and budget, will follow this path:

- Meetings with the Owner's staff to develop a program to meet their needs and objectives
- Reviewing the benefits and downside of various project delivery systems, for example, lump sum, GMP, CM
- Analyzing the program and presenting alternative feasibility studies and life cycle costing plans
- Projecting a project timeline for both design and construction
- Preparing a budget to include allowances, alternates, and contingencies
- Addressing project financing issues with the Owner

The Associated General Contractors of America (AGC) publishes a number of standard contract forms relating to design-build. Their AGC Document 410, reflected in Section 7.6.1, contains an invaluable checklist highlighting the administrative obligations of the owner and the design-builder.

7.6.1   AGC Document No. 410—Standard Form of Design-Build Agreement and General Conditions between Owner and Design-Builder *(page 7.21)*

7.6.1
AGC Document No. 410—Standard Form of Design-Build Agreement and General Conditions Between Owner and Design-Builder

(Reproduced with the expressed written permission of the Associated General Contractors of America under License No. 131.)

THE ASSOCIATED GENERAL CONTRACTORS OF AMERICA

INSTRUCTIONS FOR COMPLETION OF
AGC DOCUMENT NO. 410
STANDARD FORM OF DESIGN-BUILD
AGREEMENT AND GENERAL CONDITIONS
BETWEEN OWNER AND DESIGN-BUILDER
(Where the Basis of Payment Is the Cost of the Work Plus a Fee with a Guaranteed Maximum Price)

1999 EDITION

This edition of the Standard Form of Design-Build Agreement and General Conditions Between Owner and Design-Builder (Where the Basis of Payment is the Cost of the Work Plus a Fee with a Guaranteed Maximum Price), AGC Document No. 410 (AGC 410), is intended to be used as a follow-on document to AGC Document No. 400 (AGC 400), Preliminary Design-Build Agreement Between Owner and Design-Builder, or as a stand-alone document that addresses the entire design-build process, including the services otherwise provided under AGC 400.

This standard form agreement was developed with the advice and cooperation of the AGC Private Industry Advisory Council, a number of Fortune 500 owners' design and construction managers who have been meeting with AGC contractors to discuss issues of mutual concern. AGC gratefully acknowledges the contributions of these owners' staff who participated in this effort to produce a basic agreement for construction.

GENERAL INSTRUCTIONS

Standard Form

These instructions are for the information and convenience of the users of AGC 410, 1999 Edition. They are not part of the Agreement nor a commentary on or interpretation of the contract form. It is the intent of the parties to a particular agreement that controls its meaning and not that of the writers and publishers of the standard form. As a standard form, this agreement has been designed to establish the relationship of the parties in the standard situation. Recognizing that every project is unique, modifications may be required. See the recommendations for modifications, below.

Legal and Insurance Counsel

This Agreement has important legal and insurance consequences. Consultation with an attorney and an insurance adviser is encouraged with respect to its completion or modification.

DESIGN-BUILD FAMILY OF DOCUMENTS

In the design-build project delivery method, the owner and design-builder enter into a single contract wherein the design-builder undertakes the responsibility to provide for both the design and construction of the project in conformance with basic requirements which have been set forth by the owner. Design may be performed within the design-builder's organization, or it may be performed by design professionals under a separate contract between the design-builder and architect/engineer (AGC Document No. 420).

The AGC family of design-build standard forms has been carefully coordinated (See diagram). Use of other forms or AGC forms with different publication dates with any of this series of contract documents would require extensive modification and is not recommended.

* To order AGC contract documents, phone 1-800-282-1423. These documents are also available in electronic form through AGC DocuBuilder® software by going to www.agcdocubuilder.org.

7.6.1
AGC Document No. 410—Standard Form of Design-Build Agreement and General Conditions Between Owner and Design-Builder *(Continued)*

POST AGREEMENT SUBMITTALS AND
ADMINISTRATIVE OBLIGATIONS
AGC 410

Paragraph	Responsibility		Task (asterisk indicates task is optional)	Completed Task
	Owner	Design-Builder		
3.1.1		X	Preliminary Evaluation of Project Feasibility	
3.1.2		X	Preliminary Schedule of Work	
	X		Approve milestone dates in schedule	
		X	Recommend schedule adjustments, if needed	
3.1.3		X	Preliminary Estimate; adjustments, if needed	
3.1.4		X	Schematic Design Documents	
	X		Approve Schematic Design Documents	
		X	Identify material deviations in Sch. Design Docs, if any	
3.1.5		X	Obtain Planning Permits, if needed	
3.1.6		X	Design Development Documents	
	X		Approve Design Development Documents	
		X	Identify material deviations from earlier documents, if any	
		X	Update schedule and estimate	
3.1.7		X	Construction Documents	
	X		Approve Construction Documents	
		X	Identify material deviations from Design Dev. Docs., if any	
3.1.8.5		X	Obtain from A/E & Subs: Property rights and rights of use	
3.2.1		X	GMP Proposal	
3.2.3	X		Written comments, if any	
3.2.4	X		Accept GMP Proposal, if appropriate	
3.2.6	X		Authorize Pre-GMP work*	
3.3.1	X		Notice to Proceed	
		X	Pre-GMP list of documents applicable to authorized work, if any	
	X		Approve pre-GMP list & incorporate in Notice to Proceed	
3.3.3		X	Give all legally required notices	
3.3.4		X	Obtain building permits	
3.3.5		X	Keep detailed accounts & preserve for 3 years	
3.3.6		X	Provide periodic written reports	
3.3.7		X	Monitor actual costs; report at agreeable intervals	
3.3.9		X	Final marked up as-built drawings	
3.4		X	Prepare Schedule of Work	
	X		Approve Schedule of Work	
3.5.3			Identify safety representative	
		X	Immediately provide written accident reports, if any	
3.5.4		X	Provide copies of legally-required notices	
3.6.2		X	Report Hazardous Materials, if any, to owner/government	
3.6.4	X		Retain independent testing lab re: hazardous materials	
3.8.3		X	Obtain Certificates of Inspection, testing or approval	
3.8.4		X	Deliver warranties and manuals to Owner	
3.10	X	X	Define Additional Services before performed	
4.1.2.1	X		Info. re: physical characteristics of site	
4.1.2.2	X		Inspection & testing services during construction	

7.6.1
AGC Document No. 410—Standard Form of Design-Build Agreement and General Conditions Between Owner and Design-Builder *(Continued)*

Paragraph	Responsibility		Task (asterisk indicates task is optional)	Completed Task
	Owner	Design-Builder		
4.1.3	X		Evidence of sufficient funds committed to Project	
		X	Notice/stop work without evidence of funds*	
4.2.1	X		Provide Owner's Program	
4.3.2	X		Prompt written notice of defects/errors in requirements or Work	
4.4.3	X		Notify of Changes in Owner's Representative	
5.1	X		Propose subcontractors*	
5.3		X	In subcontracts, provide assignability	
5.4		X	Bind all subs and suppliers to this Agreement	
6.2.2	X		Direct commencement of work before insurance is effective*	
7.1.5		X	Application for Payment (Design Phase)	
	X		Within 15 days, accept or reject Application for Payment	
	X		Pay accepted amounts within 15 days of acceptance of Application	
7.1.6		X	Notice to Stop Work if Owner doesn't pay*	
8.2.10	X		Approve insurance and surety bonds	
9.1.1	X	X	Execute written Change Orders	
9.2.1	X		Issue Work Change Directives*	
9.2.2		X	Submit costs for work per Work Change Directive	
9.3.2		X	Promptly give written notice of minor changes made	
9.4		X	Within 21 days of occurrence, give notice of unknown conditions	
9.5.1.4		X	Maintain itemized account of expenses & savings	
9.6		X	Within 21 days of occurrence, make claim for add'l cost or time	
		X	Within 21 days, make claim for design costs of work not pursued	
10.1.1		X	Application for Progress Payments	
		X	Submit bills of sale and applicable insurance	
		X	Statement of funds disbursed from last payment	
10.1.2	X		Within 10 days, accept or reject App. for Payment (w/reasons)	
	X		Pay accepted amount within 15 days of acceptance of App. for Pymt.	
10.1.3		X	Notice to Stop Work if Owner doesn't pay when due*	
10.3	X		Written reasons for disapproving Application for Payment	
10.4.1	X	X	Certificate of Substantial Completion	
10.5.1	X		Before final payment, request proof bills paid*	
11.2.1		X	Obtain and maintain required insurance	
11.2.4		X	File Certificates of Insurance with Owner before Commencement	
11.4	X		Obtain Liability insurance and provide Certificate of Insurance	
11.5.1	X		Obtain and maintain Property Insurance	
11.5.5		X	Require Owner to provide copy of property insurance policies*	
	X		Give written notice before work if Owner will not have prop. insurance	
11.7.1		X	In subcontracts, require subrogation waivers	
12.1.1	X		Order suspension for convenience*	
12.2.1	X		Undertake work upon 5 days notice if non-performance*	
12.2.2	X		Terminate contract on 5 additional days notice*	
12.4.1		X	Terminate for cause on 5 days notice*	
12.4.2		X	Give 5 day notice and terminate for nonpayment*	
13.2	either	either	File mediation request*	
13.4		X	In subcontracts, require consolidation of cases	
Exhibit 1	either	either	File arbitration demand within reasonable time, if an option*	

7.7   Design-Build Projects in the Public Sector

Government agencies at all levels—federal, state, and local—are increasingly identifying design-build as a favored project delivery system. Since 1994, design-build projects totaling $2.6 trillion, spread over 25 states, have been approved by the Federal Highway Administration alone, and in the year ending 2000, $37.2 billion in public works projects utilized design-build. As of June 2004, 159 bills relating to the design-build process were introduced in state legislatures.

7.7.1   Requests for Proposals

Solicitations from government agencies for design-build projects utilize several different approaches in their selection of proposals and subsequent contract awards:

- *Direct selection.* The process where a design-builder is selected based upon definable, objective criteria such as prior experience, complete scope of work, terms, and of course, price.
- *Best value.* An award based upon a combination of qualitative evaluations and objective criteria such as price.
- *Equivalent design / low bid.* A selection of best value considering technical submissions by the design-build team followed by a critique of its proposal where respondents are afforded an opportunity to change the design and adjust the proposal accordingly.
- *Fixed-price design.* This type of RFP includes the maximum cost of the project, whereby an award will be made based on the best qualitative design proposal meeting those limits.
- *Adjusted low bid.* Upon being selected by the public agency, the price may be negotiated further with the qualified low bidder.

7.7.2   The Two-Part Public Works RFP

This two-part form of project solicitation is being used on the federal and state levels and is, in effect, a process of elimination, a method of reducing the initial bidders down to a "short" list.

Part 1 or Part A will be used to establish and evaluate the bidder's qualifications, thereby eliminating those bidders whose qualifications may not be as strong as the others. The questionnaire portion will require bidders to provide the following information:

- Bidder's technical competence and experience in the type of design and construction being contemplated.
- Documentation of past performance of the proposed design-build team to include that of the contractor, the architect, and engineers.
- The capacity of the team to meet the criteria as outlined in the RFP.
- Response to other factors that may be appropriate to the specific type of project being considered by the Agency.

Part 2 or Part B will be more pragmatic and focus on the design and budget presentations, to include:

- Submission of a technical proposal stipulating that the Agency's program will be achieved.
- Cost and pricing information are commensurate with the design parameters set forth in the technical portion of the proposal.

7.8   The Evaluation Process

Although the evaluation process is wide and varied, many public agencies use a point system, awarding points for each of the criteria contained in the RFP. The State of Maryland's Prince George's County Public School evaluation is illustrative of this weighted process where various points are awarded for experience, team resources, minority participation, and, in this case, a rather unique provision, one that allows for some innovation on the part of the bidder.

Appropriate project experience: Total 35 points (21 required as minimum). Qualifications and experience of the design-build firm to include:

1. Experience with similar design-build projects
2. Experience with other types of design-build projects
3. Management approach for the design-build delivery process
4. Experience with public school design and construction issues
5. Design and construction quality as evidenced by industry awards and recommendations
6. Related project experience
7. Specifically, State of Maryland and Prince George's County major project experience

Team resources and capacity: Total 30 points (18 points required as minimum).

1. Team resources, ability, and capacity to meet this project's design and construction requirements and to complete the project within the schedule and budget
2. Key personnel's experience with similar design-build projects and with the design-build delivery system
3. Design-build team history of working together

MBE compliance: Total 30 points (18 points required as minimum).

1. Past record of MBE participation
2. Proposed plan to achieve MBE participation in design and construction

And the unique provision of this state agency:

Educational support: 5 points (3 points required as minimum)

Experience and willingness to develop and participate in graduate school internship, mentoring, and apprentice programs with local architectural, engineering, and business science student residents.

7.9 Safety Issues

In design-bid-build projects, jobsite safety concerns are the province of the contractor, while adherence to design documents rests with the architect in his or her role of interpreter of those plans and specifications. These roles may become blurred in the administration of the design-build process. Who will be dominant in developing and administering the safety program, and how will quality issues be addressed and resolved? Both deserve special consideration.

7.9.1 The Safety Program

Owners do not want the adverse publicity associated with accidents on their projects. The design consultant and contractor have a legal and moral obligation to provide a safe jobsite. The cost of insurance or potential lawsuits can be costly, so these three reasons should be sufficient drivers to develop and administer a strong safety program.

A formalized safety program will acquaint all site managers, whether they are contractors, architects, or engineers, with the procedures to be followed:

- To provide for a safe working environment to avoid injuries, fatalities, and damage to property.

- To provide guidelines for the dissemination of safety information to comply with OSHA, both federal and state, and the Environmental Protection Agency (EPA), and to develop and implement inspection procedures to ensure compliance with these and any other involved agency safety requirements.

- To avoid the monetary penalties that will be imposed for poor safety performance, not only from government agencies but from lawsuits filed by families of injured or deceased workers and from insurance companies holding policies on the project.

7.10 Quality Issues

The blending of design and construction raises the question of how quality issues will be addressed during design and during construction. During construction the architect's historical role as the Owner's watchdog to monitor compliance with quality standards may now become blurred because the Owner may perceive the architect as serving two masters—the Owner and the design-build team.

In fact, this collaborative effort may actually increase quality, as constructability issues are discovered, addressed, and resolved during the design process, not when they occur during the hectic period of construction. That, at least, ought to be the goal.

Quality issues can be raised, reviewed, discussed, and resolved in the design phase when the contractor, using its experiences on similar projects, brings its concerns to the table rather than leaving them to be uncovered and resolved in the field.

Guidelines need to be established to resolve quality-related matters that occur in the field. Will the architect be given full authority to reject substandard or nonconforming work? How will field modifications be handled?

At the completion of the project, the design-builders must ask themselves:

1. Did the project encompass the Owner's program?

2. Did it meet the Owner's financial expectations?

3. Were the expected quality levels achieved?

The future reputation and success of the design-build team will rest on an Owner's positive response to all three of these points.

7.11   Comprehensive List of Contract Documents Issued by the Design-Build Institute of America

The Design-Build Institute of America (DBIA) has issued the following contract documents:

DBIA Document No. 501—Contract for Design-Build Consultant Services

DBIA Document No. 520—Standard Form of Preliminary Agreement Between Owner and Design-Builder

DBIA Document No. 525—Standard Form of Agreement Between Owner and Design-Builder—Lump Sum

DBIA Document No. 530—Standard Form of Agreement Between Owner and Design-Builder—Cost + Fee with an Option for GMP

DBIA Document No. 535—Standard Form of General Conditions of Contract Between Owner and Design-Builder

DBIA Document No. 540—Standard Form of Agreement Between Design-Builder and Design Consultant

DBIA Document No. 550—Standard Form of Agreement Between Design-Builder and General Contractor—Cost + Fee w/Guaranteed Maximum Price

DBIA Document No. 555—Standard Form of Agreement Between Design-Builder and General Contractor—Lump Sum

DBIA Document No. 560—Standard Form of Agreement Between Design-Builder and Design-Build Subcontractor—Cost+ A Fee w/GMP

DBIA Document No. 565—Standard Form of Agreement Between Design-Builder and Design-Build Subcontractor—Lump Sum

DBIA Document No. 570—Standard Form of Agreement Between Design-Builder and Subcontractor

DBIA Document No. 575—Standard Form of Agreement Between Design Consultant and Design Sub-Consultant (2012 edition)

DBIA Document No. 580—Standard Form of Teaming Agreement Between Design-Builder and Teaming Party (2012 Edition)

Building Information Modeling (BIM) Exhibit—This exhibit to be used when the participants have agreed to utilize BIM to assist with the design and/or planning of the project.

The Preparation and Processing
of Change Orders

8.1   Section Description

A change order is a written document, prepared by an architect, approved by an Owner, the Owner's architect, and the contractor, each of whom agree with the following conditions:

- Recognition of a change in the work
- The cost of the change and its impact on the contract sum, which will increase, decrease, or remain the same
- The impact on the contract time, which will increase, decrease, or remain the same

Although this seems like a straightforward process, it is often fraught with problems.

Before discussing the various situations generating change orders, the first order of business when considering the issuance of a Proposed Change Order (PCO), a Change Order Request (COR), or a Construction Change Authorization (CCA) is to review the contract documents. Both the contract specifications and the contract with the Owner will contain directives relating to change orders.

8.1.1   The Contract Specifications

Quite often Division 0 or Division 1 of the contract specifications includes a section on the preparation and submission of change orders dealing mainly with the process of submitting a change request to the Architect. Some specification sections include specific information such as allowable overhead and profit, or the time frame in which a general contractor must respond to an architect's Request for Quotation (RFQ) or Request for Proposal (RFP). Most deal with the number of copies to be submitted and the time allowed for architect review and response, and leave the intricacies of change order preparation to exhibits or addenda prepared by the Owner's attorneys.

A first step is to review the construction specifications to retrieve any change order information before moving on to the executed contract.

8.1.2   Defective Specifications

It is not unusual for specifications from a similar project to be used by the design consultants on the current project. But sometimes these recycled specifications may not apply to the project at hand. The author had such an experience when reviewing the specifications on a *school* project. The specifications required the installation of support brackets for wall-mounted classroom televisions; however, the specs stated that the brackets should be mounted "so the TV can be viewed by the *patient*."

Defective specifications can result from the following:

- Specifications copied from another project that don't apply to the current one (as in the preceding example)

- Specifications that have expired because of changes in the building code or industry standards (changes in AISC, ASTM, or ACI standards, for example)

- Use of generic specifications possibly obtained from one manufacturer/supplier who may no longer be in business or the use of "or equal" when we know that very few items are truly "equal" to another (possibly being equal only in the eyes of the alternate supplier).

- Specifications that are not practical, such as equipment that once installed cannot be accessed for maintenance, or concrete work installed in such a fashion that it is impossible to remove the forms without damaging the surrounding areas.

- Some specifications are not "specific" and may be subject to numerous interpretations, allowing the "interpreter of the contract documents—the Architect" to establish one that exceeds the reasonableness test.

8.2   Contract-Related Change Order Provisions

Rarely is a standard contract form executed without revisions, additions, deletions or modifications, exhibits, and exclusions, inclusions, and qualifications lists. Because each Owner-Contractor contract for construction is different, the first step in the preparation of a request for a change will involve a review of that contract to discern the portions that impact the change order process.

Relying on information or procedures from a previous project can be dangerous. Some of the more common contract provisions to look out for are:

- Time constraints on change order submissions
- Allowable overhead and profit by the General Contractor, Subcontractor, and lower-tier contractors
- The "Should Have Known" provision
- Unique provisions of a Cost Plus/GMP contract

8.2.1   Time Required for Submission of Proposed Change Orders

The contract generally sets limits on the timely presentation of the contractor's intent to submit a Proposed Change Order in support of a formal change order request.

The reference to change orders in the standard American Institute of Architects (AIA) contract form is Article 7 of AIA A201 General Conditions and will contain a provision for a timely submission of a PCO. A typical such provision will read something like this:

> The delivery of information relating to a change in contract sum or time is to be submitted by the contractor within four (4) business days, unless such timing is impractical based upon the complexity of the issues, in which case the Owner shall grant a reasonable extension of time for delivery of the information.

or

Change order proposals shall be submitted promptly and shall remain firm for a period of not less than 90 days from delivery of the change proposal. Any delay in submission will not justify or constitute the basis for an increase in the Cost of the work or Contract time.

The first "time" provision outlined here should not pose a problem for a contractor, but the second provision regarding price protection is not so innocent. The Owner is requiring the contractor to submit a price that will remain firm for 90 days. In the face of ongoing construction, work may be put in place that will require rework if this change proposal is not accepted in a timely fashion. If that is the case, the contractor may incur added costs in the form of rework or out-of-sequence work that it may not be able to recoup from the Owner.

Assume that a proposed change involves plumbing rough-ins in areas where drywall installation is proceeding rapidly. If an area must be kept open to accommodate this change, will either the plumbing or drywall subcontractor require additional money to perform this extra work if it has to remobilize its crews to that area? Even in the face of such contract language, it might be advantageous for the Project Manager, when submitting a change order such as the hypothetical one above relating to plumbing work, to insert the following provision in the PCO:

Notwithstanding the provisions of the contract, the nature of this proposed change is such that its acceptance or rejection must be made within (X) days in order to maintain the price quoted in this PCO. Due to the progress of plumbing rough-ins and close-in of the walls in the area under consideration, additional costs will be incurred by several subcontractors if they must remobilize for this work. These costs will be reflected in a revised proposal to the Owner.

8.2.2 Allowable Overhead and Profit on Change Order Work

At times, the allowable overhead and profit for change orders is included in Division 0 or Division 1 of the contract specifications, but at other times, these provisions are inserted in the contract in Article 7, when AIA contract formats are used. Some contracts will have a sliding scale for overhead and profit (OH&P). The allowable OH&P may decrease as the value of the work increases. Other provisions can limit the total percentage of OH&P including that of the prime subcontractor, second- and third-tier subcontractors, and the General Contractor.

Provisions to look for will be similar to these:

a. Overhead and profit is limited by the value of the work on a sliding scale:

For the contractor, for work performed by their own forces
Up to and including $100,000   allow 15%
$101,000 to and including $200,000   allow 10%
$201,000 and over   allow 5%
For subcontractors employed by the General Contractor this same sliding scale shall apply.

b. Limitations on general contractors, subcontractors, and second- and third-tier subcontractors.

> The undersigned (General Contractor) agrees that the total percentage for overhead and profit which can be added to the net cost of the work shall be as follows:
> For work performed by the General Contractor's own forces _____%
> For subcontracted work _____%
> For work performed by the subcontractor's subcontractor, ___% of the amount due that subsubcontractor.

c. A clause frequently regarding restrictions on overhead and profit will read:

> Total allowable overhead and profit for the entire change order is limited to ___%

This includes the General Contractor's fee plus all subcontractors.

8.2.3   The "Should Have Known" Provision

Various types of contract provisions are inserted by an Owner for the purpose of reducing the potential for change order work; one such provision could be called the "You Should Have Known" provision frequently included in a Cost Plus GMP project. These types of projects require the General Contractor to prepare an estimate based upon, generally, 65% to 70% complete drawings. The general contractor and its subcontractors must anticipate costs for work that will appear in the remaining 30% to 35% of the design. The Owner will have selected a contractor based upon the contractor's experience with similar projects and this "should have known" provision will go something like this:

> The contractor has constructed several projects of this type and has knowledge of the construction and finished product. The contractor shall immediately notify the Architect and Owner of any details that do not meet good construction practices. By proceeding with the work, the contractor indicates that all details, construction procedures and materials shown or specified in the contract documents are consistent with sound, standard, and acceptable practices.

8.2.4   Provisions Unique to a Cost Plus GMP Contract

Most GMP contracts are prepared and executed prior to the 100% completion of the plans and specifications. One of the main reasons an Owner enters into a GMP contract is that it allows them to finalize a contract sum with a builder before the plans and specs reach final completion. The contractor will base its price on 65% to 80% complete drawings and include in its estimate costs what it and the Architect perceive to be represented in the final 20% to 35%.

But there is such a thing as "design creep"—the Architect adding some minor but costly details or refining the design beyond the original concept anticipated by the contractor. For example, the lobby floor that was bluestone is now Italian marble; the 200 hollow metal door frames are now anodized aluminum.

Experienced GMP contractors include an Exclusion or Qualification list that, after some back-and-forth negotiations, becomes a contract exhibit and establishes scope to a more finite degree.

Without a detailed Exclusion or Qualifications list attached to a GMP contract as an exhibit, the contractor may be held responsible for furnishing components missing from the drawings. A "knowledgeable contractor" should have anticipated these and included them in its price.

This Exclusion or Qualifications list ought to include specific quality levels of finishes such as wall treatments, flooring, cabinetry, windows, doors, hardware—all details that will show up in the completed drawings.

A General Contractor operating under a GMP contract ought to request progress drawings from the Architect to check for conformance with the concept it envisioned in the final design. Enhancements that were not *reasonably* anticipated could be brought to the Owner's attention and resolved prior to the printing of the 100% complete documents.

8.3   Presentation of the PCO or COR Must be Clear and Concise

The presentation of a PCO lacking sufficient detail in both scope change, cost, and time extension is the basis for most Owner-builder conflicts. These problems arise because of one or more of these occurrences:

- Contractors don't clearly define the nature of the change order request and fail to provide documentation complete with drawing and/or specifications changes or a detailed explanation of the proposed change.

- The contractor has not scrutinized all accompanying subcontractor and vendor proposals to ensure that they have correctly identified the changes, their requests for an extra are justified, the costs are reasonable, and the terms and conditions of the contract such as allowable overhead and profit percentages have been met.

- The contractor does not provide cost information with sufficient detailed breakdowns that allow the reviewing party to thoroughly examine, understand, and accept all costs associated with the proposed change.

- The contractor has not provided credit for deleted items or for a reduction in the quantity or quality involved in the proposed change.

- When a time extension is required, the contractor failed to provide enough documentation to support its position.

8.3.1   Defining Scope Change (*page 8.8*)

When changes to the contract scope are requested by the Architect, accompanied by an Architect's Supplementary Instruction (ASI), revised drawings, sketches, and/or specifications, the Project Manager must review these requested changes to determine if there are additional items required to complete the work that may not be reflected in the Architect's submission. For example, a change in the structural framing drawing may impact mechanical riser locations or sizes adding costs to the estimate. Sample letter 8.3.1 represents a typical response to an ASI.

8.3.1
Defining Scope Change

Contractor's Letterhead

ABC Architects
431 Office Park Drive
Central City, Virginia 22504

Attention:

Re: (Project name and number)

Subject: ASI (insert ASI number and date)

Dear Mr. (Ms.) (   ),

We are in receipt of ASI# (number) dated (date) and wish to inform you that we consider this directive to be a change to the contract.

We will not proceed with the changes as shown unless directed. Your response as indicated below will be confirmation of your directive to proceed and the basis for establishing the cost.

With regards,

Project Manager

PROCEED AS A CHANGE TO THE CONTRACT

By:_____ Date:_____

Proceed on basis: (T&M), (Cost Plus Not to Exceed), (Lump Sum)

Cc: File

The Project Manager must review the proposed scope changes to determine whether or not a credit may be due for deleted items or because of a change in materials or equipment. Although not reflected in the Architect's directive, a thorough review of the change by an alert Project Manager could have picked this up. Brought to the Architect's attention, notification of any such credits will further the good relationship between the contractor and Owner.

8.3.2   Scrutiny of Accompanying Subcontractor Proposals

Failure by the Project Manager to review subcontractor quotes incorporated in the cost of the proposed change order is often cited as a major reason for deteriorating relations between the General Contractor and Architect. Missing credits; lack of detail in the subcontractor's quote; and errors in arithmetic, including the wrong percentage for OH&P, all point to poor project management on the part of the General Contractor.

The Owner and the Architect will accuse the Project Manager of merely "passing through" all costs presented by a vendor or a subcontractor and not scrutinizing these proposals before submitting them to the Architect. Material and equipment costs need to be checked for "reasonableness."

Labor rates, in particular, need to be reviewed. It has been this writer's experience that many subcontractors publish hourly labor rates that include "burden" items that may be inappropriate. An improper listing of labor burden items is shown in Section 8.3.3. One is rather easy to pick up on; why would the foreman's truck allowance increase by 50% when applied to premium time wages? Possibly less obvious, would General Liability add-ons per hour increase by 50% when applied to premium time work?

Section 8.3.4 provides a more definitive labor rate breakdown that clearly shows these "burden" items that increase for premium and double time and those that don't.

Even more subtle is the application of Federal Unemployment Insurance Authority (FUTA) and State Unemployment Insurance Authority (SUTA). Both have payroll limits, but these limits vary somewhat from state to state. FUTA is often limited to the first $7,000+ in wages, and SUTA may be limited to the first $10,000 or so, which equates to a worker probably meeting these limits in the first quarter of work in a new year, after which no further contributions have to be made to either the federal or state agencies. But, when reviewing subcontractor wage breakdowns for time and material work performed in July, September, and even November, subcontractors may include contributions to SUTA and FUTA even though the employer is no longer making them. Whether this is by design or inattention to the time when these costs no longer apply, the Project Manager checks these contributions at some point in the mid to later part of the calendar year. A combination of these two items can amount to a low of $2.50 and a high of $4.50 per hour. So these unnecessary add-ons can inflate the cost of a week's labor by $100 to $180, based upon a 40-hour work week. And if these burden items are erroneously increased by $1\frac{1}{2}$ or 2 for premium or double-time work, the total costs can be very high.

These are the kinds of reviews expected of an alert Project Manager and when performed will certainly enhance the Project Manager's relations with the Architect and Owner.

8.3.3   Labor Rate Breakdown with Questionable Increases for Premium Time

ABC Carpentry, Inc

Rate	Foreman	Foreman PT	Journeyman	Journeyman PT	Laborer	Laborer PT
Hourly rate	$35.00	$52.50	$31.00	$46.50	$24.00	$36.00
Truck allow	$1.50	$2.25				
Workers comp	$2.50	$2.50	$2.50	$2.50	$2.75	$2.75
Gen'l Liability	$0.57	$0.85	$0.57	$0.85	$0.85	$0.85
Total	$39.57	$58.10	$34.07	$49.85	$27.60	$39.60

8.3.4   Labor Rate Breakdown with Clearly Defined Increases for Premium/Double Time (*page 8.11*)

8.3.5   Sufficient Cost Information to Allow for a Prompt Review

How can an Architect or Owner adequately comment on a cost proposal that looked like this?

Furnish and install 50 linear feet of drywall, four 3070 doors with hardware, primer and two coats paint on door frames and drywall, stain/seal doors	$9,250.00
15% overhead and profit	1,387.50
Total cost of this change	$10,637.50

But what about this?

Furnish and install 800 square feet of 5/8" gypsum drywall on 25-gauge metal studs, 24' on center	$4,400.00
Furnish and install four 3070 20-gauge HW frames	600.00
Furnish and install four 3070 rotary sliced red oak doors	2,400.00
Furnish and install four hardware sets Type B	600.00
Apply primer, two coats, acrylic latex paint on gypsum wall and acrylic enamel on door frames	950.00
Stain wood doors, apply two coats of sealer on four doors	300.00
Total cost	$9,250.00
15% OH&P	$1,387.50
Total cost of this change order	$10,637.50

8.3.4
Labor Rate Breakdown with Clearly Defined Increases for Premium/Double Time

FEBRUARY. 23, 2006						
			Metropolitan Boston Carpenter Labor Rates			
			March 1, 2006 to August 31, 2006			
			STRAIGHT	PREMIUM	DOUBLE	
		Wage Rate	$ 32.43	$ 48.65	$ 64.86	
		Benefits:				
		Health and Welf.	$ 7.38	$ 7.38	$ 7.38	
		Pension	$ 4.25	$ 4.25	$ 4.25	
		Annuity	$ 7.26	$ 7.26	$ 7.26	
		ATF	$ 0.30	$ 0.30	$ 0.30	
		CARP L/M	$ 0.27	$ 0.27	$ 0.27	
		CITF	$ 0.04	$ 0.04	$ 0.04	
		NLMP	$ 0.02	$ 0.02	$ 0.02	
		MCTP	$ 0.30	$ 0.30	$ 0.30	
		MCAP'	$ 0.05	$ 0.05	$ 0.05	
		Payroll Taxes:				
	FICA	7.65%	$ 2.48	$ 3.72	$ 4.96	
	FUTA	0.80%	$ 0.26	$ 0.25	$ 0.25	
	SUTA	10.90%	$ 3.53	$ 3.47	$ 3.47	
	MMI	0.12%	$ 0.04	$ 0.06	$ 0.08	
		Insurance:				
		Workers Comp.	$ 4.95	$ 4.95	$ 4.95	
		Gen. Liability	$ 2.25	$ 3.38	$ 4.50	
		Umbrella	$ 0.55	$ 0.83	$ 1.10	
		Sub-Total:	$ 66.36	$ 85.17	$ 104.04	
		15% O&P	$9.95	$12.78	$15.61	
		Total:	$ 76.32	$ 97.95	$ 119.65	

Note that for the sake of expediency, we did not add insurance or bond costs (if applicable) nor did we add building permit costs. Although this is a rather simplistic example, it serves the purpose to illustrate the need for a complete breakdown of costs when submitting a proposed change order.

Owner-directed changes are rather straightforward to deal with, but other conditions that arise during the construction process are not so clear cut.

8.4   Conditions That Create Change Orders

Site work and site-related conditions account for many change order requests, generally initiated by the General Contractor. These types of change orders dealing with unforeseen subsurface conditions, differing conditions, or changed conditions are complex, because they not only often require an interpretation by all parties that is deemed "reasonable," but also often result in radically differing interpretations. These types of change order requests often lead to protracted discussions and negotiations between Owner, design consultants, and contractor that have the potential to delay the work, thereby setting the stage for a delay claim.

Because of the tendencies of these types of change orders to escalate to a dispute and quite often further to a claim and litigation, we will not deal with them in depth in this section; see instead Section 9 on "Construction Disputes, Claims, and Resolutions."

Delays created by events beyond the control of the contractor and resultant requests by an Owner to maintain the original completion schedule set up the process referred to as *acceleration*; this situation is also discussed in Section 9.

Lack of properly coordinated drawings is another problem that can become very complex; it generally surfaces when construction is proceeding at a rapid pace and the need for properly coordinated drawings becomes essential to maintaining progress. However, the widespread use of Building Information Modeling (BIM), a 3D method of design development, has had a major impact on coordination problems, greatly reducing them by virtue of its design development methodology. New Section 11, on "Building Information Modeling," delves into this process and explains how coordination problems, frequently referred to as "clashes" or "conflicts," have just about disappeared, and coordinated drawing issues have become less of a problem.

Some of the conditions set forth in the following list have the potential to provoke strong disagreements between contractor and Owner and can quickly escalate to a dispute or threat of litigation:

- Errors and omissions in the contract documents
- Unforeseen subsurface conditions
- Differing or changed conditions
- Drawing coordination problems
- Delays in the construction process, whether generated by Owner, design consultant, or contractor

- Acceleration of the project when delays occur
- Winter conditions
- Requests by the Owner or contractor to proceed with premium time work
- Latent conditions
- Incomplete design
- Owner-generated changes

8.4.1   Errors and Omissions

A standard malpractice insurance policy covering professional liability refers to errors and omissions as "errors of commission" and "errors of omission," and broaching the term "errors and omissions" to an architect is a sure way of damaging any developing relationship.

There is a fine line between an architect's responsibility for design and a contractor's responsibility to provide professional services.

Some of the more common causes of errors of omission and/or commission in the design documents are:

- Conflicts between small and large details and between the written specifications and the graphic drawings

- Boilerplate statements or details from other projects that don't apply to the project at hand

- Poor communication between design consultants, so that when a change is made by one consultant (say, the structural engineer), this change is not conveyed to others, thereby causing such problems as improperly placed floor penetrations, chases in the wrong place, etc.

- Failure to thoroughly review design development and final drawings with the Owner to ensure that its program is being met

- Allowing insufficient time for a thorough review of all design documents before the plans and specs are presented to the contractor for pricing

The above list, incidentally, was prepared by a *design professional* as reported to an editor at *ENR* magazine in 2005, to make the point that some of the problems leading to change order requests by a contractor are caused not by what architects are *doing* but by what architects are *not doing*.

Taking a reasonable (there's that word again) approach when considering a change order request for perceived errors and omissions is the proper, and more practical, approach.

Architects' malpractice insurance premiums are extremely high, and therefore architects generally carry high deductibles, which means that claims for errors and omissions will most likely be paid out of the Architect's pocket. The Project Manager should explore alternative solutions to resolving some error and omission problems such as submitting value engineering suggestions to absorb all or a portion of the costs created by these deficiencies.

8.4.2   Drawing Coordination Problems

Unless an Architect has civil, structural, mechanical, and electrical engineers on staff, it will subcontract any or all of these design responsibilities to other offices. Coordination meetings will be held by the Architect as the team leader from time to time to ensure that all systems and components are tracking properly and that problems, as they arise, will be resolved.

Without a careful and complete review of overlays and layers of design by the Architect, drawings may be released that lack responsible coordination. When this coordination process breaks down, the problems created can be minor with little or no cost implication, or they can be monumental and trigger an entire series of extra costs:

- Rework of work already in place will require remobilizing forces and create associated costs for retrofitting..
- Lowering ceiling heights or increasing the size of chases and their enclosures for MEP risers or run-outs will have associated costs.
- Delays while drawings are being coordinated and the contractor is directed by the Architect/Owner to continue working in areas not impacted by the redesign will increase costs.
- To make up for lost time, the contractor may be directed to accelerate its work, setting in motion the potential disputes that may lead to a claim.

When substantial drawing coordination issues are discovered, the Architect should be quickly notified, in writing, of the nature and extent of the problem. Section 8.4.3 deals with a general coordination problem, and Sections 8.4.4 and 8.4.5 are more specific in nature.

8.4.3   Sample Notification to Architect of Coordination Problems *(page 8.15)*

8.4.4   Sample Letter to Architect with Specific Drawing Coordination Problems *(page 8.16)*

8.4.5   Sample Letter Advising Architect/Owner of Coordination Problems *(page 8.17)*

8.4.6   Delays in the Construction Process

There are various categories of delays:

- *Excusable delays.* The contractor is granted a time extension but no monetary compensation.
- *Concurrent delays.* Delays occur, but neither the contractor nor the Owner can collect monies for damages due to these delays.
- *Compensable delays.* Delays for which either the Owner or the contractor is entitled to additional monies.

8.4.3
Sample Notification to Architect of Coordination Problems

Contractor's Letterhead

Addressed to the Architect

Attention: Responsible Individual

Re: Project Name and Number

Subject: Notification of Coordination Problems

Dear Mr. (Ms.) (   ),

During the coordination process required by (state either the contract article or the specification section dealing with the contractor's obligation to provide coordination drawings for the Architect's review), we developed a number of issues that, in our opinion, go beyond the normal coordination process.

We are in the process of defining these issues and will submit them to your office for review and resolution not later than (date).

At this time we cannot quantify their impact on cost or schedule, but we have established PCO# XX to begin to track these costs.

With Best Regards,

Project Manager

8.4.4
Sample Letter to Architect with Specific Drawing Coordination Problems

Contractor's Letterhead

Addressed to Architect

Attention: Appropriate Individual in the Firm

Re: Project Name and Number

Subject: Coordinated Drawing Concerns

Dear Mr. (Ms.) (   ),

We call your attention to (drawings in question, e.g., Drawings A1–A4 and Drawings S-1 and S-2), which in our opinion require further coordination.

Provide Examples
Example 1: The depth of beams on Drawing S-2 will not allow for a 10″ ceiling height in Rooms 100–110 on Drawing A-1
                                        or
Example 2: There are dimensional differences between structural and architectural drawings. Dimensions on drawings S-5 and S-7 do not correspond with the slab edges as shown on related architectural drawings.

State the Problem
In Example 1 above, if you will allow an 8′ 4″ ceiling height in these rooms, we will proceed accordingly. However, if you wish to maintain the 10′ ceiling height, we will need to consult with our subcontractors and your engineer to effect a solution satisfactory to all parties.
In Example 2 above, we will need confirmation of the correct dimensions and the necessary sketches to clarify any changes that may be required.

Advise of Any Potential Cost/Time Implications
Example 1: If the Architect insists on 10′ ceiling height, then after consulting with our subcontractors and field supervisors, we will advise you of the additional costs and time implications involved in effecting your request.
Example 2: When we receive the correct dimensions and related changes, we will assess the cost and time implications of this change in conditions.

With regards,

Project Manager

8.4.5
Sample Letter Advising Architect/Owner of Coordination Problems

Contractor's Letterhead

Addressed to Architect or Owner, if Applicable

Attention: Appropriate Individual

Re: Project Name and Number

Subject: (Most common coordination problems could be any one of these)

1. Between structural and architectural drawings
2. Between architectural and mechanical/electrical/plumbing riser locations or sizes
3. Between architectural and mechanical regarding plumbing, fire protection, and HVAC duct rough-ins above the ceiling

Dear Mr. (Ms)   (   ),

If no. 1, we wish to advise you of a conflict in the dimensions on drawing(s) (state the exact structural and architectural drawings).We will need to have these dimensional discrepancies resolved before proceeding with work in the affected area. Please respond not later than (date) in order to avoid any potential delays.

If no. 2, we wish to advise you of a conflict between the size and location of slab/floor penetrations for (name trade, plumbing, HVAC, etc.) as shown on drawing (structural or architectural) and [corresponding MEP drawing(s)]. We will need confirmation of the appropriate size and location for these risers. Please respond not later than (date) in order to avoid any potential delays.

If no. 3, we have completed our coordination drawings pertaining to above ceiling MEP roughs-ins and find conflicts that prevent their installation within the designated space while maintaining the required ceiling height.

We would like to set up a meeting at (time) (date) to present alternative solutions for your review, comment, and approval. Please advise if this date is acceptable so we can proceed accordingly.

With regards,

Project Manager

We will deal with the first and last categories.

Excusable delays will become important when there is a liquidated damages (LD) clause in the contract. When the contractor is bound to complete the project by the date certain in the contract or have LDs assessed by the Owner, obtaining an extension of time via a change order is very important. Excusable delays include:

- Acts of God
- Fires or other significant accidents
- Illness or death of one or more of the contractor's principals or Owners
- Transportation delays over which the contractor has no control
- Labor strikes or disputes
- Unusually severe weather

The steps to take when considering the submission of a delay change order request are:

1. Submit a written request to the Architect for a time extension (even if it appears that the project may be ahead of schedule at that time, events can alter that situation quickly).
2. Severe or adverse weather issues should be presented by comparing the unusual weather conditions against a 10-year average for the specific time of the year in the geographic location of the project.
3. If no response is received from the Architect in one week, send a follow-up letter.

8.4.7   Winter Conditions and Other Weather-Related Matters

In those geographic areas where winter work means working in subzero or freezing temperatures, some Owners prefer to exclude any costs for these winter conditions from the construction contract. Because it is nearly impossible to predict the severity of winter work, it is either excluded or some Owners prefer to have the contractor include an "Allowance" that will be reconciled as costs are incurred.

In either case, the contractor must clearly identify, isolate, and document all labor, material, and equipment costs pertaining to winter conditions. Daily Reports filed by the Project Superintendent must include a statement of weather conditions and, preferably, site temperatures at the start of the day, midday, and early afternoon to provide evidence of adverse weather conditions. The nature of the work being undertaken, the reason for temporary heat and protection, and a brief explanation of what workers, materials, and equipment will be required should be forwarded to the Architect or Owner. A copy of this memo will serve as documentation when the cost of work is submitted as a proposed change order.

In the case when winter conditions is an allowance, the Project Manager would be wise to alert the Owner when winter conditions costs have, say, consumed 75% of the allowance, anticipating that costs will exceed the allowance.

When this work is being conducted on a Time & Material basis, a cost memo to the Owner will avoid any surprises when the final proposed change order for this work is submitted.

8.4.8   Request by the Owner for Premium Time Work

A contractor may alert an Owner to the need for premium time work to avoid other increased costs and may do so by sending a letter to the Owner stipulating the reason why premium time work at this point is advisable and the approximate cost of same. Upon receiving authorization from an Owner to proceed, detailed daily labor reports and material receipts attached to that authorization will form the basis for the change order request. Section 8.4.9 shows a sample Authorization to Proceed with Premium Time Work.

8.4.9   Sample Authorization to Proceed with Premium Time Work (*page 8.20*)

8.4.10   Latent Conditions

Latent conditions can take the form of *hidden conditions* if the project involves renovating or rehabbing an existing building. The design consultants don't have X-ray vision, and until a plaster or sheetrocked wall is removed to expose the substrata, the condition or type of sub-base cannot be known. For example, the contractor, basing its estimate on a *reasonable* assumption, may find that a newly exposed masonry wall has deteriorated to the point where it may be less expensive to replace it than to repair it. Most Owners and Architects can come to some accommodation with the contractor to agree to a sum to repair/replace such a latent or hidden condition. Some subsurface conditions that remain undetected, even though extensive test borings accompanied the bid documents, may also be classified as latent conditions.

A contractor bears some responsibility when conducting a prebid inspection of an existing structure that will be updated or upgraded. The contractor needs to closely inspect some existing conditions, possibly even asking the Architect if a portion of the wall, floor, or ceiling can be demolished to obtain better insight into the makeup of the substrata. It is important that the contractor request the Architect to issue a notification to all bidders that they will be able to inspect that portion of the exposed structure. A little knowledge can be dangerous if you, as a bidder, add costs to correct a problem you uncovered and other bidders have not been made aware of that same problem.

8.4.11   Incomplete Design

Incomplete design problems can range from lack of items on a finish or door schedule to even improper identification of a building component. Room 410, as an example, may not have been included on the Finish Schedule, with the inadvertent omission not picked up by either Architect or contractor. When the finishes

8.4.9
Sample Authorization to Proceed with Premium Time Work

Contractor's Letterhead

Addressed to: Architect or Owner as Applicable

Attention: Appropriate Individual

Re: Project Name and Number

Subject: Request to proceed with premium time work

Dear Mr./Ms:

Reason No.1—Based upon a current weather forecast, it would appear that we may experience (heavy rain) (freezing temperatures) (high winds) tomorrow which could delay the installation of (concrete slabs, underground utilities, precast panel erection, etc.).

To avoid project delays due to the impending inclement weather, we request authorization to work extended hours in advance of this weather. We anticipate working a crew of (type of tradesman, number of workers) for approximately four hours each, at time and one-half (or double time) rates. Total cost of premium time work will not exceed (put in a safe number).

Reason No.2—We have received verbal authorization by (Architect or Owner) to proceed with (describe the extra work). We feel it is best to proceed with this work on an overtime basis to avoid conflicts with other trades. We anticipate that the addition of overtime labor rates will add approximately (include dollar figure) to the cost of the work which was budgeted at ($ value or work performed on regular time basis).

Reason No.3—State any other reason why it will be advantageous to the Owner to proceed with premium time work.

Your prior authorization is requested in order for us to proceed with the work on a premium time basis.

(Due to the urgency of the matter, a verbal authorization may be required, to be followed up with this letter where authorization is included, or this letter can be e-mailed to the Architect/Owner, thereby creating written documentation.)

Best regards,

Project Manager

are defined and additional costs are presented, a change order of this type might be contested by the Architect, who argues that the contractor should have picked up this omission upon drawing review. The Architect, however, will have a difficult time defending against the contractor's retort, "Why didn't you pick this up when you prepared the Finish Schedule?"

Ceiling, wall, floor, and door schedules can become a point of contention between contractor and Architect if neither has paid sufficient attention to reviewing schedules—whether they pertain to doors, windows, hardware, paint, ceilings, or flooring. Quite often, though, errors or omissions in schedules can be resolved when both Architect and contractor approach the problem *reasonably*, either through an extra-cost change order or a trade-off in some other area to cancel out these costs.

8.4.12   Owner-Generated Costs

Nearly all construction contracts contain provisions that allow an Owner to make changes to the contract scope. However, these changes must be made in a timely fashion so as not to impede the contractor's progress. One way to deal with these types of changes in scope is to provide sufficient detail to allow the Owner to understand not only the costs associated with the changes, but also the time frame in which a decision must be made so as not to affect the project schedule. It is simply a matter of adding a sentence to the change order:

> The contractor requires a decision to be made relating to acceptance/rejection of this Proposed Change Order not later than (date) so as not to impede normal progress of the project.

8.5   Types of Change Order Requests

Change order requests can be lump sum, which may be preferable, or Cost Plus where all labor, material, and equipment costs must be clearly defined and documented. On those occasions where it is necessary to commence the change order work quickly and before a lump sum amount can be established, the Cost Plus Not to Exceed approach is best.

8.5.1   Review the Contract as a First Step

Although both the lump sum and Cost Plus GMP approach seem rather straightforward, the Project Manager ought to review the contract, specifically any exhibits, to determine if unit prices apply, if there are any allowances pertaining to items in the proposed change order, or if any of the work falls within the scope of a listed Alternate.

8.5.2   Unit Price Change Orders

Contracts involving primarily civil engineering projects will generally contain a whole list of unit prices, some of which may have graduated pricing based upon quantities being charged. Vertical building contracts may include unit prices for items other than site-related costs, so it is a good idea to check the contract, once again, as costs are being prepared.

As stated previously, when buying out the job, any unit prices in the contract should be negotiated with the Owner and passed through to the appropriate vendor or subcontractor.

8.5.3   Minor No Cost/Time Impact Change Orders

Some minor changes requested by either the Owner or Architect or by the contractor may have no cost or time implication, but if they change the scope of the contract, it is best to issue a No Cost/Time proposed change order to document the fact that the scope of work has changed.

8.6   Costs to be Considered When Assembling a Change Order

When preparing a change order request in response to the Architect's directive to add five hollow metal doors and frames, it is rather obvious to include the cost of frames, doors, and hardware along with the costs to purchase, install, and paint, but did the Project Manager include the cost to unload the material and distribute it to the area in the building where it is to be installed? It is important to review all costs associated with a proposed change to ensure that they are all-encompassing. Consider the following.

8.6.1   Direct Costs

Hard dollar costs required to complete the work include the cost of:

- Contractor-owned equipment, whether active or idle, at rates no higher than published rates, including fuel costs
- Rental equipment, including replaceable accessories, such as diamond blades and jackhammer points
- Insurance premium increases
- Labor, fringe benefits, payroll taxes [*Tip:* When figuring the hourly rate for a salaried employee, remember that the employee works only 50 weeks (assuming a two-week vacation period), so divide the employee's yearly salary by 50 weeks and again by 40 hours to get the hourly rate. The same may also apply to key tradesmen such as foremen or crew leaders.]
- Material rehandling, pickup and/or delivery and distribution costs
- Fasteners (those stainless-steel Hiltis can be expensive)
- Photographs, postage, express deliveries, reproducible costs
- Safety equipment required
- Subcontractor costs
- Temporary heat and/or temporary protection
- Travel, including auto, truck, tolls, parking

- Utility costs for the building under construction (These would actually apply when change order work occurs at the end of a project and before utilities are turned over to Owner. They could also apply for Saturday and Sunday work when utilities are required and the contractor is required to pay for these off-hours costs.)
- Winter conditions or summer conditions (need to air-condition), if applicable
- Bond premium increases, building permit increases

8.6.2   Indirect Costs

Indirect costs are those that are field- and home-office-related:

- Project management costs
- Project Engineer
- Project Superintendent
- Field office expenses
- Temporary utilities
- Accounting and costs to prepare estimate

8.6.3   Impact Costs

Impact costs are those associated with changes that impact performance:

- Lost productivity due to trade stacking or other inefficiencies
- Idle equipment and idle equipment maintenance
- Lack of availability of skilled tradespeople
- Cost of disruption to the orderly flow of work
- Cost to work out of sequence
- Cost of extended warranties on equipment installed during the project that may expire before the standard one-year warranty

8.7   The Construction Change Directive (CCD)

This somewhat different approach to the preparation of a change order first appeared in the 1987 edition of AIA General Conditions Document A201 and since that time has demonstrated clear advantages to the Owner, Architect, and contractor.

8.7.1   Purpose of the Construction Change Directive

The CCD allows the Architect to authorize extra work when there is either no time to compile and submit a lump sum price or when there is a lack of agreement over the terms of the proposed change order, which usually translates into disagreement over cost.

As stated in Article 7 of AIA Document A201, the contractor is authorized to commence the extra work and total costs are to be determined by one or more of the following:

1. A mutually accepted lump sum price.

2. Using applicable unit prices contained in the contract.

3. Costs mutually acceptable to both parties—in effect, a negotiated sum.

8.7.2   How a CCD Is Initiated

An Architect or Engineer can initiate a CCD by sending a directive to the contractor setting forth the work to be performed and the documented "costs" that will be acceptable for inclusion; this directive may merely refer to Article 7.3 of AIA Document A201, which stipulates the cost to be reimbursed under a CCD. If an AIA document is not being used, it is best to refer to Article 7.3 as a basis for costs due for reimbursement.

A General Contractor can initiate a CCD by sending a letter to the Architect. Section 8.7.3 provides an example.

8.7.3   Sample Letter Initiating a Construction Change Directive (page 8.25)

8.7.4   Allowable Costs in a CCD

Article 7.3.6 AIA A201 states acceptable costs, when properly documented, as:

- Cost of labor, including all appropriate labor "burden" items
- Cost of materials, equipment, and supplies, including cost of transportation to the site
- Rental costs of machinery and equipment, whether rented from the contractor or a rental company. The cost of hand tools is excluded.
- Cost of bond premiums, if applicable, insurance, permit fees, sales and use taxes related to the work
- Added costs of supervision and field personnel directly involved in the change order work

8.7.5   Advantages of the CCD

There are many advantages to a contractor when work is authorized via a CCD:

1. Work can proceed immediately upon acceptance of the CCD, thereby reducing the possibility of the added cost, or dispute over costs to remobilize or resequence crews.

8.7.3
Sample Letter Initiating a Construction Change Directive

Contractor's Letterhead

Addressed to: Architect

Attention: Appropriate Individual in the Firm

Re: Project Name and Number

Subject: Proceeding with Requested Change Order Work via a CCD

Dear Mr. (Ms.)   (   ),

We refer to (the document from the Architect requesting the change order/extra work).

Inasmuch as there is no agreement on the cost of this work, we request authorization to proceed on the basis of a Construction Change Directive (CCD).

Upon receipt of your authorization, we will commence the work (as outlined above or in a brief description).

With regards,

Project Manager

Proceed with Work as a Change to the Contract

By:_____   Date:_____
    Architect

Proceed on basis of a CCD

2. As long as all costs are fully documented, there should be less dispute over the final cost of the extra work.

3. The CCD allows for reimbursement of supervision and field office personnel costs, costs that many architects dispute when following the conventional PCO approach.

4. And another important point, any costs not in dispute can be included in the current requisition, an option not usually allowed when the conventional PCO route is taken. Quite often a subcontractor's payment for extra work will be held up because the Owner disputes the cost of the carpentry work in the change order. The CCD allows undisputed costs to be requisitioned.

8.8   Roadblocks to Acceptance of Change Orders

What should be a rather straightforward approach to the submission, review, and acceptance of a change order is often contentious, leading to an atmosphere of distrust between the Owner and contractor. Viewing the change order process from the perspective of an Owner, Architect, and Engineer may shed some light on the objections frequently voiced by those entities. The builder ought to keep these perspectives in mind when encountering opposition to any proposed change order request.

8.8.1   The Owner's Perspective

Owners of construction projects select a team of architects and engineers to develop a set of plans and specifications to meet their building program goal. In their pro forma financial analysis, the cost of design and construction and a contingency for unforeseen events will be included. Unless the Owner adds or deletes scope of work after a contract is awarded to the General Contractor, the Owner's question will be, "Why is the contractor sending me a change order?" For some obvious and minor omissions or errors in the contract documents, the Owner may have a valid point, but when serious coordination problems arise, or blatant drawing or specification deficiencies are discovered, the contractor may have a valid right to request an increase in the contract sum and completion time. Although some contractors have exploited the change order process to seek short-term profits at the expense of long-term reputation, most contractors would rather avoid change orders altogether because of their generally disruptive nature.

But the Owner has a point, and the contractor's response to "Why is the contractor sending me a change order" must be very explicit in detailing the rationale behind the request, accompanied by reasonable, complete, and easy-to-review costs.

8.8.2   The Architect's and Engineer's Perspective

The preparation of today's project plans and specifications is very complex, not only because of sophisticated structural and mechanical systems, but also because of the network of building codes and local and federal laws and regulations that require compliance. The perfect set of plans and specifications will be difficult to produce, and the standard contract documents make reference to this fact. Minor errors and omissions are expected, but many of the problems that morph into change orders relate to lack of proper review of the drawings before being released for bid—clearly not a contractor responsibility.

But consider the Architect's view. The Architect wishes to develop or maintain a good relationship with the client, and that client will ask the same question, "Why is the contractor asking for an extra; why did you miss this on the drawings?"

Reasonableness must prevail with all parties: the Architect must recognize that he or she owes the contractor a "reasonably" complete set of documents, the contractor must not expect a perfect set of plans and specs, and both ought to work out their problems accordingly.

The contractor can often mitigate some of these extras by developing a few value engineering suggestions to offset the added costs of the extra. By requesting their subcontractors to look for any VE recommendations, the contractor will have made an effort to assist the Architect in dealing with any unexpected added costs.

8.8.3   The Contractor's Perspective

The contractor, particularly in a competitive open bid situation, will have based its estimate on its interpretation of the requirements of the plans and specifications and also on the bidding philosophy of its competitors. Standard AIA contract provisions require the contractor to review the plans and specifications and notify the Architect of any errors, inconsistencies, or omissions; however, the contractor is not required to ascertain that the plans and specs meet applicable building codes. When working in a competitive environment, though, there is a fine line as to what to *assume* and what to *include* in the estimate.

Working in a very competitive industry, contractor profit margins are slim. According to the Construction Finance Management Association's 2006 *Construction Industry Annual Financial Survey*, a composite of 502 contractors provided detailed financial information responding to their survey. Among those industrial and nonresidential contractors responding, Best in Class reported 6% gross margins, operating margins of 1.3%, and net margins before taxes of 1.6%. These figures don't leave much room for error and do not allow a contractor to absorb many unanticipated costs.

If more Owners were aware of these low margins, they might be more sympathetic to some of the contractor's change order requests. Major omissions, inconsistencies, or errors in the bid documents should rightfully not be the responsibility of the contractor to absorb.

8.8.4   The Solution

One word, *reasonable,* is the key to the resolution of many disputed change orders. Can the Owner, Architect, and contractor agree that problems arising from deficiencies in the contract documents and the costs to correct them as proposed by the contractor are reasonable?

Can *reasonable* tradeoffs offset the cost of any proposed extras? Can a material or product be substituted, the savings of which would partially or fully offset the cost of the extra?

Two legal terms can be offered if there is reluctance to approve a reasonable request for an extra: "quantum meruit" and "unjust enrichment."

8.8.5   Quantum Meruit and Unjust Enrichment

Quantum meruit (pronounced "quantum mare-o-it") is referred to as a "quasi-contract" method of recovery of costs associated with change order work. It is called quasicontract because it does not relate to specific contract terms or language, but rests upon the principle that an Owner receives value from the contractor's actions. The contractor must prove that the Owner has benefited from the extra work that was completed and accepted. The principle of quantum meruit is based upon the concept that an Owner is obliged to pay a contractor for the benefit received by the incorporation of work into the project.

The term "unjust enrichment" might easily be explained as "you can't get something of value for nothing." If the contractor proceeds with the extra work based upon a good faith verbal commitment from either the Architect or Owner and either of these entities denies making such a verbal agreement, the Owner will have been clearly *enriched* by this contract modification because value has been added to the project. The Owner may argue the amount of the "value," but the Owner will have a difficult time disagreeing that some value has been added to the project.

Change orders are nearly unavoidable, and when the test of "reasonableness" is applied by all parties to the contract, a smoother flow of change order requests should follow.

8.9   Protocol for Change Orders, Premium Costs, Winter Conditions

Change Orders

1. Each proposed change order is to contain a brief explanation of the nature of the change and who initiated the request (Owner, A/E, contractor). Attach all supporting documentation (e.g., letter from Owner, SK from A/E, request from subcontractor, etc.).

2. If the scope of the work is increased or decreased, state the prior condition and proposed condition and the delta between prior and proposed.

3. All costs submitted by the contractor for self-performed work and costs submitted by subcontractors shall be broken down into labor (hours x rate), materials (number if applicable, lineal or square feet if applicable), and cost.

Overhead and profit to be added to the "cost." Check contract to confirm percentage OH&P.

4. Equipment, if employed. Indicate whether owned or rented by contractor. List the number of hours/days x applicable rate. Provide receipt for delivery to jobsite and receipt for pick-up.

5. Review of subcontractor proposals shall include scrutiny of labor hourly rates to ensure that these rates and burden are proper and reasonable. When premium time rates are being charged, verify that the upcharge is correct with respect to labor rate and applied burden. When FUTA/SUTA limits have been reached, hourly rates should reflect deletion of this item from the labor burden computation.

6. If work is being done on a Time & Material (T&M) bases, follow the procedures indicated below.

7. If requested by Owner, allow Owner's representative to be present when the change order negotiations with subcontractor and/or vendor involving change order work are taking place.

Time and Material Work Authorized by Owner

1. When the Owner has authorized work on a T&M basis, the contractor's supervisor is to obtain daily tickets for all T&M work, including self-performed work. Each ticket should include worker's name, trade category (carpenter, laborer, etc.), number of hours worked, and task performed. Each ticket must be signed and dated by the contractor's supervisor; this validates the contents. Receiving tickets for all material are to be signed by the contractor's supervisor and attached to labor tickets.

2. For subcontracted work, daily tickets should be submitted by each subcontractor, listing tradesman by name and category (if this is union labor, list by apprentice, journeyman, etc.). The ticket should state the number of hours each person worked and the task(s) performed. This ticket is to be signed by the subcontractor's foreman and by the contractor's supervisor. Receiving tickets for all materials are to be signed by the subcontractor, confirmed by the contractor's supervisor, and attached to the labor tickets.

3. Equipment, either owned or rented, requires a ticket indicating date/hour brought on site, when use commenced, number of hours employed, and date/time when use ceased. The hourly rate is to be included on the receipt, which must be signed by the contractor's supervisor.

Winter Conditions (if applicable)

1. Indicate the operation taking place requiring winter-condition work.

2. Provide a log with temperature readings at 7:00 A.M., noon, and 2:00 P.M.

3. Provide daily tickets for labor as outlined earlier in this subsection.

4. Provide a list of materials used and type of fuel consumed.

5. Provide a list of equipment used, number of hours each piece was in operation, and the hourly rate for each piece.

6. All such tickets are to be signed by the contractor's supervisor.

7. The contractor should accumulate and present all tickets to the Owner on a weekly basis, with a running total for winter conditions and the task requiring winter conditions (e.g., concrete, taping of sheetrock, painting).

Construction Disputes, Claims and Resolutions

9.1   General

At times, even the best efforts put forth by a Project Manager or construction executive fail to resolve a disagreement with an Owner. Disagreement may concern the cost of the work, or whether the scope of the construction contract has been exceeded or reduced, or whether problems were uncovered in the plans and specifications as construction accelerated. Such disagreements can quickly escalate to disputes and the filing of claims, if not dealt with quickly by reasonable people.

Before the increased use of Building Information Modeling (BIM) design software, many disputes occurred because of drawing coordination or "conflict" problems. Because some drawings were not properly coordinated between architectural, structural, and mechanical and electrical plans, conflicts between various systems had to be brought to the Architect's attention, often as installation was taking place during construction. Gas or water piping locations, if not changed, would bump into ductwork or a steel beam; there were instances where the mechanical and electrical work would not fit within the space allotted above the ceiling or in a vertical chase. BIM eliminated many of these kinds of coordination and conflict problems that often resulted in disputes and claims.

However, two of the lingering problems that BIM cannot entirely solve involve site work, most often issues involving differing or changed conditions and unforeseen subsurface conditions. Many of the construction disputes and claims we see today relate to these two conditions. Test borings, test pits, and geotechnical investigations and reports can't cover an entire construction site, so a certain amount of guesswork remains in every contractor's excavation and site work estimate.

Other owner-contractor disputes and claims can vary from problems occurring during the bidding process, particularly on public works projects where there is little or no ability to commit to the give-and-take required to successfully negotiate differences as in the private sector.

The court decisions referenced throughout this chapter may provide the background for some legal decisions and how the legal community and the courts view various disputes and claims, but they are not meant to take the place of advice obtained from your company's legal representation in matters relating to a specific claim.

9.1.1   State of Global Construction Disputes

The EC Harris Company, a global built asset consultancy company headquartered in London, England, reported on May 28, 2012, that global construction dispute values fell in 2011 in most parts of the world—except in the Middle East, where the value of the average claim more than doubled from $56.25 million in 2010 to $115 million in 2011. Harris indicated that this increase was due in part to the fact that many mega-projects were nearing completion; the failure to properly administer those contracts was the most common cause of these disputes.

In the United States, Harris discovered that the average value of a construction claim *decreased* between 2010 and 2011, falling from $64.5 million in 2010 to $10.5 million in 2011. According to Harris, this reduction in claims was due, in part, to two factors: increased emphasis on avoiding and/or mitigating any claims arising in both the private and public sectors, and a generally depressed construction market.

9.1.2   Top Five Causes of Disputes

The top five causes of disputes in the United States, as found by the Harris study in 2011, were:

1. Ambiguities in the contract plans and specifications.
2. Incomplete design information.
3. Conflicting party interests.
4. Failure to make interim awards on extensions of time and provide adequate compensation when added time was recognized.
5. Failure to properly administer the contract (a catch-all category that could cover a wide range of sins committed by any or all parties).

9.1.3   Party-to-Party Negotiations Work

Most interestingly, Harris found that party-to-party negotiation resolved many of the disputes and claims (the second and third most popular resolution methods were mediation and arbitration). The study also found that more emphasis was placed on tackling and resolving issues before each party's position hardened and the dispute escalated into litigation.

The new contract formats developed by the American Institute of Architects (AIA) and the Associated Contractors of America (AGC), in conjunction with input from other construction/design and engineering associations, stress collaboration among all parties to the construction process. The facts, as developed by EC Harris, seem to support the effectiveness of this collaborative environment.

9.2   When Negotiations Fail

There will be times when negotiation has run its course and either one party or both decide to ratchet up the resolution process. Resolution of construction claims and disputes can be categorized into five methods:

1. Utilizing a contract administrator, generally specified in the contract as the Architect, to bring all parties together and present their informed decisions on how to resolve the issue at hand. Because the Architect has been engaged by the Owner, the contractor may be of the opinion that an impartial resolution will be difficult to achieve. Experience has shown that everyone must approach resolution fairly and equitably.

2. The most tried-and-true method of resolving a claim is direct negotiation among all parties who sincerely have as their goal an equitable resolution. It is often said that when each party walks away from a settled negotiation feeling that it has given more than it has received, it is a successful negotiation.

3. Mediation. In this process, a disinterested third party is engaged to achieve resolution.

4. Arbitration. Many contracts include provisions for *binding* arbitration, in which the arbitrator's decision is final; if that decision is not complied with, the aggrieved party can obtain the necessary judgment from a court to enforce the arbitrator's decision. The American Arbitration Association (AAA) has quite a bit of information explaining the arbitration process.

5. Litigation. When all other means of resolution fail, litigation is the final step in settling a claim.

9.3 Some Disputes Occur in the Bid Proposal Process

Disputes can arise even before a contract is awarded, especially when submitting bids in the public sector. Private-sector work allows an Owner to select a contractor even though that contractor may not be the low bidder, or may lack some minor compliance with the requirements of the Request for Proposal (RFP). Quite often RFPs in private-sector work include the phrase that an award will be made to the "low responsible bidder," thus allowing the Owner to determine who has the most competitive bid, or "best value," or who may, in the Owner's opinion, be best to work with. The Owner, being subject to no statutory regulations to prohibit it from bypassing the low bidder, can choose to contract with any of the bidding entities. This is not so easy in public works projects.

9.3.1 Public Works Procedures

In public works RFPs, there are any number of procedures that must be followed to have the submitted bid considered acceptable by the agency:

- The bid must be signed by a company officer.

- Bid bonds are generally required in a specific dollar amount and are forfeited if the selected contractor thereafter elects not to accept a contract presented to it. A certified check or letter of credit is often accepted in lieu of a bid bond.

- Submission of public proposals generally requires that the sealed bid be submitted at a specific date, time, and place. Bids, which are usually time-stamped to ensure compliance with the agency's requirements, can be justifiably rejected if they fail to meet these submission requirements—but they need not necessarily always be rejected.

9.3.2 Late Bids Rejected?

How many times has a project manager dashed to the bid opening office hoping to get there on time and get a bid stamped before the deadline? But suppose that heavy traffic caused by road repairs, or some other unanticipated event, precludes the company's bid from being stamped in, say, precisely at 10:00 A.M; instead, it is officially accepted at 10:10 A.M. Will this bid be automatically rejected? Not necessarily.

9.3.3 Challenging a Bid Rejection

If the local authority rejects the bid because it was submitted late, don't despair. If the late submission was not due to fraud, collusion, or intent to deceive, the Project Manager should state verbally, in the presence of witnesses (which include the other bidders), that the company will file a written protest with the public official for refusal to accept its bid. The Project Manager should remain at the bid opening, taking notes as competitor's bids are opened, and document any other comments of interest.

9.3.4 In the Public Interest

If your rejected bid is determined to be the low bid, and if the project is important to your company, a formal protest and challenge on the basis of public interest can have an effect. Public agencies do have some leeway if they deem it to be in the *public interest* to accept the low bid from a responsible contractor even though submission of the bid was 10 minutes late. The agency may cite this late submission as a "minor" discrepancy.

9.4 Promissory Estoppel

Let's suppose that in the rush to assemble a bid, a subcontractor phoned in its proposal just as you were preparing the bottom line on your way out the door, but did not follow up with a written confirmation. And let's suppose that your company was awarded the work, after which that subcontractor called to say it made a mistake and wishes to withdraw its proposal. Without a written confirmation of the subcontractor's bid, can anything be done? In some cases the answer is yes.

9.4.1 Promissory Estoppel Principle

The principle of *promissory estoppel* may apply and render an "oral contract" enforceable. Under this doctrine, the subcontractor may be held responsible for any damages incurred by the General Contractor, and be required to pay the General Contractor the difference between its (oral) proposal and the next highest subcontractor's bid. A case in point is discussed in subsection 9.4.2.

9.4.2 Challenges to Oral Bid Withdrawal

The case of *Bridgeport Pipe Engineering v. DeMatteo Construction Company,* 159 Conn. 242, 244 (1970), involved just such a matter: An HVAC subcontractor that phoned in a bid later declined to enter into a contract. The Connecticut Supreme Court ruled that the telephone bid was an oral contract and the General Contractor accepted that offer when it became the successful bidder. Another case, *H.W. Standfield Construction Corp. v. Robert McMuller & Sons, Inc.,* 92 Cal. Rptr. 669 (1971), presented a similar situation: The general contractor had relied on the promise of a painting subcontractor to do the work for a stipulated sum, but the painter later reneged and refused to do the work at the price quoted initially. The California Supreme Court upheld a ruling in Standfield's favor.

9.5 Other Contentious Disputes

Three types of claims were not included in Harris's "Top 5 of 2011" in the United States. Some of the most contentious disputes and claims are:

- Delay claims and ensuing lost productivity
- Differing, changed, or concealed conditions
- Acceleration

9.5.1 Delay Claims

Delay claims fall into one of three categories:

- *Excusable delays.* For an excusable delay, a contractor is granted a time extension but no monetary compensation.
- *Concurrent delays.* Delays caused by both contractor and Owner deny both parties damages for these delays.
- *Compensable delays.* Delays for which either the Owner or the contractor will be entitled to additional compensation.

9.5.2 Delays and Lost Productivity

Delay claims also create an environment in which two other situations may arise: a contractor's claim for lost productivity due to the delay, and the Owner's demand that the contractor complete the project in accordance with the original completion date. In other words, the Owner demands that the contractor accelerate the work.

9.5.3 Sample Letter Advising of a Delay (*page 9.8*)

In this sample letter, the delay is due to holdup of shop drawings or nonresponse to a Request for Information (RFI). However, it can easily be adapted to other types of delays.

9.5.3
Sample Letter Advising the Architect or Owner of a Delay Due to Shop Drawings or Nonresponse to RFI

Contractor's Letterhead

Addressed to Architect or Owner

Attention: Responsible Individual

Re: Project Name and Number

Subject: Notification of a Delay

Dear Mr. (Ms.)   (   )

(State reason for delay request, e.g.)

1. Delay in return of shop drawing—note drawing, date sent, still open as of this date.
2. Delay in response to General Contractor's request for information—note RFI number, date issued, still open as of date.

These delays commenced on (date) and
If reason 1 is selected: have impacted the release of (name equipment or product).
If reason 2 is selected: have delayed the progress of work in place until we receive a response to that RFI.

Upon receipt of the (shop drawing) (RFI), we will assess the cost and time implications relating to this delay and will issue the PCO# (or COR) to track costs incurred.

With regards,

Project Manager

9.5.4   Contractor's Claim for Lost Productivity

Many situations can contribute to lost productivity:

1. Delays in responding to a contractor's Request for Information (RFI), thereby forestalling work in a specific area and having to shift forces to another part of the project.

2. Delay in approving a change order that will, in turn, delay the purchase of materials and/or equipment required to maintain the contract schedule.

3. Interruption of the schedule by an unanticipated excusable or compensable delay, which, in turn, affects the orderly sequence of construction activities as reflected in the approved or acknowledged schedule.

9.5.5   Contractor to Prove Loss of Productivity

The contractor now must prove that any of the events discussed earlier caused a loss of productivity and therefore affect the contractor's costs to complete the project, including, but not limited to, an extension of the approved construction schedule. Depending upon the nature of the delay or interruption, the costs to get back on track may be substantial, and the Owner and its design consultants will surely want a detailed explanation and accounting of those costs.

9.5.6   The Measured Mile

One approach to defining the costs for lost productivity is known as the *Measured Mile*. This procedure compares productivity initially estimated and achieved during periods in the project not adversely affected by events creating the claim with productivity during the affected period. The costs estimated and incurred during the nonimpacted period of performance are referred to as the Measured Mile, against which the impacted costs will be compared.

Three Components of the Measured Mile Approach

1. The budgeted cost of the work scheduled (BCWS)
2. The budgeted cost of the work performed (BCWP)
3. The actual cost of the work performed during the affected period (ACWP)

9.5.7   Challenges to the Measured Mile

The challenges to the Measured Mile approach to lost productivity are considerable, so rebuttal must begin when the Project Manager or field Superintendent first becomes aware of *the potential* for an increase in costs due to the impending delays in the schedule.

Documentation at this point, in anticipation of the problem, is one of the keys to the success of any ensuing claim. This is not to say that the Daily Reports produced prior to those delays should be scant of key information relating to the potential delay(s). If in doubt, put it in the Daily Report. This document can never be criticized for containing too much detail.

9.5.8   Subcontractor Support

Affected subcontractors may also be a source of important information, as they usually have their own daily reports reflecting hours generated for specific operations and the quantity of work set in place. These unit costs or cost per operation can be compared with their project budgets and with costs on prior, similar projects. As an example, lost productivity in structural erection may be documented by the steel subcontractor's estimate, accompanied by the actual costs to unload, set in place, and bolt up or weld components during the "lost productivity" period. Historical costs on similar, prior operations should be included to reflect a *pattern* of typical costs for projects of the same nature and scope.

9.5.9   Example of the Measured Mile Approach to Lost Productivity

A concrete subcontractor may have project after project square foot costs to place and finish concrete slabs on grade or suspended slabs. Let's say its costs over a dozen recent projects range from $6.50 per square foot to $7.25 per square foot. The cost to install a slab involved in a "lost productivity" claim was, say, $10.25; therefore, using the average between the two square foot prices from previous projects as $6.88 would place the lost productivity costs at $3.37 at a minimum. This type of documentation is essential in developing a lost productivity claim.

9.5.10   Two Court Decisions Providing Further Validation of the Measured Mile Approach

In a case in the Commonwealth of Pennsylvania, *James Corp. v. North Allegheny School District,* 938 A.2d 474 (Pa. Commw. Ct. 2007), the school district challenged the Measured Mile approach presented by the expert consultant hired by James. The court noted that the expert consultant did not have to present costs with "mathematical precision," but merely had to present them as a "reasonable basis" for the Measured Mile calculation.

This was an important point. Equally important was a contract provision in the contract between James and the school district that required a 21-day notification of any event giving rise to a claim. James did not file its claim within that 21-day period. The court said that "to adopt the severe and narrow application of the notice requirements … would be out of tune" with an equitable solution because the district continually changed the project schedule and the contractor's resequencing of work, and the school district was actively involved in the project on a day-to-day basis.

9.5.11   Subcontractor's Measured Mile Approach

In the case of *Bell BCI v. United States,* 81 Fed. Cl. 617 (2008), the government made about 200 modifications to the building's HVAC ductwork system. Bell's subcontractor filed a claim for loss of productivity and Bell, in turn, filed a claim on behalf of this subcontractor, claiming that the subcontractor had compared its production rates during the earlier nonimpacted period with the cost of work that had been impacted by the many changes. The court accepted Bell's and its subcontractor's use of the Measured Mile approach as a proper method to recover costs.

9.6   The Total Cost Approach

This method of presenting a claim for lost productivity, as opposed to the Measured Mile approach, uses the total cost of the project, impacted as it was by those events that resulted in lost productivity. Of course, this claim can only be assembled once the project has been completed and the contractor has submitted the final cost of the project (as opposed to the original project budget). It is rather easy to see why this method has not been highly regarded. First, the contractor's initial estimate or budget may have not been correct. Perhaps the contractor got the job initially because its estimate was erroneous. Second, perhaps the contractor did not work efficiently throughout the project and is using the lost productivity claim to recover money it doesn't deserve.

9.6.1   Contractor Proof Requirements

Before a Total Cost approach can be used to back up a lost productivity claim, the contractor must prove that:

- The losses incurred could not be substantiated with any degree of certainty due to the complexity of the situation that created these losses.
- The contractor's bid/estimate was realistic.
- The contractor's final costs were reasonable based upon the conditions encountered.
- The contractor did not share in the responsibility for these added costs or expenses (i.e., overruns were not the result of the contractor's poor performance).

9.6.2   Court Case Using Total Cost Approach

In a California court case, *Peterson v. Container Corp. of America,* 218 Cal. Rptr. 592 (Cal. Ct. App. 1985), Petersen was engaged to construct a paper mill modernization project on the basis of a Cost-Plus-a-Fee not to exceed a Guaranteed Maximum Price (GMP). The Owner indicated that the construction drawings would be completed within two to three weeks of a construction start; however, the revised drawings were not completed until *one year* after the construction start. Changes made by the Owner were flying left and right and the contractor claimed that it could not keep track of all of them. Several hundred other changes were required to correct defects in those final drawings.

The Owner and the contractor engaged a referee to review the facts of the case. The referee ruled that the Owner had abandoned and breached the contract and therefore the contractor was entitled to be reimbursed for the value of its work on a "total cost" basis.

The appellate court agreed with the referee's decision, adding that the excessive number of changes could constitute an abandonment or breach of the contract because the number and frequency of changes *were inconsistent with the original scope of the work*. The court went further and ruled that the contractor was unable to segregate the costs associated with the multitude of changes, so the Total Cost approach was the only reasonable means of pricing this added work.

9.7   Differing Conditions

In Section 8, we discussed the process of preparing a change order relating to differing conditions. Although the contractor may follow all of the guidelines for the preparation and submission of such a change order, the review and approval of a differing conditions change order will be a formidable challenge. The Project Manager will usually encounter more than a little resistance to this type of change order, no matter whether the contract is a public- or a private-sector contract.

9.7.1   Stop Work When a Differing Conditions Claim Is in the Offing

First of all, **read your contract with the Owner**, paying particular attention to the provisions concerning notice when unknown or concealed conditions are encountered. These provisions may vary considerably depending on whether the project is private or public.

In many cases, private-sector contracts use or are based on the AIA A201—General Conditions contract (2007 Edition), specifically Article 3.7.4—Concealed Conditions; or Article 3 of ConsensusDOCS® 200 (2011 edition, revised April 2012). Please read this document thoroughly!

9.7.2   Sample Letter Regarding Change in Conditions (*page 9.13*)

9.7.3   AIA Contract Language Regarding Changed Conditions

The article in the AIA General Conditions document referring to differing or changed conditions requires the contractor to promptly provide notice to the Owner and the Architect **before conditions are disturbed**, and in no event later than 21 days after the first observance of the conditions described as follows:

1. Conditions on the site that are subsurface or otherwise have concealed physical conditions that differ materially from those indicated in the contract documents.

2. Unknown physical conditions of an unusual nature that differ materially from those ordinarily found to exist, and generally recognized as inherent in construction activities of the character provided for in the contract documents.

9.7.2
Sample Letter Regarding Change in Conditions

Contractor's Letterhead

Addressed to Architect or Owner (if applicable)

Attention: Responsible individual

Re: Project Name and Number

Subject: Change in Conditions

Dear Mr./Ms.:

This Change in Condition Notice has been issued to inform you of the potential cost and schedule impact due to the Design Team's (select one of the following)

1. Response to RFI Number (?)
2. Issuance of ASI Number (?)
3. directive from the Architect dated (date)

Please note that this (RFI, ASI, directive) constitutes a change to the contract documents or may modify work that has already been put in place per previous contract or modified directives issued from your office.

Since time is of the essence, we will proceed with this change as described in the (RFI, ASI, directive) unless we are directed otherwise,

We have established PCO# (?) to track costs related to this issue.

With regards,

Project Manager

9.7.4   Architect's Role in Determining Changed Conditions

The pertinent AIA General Conditions contract article continues by stating that the Architect may determine that these conditions differ materially and will cause an increase or decrease in the contractor's cost of, or time required, to fulfill the contractor's performance of the work. The article further states that if the Architect deems these conditions not materially different from those indicated in the contract documents, the Owner and contractor will be notified that no change in the terms of the contract are warranted. If that is the case, the contractor can proceed to press its case in accordance with Article 15 dealing with disputes and claims.

9.7.5   ConsensusDOCs® Differing Conditions/Concealed/Unknown Conditions Subsection (*page 9.15*)

9.7.6   Public Sector Differing Conditions Requirements

Public contracts may follow the guidelines incorporated in the federal government's differing site conditions provision, 48 C.F.R. 52.236-2:

(a) The Contractor shall promptly, and before such conditions are disturbed, give a written notice to the Contracting Officer of (1) subsurface or latent physical conditions at the site which differ materially from those indicated in this contract, or (2) Notice of unknown physical conditions at the site, of an unusual nature, which differ materially from those ordinarily encountered and generally recognized as inhering in the work of the character as provided for in the contract.

(b) The Contracting Officer shall investigate the site conditions promptly after receiving the Contractor's notice. If the conditions do materially differ and cause an increase or decrease in the Contractor's cost, or time required for, performing any part of the work under this contract, whether or not changed as a result of the conditions, an equitable adjustment shall be made under this clause and the contract modified, in writing, accordingly.

(c) No request by the Contractor for an equitable adjustment in the contract under this clause shall be allowed, unless the Contractor has given the notice required; provided, that the time prescribed in the contract for giving written notice. However the time required may be extended by the Contracting Officer.

(d) No request by the Contractor for an equitable adjustment to the contract for differing conditions shall be allowed if made after final payment under this contract.

9.7.5
ConsensusDOCs® Differing Conditions/Concealed/Unknown Conditions Subsection

Release Notes - April 2012

(Only applicable to the 2012 updated editions which will be delivered exclusively through the ConsensusDocs New Technology Platform. These revisions DO NOT APPLY to 2011 and older editions of ConsensusDocs delivered through DocuBuilder.)

I.   Global Change to Differing Site Conditions/Concealed or Unknown Site Conditions

Subsection (b) has been modified to read "unusual *and* unknown." The previous editions had "unusual *or* unknown." Below is how ConsensusDocs 200 (2011 edition; revised April 2012):

> 3.16.2 CONCEALED OR UNKNOWN SITE CONDITIONS If the conditions encountered at the Worksite are (a) subsurface or other physical conditions materially different from those indicated in the Contract Documents, or (b) unusual and unknown physical conditions materially different from conditions ordinarily encountered and generally recognized as inherent in Work provided for in the Contract Documents, the Constructor shall stop affected Work after the condition is first observed and give prompt written notice of the condition to the Owner and the Design Professional. The Constructor shall not be required to perform any Work relating to the unknown condition without the written mutual agreement of the Parties. Any change in the Contract Price or the Contract Time as a result of the unknown condition shall be determined as provided in article 8.

A similar revision was made in the following 2011 document editions:

> ConsensusDocs 205, section 5.11;
> ConsensusDocs 300, section 13.14;
> ConsensusDocs 410, section 9.4;
> ConsensusDocs 500, section 3.18.2;
> ConsensusDocs 510 section 3.10; and
> ConsensusDocs 750 section 7.3.

Note that the 2012 editions of the following documents also contain this language:

> ConsensusDocs 235, section 6.11;
> ConsensusDocs 415, section 8.5;
> ConsensusDocs 450, section 7.3; and
> ConsensusDocs 460, section 7.3.

Source: By permission: ConsensusDOCS™—copyright license No. 0464.

9.7.7 Five General Rules to Follow

It is extremely important to follow these five general rules, which appear in the AIA and ConsensusDOCS® contracts, as well as federal government contracts, when differing conditions are claimed by the contractor:

- Stop working in the affected area ASAP!—even if the condition is not immediately identified as a differing or changed condition, but you merely suspect that it might be so. The next order of business is to preserve that area as it is/was discovered. The Daily Log should reflect the work stoppage and the stage the excavation was in; photographs should be taken and noted in the Daily Log.

- At this point, immediately notify the Owner or its representative. If the Owner's representative is on site, have him or her inspect the location of the condition and provide written notification to document your conversation with that on-site representative. If that representative is off-site, take the appropriate photos and advise the representative (via e-mail or fax) of the time and date that work was stopped in the area in question, and that it is your opinion that the conditions encountered differ materially, and so on. If the condition was brought to the Project Manager's attention by an excavation subcontractor, have that Subcontractor prepare a notice to the General Contractor or Construction Manager, outlining its position in this matter, and then attach that document to the one the Project Manager is transmitting to the Owner.

- Request that the Owner reply with its instructions on how the Owner wishes you (the Project Manager) to proceed. The flow-through provision in most contracts between Owner and General Contractor will provide the basis for passing this directive onto the excavation Subcontractor, if that entity was the originator of the condition.

- If the Owner's or Architect's instructions are to cease work in that area until further notice, document these instructions as to the time and date that the Stop-Work Order was issued, the exact operation that was taking place at the time the Stop-Work Order was issued, and an indication that a notice to proceed with the work in the affected area, unless received by "X" date, may be cause for a delay in the project. A portion of the site plan where this condition exists, marked accordingly, should also be prepared and placed in the project file.

- If the Owner fails to respond within a reasonable period of time, send another notice indicating that further delays may impact the construction schedule and continue to do so daily until further instructions are received.

9.8 Type I and Type II Differing Conditions

In the case of differing conditions, any claim may be placed into one of two categories: Type I or Type II conditions. Type I concerns erroneous contract indications; Type II concerns conditions unknown to the Owner.

9.8.1 Type I: Differing Conditions Claims

For a Type I differing conditions claim, a subsurface or latent condition must have been encountered that differed materially from what was indicated in the bid and/or contract documents. That is, the bid/contract would have contained some indication of the expected conditions and the actual conditions must have varied from that indication. The contract documents could have expressed the anticipated conditions in the bid/contract by virtue of the following:

- Boring logs and the locations of the borings on the site plan
- Subsurface and/or geotechnical reports
- Ground water levels
- Foundation investigation reports

Type I claims generally arise when the Owner does not conduct an adequate subsurface investigation and prepares the bid documents based upon *assumptions* of the nature of the subsurface conditions.

9.8.2 Type I: Claim Court Ruling

The case of *H.B. Mac, Inc.*, 153 F.3d 1338 (Fed. Cir.1998), reveals how important it is to visit a site before submitting a bid—and also not to rely completely on the information contained in boring logs. H.B. Mac did not visit the site prior to submitting its bid, and instead relied on eight soil borings that were part of the bid package. The contract included construction of two buildings 300 yards apart. The contractor assumed that the subsurface conditions revealed by the test borings were representative of both sites, even though all the borings were taken on only one of the sites. When excavation commenced, the contractor found that the actual conditions were significantly different; it had to dewater and install some piles, and as a result incurred delays. H.B. Mac claimed a contract adjustment for Type I differing soils conditions. The court said that the test borings could not be reasonably interpreted to include a building 300 yards away in an area that was geologically diverse.

9.8.3 Timely Notification of Claim

Proper notification of a claim for Type I differing conditions is exemplified in the case of *City of Savannah v. Batson-Cook Co.,* 310 Ga. App. 878, 714 S.E.2d 242 (2011). The engineer reports provided by the city to the contractor revealed the existence of soft clay in one corner of the proposed parking garage. The excavating subcontractor employed by Batson-Cook said that the material encountered was materially different from the clay noted in the engineer's report and requested additional money to redesign the support system. The case went to trial and the contractor was awarded a total of $17 million. The city did not challenge the finding of differing conditions, but stated that the contractor had not given the required 21 days formal notification after observing the differing condition.

The court denied the city's claim, indicating that the city had received *timely* notice via e-mails between the contractor, subcontractor, engineer, and the city.

9.8.4   Type II: Differing Conditions Claims

A Type II differing conditions claim involves physical site conditions that are unusual, unknown, or *materially different* from those that would ordinarily be encountered and that the contractor could not reasonably have anticipated from a site inspection, and of which the contractor did not have knowledge from any other source (presumably scant or nonexistent information furnished by the Owner in the bid documents) or prior knowledge of the specific site.

9.8.5   Type II: Actual versus Contract Interpretation

A Type II claim invites comparison between the *actual* conditions and what the contract could have led the contractor to *reasonably* expect.

9.8.6   Importance of the Geotechnical Report

In a decision reported by the Armed Services Board of Contract Appeals (*Mass Construction Group, Inc.,* ASBCA No. 55440), the contractor claimed that it had encountered Type II differing site conditions, namely "unanticipated groundwater which flooded the footing excavations." The ASBCA denied the contractor's claim for the following reasons, which could apply to any number of situations encountered by a contractor that does not thoroughly review the geotechnical report included in the RFP:

1. The bid documents advised the bidders that subsurface data was available for review at the government's contracting office.
2. The subsurface data indicated that a groundwater table existed between 3-1/2 and 6 feet below the surface elevation.
3. The contractor did not review any subsurface data before submitting its bid **and did not attend the prebid site visit.**
4. The site conditions the contractor encountered were similar to the subsurface drilling logs the contractor failed to review.

Lessons Learned.   The lesson learned from this Type II claim is: review all subsurface data thoroughly and attend any and all prebid site visits. It seems as though these two actions would have been taken by any *reasonable* contractor, but in this instance the contractor failed to do so—to its detriment.

9.8.7   Complexity of Filing Type II Claims

The case of *Hydro-Dredge Corp. v. Corp of Engineers,* provides another lesson in the complexities of filing a Type II claim. Hydro-Dredge, which had a dredge work contract in the Hyannis (Massachusetts) harbor, claimed that the thick eelgrass encountered was a Type II differing site condition; it also claimed that the discovery of a sunken sailboat was such a condition. The Board of Appeals found that the amount of eelgrass was unknown and unusual, but stated that the sunken sailboat should have been anticipated (possibly because Hydro-Dredge was working in a harbor where yachts are moored?).

A contractor claiming a differing site condition is confronted with a heavy burden of proof. The contractor's unfamiliarity with the area doesn't seem to constitute a strong defense. The proof required for relief in a Type II claim is, as one expert put it, "amorphous and vague, and without a reference point in the contract."

9.9   Owner's Obligation to Disclose Superior Knowledge

Another consideration when differing conditions claims are possible involves information that the Owner might have possessed regarding subsurface conditions, but failed to pass on to the bidders.

In almost any contract, the Owner's obligations include:

- Not to withhold vital knowledge of a fact or event that could affect the bidder/contractor's performance or costs.

- An assumption that the bidder/contractor had no knowledge of this fact or event and therefore had no reason to obtain such information.

- Contract specifications that do not mislead the contractor nor rest upon a fact or event known by the owner but withheld from the bidder/contractor.

- The owner must reveal all pertinent information.

9.9.1   Principles Involving Proper Disclosure to Bidders

What constitutes proper or improper disclosure of subsurface conditions in the Owner's bid/contract documents?

1. The courts have confirmed that Owners, be they public or private, have a duty to disclose to the contractor pertinent information that the Owner possesses or knows is available if such information could reasonably be expected to have a material effect on bidding or construction of the project.

2. Ambiguous disclaimers are frowned upon and often ruled unenforceable by the courts.

3. General disclaimers are also frowned upon and often ruled unenforceable by the courts.

4. Narrowly tailored, specific disclaimers may be accepted by the courts.

5. The Owner assumes the risk in the following situations:

   a. If pertinent subsurface information is withheld.

   b. If inaccurate test data or results are provided in the bid/contract documents.

    c. If bidders are not provided either adequate subsurface information or a reasonable opportunity to conduct subsurface conditions.

    d. If defective plans/specifications are provided (although the defects may constitute a separate and different legal issue).

6. Information available to the bidders should have been reviewed, analyzed, and interpreted by qualified geotechnical engineers or engineering geologists.

9.9.2 How Do Owners Respond to a Contractor's Claim for Differing Conditions?

By viewing the questions an Owner may consider as it reviews the contractor's differing conditions change order, the contractor may gain some insight into the necessary components of that claim. What will the Owner zero in on? The following points may be of value in that regard:

1. What provisions in the construction contract pertain to notification of a claim, in general, and a differing conditions claim in particular?

2. If the contract contains a Site Investigation clause, did the contractor perform a *reasonable* inspection? Did the contractor provide any written documentation of this inspection?

3. Has the contractor submitted a proper claim, if the contract document differentiates between a Type I and a Type II claim?

4. Does the contractor's interpretation of the contract provisions regarding differing conditions appear to be reasonable?

5. Did the contractor rely on information contained in the bid documents and incorporated into the contract documents to justify its claim? Was this information in sufficient detail to *reasonably* present the existing site conditions?

6. Did the contractor have any prior knowledge of the site's condition (e.g., had the contractor worked on or near this site on previous construction work)?

7. Did the contractor rely on the total information available in the contract documents or only in a portion to support its claim?

8. Is there any exculpatory language in the contract for construction, or disclaimers that can affect this claim?

9. Would these disclaimers or other provisions of the contract be enforceable in the venue in which the contractor's claim is being presented?

10. Is this claim actually, a claim for differing conditions, or does it appear to be a claim to cover an unrelated cost the contractor has incurred?

11. Could this claim be construed as the contractor attempting to recover losses due to its misreading/understanding of the bid documents?

12. If there is validity to the contractor's claim, has the contractor adequately and accurately set forth the costs associated with the claim?

13. Since the burden of proof is on the contractor to prove its claim, has it provided evidence to support its claim; that is, to support the costs included in its claim?

14. Has the contractor complied with all contractual prerequisites to establish entitlement to the claim?

15. Has the contractor submitted sufficient proof of damages, costs, inefficiencies, time delays, contract completion extension?

16. A review of the contractor's recordkeeping should reveal adequacy, completeness, and applicability relating to the claim.

17. If loss of efficiency is being claimed, is the benchmark upon which these inefficiencies are based correct and proper? Was there loss of productivity caused by factors unrelated to the differing conditions claim?

18. The contractor should be questioned about the extent of the amount of soil that is claimed as "differing" from what was expected—was it 10%, 25%, etc.? What was the condition of the immediate adjacent soils?

19. Did the benchmark period include both high- and low-productivity excavation? What was the amount of equipment downtime, if any, and was it typical for this type of operation?

20. With respect to costs, did the contractor include lower productivity due to the normal start of an operation, the learning curve on a particular piece of equipment, or unexpected or unforeseen adverse weather?

21. In reviewing the contractor's unit costs as a benchmark, compare with costs from an estimating service, or your design consultant's experience.

22. Analyze costs in other periods of the contractor's work progress on the site to determine if the costs associated with its claim appear to be representative costs.

23. Is the Measured Mile approach being utilized in the preparation of the contractor's claim, and if so, are the substantiating costs properly documented?

24. In addition to loss of productivity, expect to see contractor costs associated with trade stacking, stop-and-start work flow, unproductive time (waiting for men/materials), and impact on other trades.

25. Expert assistance in evaluating the claim may be required; if so, seek assistance from claims and legal experts, engineering, and construction managers acting as an Owner's agent, among other possibilities.

26. Do not dismiss seeking an alternate solution to the claim in lieu of pursuing litigation.

9.10   Delays and Potential Acceleration Claims

As previously discussed, delay claims take one of three forms: *excusable*, for which the contractor is granted a time extension to the contract but no monetary damages; *concurrent*, for which neither contractor nor Owner can collect damages for delays; and *compensable*, for which either Owner or contractor is entitled to additional monies.

We will deal here with the excusable/compensable delay and the contractor demand for costs to accelerate the project that sometimes ensues in these situations. The latter eventuality could arise when the Owner recognizes that a delay has occurred and that the delay impacts the schedule, but still requires the contractor to complete the project in accordance with the initial contract completion date.

When pursuing a delay claim, the burden of proof is on the contractor to show cause: that the Owner's actions or inactions affected work activities as set forth on the contractor's approved CPM schedule. Many of these CPM activities include "float time," but others do not. If activities having no float are not performed on time, they will most likely affect the ones that do and either substantially reduce that float time or absorb it all.

9.10.1   As-Planned versus As-Built

The delay claim involves a comparison of the initial, detailed, as-planned, CPM schedule incorporated into the construction contract with the current "as-built" schedule. This requires the CPM schedule to be updated on a regular basis; this is even more important when events on the horizon or actually taking place indicate that "as-planned" activities are being impacted by actions or inactions of the Owner. However, this initial schedule must be reasonable. Thus, it must be prepared with a great deal of thought and input from the contractor's database, and supplemented by subcontractor and vendor participation, because it will be put under a magnifying glass if a dispute or claim arises.

9.10.2   Validating the Importance of an Accurate CPM Schedule: Look at It Again

Because most delay claims will be supported by the impact on the "contract" CPM schedule, it may be worthwhile reviewing some of the important components of that schedule. Does your CPM schedule reflect attention to the following?

1. Does the schedule include sufficient detail to be effective in reflecting the dependence of key activities on predecessors and successors?

2. Are activities for submittals and submittal review time included?

3. Are "windows" included for the delivery of Owner-supplied materials and equipment?

4. Are all activity relationships complete, valid, and confirmed by subcontractors and vendors?

5. Are activity durations reasonable and, if the work is subcontracted, validated by those subcontractors?

6. How were these activity durations validated: by historical data, productivity calculations, or seat-of-the-pants estimates? (Better to know now than when documentation is required to support your claim!)

7. Is the critical path reasonable, and are critical activities clearly identified?

8. Do any conflicts appear between concurrent activities?

9. Is adequate and reasonable float time included for key activities and Owner-supplied materials and equipment?

10. Is there an activity for Test—Adjust and Balance (TAB) and other close-out operations and procedures? Is sufficient float time included in case the TAB process becomes complicated?

11. Is there an activity for "Substantial Completion"? Review the contract with the Owner to ensure that the schedule's definition of substantial completion includes the prescribed activities and associated duration time.

12. Has a routine been established to ensure that this schedule will be updated?

9.10.3   Prompt Updating of the CPM Schedule

A case in point may be found in *George Sollitt Construction Co. v. United States*, 64 Fed. Cl. 229 (2005). George Sollitt Construction claimed that three circumstances were chargeable to the Navy while Sollitt was building, in three phases, a new pump house, two other buildings, and an addition to a third. All three delayed the completion of Phase I construction. Sollitt also claimed that seven other circumstances delayed the completion of the second and third phases. Although Sollitt was awarded $551,056 for miscellaneous contract entitlements, its claims for delays were dismissed by the court. The court ruled that the contractor should have updated its CPM schedule immediately after each purported delay; this would have highlighted the differences between the initial CPM schedule and the revised schedule reflecting each of these delays. The court stated that "for some months at the beginning and end of the project [Sollitt] failed to provide monthly CPM schedule updates, a contractual requirement." Some of the schedule updates lacked specific information about the start and end dates of certain work activities on the critical path. These activities were reported only as to their percentage of completion. In other instances, when specific activity sequences could be determined from the schedule updates, the dates were inaccurate.

9.10.4   Interdependence of Related Activity Delays

The contractor must present its case for delays by expressing the interdependence of a delay in one activity and the impact of that delay on subsequent activities. For example, a delay in the installation of in-wall electrical conduits will delay the installation and taping of drywall, which in turn may delay painting or vinyl wall covering applications, which in turn could possibly delay installation of floor coverings (vinyl tile, carpet).

9.10.5   Key CPM Documents to Be Presented with a Claim

Five major CPM schedule documents should be presented with the delay claim:

1. The as-planned schedule is prepared in accordance with the contract documents.
2. The as-built schedule reflects the actual sequence.
3. A series of adjusted schedules explaining the sequence of events that transformed the as-planned schedule into the as-built schedule. The first such adjusted schedule should incorporate the first delay and its impact on the schedule and estimated damages. This first adjusted schedule becomes the benchmark for subsequent delays. This process of creating further adjusted schedules should occur for each subsequent delay that transpires.
4. The as-projected schedule will show the anticipated adjusted project completion date.
5. If the project has been completed, an Impact Schedule can be developed to reflect all compensable and excusable delays.

9.11   Acceleration: Recognizing a Delay but Requiring the Contractor to Meet the Initial Contract Schedule

When a contractor has incurred an excusable or compensable delay, the contractor is entitled to recover the added costs resulting from the delay *or* an extension of time to complete the contract scope of work and the added costs associated with that time extension. Two conditions can now face the contractor: (1) The Owner may recognize the delay but require the contractor to complete the project per the initial completion schedule at additional cost. (2) The Owner does not grant the contractor's request for an extension of time and the contractor must accelerate its performance to attain completion per the initial schedule. This latter is a *demand for acceleration*.

9.11.1   Five Components of an Acceleration Claim

There are five elements of a claim for acceleration, all of which the contractor bears the burden of proof to establish:

1. An excusable delay has occurred.
2. Timely notice of the delay to the Owner, along with a request for a time extension, is presented, in writing, from the contractor to the Owner.
3. The time extension is denied and the Owner requires the contractor to complete the project by the initial project completion date.

4. The contractor notifies the Owner that this directive to complete the project per the initial schedule is considered a constructive change.

5. The contractor incurs extra costs to meet this accelerated schedule; such costs may include:

    a. Resequencing work activities

    b. Increasing the labor force by working overtime or adding new shifts

    c. Stacking of trades and subsequent loss of labor efficiency

    d. Additional supervisory costs, whether added to the normal shift or added to the second or third shifts

    e. Adding extra equipment (owned or rented)

    f. Expediting equipment and/or material deliveries

    g. Increased overhead costs (office trailer rental, computer/cell phone costs, added utility costs)

9.11.2 Sample Letter Responding to an Architect or Owner's Request to Accelerate (*page 9.26*)

9.11.3 Contractor Responsibility

The contractor also bears the responsibility to show whether:

- The delay leading to the acceleration demand was, in effect, excusable.

- The delay was or was not concurrent with other events.

- The contract includes or does not include a "no damages for delay" clause.

- The contractor followed the contract requirements for notification of a time extension.

9.11.4 Documentation

Documenting the claim will require the review and inclusion of these important elements:

- Notes and meeting minutes

- Correspondence between all affected parties

- Detailed job costs

- Initial budget and schedule of values

- Daily Log books and Daily Reports

9.11.5 Disputed Constructive Acceleration

The basic definition of constructive acceleration is when the Owner fails to grant any time extensions to the contractor. *Disputed* constructive acceleration occurs when the Owner grants a time extension that is less than the contractor

9.11.2
Sample Letter Responding to an Architect or Owner's
Request to Accelerate

+--+
| Contractor's Letterhead |
+--+

Addressed to Architect or Owner (as applicable)

Attention: Responsible Individual

Re: Project Name and Number

Subject: Request to Accelerate the Work

Dear Mr. (Ms.)   (   ),

Upon direction received from your office on (date), we have been authorized by (individual giving authorization) to accelerate the progress of the project in order to maintain the contract complete date of (insert date).

To that end we have authorized our subcontractors to expedite their work by increasing their workforce and/or adding additional shifts to extend working hours to include premium time. We will submit weekly costs associated with this acceleration effort which will commence on (date).

Although difficult to ascertain at this time, we believe that the total cost to accelerate will be in the range of (insert dollar figure such as $750K to $1 million). This is not to be considered as a maximum upset number, but merely an estimate.

Unless we hear from you to the contrary, we are proceeding on that basis.

With regards,

Project Manager

has requested. If this should occur, the contractor must first prove not only that it was entitled to a time extension, but also that the time extension granted by the Owner was less than the extension the contractor felt was justified. In such a situation, the contractor must give the Owner written notification that the time extension is inadequate and that the contractor will seek additional compensation for the acceleration effort.

9.12   Claiming Extended Home Office Overhead

When contractors prepare claims for extended overhead, they often include both field and home office overhead. A standard formula for determining extended office overhead was known as the Eichleay Formula, so called because it was first articulated in the 1941 lawsuit *In re Eichleay Corp.*, ASBCA No. 5183, 60-2 BCA 2688. In that case, Eichleay, the contractor, won damages for extended office overhead. Until 1997, when a federal court decision was thought to have sounded the death knell for use of the Eichleay Formula, this formula was used to calculate the cost of a contractor's home office overhead when a claim for a delay and subsequent time extension was granted.

9.12.1   The Eichleay Formula

The Eichleay Formula was rather straightforward. Let's assume a situation in which a $2 million contract involves a delay claim and the contractor incurs 45 compensable calendar days' delay. During this period, that same contractor has four other projects underway, all five totaling $10 million. The contractor's home office overhead during this period is $250,000. The duration of the project subject to the claim, including delays, totals 265 days. Home office overhead damages using the Eichleay Formula are as follows.

1. Contractor's total volume at this time: $10,000,000 divided by $250,000 (total office overhead) = $40,000 overhead per project.

2. Allocable overhead of the project in hand divided by the number of compensable delay days equals the daily allocable overhead rate:

   $40,000 divided by 265 days = $150.94

3. Extended home office overhead damages are calculated by multiplying the number of compensable delay days by the daily allocable overhead rate:

   $150.94 per day × 45 days = $6,792

9.12.2   Is Eichleay Dead?

Although a 1997 federal court decision was thought to have "killed" the Eichleay Formula, along with further claims calculated according to that formula, a case decided the next year (*West v. All State Boiler* [1998]) clarified some of the points in the 1997 case, and the Eichleay Formula was given new life.

In 1994, in *Wickham Contracting Co v. Fisher*, 12 F.3d 1574, 13 F.P.D. 1, 18 C.C. 121 (Fed. Cir. 1994), the court said that the Eichleay Formula was the "exclusive method of calculating unabsorbed office overhead." Since the late 1990s, several appeals seem to float through the federal court system annually concerning use of the Eichleay Formula for determining extended office overhead. So, if for no other reason than that it is still litigated, today's project managers should be aware of this controversial formula.

9.13   Alternatives to Litigation: Mediation, Arbitration, Dispute Resolution, Fact Finding

9.13.1   Mediation

Mediation is a process in which the aggrieved parties meet with a mutually elected impartial and neutral moderator who assists them in the negotiation of their differences. There are many reasons for pursuing mediation to resolve a potential claim:

- Mediation is neutral and fair. The parties involved have an equal say in the terms of the settlement. There is no determination of guilt or innocence in the process, merely resolution of the matter at hand.

- Mediation generally occurs early in the dispute process, before the participants' positions become hardened. Many mediation sessions are completed in one day. Legal representation is optional but not required.

- Mediation is confidential. All parties sign a confidentiality agreement stipulating that any information disclosed during the mediation will not be revealed to anyone.

- Mediation can avoid expensive litigation.

- Mediation fosters cooperation because it involves a problem-solving approach to the resolution of disagreements.

- Mediation provides a neutral and confidential setting in which all parties can discuss their problems and views on the matters in dispute. This sharing of information leads to a better understanding of other parties' positions.

- The neutral party (the mediator) assists each party in reaching a voluntary, mutually beneficial resolution.

- An independent survey revealed that 91% of all respondents would use mediation again.

How it Works.   Mediation usually commences with a joint session of all parties, conducted by the mediator, who also sets the agenda. This allows all parties to become familiar with the process and ascertain the position of each party involved. This joint session is followed by private caucuses where the mediator meets with each party and allows that party to explain and enlarge upon its position and its goals. The mediator may ask questions to further understand each party's position; in so doing, the mediator may lead that party to realize that perhaps its position is not as strong as originally perceived. By raising these questions with the parties individually, they may see their positions differently than initially viewed, leading to a willingness to compromise and achieve resolution.

9.13.2   Arbitration

Arbitration is a process for resolving disputes in which the arbitrator reaches a decision that can be either *binding* or *nonbinding* depending on the contract provision requiring arbitration. Except in certain fairly rare circumstances, the decision in binding arbitration is not challengeable by the damaged party, although that party may try to seek a judgment from a court. If the arbitration is nonbinding, the parties may accept the arbitrator's final decision, or reject it and decide to pursue litigation or use other means to resolve their differences.

Arbitration is a formal process whereby a single arbitrator or a board of arbitrators listen to the aggrieved parties present their cases. Presentations may be made either by the principals of each party or by their attorneys. After hearing all the facts, the arbitrators retire and return with their written decision in the matter at hand.

9.13.3   Alternate (or Alternative) Dispute Resolution (ADR)

The process known as alternate or alternative dispute resolution is kind of a catch-all for settling disputes without resorting to litigation. The broad ADR category includes mediation and arbitration, and goes further to include *negotiation,* in which the aggrieved parties resolve their differences in an informal environment, getting together and ironing out their concerns and proposing solutions to end their dispute.

9.13.4   Dispute Resolution Board

Dispute Resolution Boards (DRBs) have been used for years. The inclusion of a DRB process in the contract between constructor and Owner has allowed processing of thousands of disputes since the introduction of DRBs in the early 1970s. This method is still widely used today.

The DBR process requires a board, typically composed of three members appointed by agreement of both parties. These board members periodically meet at the construction site, so they are familiar with the events taking place there. The dispute is presented at a DRB hearing. Because the members of the board are cognizant of the unfolding facts that have led to the disagreement, they are in a better position to hear each party's approach to the dispute and present their recommendation based on knowledge gained during their periodic attendance at the construction site meetings.

DBR rulings can be either binding or nonbinding, depending upon the contract language.

Nine Key DRB Components.   The Dispute Resolution Board Foundation lists nine components that contribute to DBR hearing success:

1. All three members of the DRB are neutral and are subjected to the approval of all parties to the dispute.
2. All members sign a three-party agreement obligating them to serve all parties equally and fairly.
3. The fees and expenses incurred for the DRB are shared among the parties.
4. The DRB is organized at the project's inception before any disputes have occurred.
5. The DRB keeps abreast of the job progress and development by periodic site visits and review of pertinent documents.
6. Any party can refer a dispute to the DRB even if provisions for doing so are not included in the construction contract.
7. An informal but comprehensive hearing is convened promptly upon determination to proceed with a DRB hearing.
8. The written recommendations or decisions of the DRB are not binding on either party unless so mandated in the construction contract. These recommendations/decisions are generally admissible as evidence, to the extent permitted by law, in a subsequent arbitration or litigation procedure.
9. DRB members are absolved from any personal or professional liability arising out of their DRB activities.

9.13.5   Fact Finding

This method of dispute resolution involves a third-party investigation by an experienced, independent neutral party to establish a factual account of the dispute. The process can produce its findings with or without a recommended settlement. It may involve assessing the true value of the claim or the cause of the dispute. The neutral investigator will issue a report of his or her findings, and this report is usually used to effect a settlement of the claim. There is value to this method of settlement when both parties agreed on a framework for settlement but have not been able to arrive at its value. The neutral investigator's evaluation of the true cost may be all that is needed for both parties to finally resolve their differences.

9.14   A Checklist of Documents to Be Assembled When Preparing to Institute a Claim

Contract Documents

☑ The contract documents themselves, including all approved change orders, disapproved change orders, and all correspondence relating to the denied change orders.

☑ All subcontract agreements and any approved change orders, denied change orders, and the correspondence relating to change orders.

☑ All purchase orders for materials and/or equipment.

☑ The contract specifications, including any modifications, addenda, Architect's Supplemental Information (ASI), Architect's clarifications, and bulletins.

☑ The contract drawings, including all revisions, sketches (SK), and ASIs.

Bid Documents

☑ Initial estimate, changes to the estimate, subcontract proposals, vendor proposals, telephone or e-mail bids from subcontractors/vendors.

☑ The formal bid, including all papers and documents utilized in the calculation and formation of the bid.

☑ Changes to the bid documented with the reason for the changes.

☑ The initial schedule, whether bar chart or CPM, submitted with the bid and all backup material pertaining to the preparation of the schedule, including a shop drawing submission schedule if required in the bid documents (these are sometimes not required unless or until the construction contract includes that provision).

☑ Correspondence from the Owner or the Architect related to items to be included/excluded from the bid or other instructions not included in the written Request for Proposal (RFP).

Schedules

☑ The initial CPM schedule, the schedule submitted with the contractor's bid, and the schedule included in the contract as the original accepted baseline schedule by all parties.

☑ All updates, including a narrative of other documentation that accompanied the reason for changes to the schedule.

☑ All one-, two-, and three-week look-aheads presented at the project meetings and documented in those meeting minutes.

☑ Any recovery schedules and/or schedules with worksheets prepared for a proposed acceleration of the work.

Project Logs

☑ Shop Drawing Submittal log, including dates submitted, dates outstanding, date returned, and A/E comments.

☑ Request for Information (RFI) logs.

☑ Request for Clarification (RFC) logs.

☑ Log of e-mails or faxes to design consultants and/or Owner.

☑ Change Order log.

☑ Log of backcharges with appropriate backup material.

☑ Correspondence log of communications to design consultants and/or Owner.

Daily Logs and Reports

☑ The Daily Reports, whether maintained electronically or in handwritten form. Electronic files are to be supplement by hard copies.

☑ Inspection reports, whether from local building officials or Owner-supplied inspection services. These reports, such as concrete field inspection, soil compaction, and various lab reports are often maintained in separate files relating to the particular subject.

☑ Photographs documenting an event appearing in the Daily Log or Daily Report.

Meeting Minutes

☑ Minutes of regularly scheduled Project Meetings.

☑ Minutes of meetings with design consultants and Owner representative that are not a part of the regular Project Meeting minutes.

☑ Minutes of subcontractor and vendor meetings.

☑ Minutes dealing with schedule updates, whether with the design consultants, Owner, subcontractors, or vendors.

☑ Special meetings dealing with Requests for Information (RFIs) or Requests for Clarification (RFCs) or other issues of importance.

Other Correspondence

☑ Correspondence with the design consultants and/or Owner arranged chronologically, which should have been kept up to date by field office personnel.

☑ Subcontractor and vendor correspondence files, organized chronologically, which also should have been kept up to date by field office personnel.

☑ Correspondence from government agencies, such as field inspections, safety (OSHA) inspections/approvals, notice of violations.

Project Manager's Narrative of Events

☑ A history of the events that led to the preparation and submission of the claim, in story form: what happened, when did it come to the Project Manager's attention, how it progressed and morphed into a claim.

☑ Include a narrative of the disruption caused by the event, the delays that occurred, and the impact of these delays on the project's progress. This should be a document that someone who is not familiar with the claim could read and get a clear picture of what happened.

☑ This narrative should include dates, and cross-references to key documents such as date when the Owner was apprised of the problem, stop-work orders, and correspondence from affected subcontractors and vendors. Be specific!

Claim for Extended Overhead

☑ Actual field costs incurred during the extended period, documented by invoices (such as for trailer rental, field office supplies, utility costs, cleaning services, computer rental, and Internet access fees).

☑ Salaries of the Field Superintendent and any salaried or hourly foremen, to include actual payroll records and documentation of all fringe benefits.

☑ Cost of extended site security measures.

Progress Schedules and Funds Analysis

10.1   Managing Schedules

10.1.1   Concepts and Section Description

Planning is determining the activities to be performed, along with their respective durations, and arranging them in proper sequence or otherwise defining their relationships. Placing the plan on a calendar makes it a schedule.

This Section presents an overview of the primary types of schedules and their uses, suggestions to facilitate the planning function, the rights of the Contractor with respect to end dates, and the responsibilities of all parties to the scheduling effort. It then deals with the construction and use of cash-flow projections.

Throughout the section, the focus is on the logistics—the management—of the schedules, including:

- Confirming the legitimacy of the original plan
- Securing commitment from all those who must carry it out
- Determining and implementing appropriate action in all cases where performance is or may be slipping

10.1.2   Construction Tool or Contract Compliance?

Most specifications will include some requirement that the Contractor provide a schedule for the respective project. Primary reasons for the Owner's and design professionals' interest include:

- Assurance that a logical plan is indeed in place
- Some idea of the anticipated cash flow of the project (to help plan for their own needs)
- Documentation of the Contractor's plan and progress as a recordkeeping device, and as possible substantiation of or defense against claims for delay costs and consequential damages

It is becoming more common to see specifications that include increasing elaboration in the schedule requirement, apparently to assure the Owner that appropriate planning and scheduling is actually being done by the Contractor. It is an acknowledgment by the industry that too many Contractors do not normally conduct this effort adequately—and some don't do it at all—despite the fact that the effort is actually crucial to the timely, cost-effective completion of a project in a manner that will best avoid the extreme costs and liabilities associated with late completion.

Even basic scheduling efforts will satisfy most specification requirements. The point, however, is that project planning and scheduling must be done by the Company correctly and consistently because it is *the* most important function that project personnel can perform for the Company's sake. Realize now that from the schedule all else flows. Acknowledge that without an adequate scheduling effort, purchasing, correspondence, submittals and approvals, accounting for changes, and dealing with each day's decisions become random, uncoordinated, and therefore very inefficient (and possibly ineffective) activities.

The success of the scheduling effort is not so much related to the complexity of the scheduling mechanism itself or to the system selected. Whichever scheduling method is selected, success in the effort is much more related to the fact that you do it, do it well, do it consistently, and do it for the life of the project. Keep the effort foremost, and you will be rewarded with:

- Superior knowledge of your own projects in ways that you may not have thought possible
- The ability to define, track, and determine the effects of history on the project
- A stable of subcontractors and suppliers who will take your requirements seriously—first because they see that *you* take them so seriously
- Owners and design professionals who respond to the needs for proper actions on their part, because they see your ability to organize and present the effects of late or inappropriate action

Finally, it is perhaps most important to realize that schedules are there as tools to aid in making decisions—not to make the decisions for you.

10.1.3 Company Responsibilities

The nature of schedules is such that they should have been constructed *yesterday*. On many projects, certain relationships are commonly known or otherwise become apparent.

Example

1. The long lead item is hollow metal frames.
2. They must be coordinated with the finish hardware, purchased, submitted, and approved before fabrication can proceed.
3. The finish hardware must be finalized and approved before the hollow metal can proceed (and possibly even before it can be purchased).
4. The very first item necessary to maintain the schedule therefore turns out to be finalizing the finish hardware.

If these relationships were not known, the immediate hardware priority would be overlooked. The project would move along with those activities that are obvious for the moment (concrete, masonry), but would stop suddenly when the hollow metal frames do not arrive (and worse, when it is realized at that point that they are still weeks away…).

A comprehensive scheduling effort requires an intensive review of all construction details and component relationships to complete it properly. In such a process involving the key project participants, the obvious and subtle relationships that must be placed in proper sequence will become disclosed. The risk of being bitten by an overlooked detail—particularly in the project's early stages—will be greatly reduced.

It is the Project Manager's responsibility to organize and finalize the development of the complete baseline schedule for each project under his or her responsibility. Contributions to the complete effort should be made by:

- All those who prepared any portion of the original bid or component estimates

- The Project Engineer and the Site Superintendent

If possible, it is also wise to involve key subcontractors or trade contractors to improve the quality of planning information. These can be the ones with whom relationships exist and who are trusted, or those who provided subcontractor and trade contractor estimates to the Company at the time of bid or proposal preparation.

Updating the schedules is *everyone's* responsibility. The Project Engineer will be responsible for actually producing the finished schedule update and transmitting correspondence, but the information necessary for the update will be generated *daily* by the Project Manager, the Superintendent(s), the Project Engineer, and everyone else with knowledge of any project effect. *These pieces of information must be recorded on the posted schedule as they occur,* for later consolidation into the finished update.

10.1.4 Subcontractor and Trade Contractor Responsibilities

The Pass-Through Clause (see Section 3.5.8) ties each Subcontractor directly to the complete specification requirements for scheduling as they relate to the work of each respective Subcontractor. Included here are the general responsibilities for items such as:

- Timely compliance of all work

- Providing adequate labor

- Performing the work in a manner that will not interfere with or otherwise delay the orderly sequence of work by others

Further included is the implied responsibility for the Subcontractor to meet requirements in a manner that will not cause the total contract time to be exceeded.

In addition to these very general considerations, the subcontract itself should include adequate and specific scheduling responsibilities, including ideas such as the following:

- The fact that time is of the essence to all dates and schedules.

- The acknowledgment that all schedules change continually. The Subcontractor must accordingly be aware of all *current* schedule requirements as they may have changed, and comply with them in every respect.

- The agreement to adjust manpower, equipment, overtime, and Saturday, Sunday, and holiday work as necessary to meet all schedules.

Beyond these stipulations, the subcontract should also contain an adequate Acceleration Clause. This language would give the General Contractor the right to accelerate the work of any particular Subcontractor by specifically directing the addition of labor, equipment, overtime, or Saturday, Sunday, and holiday work whenever it becomes apparent to the General Contractor that a respective portion of the work is not likely to be completed on time or as promised. The clause can then go on to state that:

- If the acceleration is not due to the fault of the particular Subcontractor, that Subcontractor may be reimbursed for the acceleration costs.

- If it is due to the fault of the Subcontractor, the Subcontractor will remain responsible for all acceleration costs.

- However, in either case, the Subcontractor *cannot refuse* to accelerate as specifically directed.

Finally, some condition (such as tying nonperformance directly to the liquidated damages provision of the general contract) should be considered to be assigned to the respective Subcontractor's failure to accelerate when so directed.

With all these rights, responsibilities, and tools, the management of the Subcontractors' portions of scheduling responsibilities will boil down to the amount of direct and specific effort made by the Project Manager, Project Engineer, and Site Superintendent(s) in involving the Subcontractors *daily*.

10.2   Schedule Types and Uses

10.2.1   General

Schedule types vary greatly in their:

- Simplicity or complexity
- Levels of detail
- Ability to actually display the plan (How did we decide to place that activity there?)
- Visibility (Can we see and understand the information?)
- Recordkeeping ability (What actually happened?)
- Ability to display cause and effect

Each schedule type has its purpose. Each is strong in one or a few of the ideals listed and weak in others.

Schedule selection criteria should be based upon the actual needs of the respective parties. Examples may include the following:

- Building committees or the Owner's finance committee may only want to see a bar chart to get a general sense of the project.

- Although you've just selected the most expensive scheduling software on the market, are you really interested in "resource leveling," particularly if you subcontract most of your work? If so, do you really think that your project people will be conducting the operation consistently for the life of the project, and you will be able to manage appropriate Subcontractor responses effectively?

- Do you really need to break your schedule into 2000, 1000, or even 500 activities? Are you trying to micromanage the work from your office miles away from the site? What do you think the probability is of conducting a greatly detailed sequence of activities exactly as planned—one year from now?

These general ideas must be decided upon before the scheduling method—or combination of scheduling methods—is selected. The description that follows will help guide those decisions.

The information in this section is a guide. It is not sufficient by itself to develop complete proficiency in any particular scheduling method. For that, more time and effort will be required, as well as the complete materials needed for this purpose.

10.2.2   Bar Charts

Description

1. The simplest of all methods: activity descriptions are placed in a column on the left, along with other desired relevant information (budget cost, for example).

2. A calendar is placed along the top of the chart, extending for the duration of the project.

3. A line or "bar" is placed alongside each activity's description in the time area to correspond with the calendar above.

Example

10.2.2
Example Bar Charts

DAY	1 2 3 4 5 6 7 8 9 10 11 12 13 14 15 16 17 18 19 20 21 22 23 24 25 26 27 28
Foundation	--------------------------------------
Masonry	--------------------------
Roof Framing	------------------------
Roof	----------
Slab On Grade	---------
MEP Underground	---------------
Complete MEP	-------------------------
Subst. Completion	◆

Advantages

1. It is simple to prepare.

2. It displays anticipated time frames of major activities visually.

3. It is easy to understand; can be a good "communicator" to large groups such as building committees.

4. It is historically accepted—used by most Superintendents and project personnel since the beginning of time; accepted by many design professionals and Owners.

5. Activities are normally listed in order of specification section. It is easy to correlate with the Schedule of Values and other payment records.

Disadvantages

1. It is most often oversimplified in its approach, not providing enough detail to use as an actual management tool.

2. It does not display the activity relationships. The schedule preparer *intuitively* considered these, but the logic is lost. This problem is emphasized because the activities are normally listed in order of specification sequence, as opposed to any logical sequence.

3. It is seldom updated or corrected.

4. "Updating" can only record a greatly simplified version of history—only when a respective activity actually started and ended. It cannot:
   - Display cause and effect
   - Forecast effects of current (good or bad) activity status
   - Accommodate changes in any way

5. Any efforts by project people to reschedule or accommodate any changes at all from the original program are completely left to the memories of those conducting the change.

6. You have no ability at all to communicate changes and their effects—not just to the Owner or Architect, but to your own people.

Conclusion. A bar chart can be a quick, convenient *preliminary* planning tool that can give a good idea of the general project parameters at the onset of a project, and as a first guide to a more useful scheduling method. It is also a good presentation vehicle that can place complex relationships into understandable form for large groups. It is not, however, an acceptable tool to manage a project by itself.

10.2.3 CPM/PDM *(page 10.9)*

Description. CPM, or critical path method, has become an overused buzzword in the industry. It is usually (but not necessarily) a computer-generated schedule that accommodates each activity's start and completion points, duration, dependencies on previous activities, and relationships with succeeding activities. The activity relationships are generally defined into the computer, and the computer is given the job of figuring out the resulting "network"—where each activity will actually fall in its complete set of relationships and on the calendar. As the job progresses, information regarding actual start and completion dates, durations, logic corrections, changes, additions, and deletions are added to each activity's information, and an update or "reschedule" is generated by the computer. Displays are either "activity-on-arrow" or "activity-on-node."

PDM, or precedence diagramming method, is an "enhanced" version of CPM that adds an attempt to provide probabilities of the success of various baseline and update schedule scenarios. PDM may still be used in other industries such as aerospace, chemistry, and so on, but has generally been abandoned for construction because of its complexity that is wholly unnecessary for our needs. For this reason, the remaining discussion will focus on CPM only.

Example

10.2.3
CPM/PDM Example

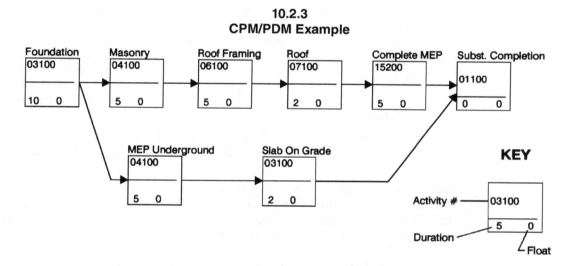

Advantages

1. It can be developed to any level of detail. Originally developed as a method to accommodate extremely large or complex schedules (building a missile system, for example), CPM can handle more than adequate levels of detail needed for any type of construction project. The schedule size is limited only by computer memory, which is becoming cheaper every day.

2. It can accommodate any number of changes to the program—such as simple corrections to logic, or change orders, delays, etc.

3. It will display all activity relationships that define the original and changed schedules. The logic of all "before and afters" is left intact.

4. There is great flexibility for adjustment throughout the baseline schedule development.

5. Computer programs can provide many types of reporting to managers, which can be tailored to exactly how each manager wishes to run his or her projects.

Disadvantages

1. It requires a computer, specialized software, and basic computer familiarity. Specialized training used to be a significant drawback. These days, however, the software is very straightforward in its use, and very "friendly." A little initiative and a dose of faith (if you are not familiar with computers) should be all that is necessary to be trained or to train yourself in a reasonably short period of time. This "disadvantage" is becoming less so each day.

2. It has very poor visibility. Unless combined with the "logic diagramming method" as described below or something similar, CPM diagrams themselves display no relationship to time. An activity requiring 5 days to complete will take up as much space on the diagram as one requiring 100 days to complete.

   Even the most accomplished CPM purist can therefore not get any sense of where the schedule really is by simply viewing it. The information must be studied, clearly understood, and visualized by the reviewer. All analysis is accordingly left to straight consideration of tedious lists of dates. This is by far the most serious disadvantage for any manager or communicator.

3. It must be updated completely—from start to finish and in all its detail—each time that it is updated. This can be a large task, which may become compromised in practice.

4. Each update generates essentially an entirely new schedule. The set of original activity targets, dates, and milestones can be too easily (if not conceptually) lost after just one or two updates if the manager is not careful to take definite steps to prevent this from happening.

Conclusion. CPM logic is the basis for all effective scheduling methods. Its earlier forms used on computers with limited graphics capacity were more prone to the "visibility" disadvantages just discussed. Later forms combine the CPM logic with the improved visibility of logic diagrams. These combined forms are the most effective as management tools.

10.2.4   Program Evaluation and Review Technique (PERT)

The PERT scheduling technique was developed in the 1950s when the U.S. Navy's Polaris submarine project was in full swing. Having thousands of subcontractors and suppliers engaged for this complex project meant having to produce a schedule that could deal effectively with thousands of activities spanning a multiyear time frame. PERT shares many similarities with CPM: It is based on the concept that before an activity can proceed, its predecessor activities must have been completed. There can be multiple predecessor events and multiple successor events. A PERT event cannot start until the predecessor event(s) has (have) occurred. It also has a network diagram with a sequence of activities and time frame for activity start and completion. The network consists of *activities*, represented by arcs (lines), and milestones, represented by nodes (circles).

PERT looks at anticipated completion of events somewhat differently from CPM in that it considers four potential completion dates:

O: Optimistic time; the minimum time required to complete a task.

P: Pessimistic time; the maximum time required to complete a task, anticipating that everything will go wrong.

M: Most likely time; a best estimate to complete a task based upon everything going right.

TE: Expected time; taking into the fact that not everything goes right and not everything goes wrong, a formula is developed as follows:

TE = (O + 4M + P divided by 6 – the expected average time the

task would take if repeated a number of times over an extended period of time

A typical PERT diagram would be similar to that in the following figure, which shows a seven-month project with five milestones numbered 10 through 50 containing six activities A through F, allowing from one to four months for completion. This example does not indicate whether "t" is actually TE, but merely illustrates the diagrammatical composition of a PERT schedule.

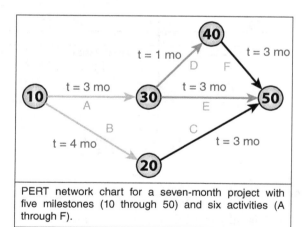

PERT network chart for a seven-month project with five milestones (10 through 50) and six activities (A through F).

It is doubtful that a Project Manager working on a residential, commercial, institutional, or public works project will ever encounter a PERT schedule, but all project managers should have some familiarity with the concept and process.

10.2.5 MOST

Description. "MOST," or management operation system technique, is a scheduling method originally developed as a companion to PERT and CPM before computer graphics become available, specifically to add visibility to CPM schedules. As developed for construction, it quickly evolved into a stand-alone system with major advantages.

Before the availability of computer graphics that could generate logic diagrams, MOST began as a manual drafting of what the industry later dubbed the "logic diagram." The MOST baseline schedule therefore appears nearly identical to the logic diagram.

MOST differs from all other scheduling methods in its unique ability to record all project effects in a way that is almost unbelievably clear, to pinpoint complete accountability for all project effects, and to reschedule without redrafting or replotting. This last feature is the mechanism that allows the MOST schedule system to be used effectively without any computer. Updating it is very easy, and is accomplished in less time than for the other methods.

Example

10.2.5
MOST Example

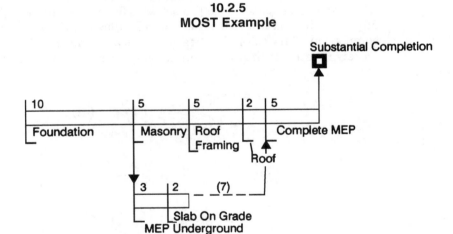

Advantages

1. It has all the advantages of the logic diagram.

2. It requires no computer. A computer is effective during the development of the baseline schedule because of the convenience for modification. However, a computer is not at all necessary.

3. It is possible to reschedule without redrawing or replotting. Reschedules and updates do not require that the schedule be redrafted. This feature allows all updates to be conducted manually.

4. The entire schedule or only the portion of immediate interest can be updated.

5. It always retains all original activity plans and milestones visibly. The current status is always placed against the *original* program to display exactly what is going on.

6. It greatly simplifies the otherwise complex history of detailed effects on activities and sequences. It keeps names pinned to effects, and it specifically quantifies the effects.

Disadvantage.   The updating method will not computer-generate reports.

Conclusion.   Probably the most effective combination of schedules is to:

1. Use a computer with CPM/logic diagramming abilities to develop the original baseline schedule and, if a plotter is available, to plot the baseline schedule to save drafting time.

2. Once the final baseline is determined and either drafted or plotted, use the MOST technique for all updating, reporting, and presenting.

10.3   Schedule Preparation and Development

10.3.1   General Procedure

1. The Project Manager will direct the development of the baseline schedule and oversee its periodic updates. The Project Manager, the Project Engineer, or the Company Scheduler can construct the actual document; this will be done with information coordinated from many sources, including:
   a. The project team's sense of specific logic sequences and the direct application of relevant experiences
   b. The bid or proposal recapitulation, including detailed cost estimates
   c. Schedule considerations in the original estimates
   d. The plans, specifications, bid documents, and referenced standards

2. As the schedule draft is developed, all activities must be studied for their specific relationships with contiguous work. Each activity must be identified or coordinated with:
   a. Its specification section number and description
   b. The specific details for the specific project (subtle differences can dramatically alter "familiar" sequences)
   c. Complete, accurate item descriptions
   d. All submittal or approval constraints
   e. The extent of required submittals, anticipated time to prepare and submit
   f. The estimated time for design coordination and approvals
   g. Material fabrication and delivery times after receipt of approved submittals
   h. Erection and installation times
   i. Specific erection and installation sequences and constraints

     j. The need for any "built-in" items by others before or with the item itself

     k. Items required to be installed before or coordinated with

     l. Manner of erection and installation—steady sequence or intermittent

3. To the complete extent possible, involve key Subcontractors or trade contractors as early as feasible in the rough draft development. At the very least, use their specialized experiences if the bid packages have not yet been sublet, or incorporate actual subcontract commitments if they have been. This will help improve the accuracy of the decisions made with respect to the considerations of item 2. Wherever possible, allow for slightly more time than a particular Subcontractor says it "needs" to complete an activity. This can be tightened up later in the final schedule development if need be.

4. Identify and consider the timing of all key items, including:

     a. Contract start date

     b. Actual project start date (if different)

     c. Anticipated shut-down periods

     d. Specific contract, site, major sequence, and weather constraints

     e. Characteristics of major components, such as:

        (1) Delivery and erection of superstructure

        (2) Delivery of switch gear, light fixtures, etc.

        (3) Delivery of key HVAC equipment

        (4) Delivery of unusual components that dramatically affect sequences (for example, spandrel glass roof)

     f. Project milestones, such as:

        (1) Roof complete

        (2) Building weathertight

        (3) Permanent power

        (4) Permanent heat

        (5) Substantial completion

        (6) Punchlist period

        (7) Turnkey date

5. As the draft nears completion, conduct a preliminary confirmation meeting with all key players signed on to the project by then:

     a. Review the details of the entire draft with everyone together.

     b. Negotiate all sequences to the agreement of everyone, and secure confirmation of their respective intentions to proceed as coordinated.

     c. Take attendance. The resulting final schedule, along with an appropriate confirming letter of transmittal, will normally be sufficient to serve as the record of the meeting.

6. Finalize the complete schedule and distribute it in accordance with Section 10.4.

10.3.2 Schedule versus Contract End Dates

The schedule and contract end dates will not necessarily be the same. The contract will require completion of the schedule within a certain number of working or calendar days, but this does not require that your schedule be *extended* to this complete period if it does not need to be. The two situations then will be:

1. A schedule draft exceeding the contract time
2. A schedule requiring less than the contract time allowed

A Schedule Draft Exceeding the Contract Time.   In practice, if the first baseline schedule draft exceeds the contract time period, it must be massaged and shortened to fall within the allotted contract time. If efforts to accomplish this are foiled by some specification requirement that is beyond your control, an extension of the contract time (along with appropriate compensation) may be in order.

Examples of this kind of effect include:

- Specified equipment that is not available to be delivered to the project within the required time period
- Areas of the project that will not become available to the Contractor to work in sufficient time during the contract period
- Unusual sequences or events that could not have been reasonably anticipated at the time of bid

The Owner and the design professionals had an obligation, throughout the preparation of the contract documents, to provide a specification that is consistent— one that is *achievable* within the allotted time, notwithstanding the fact that there may be several ways to build the same project. In addition, it is the design professional who carries the *implied warranty* that this is the case. If subsequently it is discovered that such performance is impossible, it will be the designer who will bear the responsibility. Refer to Section 3.5.13, Impossibility and Impracticability, for related discussion.

A Schedule Requiring Less than the Contract Time Allowed.   If it is reasonably demonstrated that the schedule is achievable in less time than that allowed by contract, the Contractor has a right to so complete it.

If there is *very* clear and complete (and fairly elaborate) language in the contract as to the Owner's lack of ability and/or intention to take the project any sooner, there may be a basis for the Contractor to "carry" the project unoccupied to a certain degree. But this case is extremely rare, and even then suspect.

If the true schedule that is being used to manage the project is shorter than the contract time, the Owner, design professionals, Subcontractors, or anyone else may not interfere with it. They will be just as responsible for a delay in the *schedule* end date as for a delay in the contract date.

Know this condition, and tolerate no argument to the contrary. Owners who have the delay-claim experience (many public entities, for example) will argue energetically that your schedule *must* correspond to the contract time. Realize that these points are attempts to cleanly sidestep complete responsibility for at least the time difference between the schedule end date and the contract date if a delay situation develops.

10.4   Baseline Schedule Distribution and Final Confirmation

10.4.1   Final Baseline Schedule Confirmation

By the time the schedule is ready to be released (assuming the effort has been conducted as early as possible), many (but not all) of the major Subcontractors and suppliers should have participated. The Owner and the design professionals will not have participated (unless serious problems became apparent that required design changes), and many major and minor Subcontractors and suppliers for various bid packages will not yet have been signed on. For these items, extreme care should have been taken to verify the validity and reasonableness of all planning information, so that problems and surprises with new constraints disclosed when those items are finally purchased are minimized.

The baseline schedule should accordingly be confirmed to the fullest extent possible before eventual submission to the Owner as the complete document against which all project performance is to be measured. Accordingly, the sequence will be as follows:

1. Transmit the schedule to all who participated in its development in order to secure final confirmation of all information. Tolerate only the most serious pieces of new information that cannot be modified at this point if anyone wishes to change previously coordinated information.

2. Transmit the final confirmed schedule to the Owner, first in accordance with any specification requirement, and then with any Company requirements.

3. Use the finalized schedule as the latest specific subcontract or trade contract requirement for each new vendor signed on to the project.

10.4.2   Sample Letter to Subcontractors and Suppliers Regarding Baseline Schedule Confirmation (page 10.17)

The sample Letter to Subcontractors and Suppliers Regarding Baseline Schedule Confirmation is to be used to transmit the baseline schedule to all those who participated in the schedule's development or have otherwise been signed on to the project as of the schedule release date. Its purpose is to secure positive confirmation from all key project participants that the specific performance requirements have been directly considered and that they agree to adhere to those requirements in every respect.

10.4.3   Sample Letter to Subcontractors and Suppliers Regarding Baseline Schedule Final Release (page 10.18)

The sample Letter to Subcontractors and Suppliers Regarding Baseline Schedule Final Release is to be used to transmit the final schedule to all parties after all

10.4.2
Sample Letter to Subcontractors and Suppliers
Regarding Baseline Schedule Confirmation

Letterhead

To: 1) _____   Date: _____

2) _____

3) _____   Project: _____

4) _____

5) _____   Project No. _____

6) _____

7) _____   Subj.: Confirmation of Baseline

8) _____   Construction Schedule

9) _____

Gentlemen:

Attached are two copies of the Baseline Construction Schedule finalized as a result of the schedule coordination meeting conducted on _____, 20____, indicating your general items of work and their relationship with the project. Note that the omission of any items required by your respective contracts does not relieve you of the requirement.

Please review the schedule, specifically considering all that is necessary to achieve the performance results indicated for the complete project, including:

*   Submittal dates and durations.
*   Material fabrication and delivery times.
*   Activity durations.
*   Dependencies (work required before your work can proceed).
*   Logic (correct sequences).
*   All other relevant considerations.

If modification to the schedule as it is represented is necessary, note same in red and return one copy to my attention for approval. If the schedule is acceptable as it is, please so confirm by signing one copy noting "approved," and return it to my attention.

Your response is required by _____, 20 ____.

Thank you for your cooperation.

Very truly yours,

COMPANY

_____
Project Manager

cc: File: Baseline Sched. w/att.
                CF

10.4.3
Sample Letter to Subcontractors and Suppliers
Regarding Baseline Schedule Final Release

Letterhead

To: 1) _____   Date: _____

2) _____

3) _____   Project: _____

4) _____

5) _____   Project No. _____

6) _____

7) _____   Subj.: Baseline Schedule

8) _____   Final Release

9) _____

Gentlemen:

Attached is the final Baseline Construction Schedule for the project.

This schedule will be periodically updated by this office to reflect current progress, and to identify changes to the program that may become necessary to maintain critical dates.

Per your subcontract, it is your responsibility to be aware of all current schedule conditions *as may be so changed from time to time*, and to comply with them in every respect. Failure to review all current information and to be aware of all current requirements will in no way relieve you of the responsibility for the information, obligations, and performance requirements.

Please acknowledge receipt of this schedule by signing below, and returning this letter to my attention.

Your response is required by _____, 20 _____.

Thank you for your cooperation.

Very truly yours,

COMPANY

_____

Project Manager

cc: File: Baseline Sched. w/att.
               CF

confirmation efforts have been completed. It is therefore to be used to send the final schedule requirements to:

1. All vendors who have appropriately responded to previous coordination efforts.

2. All vendors who have not responded to the sample Letter to Subcontractors and Suppliers Regarding Baseline Schedule Confirmation of Section 10.4.2 by the date required.

3. All vendors signed on to the project after the schedule was confirmed.

10.5 The Cash-Flow Projection

10.5.1 General

The cash-flow projection is the monthly and cumulative description of the value of the work that is anticipated to be completed within each respective payment time period. It relates the approved Schedule of Values directly to the finalized baseline construction schedule to result in a schedule of planned billings for the life of the project.

Its accuracy is directly dependent upon:

- The accuracy of the general Schedule of Values
- The degree of direct correlation of the general Schedule of Values with those of each major Subcontractor and supplier
- The accuracy of the baseline construction schedule
- The degree of direct correlation of the construction schedule with the general Schedule of Values (and the corresponding lack of subjectivity in relating the two)

If the accuracy and direct correlation of all these documents can be maintained:

- The cash-flow projection will be realistic to begin with.
- The relation of cash progress with schedule progress will be accurate.
- Subsequent cause-and-effect and even consequential damage analysis will remain valid.

10.5.2 Preparation

Preparation is most straightforward if care has been taken to relate the baseline schedule directly to the general Schedule of Values. This itself is more readily achievable if each has been related directly to the project specification in the first place.

In any case, the complete procedure boils down to:

1. Assigning an appropriate cost to each scheduled activity.
2. Distributing that cost appropriately within the activity duration (as recognized to be billable).
3. Relating those distributed costs to the respective payment periods in which each falls.

The procedure is essentially the same, whether a computer schedule is used or the information is calculated manually. Computer programs will often have a mechanism to allow assignment of cost to an activity, and also allow some degree of control as to the placement of cost in an either distributed or point-cost manner. If not, the computer schedule plot can be used as it is, with the cost-plotting procedure conducted manually.

In either case, the critical issue is keeping complete and accurate notes as to the actual estimates, determinations, and procedures used throughout the preparation exercise in order to substantiate your work if the projection should ever be questioned at a later date.

10.5.3 Procedure

The procedure described is to construct the projection manually. The principles are identical to those applied to any computer program, and should be treated accordingly.

1. Assuming a time-scaled schedule, draw vertical lines through the schedule at each payment period (presumably months). Note that this step is not necessary with computer-generated cash projections.
2. Assign costs for all schedule activity items and material deliveries.
   a. For each activity and material delivery item, insert the total cost estimate as your reference for eventual distribution along the respective activity.
   b. Keep this total cost distribution as closely related as possible to the approved Schedule of Values.
3. Distribute activity costs as appropriate for each specific item.
   a. For on-site activities, it will generally be most appropriate to distribute the total activity's cost evenly over the duration of the activity.
      (1) With the exception of unusual or very long (and possibly oversimplified) items, this method will be sufficient for the purposes of the projection.
      (2) If there is a significant reason why a relatively even distribution of costs will not be appropriate, consider breaking the activity into those components that define the detail more correctly and assign the separate costs accordingly.
   b. For computer programs, it will normally be sufficient to simply indicate this distribution instruction. For manual preparation, simply prorate the

percentage of cost to the same percentage of the actual activity time falling in each payment period.

4. Distribute material delivery costs.

   a. Most contracts provide that materials will be paid for only when delivered to the site. If the prescribed conditions are different, so be it.

   b. While it is true that other payment conditions may apply in problem circumstances, they are normally there to deal with special situations that may develop. They accordingly have no place in a projection that does not directly anticipate such a problem. If you are aware of payment conditions that allow for off-site payment, your schedule should have a "delivery/storage" bar or item that specifically shows it. It can then be treated in the same manner as all other delivery items.

   c. Because of the foregoing conditions, it will then be appropriate to assign 100% of the complete material cost for the item to the last day of the material delivery schedule time bar or item. In other words, all of it will be billable when all of it arrives at the jobsite.

5. Consolidate all costs. Total all costs for each payment period, and enter the value at the bottom of the column.

6. Apply overhead, supervision, profit, and all other direct and soft costs either distributed evenly over the entire schedule or as otherwise appropriate. This step can be omitted if these cost factors have already been considered in the preparation of the individual activity costs as "complete" costs in the preceding steps.

   a. Some costs such as "jobsite overhead" are more likely to be accurate if spread evenly per month (trailer rental, for example).

   b. Some costs are more appropriate to be period-specific (bond premium costs falling in the first month, for example).

   c. Some cost categories can go either way. (Profit can be treated as applying strictly to the same percentage as work complete, or it can be more directly related to the specific bid packages.)

7. Add step 6 overhead, supervision, and profit to the total direct costs of step 5 to arrive at the total receivable values for each payment period. The cash-flow projection itself is not complete.

The simplified illustration given in Section 10.5.4 shows the specific steps as described. This illustration is complete; more detailed schedules only have more items.

10.5.4   Example Cash-Flow Preparation Worksheet *(page 10.22)*

10.5.5   The S-Curve

This section describes the creation of the S-curve and includes basic direction on its uses and purposes. You should be aware that very elaborate analyses of total project performance and the ability to demonstrate many direct cause-and-effect relationships are possible with the S-curve, but these detailed developments are beyond the scope of this Manual.

10.5.4
Example Cash-Flow Preparation Worksheet

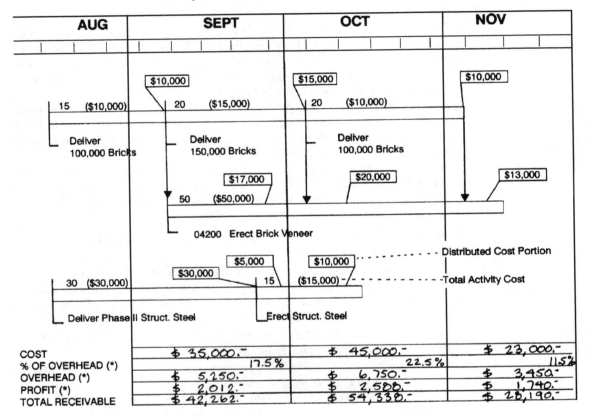

	AUG	SEPT	OCT	NOV
COST		$ 35,000.⁻	$ 45,000.⁻	$ 23,000.⁻
% OF OVERHEAD (*)		17.5%	22.5%	11.5%
OVERHEAD (*)		$ 5,250.⁻	$ 6,750.⁻	$ 3,450.⁻
PROFIT (*)		$ 2,012.⁻	$ 2,500.⁻	$ 1,740.⁻
TOTAL RECEIVABLE		$ 42,262.⁻	$ 54,330.⁻	$ 28,190.⁻

(*)   For this example:
Total Project Cost is $200,000
Overhead is $30,000
Profit is 5%, and distributed evenly

The S-curve is a cumulative plot of the cash-flow projection as developed in Section 10.5.3. In order to create it, simply:

1. Cumulate each month's total value with the combined values of all preceding periods.
2. Plot on graph paper to create the S-curve.

In the example given in Section 10.5.4, and assuming September in the example is the project's first period, the information would be as follows:

	Period 1 September	Period 2 October	Period 3 November
Total receivable	$42,262	$54,338	$28,190
Cumulative totals	$42,262	$96,600	$124

The S-curve plot for these three periods would look like the following:

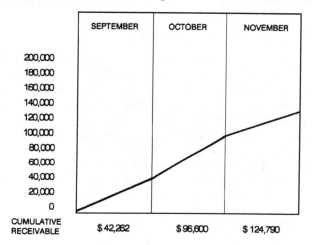

In this simplified example the "curve" is abrupt, with no distinct shape. In an actual, detailed analysis of an entire project spanning many time periods, the curve would begin shallow, steepen during the middle periods, and then level off again in the ending periods. This is a common characteristic that basically represents slower "start-up" activities in the beginning, the project maturing into its production phase of high activity, and closing off with slowing activities and with finishes that are not as high in cost relative to the complete project. It is this characteristic that will cause the actual curve to approximate the shape of an S. An actual S-curve for an entire project may look generally like the following:

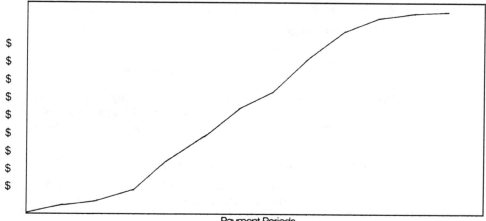

As a footnote, because this shape is so characteristic of most conventional schedules, if you should note that the shape of one of your curves is dramatically different, it may be an indication that the schedule itself is not realistic or is inaccurate. There may turn out to be a perfectly good explanation, but in such a case the S-curve analysis will at least have forced a review to confirm the validity of the original schedule. Most often, however, such a reconsideration would disclose an error.

10.5.6   Comparison of Actual to Projected Cash Flow

The first use of the S-curve is to allow planning for the cash flow for each project as specifically related to planned progress, and then monitoring such performance on an ongoing basis. As part of that function, actual applications for payment should be compared to those planned—as they are occurring. If this simple procedure is conducted for all projects on an ongoing basis, total Company cash flow becomes an item that is planned and managed, not something that becomes known as it happens (for better or worse).

In order to do this, plot the values of each requisition (contract value without approved change orders) on the same plot as the original S-curve. The relationships will immediately become apparent:

- If you are reporting the project "on schedule," the actual requisitions should be very closely paralleling the projected curve.
- If the project is being reported slightly behind schedule, the curve should be falling off slightly.
- If the project is being reported ahead of schedule, the actual payment line should be falling above the planned curve to the same degree.

If this direct one-to-one relationship between the reported schedule activity progress and actual billings is not apparent, something is wrong with one or more of the following:

1. The baseline schedule is inaccurate.
2. The cash-flow projection was prepared incorrectly.
3. Schedule progress reporting is not correct.
4. Billings are not being conducted correctly.
5. Billing *approvals* by the Owner or the Architect are not correct.

If the schedule is truly being used to manage the project, the situation as it exists will become apparent to management as soon as it is occurring—in plenty of time to do something about it before it gets out of control.

10.5.7   Comparison of Cash Progress with Activity Progress

The S-curve can be very useful in validating the original schedule and supporting the schedule updates. The basics of the subject are reviewed here, but the very detailed development of the techniques is beyond the scope of this Manual.

Curve Shape and Slope.   The first analysis has to do with the shape of the curve itself. The slope, or steepness, of the curve is actually the rate of progress of the work. The steeper the slope, the higher the billings, the more work is being represented to be complete.

The S shape is the result of the characteristic work flow of a project, which includes:

1. A mobilization and start-up period (shallow curve slope).
2. Work increasing and developing to the "pace" of the project (steepening curve slope).
3. A general slowing down of the rate of progress, and work being done on finishes— work that is lower in cost relative to the complete project (flattening curve slope).

As mentioned previously, if the curve does not assume the characteristic S shape, there is a good chance that something may be wrong with the planned schedule. Such common problems include:

1. The curve begins too steeply, indicating that too much work is anticipated too soon in the very beginning of the project.
2. The shape varies greatly among payment periods in the job's pace cycle, indicating billings that are jumping around when progress should actually be steady.
3 The curve ends too abruptly instead of flattening out. This can mean that you are expecting the project to run to its end at the rate of its pace cycle and then just stop—possible but not probable.

There may be good explanations for these effects (such as coincidental delivery of a large number of expensive equipment items that causes the projected value of a certain period to jump dramatically). The point, however, is simply that if the curve varies dramatically from the expected shape, you must either confirm that there are appropriate reasons, or fix the mistake immediately.

Comparison of Planned Curve with Actual Curve as a Means of Validating the Original Schedule Itself.   In a delay situation, the original schedule will become the subject of an extreme amount of criticism. It will be accused of being too aggressive and not properly prepared in the first place. In the worst conditions, it might even be accused of being intentionally prepared in an overly aggressive manner, specifically to maximize future delay claims.

If such a situation develops in which the project was delayed, the cause of the delay is eventually resolved, and work resumes, the curve can go a long way in substantiating the validity of the planned program.

In such a delay situation, the planned curve for the period will typically be steep (indicating a good amount of planned progress), and the effects of the delay will flatten the curve of the actual payments (indicating that less work was done than planned). When work resumes, the actual curve should then approximate the original curve of the preceding delay period. If it does, the confirmed actual rate of progress (steepness of the actual curve) that matches planned curves for the same kind of work will directly confirm that the original planned work was absolutely achievable in the first place. It will nail down the fact that the planned schedule was realistic, and therefore the resulting delay calculations are accurate. All other periods having actual curves that approximate the planned curve will add to the substantiation of the original schedule as a valid program.

As a final note, the most direct comparison is of actual contract work to planned contract work—without changes and change orders. You may therefore find it helpful to plot three actual curves:

1. Actual billings of contract work only.

2. Actual billings of change order work.

3. The combined curve of actual contract work plus changes.

In a fine-tuned analysis the curves will have different uses. Even with no specialized training, however, project people will be surprised at the amount of information that will jump off the paper while they are just reviewing the plots.

10.6   Schedule Updating Considerations

10.6.1   General

The actual updating technique will vary depending upon the scheduling method selected. Bar charts only provide for the recording of the passage of time—as it passes. Other methods will allow for recording and accounting for:

- Actual activity start dates
- Actual activity completion dates (substantial completion of enough work to allow following trades to proceed)
- Time remaining to complete activities that have started but are not yet complete (as of the update issue)

Beyond this, scheduling methods and computer programs vary in their abilities and procedures for accommodating changes, adjustments in schedule logic, and so on.

10.6.2   Scheduling Approach

The important ideas for managers to focus on continually include the following:

1. Throughout all schedule updating efforts, consider not so much "how much time has gone by" as "*when* will it be *done.*"

2. Involve *all* Subcontractors and suppliers on a continuing basis. Force them to provide complete information, and in writing if necessary.

3. Involve the Owner and design professionals. Develop a routine for transmitting schedule updates, reporting the status of the project, indicating the effects of everyone's actions and inactions, and describing "get-well" plans.

4. Use your schedule as *the* agenda at all job meetings. Let everyone see that you are *always* considering the scheduling effects of all occurrences and decisions, and displaying those effects on paper.

5. Develop a complete lack of tolerance for *any* schedule slippages. Be forceful. Don't risk the development of even the slightest bad habits on the part of any project participant.

Use the sample forms and letters that complete this section, but be creative. Use them as a guide from which to develop detailed accountability systems that are closely coordinated to the specific scheduling system used by the Company.

10.6.3 Sample Schedule Analysis/Evaluation Report *(page 10.28)*

The sample Schedule Analysis/Evaluation Report that follows can be used as the routine correspondence that:

1. Transmits the schedule or schedule update to the Owner and design professionals.

2. Distributes all current schedule information to each key Subcontractor.

3. Reports the current status of the project as of the date of the update.

4. Includes anticipated "get-well" programs intended to bring identified deficiencies back into line.

In preparing the report:

1. Include names, companies, and dates. Be as specific as possible.

2. In the "Remarks" section, indicate all new constraints and conditions:
   a. If the project is doing well (on or ahead of schedule), meaningful remarks might include:
      - The level of confidence by management in the ability of the project to maintain such performance
      - Specific critical areas that require direct attention to maintain the status
      - Rough extrapolation of the performance into the future periods, if other known conditions exist whose effects have not yet been felt on the project
      - Any significant facts, proceedings, etc., that might impact on the status report of the *next* or other future update(s)

10.6.3
Sample Schedule Analysis/Evaluation Report

Date: _____   Project: _____ #: _____

To: _____   Report Date: _____ Rev. #: _____

_____

_____

ATTN: _____

# COPIES	REV #	REVISION DATE	DESCRIPTION / TITLE	BASELINE COMPLETION	REVISION COMPLETION	DEVIATION (+ /-)	
							WD CD

As of _____, 20 _____, this project is _____ Working Days (Calendar Days) ahead
of (behind) schedule, primarily due to:

_____

_____

_____

_____

_____

_____

_____

_____

_____

_____

_____

_____

Remarks:

_____

_____

_____

_____

_____

_____

_____

_____

_____

_____

_____

_____

_____

_____

cc: Jobsite w/att.   Signed:

_____   _____

_____   _____

File: Sched. Rev #_____ w/att, CF

b. If the project is in an unacceptable condition (behind schedule), include detailed outlines of:

- All confirmed commitments, promises, etc., relative to any "get-well" programs devised to this point
- Nonperformance, inattention, inappropriate action, etc., on the part of any Subcontractor, supplier, design professional, the Owner, or whomever (be specific, name names)
- Previously "noncritical" items that have now become critical
- Specific actions required by whomever

   Use names and dates. Pin it down. Refer to all appropriate correspondence, meeting minutes, and so on to support your remarks with the more detailed project record (and make subsequent research much easier).

3. Distribute copies of the completed report to:
   a. All parties referenced in any way in the "Status" or "Remarks" sections
   b. The Owner
   c. The Architect
   d. All major Subcontractors

   The periodic distribution of such complete information will generally fulfill most procedural requirements that will support any requirement for timely and complete notice.

10.6.4   Sample Delay Letter #1 to Subcontractors *(page 10.30)*

The sample Delay Letter #1 to Subcontractors can be used as a first formal notification to a Subcontractor that its work is falling off and that the schedule is accordingly in jeopardy. It:

1. Notifies the offending party that current progress is unacceptable and is adversely affecting the project.

2. Indicates the specific problem activities.

3. Notifies of responsibility for the resulting effects.

4. Requires specific written confirmation of new commitments necessary to regain the schedule.

   A supply of these form letters should be on hand and completed *as each schedule update is being conducted.* When in doubt, send it out.

10.6.5   Sample Delay Letter #2 to Subcontractors *(page 10.31)*

The sample Delay Letter #2 to Subcontractors is to be used in those instances where a chronically deficient Subcontractor continues to be unresponsive to you and to the project requirements. It:

1. Acknowledges that no acceptable response has been received to Letter #1.

2. Invokes the subcontract acceleration provisions directly.

10.6.4
Sample Delay Letter #1 to Subcontractors

	Letterhead

To: _____   Date: _____

   _____   Project: _____

   _____   _____

   _____   Project No.:_____

Attn: _____   Confirmation of Fax:
                                        Fax #: _____

SUBJ: Anticipated Project Delay

Mr.(Ms.) (   ),

At your current rate of progress, it is apparent that the schedule completion dates for the following activities will be missed:

Activity   Scheduled Completion

_____   _____

_____   _____

_____   _____

_____   _____

_____   _____

If these completion dates are not regained, you will directly interfere with and delay all activities scheduled to immediately follow, thereby directly delaying key milestones and completion dates. In this event, you will be held responsible for all associated costs and other problems resulting.

Please submit your written statement to my attention by _____, 20 _____, specifically advising of how you intend to regain the schedule. Include:

    1.   Number of workers per day.
    2.   Number of regular and *overtime* hours per day planned.
    3.   Number of days for each subactivity.
    4.   Specific measures being taken to expedite all necessary material deliveries.

Your complete written response is critical to the orderly completion of the project without serious incident. Your cooperation is appreciated.

Very truly yours,

COMPANY

_____   cc: Jobsite
Project Manager   cc: File: Vendor File: _____
                                                            Sched. Rev #_____, CF

10.6.5
Sample Delay Letter #2 to Subcontractors

+--+
| **Letterhead** |
+--+

(Date)

To: (Subcontractor)   Confirmation of Fax:
                                            Fax #: _____

                                            CERTIFIED MAIL
                                            RETURN RECEIPT REQUESTED

RE: (Project)
           (Company Project #)

SUBJ: Project Delay and Directed Acceleration

Mr. (Ms.) (   ),

As of this date, we have received no appropriate response to our letter of (Date) regarding your anticipated project delays. Since that date, your work has continued to progress unacceptably, causing additional interferences.

Per Paragraph (   ) of your subcontract, you are hereby directed to immediately accelerate your work by adding all labor, material, equipment, overtime, Saturday, Sunday, and Holiday work as necessary to regain the schedule.

Your failure to comply with this directive will be considered a material breach of your subcontract. Appropriate action will then be taken by this office, which may include termination of your subcontract.

Please deliver your written confirmation to this office by or before (Time) on (Date) that you are either mobilizing to comply with this directive, or that you specifically refuse to comply.

No response by the date and time indicated will be construed as your notification by default of your refusal to comply. In that event, steps will be taken to expedite your work by other means. Your company will then be backcharged for all costs incurred, including but not limited to procurement and coordination time and effort, supervision, overhead, and profit. You will also be held responsible for all project delay costs and consequential effects.

Very truly yours,

COMPANY

Project Manager

cc: Jobsite
       File: Vendor File: (   )
                Sched. Rev. # (   ), CF

3. Notifies of the serious consequences of the continued inappropriate action.

4. Again requires a written statement of planned correction.

5. Notifies of at least the pending backcharge resulting from completing the work by other means.

The unfortunate truth is that if this letter becomes necessary, there are sure to be other problems with that Subcontractor. The Project Manager should at least be consulted before the letter is sent out, and it is probably a better idea for the letter to go out with the Project Manager's signature. In any event, if things have deteriorated to this degree, there should undoubtedly be other communications going on to bring such a serious problem under control. If all this is happening correctly, this Letter #2 becomes an important supplement, rather than the only communication. For its best effect, it should be used as a word-for-word letter.

Building Information Modeling

11.1   Section Description

Building Information Modeling (BIM) is a revolutionary software program that allows design consultants (Architects and Engineers) to create their structure's design in three dimensions. Since its early introduction into the design community in the 1970s, adoption and use by the architectural and engineering professions has surged from about 28% in 2007 to approximately 71% in 2012.

The concept's benefits to the construction industry are many. Both project managers and project superintendents who may not be familiar with BIM or working with computer-aided design systems may need an introduction to the process.

The Architect creates a three-dimensional building envelope design, which is passed through to the Structural Engineer and the MEP Engineers in basically the design development stage, for review, comment, and (perhaps) modification. This process brings constructability issues and coordination problems to the forefront while there is the opportunity to correct them in the design stage rather than in the field. Clashes and conflicts—between structural, HVAC, plumbing and electrical systems, vertical and horizontal piping, ductwork, conduit runs, and the building's structural design—are caught and corrected as the design progresses, not in the field where construction must come to a halt while the design team figures out how to make everything fit in its allotted space.

An added benefit of detecting and curing these coordination problems during design is that it allows mechanical, plumbing, and electrical subcontractors to prefabricate some assemblies with assurance that they will fit when installed.

11.2   Why Contractors Value BIM: The 3D Approach

Contractors are familiar with the specification requirement to prepare coordination drawings in the two-dimensional (2D) design world. This requirement is intended to avoid clashes and conflicts by detecting them as the various trades prepare their own drawings for an overlay by the Project Manager. A typical section of these coordination requirements might read as follows:

The Contractor shall coordinate construction operations included in the various sections of the specifications to assure the efficient and orderly installation of all parts of the Work. . . .

Reproducible drawings showing work with horizontal and vertical dimensions to avoid interference with structural framing, ceilings, partitions, equipment, light, mechanical, electrical, and conveying systems, and other services, shall include:

1. In and above suspended ceilings

2. Within partitions and walls

3. Within chases

4. In mechanical spaces

5. In electrical spaces

Prepare coordination drawings (individual drawings are submitted by the appropriate subcontractor).

The General Contractor prepares this "coordination" drawing pointing out the areas of concern. For example, some of the work does not fit above the ceiling, or pipes and ductwork are obstructed by the structure's concrete or steel components. When such problems are identified, the GC must then await direction from the Architect on how to proceed. Some ceilings may need to be lowered a bit, or steel beam shapes and sizes changed, or partitions or chases enlarged, or duct configurations changed.

Unfortunately, such conflicts between systems are not always 100% identified during design; often they go unnoticed until the situation arises in the field. At that point, work stops; the contractor waits for direction from the Architect; changes, if required, are submitted to the various subcontractors; and change orders are issued. Potential additional work, delays, and loss of productivity due to work stoppage are almost inevitable—and thus claims are born.

The BIM process allows the designers to spot these "conflicts" or "clashes" (as they are referred to) during the design process and make the necessary changes to the drawings. The figure in 11.2.1 illustrates how BIM enabled a designer to include sleeves through a series of grade beams when it appeared that a pipe's path could not be altered. The figure in 11.2.2 illustrates, again, how a plan view alerted the designer to a clash with a continuous footing; the 3D depiction allowed the designer to cure the conflict by creating a sleeve through the footing to avoid the clash.

In 2012, according to a McGraw-Hill survey of BIM users, document errors and omissions were reduced by 57% when this 3D system was employed. Contractors reported that rework was reduced by 65%. Slightly more than one-half of the contractors surveyed reported that there were 55% fewer drawing errors and omissions when BIM was used. Also, 53% of the contractors reported reduced project duration and increased productivity, presumably because little or no time was lost in unraveling coordination problems.

11.2.1 Sleeving through a Grade Beam (page 11.4)

This figure shows how sleeving through a grade beam resolved a clash. The use of BIM revealed the clash in the "drawings" and prevented it from getting through to be discovered later in the field.

11.2.2 Sleeving through a Foundation (page 11.5)

In this instance, BIM enabled the designer to identify and resolve a clash in underslab drainage that conflicted with footings. Again, this problem was revealed in the "drawing" stage, and prevented the error from getting through to be discovered later in the field.

11.2.1
Sleeving through a Grade Beam

SLEEVING THRU GRADE BEAMS

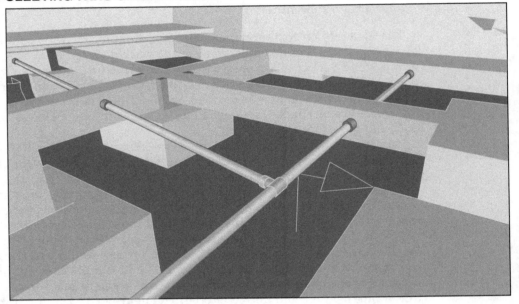

Source: The Albert Sherman Center, University of Massachusetts Medical School project.

11.3   How Does BIM Impact the Role of the Project Manager?

The entire BIM process is one of collaboration among Owner, design consultants, and, more often than not, the General Contractor/Construction Manager. The change from working in a two-dimensional drawing environment to a three-dimensional one requires new skills for dealing with changes in software and a new way of looking at time and materials allocations. Those contractors and project managers who have attained the necessary skills to work in a BIM environment will certainly have a leg up on the competition.

Because it will be necessary to import or export design data, the contractor and the other team members—subcontractors and vendors—will need to have, and be working from, the same software package. The construction Project Manager assigned to a BIM project must become familiar with the use of that software and may have to attend one of the many courses offered by various organizations to obtain this training.

11.2.2
Sleeving through a Foundation

UNDERSLAB DRAINAGE CONFLICTING WITH FOOTINGS

	*UMMS- Albert Sherman Center - Constructability Report*
ID	**ID-004**	**Trades Involved**	**SP**	**Severity**	**Medium**
Description		Clash between 8" diam PVC pipe sleeve (top of elv = 428'-10", bottom of elv = 428'-2") and CIP continous footing 50" × 96" × 30" bottom of elv = 426'-0".
Floor	**FLOOR 1**	**Grid Reference**	**E/8-9**	**Sheet Reference**	**S-1.0.00**

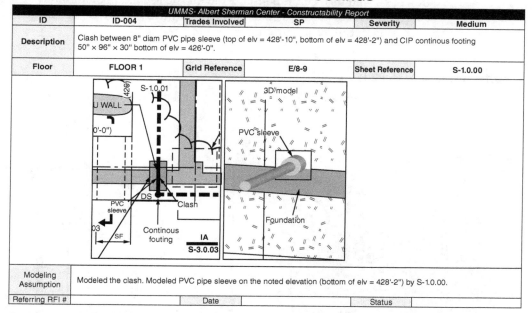

Modeling Assumption	Modeled the clash. Modeled PVC pipe sleeve on the noted elevation (bottom of elv = 428'-2") by S-1.0.00.
Referring RFI #	**Date** **Status**

Source: The Albert Sherman Center, University of Massachusetts Medical School project.

The 3D model developed by the design team will certainly have value to the construction team, including the subcontractors and vendors. The 3D design process will also be of value to the General Contractor or Construction Manager's Estimating Department, as the process allows users to develop quantity take-offs as a byproduct when the building components have been completed.

When general contractors or construction managers (and their subcontractors and suppliers) are brought into the project during design development, their experience and knowledge will be welcomed. These contractors can suggest simple or not-so-simple changes to the model and help fine-tune the design. The contractor may also comment on design details and their impact on schedule and cost.

Project managers may find themselves working in multiple views of the same detail expressed in 3D, plan view, elevation, and section (see Section 11.3.1).

Or they may be presented with several design options for their review and comment. For an example of comparing options, Section 11.3.2 shows a building entry detail in which a full-length column from grade to the second floor is shown, as well as a column supported on the second floor, possibly by a perimeter beam at that second-floor location. Can the contractor comment on the preferable detail, costs, productivity, etc.?

11.3.1 Working in Multiple Views

The figure shows a 2D plan, elevation, and section and the 3D presentation.

WORKING IN MULTIPLE VIEWS

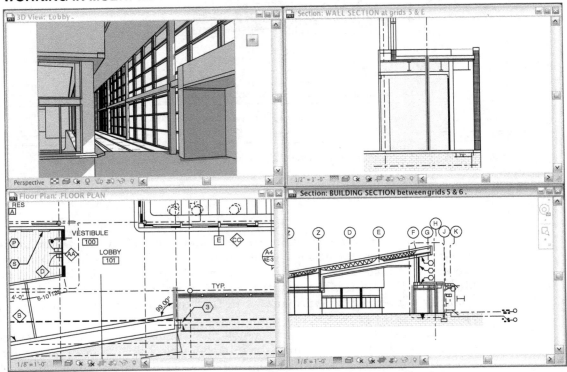

Source: Dekker/Perich/Sabatini presentation at 2011 CSI Southwest Region Conference.

11.3.2   Comparing Options

This 3D presentation allows comparison of two approaches to a building entry design.

Comparing Options

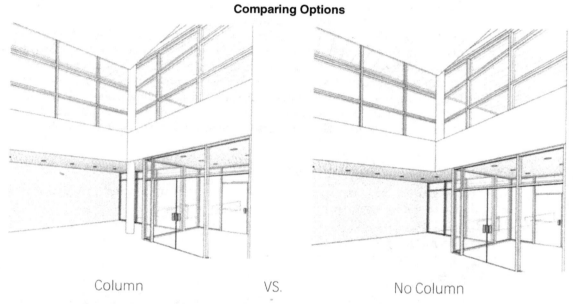

Column   VS.   No Column

Source: Dekker/Perich/Sabatini presentation at 2011 CSI Southwest Region Conference.

11.4   Building Information Modeling: 3D, 4D, 5D

The three-dimensional depiction of a building was the first iteration of the BIM process, but two other "dimensions"—4D and 5D—have since been added. Software companies recognized that value would be added for design and construction professionals by advancing beyond that third dimension.

11.4.1   The 4D Aspect of BIM

The use of 3D in developing a building's design pretty much follows the sequence in which conventional building construction takes place. The Civil Engineer will be assigned the responsibility to rework the topography of the site to accommodate the new structure, design the underground utilities in consultation with the Mechanical Engineer, and develop the site improvements (roadways, parking areas, walks, site structures, etc.). The Structural Engineer will commence development of the structural system based upon the foundation type and the Architect's design of the building envelope. The MEP engineering group will design the size and location of the building's plumbing, HVAC, electrical, and communications systems.

Why not utilize this software, with a little tweaking, during these phases of development to create a virtual schedule? This is the 4D approach, in which development of a schedule graphically incorporates the time frame for each of these construction phases: site grading, foundation work, steel frame erection, building envelope, and the site and interior utilities to service the building's plumbing, HVAC, and electrical systems.

An example of use of this 4D design tool to produce a virtual schedule is a hospital built in Worcester, Massachusetts: The Albert Sherman Center. By adding time sequences to the operations as developed, and including a start and completion date for each phase, a virtual schedule was created. Brought up on a computer screen at the jobsite, managers could merely look out the window and see if the actual progress matched the stage of construction depicted on the "virtual" schedule.

The following schematic drawings are examples of this 4D approach.

11.4.2   Building Platform Brought Up to Grade

This figure reveals the building platform brought up to grade, ready for foundation work to commence.

SITE PLANNING

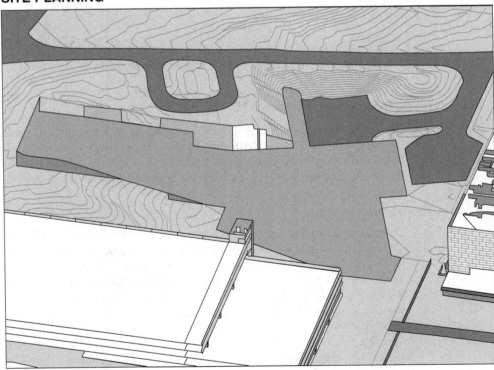

Source: The Albert Sherman Center, University of Massachusetts Medical School project.

11.4.3   The Core and Shell Structural Steel Framework in Progress

This image shows the core and shell structural steel framework in progress.

CORE/SHELL STEEL STRUCTURE

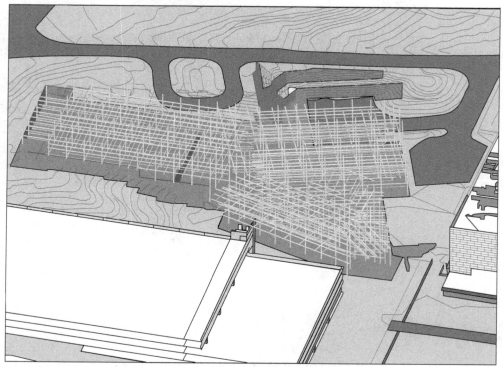

Source: The Albert Sherman Center, University of Massachusetts Medical School project.

11.4.4   Roof Steel and Floor Slabs, Partial (*page 11.10*)

The roof steel is not complete in this image, but floor slabs have been poured, the roof is on one portion of the building, and the building envelope is in progress in that same area.

11.4.5   Building Envelope with Floor Slab and Stairwells
(*page 11.11*)

The building envelope, continuing floor slab work, and stairwells up to the penthouse level on the main building are displayed in this image.

11.4.4
Roof Steel and Floor Slabs, Partial

CORE/SHELL

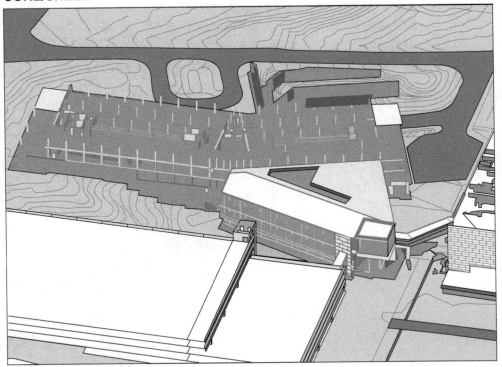

Source: The Albert Sherman Center, University of Massachusetts Medical School project.

11.5   The 5D Aspect of BIM

As a design is being developed—say, for the structural steel system—another "dimension" can be added to the process: the ability to create a list of materials and quantities. As each piece is sized and dimensioned, unit costs for each component (connections, bearing plates, etc.) can be added to create a credible cost estimate. That is the capability of what is called the *fifth dimension*. For example, in the case of a steel structure, a "cutting list" can be prepared as the steel members are selected, sized, and dimensioned. A quantity take-off can be concurrently produced during the process and, if a structural steel contractor has been engaged, a reliable estimate can be assembled. The Subcontractor can submit this cutting list to the mill and obtain a place in the mill's production cycle. A bill of quantities can be prepared for other trades as well: perhaps

11.4.5
Building Envelope with Floor Slab and Stairwells

CORE/SHELL

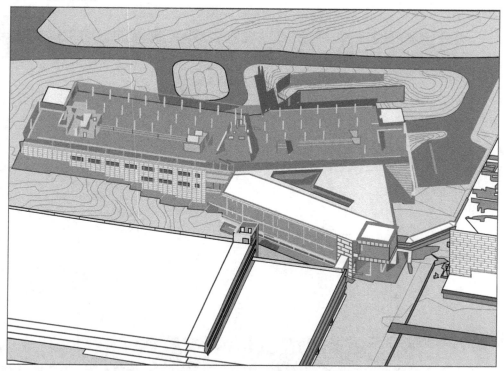

Source: The Albert Sherman Center, University of Massachusetts Medical School project.

miscellaneous metals, or fireproofing of steel members. An illustration of a 5D quantity take-off is shown in Section 11.5.1.

If a Construction Manager or General Contractor has been selected by the Owner during this modeling phase, fairly accurate project costs can be developed as design progresses, utilizing the CM or GC's cost data base and costs received from the subcontractors and vendors. If costs in a particular trade are exceeding the budget, design modifications can be made to bring the project back on budget.

Building Information Modeling, in all of its aspects, is proving itself to be a very effective tool. Part of its value lies in its ability to foster collaboration among all parties to the design-build team in increasingly harmonious relationships.

11.5.1 A 5D Take-Off

In this illustration, the take-off portion contains three columns from left to right: Description (of the material), Unit (square foot, lineal foot, cubic yard), and Value.

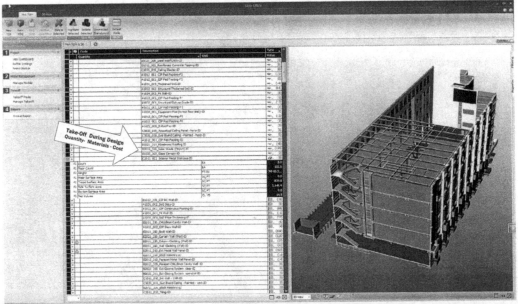

Source: By permission: Trimble Navigation Ltd., Sunnydale, CA.

Green Buildings and Sustainability

12.1   Introduction to Green Building and LEED

The process of producing an environmentally friendly structure, in both design and construction, was the basis for the creation of a rating system that would award "points" for accomplishing certain criteria and achieving specific goals. This program, developed by the U.S. Green Building Council (USGBC) just after it was founded in 1993, was entitled Leadership in Energy and Environmental Design (LEED). What started out as a small movement by the USGBC in 2000 has burgeoned to approximately 7,500 LEED-certified buildings in 2010. Adding 1 billion square feet in 2010 alone, this represented a growth rate of 14% over previous years.

12.1.1   Definition of Green Building

The U.S. Environmental Protection Agency defines *green building* as follows:

Green building is the practice of creating structures and using processes that are environmentally responsible and resource efficient throughout the building's life-cycle from siting to design, construction, operation, maintenance, renovation and deconstruction. This practice expands and complements the classical building design concerns of economy, utility, durability and comfort. Green building is also known as a sustainable high performance building.

12.1.2   The Eight Components of Green Building

According to the U.S. Environmental Protection Agency (EPA), the eight components of green building are:

1. *Energy efficiency*, which includes the EPA's Energy Star program; renewable power sources (solar, wind, wave); heat island reduction (reducing the heat island effect created by asphalt and other hard-surfaced paving); and energy-efficient roofs, both cool roofs and green roofs.

2. *Water efficiency,* which includes promotion of water-efficient appliances and reuse of waste water (both storm water runoff and treated, recycled water).

3. *Environmentally preferable building materials,* which includes recycling of demolition and other contraction debris/waste materials.

4. *Waste reduction,* which encourages projects that reduce, reuse, or recycle waste materials, as well as the development of environmentally friendly solutions such as large-scale landscaping designed to preserve natural resources.

5. *Development of a Lifecycle Building Challenge program* that will facilitate disassembly of material for reuse to minimize waste.

6. *Toxics reduction* by substituting safer chemicals and educating users on the health and safety values of certain materials, such as spray polyurethane foam.

7. *Indoor air quality*, also referred to as "Indoor Environmental Quality," that offers equipment, tools, and programs to protect occupant health, promote comfort and productivity, and enhance the durability of structures.

8. *Smart growth and sustainable development* that promote greenscapes to preserve natural resources, reduce runoff of pollutants from urban environments into streams and rivers, and encourage new approaches to storm water management and environmentally responsible development and reuse of contaminated sites.

12.2   How Does LEED Work?

LEED is a point-based system in which points are awarded for complying with, and achieving, various environmentally sensitive design and building components. These components include:

1. Sustainable sites

2. Water efficiency

3. Energy and atmosphere

4. Materials and resources

5. Indoor environmental quality

LEED certifications are divided into five levels of compliance:

- Certified—requires 40-49 points
- Silver—requires 50-59 points
- Gold—requires 60-79 points
- Platinum—requires 80-110 points

LEED certification is available for multiple building types, including:

- New construction and major renovation
- Existing construction
- Commercial interiors
- Core and shell
- School
- Retail establishments
- Health care facilities
- Residential
- Neighborhood development

12.2.1   A Typical LEED Checklist

This example checklist is for new construction and major renovations.

12.2.1
A Typical LEED Checklist

Project Name
Date

LEED 2009 for New Construction and Major Renovations
Project Checklist

Sustainable Sites — Possible Points: 26

		Y ? N
Prereq 1	Construction Activity Pollution Prevention	Y
Credit 1	Site Selection	1
Credit 2	Development Density and Community Connectivity	5
Credit 3	Brownfield Redevelopment	1
Credit 4.1	Alternative Transportation–Public Transportation Access	6
Credit 4.2	Alternative Transportation–Bicycle Storage and Changing Rooms	1
Credit 4.3	Alternative Transportation–Low-Emitting and Fuel-Efficient Vehicles	3
Credit 4.4	Alternative Transportation–Parking Capacity	2
Credit 5.1	Site Development–Protect or Restore Habitat	1
Credit 5.2	Site Development–Maximize Open Space	1
Credit 6.1	Stormwater Design–Quantity Control	1
Credit 6.2	Stormwater Design–Quality Control	1
Credit 7.1	Heat Island Effect–Non-roof	1
Credit 7.2	Heat Island Effect–Roof	1
Credit 8	Light Pollution Reduction	1

Water Efficiency — Possible Points: 10

Prereq 1	Water Use Reduction–20% Reduction	Y
Credit 1	Water Efficient Landscaping	2 to 4
Credit 2	Innovative Wastewater Technologies	2
Credit 3	Water Use Reduction	2 to 4

Energy and Atmosphere — Possible Points: 35

Prereq 1	Fundamental Commissioning of Building Energy Systems	Y
Prereq 2	Minimum Energy Performance	Y
Prereq 3	Fundamental Refrigerant Management	Y
Credit 1	Optimize Energy Performance	1 to 19
Credit 2	On-Site Renewable Energy	1 to 7
Credit 3	Enhanced Commissioning	2
Credit 4	Enhanced Refrigerant Management	2
Credit 5	Measurement and Verification	3
Credit 6	Green Power	2

Materials and Resources — Possible Points: 14

Prereq 1	Storage and Collection of Recyclables	Y
Credit 1.1	Building Reuse–Maintain Existing Walls, Floors, and Roof	1 to 3
Credit 1.2	Building Reuse–Maintain 50% of Interior Non-Structural Elements	1
Credit 2	Construction Waste Management	1 to 2
Credit 3	Materials Reuse	1 to 2

Materials and Resources, Continued

		Y ? N
Credit 4	Recycled Content	1 to 2
Credit 5	Regional Materials	1 to 2
Credit 6	Rapidly Renewable Materials	1
Credit 7	Certified Wood	1

Indoor Environmental Quality — Possible Points: 15

Prereq 1	Minimum Indoor Air Quality Performance	Y
Prereq 2	Environmental Tobacco Smoke (ETS) Control	Y
Credit 1	Outdoor Air Delivery Monitoring	1
Credit 2	Increased Ventilation	1
Credit 3.1	Construction IAQ Management Plan–During Construction	1
Credit 3.2	Construction IAQ Management Plan–Before Occupancy	1
Credit 4.1	Low-Emitting Materials–Adhesives and Sealants	1
Credit 4.2	Low-Emitting Materials–Paints and Coatings	1
Credit 4.3	Low-Emitting Materials–Flooring Systems	1
Credit 4.4	Low-Emitting Materials–Composite Wood and Agrifiber Products	1
Credit 5	Indoor Chemical and Pollutant Source Control	1
Credit 6.1	Controllability of Systems–Lighting	1
Credit 6.2	Controllability of Systems–Thermal Comfort	1
Credit 7.1	Thermal Comfort–Design	1
Credit 7.2	Thermal Comfort–Verification	1
Credit 8.1	Daylight and Views–Daylight	1
Credit 8.2	Daylight and Views–Views	1

Innovation and Design Process — Possible Points: 6

Credit 1.1	Innovation in Design: Specific Title	1
Credit 1.2	Innovation in Design: Specific Title	1
Credit 1.3	Innovation in Design: Specific Title	1
Credit 1.4	Innovation in Design: Specific Title	1
Credit 1.5	Innovation in Design: Specific Title	1
Credit 2	LEED Accredited Professional	1

Regional Priority Credits — Possible Points: 4

Credit 1.1	Regional Priority: Specific Credit	1
Credit 1.2	Regional Priority: Specific Credit	1
Credit 1.3	Regional Priority: Specific Credit	1
Credit 1.4	Regional Priority: Specific Credit	1

Total — Possible Points: 110

Certified 40 to 49 points   Silver 50 to 59 points   Gold 60 to 79 points   Platinum 80 to 110

Source: greenexamguide.com

12.4

12.3   Public- and Private-Sector Green Buildings

The public sector has embraced LEED. According to USGBC figures, government-owned or -occupied LEED buildings constitute 27% of all "Green" projects. The federal government owns 828 certified projects, and another 3,900 are awaiting certification. State governments have 911 certified Green projects and nearly 2,000 awaiting certification as of this writing.

12.4   Costs and Financial Aspects of
Green Buildings

Various studies have been conducted around the country to discern the cost of going green. In 2003, a California study, *The Costs and Financial Benefits of Green Buildings*, led by Greg Kats of Capital E, a Washington, D.C. consulting firm, concluded that the average premium cost for the 33 buildings studied was slightly less than 2%, equating to about an additional $3.00 to $5.00 per square foot.

A report by Turner Construction Company in 2005, issued after a study of 30 green-built schools, revealed that green schools cost less than 2% more than conventional school construction. However, this extra cost is potentially offset by other considerable financial benefits of green building, such as reducing occupant health problems, delivering lower operational costs, and resulting in enhanced student learning.

Davis Langdon, a global construction consultancy firm, conducted several green building cost studies over the years. In a July 2007 study the firm conducted, it investigated 221 academic, laboratory, library, and community buildings and health care facilities, and "suggested that the cost per square floor for buildings seeking LEED certification falls into the existing range of costs for buildings of a similar program type." Thus, it appears that prudent and careful "green" planning and design do not necessarily increase per-square-foot costs.

Another study by the City of Seattle in 2000 reported that the incremental cost to achieve LEED Silver standards across all projects was 1.7%. The premium cost equated to between $3.00 and $5.00 per square foot, similar to the premium found in the California study.

A Davis Langdon "Cost of Green in NYC" study from the fall of 2009, which analyzed high-rise residential buildings and compared LEED and non-LEED construction costs (see the following tables), added that soft cost increases were not substantial. The median cost of LEED design fees was $0.56/sf, the median cost of LEED documentation was $0.30/sf, and the median cost of commissioning was $1.55/sf.

Construction Costs per SF for High-Rise Residential Buildings in New York City

	All	Non-LEED	LEED Certified	LEED Silver	LEED Gold	LEED Platinum
Average	$438	$436	$315	$467	$433	$463
Median	$431	$407	$315	$439	$440	$463

Construction Costs per SF for Commercial Interior Projects in New York City

	All	Non-LEED	LEED Certified	LEED Silver	LEED Gold	LEED Platinum
Average	$197	$204	N/A	$156	$330	$100
Median	$160	$163	N/A	$158	$244	$100

Note: No explanation was given for the disparity between LEED Platinum costs for interiors other than commercial interior projects, nor was a reason given for why LEED Certified costs for high-rise residential buildings were less than "All."

12.5   The Learning Curve for Project Managers and Construction Managers

Project Managers and Construction Managers desirous of learning more about green construction might tap into GREENBUILD, an annual international conference and exposition started in 2002. GREENBUILD is now held in various locations throughout the United States, offering seminars, industry exhibits, and renowned speakers.

For Project Managers or Construction Managers planning work on LEED projects, here are some tips to follow that are unique to these types of projects:

- Establish a LEED documentation center where all of the latest applicable LEED resources are stored. This center should include:

  1. The construction waste management plan
  2. Areas set aside for recycling and a concise list of the materials that are and are not to be recycled
  3. The indoor air quality management plan that will be referred to during commissioning
  4. The Green Interior Design and Construction Reference Guide (obtainable from the design consultants)
  5. A current project scoreboard of goals, both achieved and yet to be achieved

- The construction schedule should include sufficient float or contingencies if the completion date (and final payment) has any tie-ins to federal, state, or local subsidies or tax credits.

- Fully understand what you are building. If certain products are unfamiliar, contact the manufacturer or a manufacturer's representative and request a meeting with its technical people to learn as much as you can about installation and function.

- Review environmental standards with vendors and subcontractors to ensure that they fully understand the parameters with respect to product installation and service performance.

- Take photographs or videos during construction to document achieved sustainability goals; also use photos or videos to document compliance with a manufacturer's product installation procedures.

- Punchlist work must also meet the project's sustainability criteria.

- Periodic and final cleanup must meet green standards for cleaning practices and waste disposal procedures.

12.6   Sustainable Construction

Sustainability is the movement to sustain economic growth while maintaining the long-term health of the environment. When this term is applied to the design and construction industry, it means creating designs that balance the short-term goals of a construction project with the long-term goals of efficient operating systems that protect the environment and nature's resources.

For example, recyclable steel used to manufacture metal studs for framing has the potential to reduce our dependence on wood for that purpose. Manufactured or engineered wood products such as oriented strand board (OSB) make use of wood chips and scraps; medium-density fiberboard (MDF) uses wood fibers to create a whole host of products for cabinetry, doors, and furniture.

Sustainable buildings represent a holistic approach to construction, combining the latest technology with nature to enhance the building's efficiency rather than detract from it; using natural light where possible to reduce dependence on artificial light; orienting the building to reduce both heating and cooling loads; and using storm water to water landscaped areas, thereby reducing overall water consumption.

Sustainable building guidelines include the following:

- Use a multidisciplinary, integrated approach to design that invites each discipline to participate in the design process.

- Simple is preferred instead of complex.

- The overarching framework for this type of project should reflect a respect for nature, neither harming it nor depleting it.

- Life-cycle costs must be an integral part of the design concept. Although initial costs may be high, long-term costs may overbalance those high first costs, particularly when it comes to operating cost efficiency.

- When selecting building materials and mechanical systems and appliances, choose those that require minimal energy to create or operate.

- Maintenance of the structure ought to be considered when both design and construction are being considered.

- Attempt to build with local materials whenever possible to reduce transportation costs.

- Design with an eye to passive strategies such as building orientation, over-hangs and sun shades, thermal mass, and natural lighting.

12.7   Basic Sustainable Construction Goals
for Site and Building

Some of these procedures can be employed on construction sites that are not necessarily designated "sustainable," as they actually represent a "best practices" approach for any contractor.

Site work goal. Meet or exceed standards for erosion control and sedimentation control by:

- Preventing loss of soil during construction due to storm water runoff and wind erosion
- Preventing siltation of existing storm sewers and streams
- Protecting the topsoil stockpiles for reuse, or modifying existing soils to meet topsoil standards

Site utilities goal. Reduce storm water runoff and reuse by accomplishing the following:

- Minimize or strive for total elimination of storm water runoff by instituting infiltration swales and basins to reduce impermeable surfaces instead of installing detention ponds.
- Retain and recharge existing water tables by minimizing disturbances, saving trees and natural vegetation, and supporting and enhancing natural landforms and drainages.
- Store roof runoff for future use as gray water or reclaimed water.
- Install a small-footprint state-of-the-art treatment plant on-site to recycle water for irrigation use.

Open space and landscaping goal. Protect and restore existing vegetation by:

- Protecting trees. This not only enhances the site's value, but provides some cooling benefits.
- Using indigenous landscaping that supports wildlife and biodiversity and does not require the level of irrigation that some new ground cover requires. It also eliminates the need for chemical treatment of these areas.
- Minimize pesticide use by installing weed cloth, mulches, and dense plantings.

Circulation and transportation goal. Improve circulation, and decrease the need for private transportation by:

- Tying the development or building to transit modes and emphasizing alternatives such as organized car pooling, water taxis (if applicable), buses, and car sharing.

The building. During construction, reduce waste by:

- Diverting at least 75% of construction waste, demolition debris, and land clearing from disposal as landfill.

- Deconstructing all existing structures with substantial recoverable materials and disposing of them to recyclers.
- Adjusting new site contours to provide a balanced site.
- Modifying nontopsoil soils to acceptable topsoil requirements.

12.8   The National Renewable Energy Laboratory (NREL): Lessons Learned

NREL conducted a study of a 13,600 square foot (1260 square meter) model building constructed at Oberlin College in Oberlin, Ohio, in 2000 to better understand the impact of an environmentally designed building that would be not only energy efficient but also able to export energy to the local utility company. The building used passive solar devices, geothermal heat pumps for heating and cooling, and roof-mounted photovoltaic (PV) cells to generate electricity. The NREL study monitored the completed structure from 2000 to 2003 and came up with the following findings that are more generic in nature than site specific.

- PV systems must be engineered to minimize transformer balance and system losses.
- PV systems may not significantly reduce the building demand.
- During summer months, large PV systems in commercial buildings, on average, can export electricity from 8:00 A.M. to 6:00 P.M.
- Control design must be completely integrated using the full capabilities of the equipment in the building, such as CO_2 sensors, motion sensors, and thermostats. A balance must be achieved between human operations and automation.
- Dark-colored ceilings must be avoided to take full advantage of daylighting and uplighting.
- Daylighting sensors are needed in all daylit areas; it is not sufficient to rely on manual controls.
- Daylighting must be designed into all occupied areas with consideration of any additional heating and cooling loads imposed upon the building.
- Overglazed areas such as an atrium can provide abundant daylighting but can also impose additional heating and cooling loads.
- Specification for heat pumps must work with appropriate groundwater temperature.
- Electric boilers can be employed as backup sources if used sparingly. Controls and staging are essential for the integration of limited-use systems such as these.

Index